Discrete Mathematical Structures

Discrete Mathematical Structures

Mario Benedicty
University of Pittsburgh

Frank R. Sledge
Center for Naval Analyses
Alexandria, Virginia

Harcourt Brace Jovanovich, Publishers
and its subsidiary, Academic Press
San Diego New York Chicago Austin Washington, D.C.
London Sydney Tokyo Toronto

ISBN: 0–15–517683–8
Library of Congress Catalog Card Number: 86–70556

Printed in the United States of America

Preface

Our goal in this book is to provide material for one or two introductory courses for computer science and mathematics majors. *Discrete Mathematical Structures* is written for students who know some high school algebra, can manipulate formulae and symbols, and wish to develop problem-solving skills that have a broad range of applications. Topics in discrete mathematics provide an ideal context for learning such skills and for applying them to inherently useful situations.

The order of presentation is a consequence of our pedagogical criteria. The first of these is that abstraction is a lengthy procedure: compare several different situations that have something in common, take out what is useful, leave behind everything that is not essential. As a consequence, we examine each idea at least three times. First, we aim at the substance and not at the form, starting from familiar and easy material. For instance, at this stage a "combination" is not much more than the description "no repetitions, order is irrelevant." Second, we present the same idea more formally: combinations become subsets or characteristic functions. We emphasize the need for formalization and abstraction in order to transfer good sense thinking to a more sophisticated level of problem solving, to see different approaches to the same problem, and to apply the same approach to different problems. Third, we use the idea in applications, using examples and illustrations from real-life, state-of-the-art problems. To continue with the example of combinations, they are used in graphs, in coding theory, and in Boolean forms.

Our second criterion is that most of the material we cover is based exclusively on good sense, clear thinking, and good syntax. We stress syntax in general and mathematical language in particular (for instance: "$a < b$" is as solid a sentence as "Jim is younger than Jane"). New vocabulary is introduced gradually, especially in view of the fact that prevailing usage is not yet as consistent or well established as it should be. As the presentation progresses, we assume (or presume) that students will be able to understand and use a progressively more formal, compact, and sophisticated language.

Third, we try—at all levels—to be pragmatic, to emphasize counting, to make connections between different concepts, to use visual representations (figures, diagrams, and tables) as an aid to the understanding of concepts, abstractions, and applications.

Fourth, we cover most of the traditional subfields of discrete mathematics in a progression that is compatible with our criteria. The book is roughly divided into two parts (see the diagram in the Note on Course Design below). The first part (Chapters 1–9) comprises basic material, although a few simple applications are included. Nearly everything in the first part is needed for the second part. The second part (Chapters 10–15) is devoted mainly to applications of the basic ideas. It includes some material that may be considered traditional but also a few "side trips" (some simple, some sophisticated) in directions that may interest students or instructors who want something more challenging. The chapters in the second part are quite independent of each other and allow an ample selection according to the preferences of instructors and students.

Fifth, we place no stress on computer programs, and we do not consider any of them in the first few chapters. The reason is simple: in its early stages computer programming is easy and mechanical; but serious programming (including the use of existing software) is a sophisticated process that requires a good understanding of programming techniques and of the concepts involved. Starting with Chapter 4, in the belief that a different perspective can contribute to a better understanding of the ideas, we introduce several tasks that could involve the computer.

We express our thanks to the many persons and institutions whose names appear in this book, directly or indirectly, and in particular to Werner Rheinboldt for personal encouragement and constructive criticism; to Willy Brandal, University of Idaho; Dennis Broline, University of Evansville; William Bonnice, University of New Hampshire; Donald Bushaw, Washington State University; Steve Fisk, Bowdoin College; Charles Franke, Eastern Kentucky University; Curtis C. McKnight, University of Oklahoma; Samuel Shore, University of New Hampshire; Tina Straley, Kennesaw College; Peter Tannenbaum, California State University, Fresno; Thomas Upson, Rochester Institute of Technology; and Jan Vandever, South Dakota State University, for their suggestions during the review process; to Richard Wolf and Carol Beringer for their technical advice during the preparation of the manuscript; and to Wesley Lawton, Ted Buchholz, Richard Wallis, Jack Thomas, Karen Bierstedt, and Lenn Holland for their editorial contributions. But we are especially grateful to a large group of young men and women whose names do not appear: our students, those who studied from the preliminary version of this book and those who will study from this version.

Mario Benedicty
Frank R. Sledge

A Note on Course Design

Discrete Mathematical Structures will accommodate several different course designs: a one-term or a two-term course at the freshman level, a one-term course at a more advanced level, and a course with emphasis on pure mathematics.

A One- or Two-Term Course at Freshman Level

The first term might include the first part of the book (Chapters 1–9) with omissions and with selected topics from the second part. The second term would cover the rest of the text.

A One-Term Advanced Course

A one-term course in the sophomore or junior year would omit Chapters 1–9 (or devote just enough time to them to develop uniformity of vocabulary and notations) and cover the second part of the book more deeply. The selection depends, of course, on the students' level of preparation and the goal of the course.

A Course for Mathematics Majors

Mathematics majors may need a course that places more emphasis on the theoretical parts of the subject matter and less on applications, especially computer-oriented applications. This can be accomplished by omitting Sections 2.5–2.9, 3.3, 3.4, 6.6, 6.7; studying in detail Chapters 5 and 7–10; and selecting some "pure" topics from Chapters 11–15 (for instance, expressions, graphs, homological properties, and planarity of graphs).

The following diagram suggests the interrelations among topics. Once the goals of the course are located in the diagram, tracing back through the prerequisites will indicate what is needed to reach those goals. Notice that there is much independence within the second part (Chapters 10–15). The sections in each of Chapters 13–15 flow in linear order, and none is needed for other chapters.

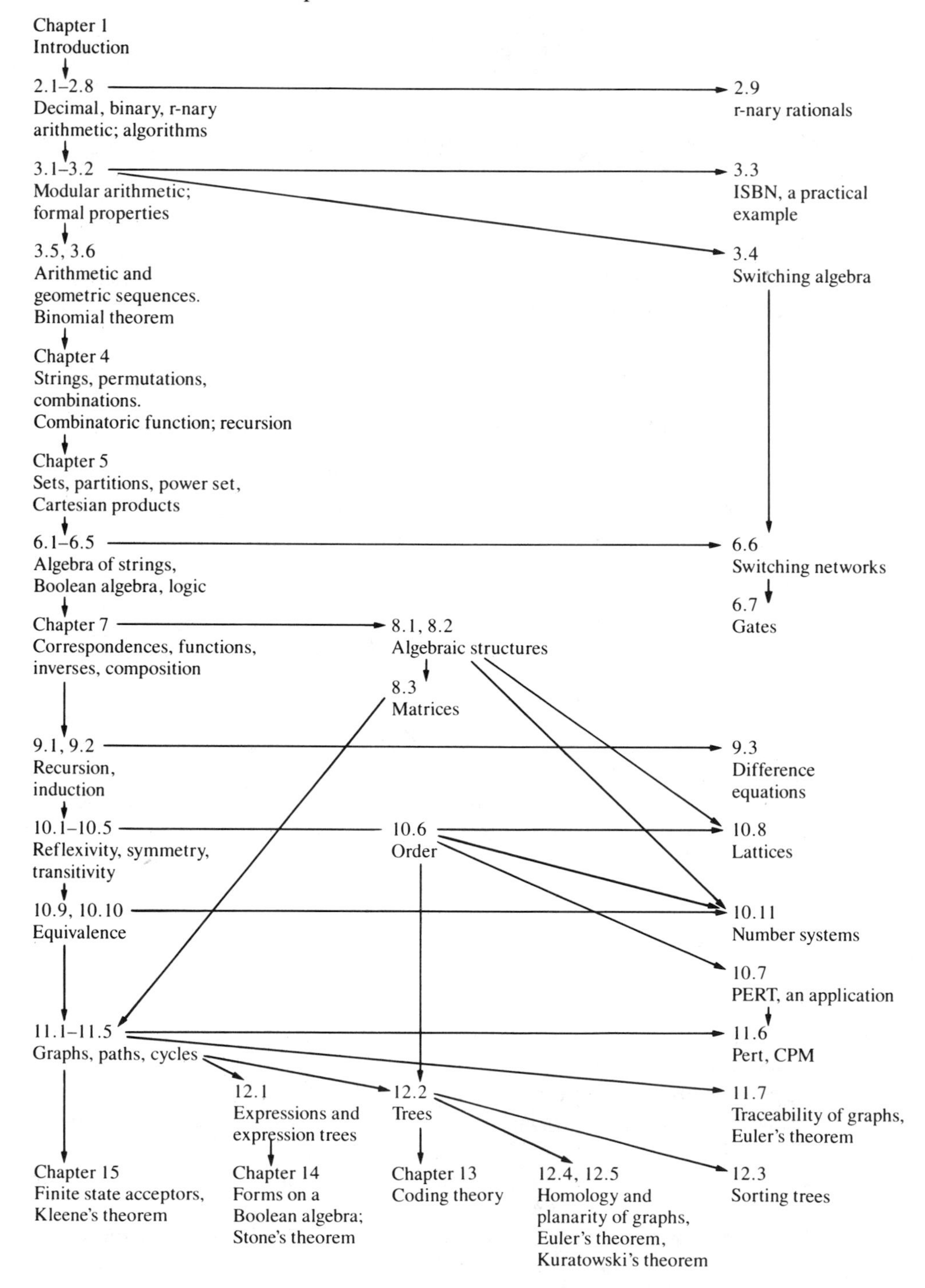

Contents

Discrete Mathematical Structures

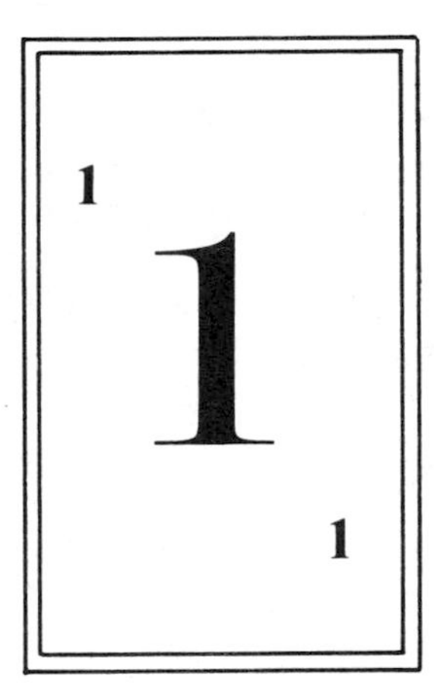

Introduction

1.1 Games: Objects and Rules

Bertrand Russell once defined mathematics as "a game played with symbols." He was not joking. Modern mathematics has been instrumental in exploring distant galaxies and the insides of atoms, in producing huge skyscrapers and miniature computers, in shaping science and commerce and sometimes even art. Yet for all of this vast technology, in essence, mathematics itself resembles nothing so much as a game. The parallels are profound, and in beginning our study of mathematics, we should first look at just what it is that constitutes a game, and what it means to play games with symbols.

Every game contains two basic components: *objects* and *rules*. In chess, the objects are a chessboard and thirty-two chess pieces; in baseball, the objects are a baseball, a bat, a field, home plate, and three bases; in a card game, the objects are the cards of a deck. The rules dictate what to do with the objects. They are seldom simple, and, most important, they indirectly produce more objects and additional rules that are implicit.

A first complication implicit in the rules is that they produce more objects. In baseball, a "ball" or a "strike" is an abstract object that is produced when the batter fails to get a "hit." In chess, a "checkmate" is an abstract object that is produced when the pieces are in a very special mutual arrangement. In tic-tac-toe, we have a somewhat extreme situation: starting only with the grid, the objects of the game—the Xs and Os—do not exist at all at the beginning, but are created as the game is played.

A second complication is that the stated rules produce additional, implicit rules. An example occurs in a *round-robin* tournament among twenty tennis players. The rules state that each of the players is to play exactly one match with each of the other players. But the rules do not explicitly state how many matches are to take place (the answer, as we will see later, is 190). Conversely, in tennis or baseball, the rules regarding the end of the game do not place any strict time limit. The theoretical implication is that the game could go on indefinitely.

A third complication is that, as any sports fan knows, the rules have a tendency to produce a vocabulary all their own, one that can be confusing to the newcomer. Imagine trying to explain to a Martian the meaning of the word "ball" in baseball. "How is it," asks ZQXP-351, "that a batter can have one strike and three balls? There's only one ball; the pitcher throws it." You see the problem. Moreover, these vocabularies are not distinct. Both a baseball player and a bowler would instantly recognize the word "strike," but with very different feelings about getting one.

A fourth—and, for the time being, last—complication: The same objects can be used with different sets of rules to produce different games. With the same deck of cards, you can play bridge or solitaire (in fact, many versions of each); with the same pair of dice, you can play craps or randomly produce the numbers 2 to 12 (though not all with the same probability).

In summary,

- A game comprises objects and rules.
- Playing the game means applying the rules to the objects.
- Applying the stated rules to the objects produces additional objects, additional rules, and vocabulary.

1.2 Mathematical Structures

A **mathematical structure** is a game in which the objects are not physical. That is, a mathematical structure is made up of abstract objects (or symbols that represent them) and rules. Studying and using the structure means applying the rules to the objects. As a consequence, the study or use of a mathematical structure will produce additional objects and rules, and, out of practical necessity, the vocabulary that goes with them. You are already familiar with at least two mathematical structures, the natural numbers and the integers.

The objects of the **natural numbers** are the "counting numbers," including zero; that is, 0, 1, 2, 3, 4, ... The rules of the natural numbers are those you learned in childhood: 1 comes after 0, 2 comes after 1, 3 comes after 2, and so on until all natural numbers are reached; nothing comes before 0. Some of the implicit objects are even and odd numbers, prime numbers, squares, and so on. For implicit rules, we have addition, multiplication, order, and the like. How is addition, for instance, an implicit rule? One simple way to define addition is by "counting up." As an example, if you want to calculate $5 + 3$, you may start from 5 and count up three times: 6, 7, 8. The same rules that tell you that 6 comes after 5, 7 comes after 6, and 8 comes after 7 tell you also, after some work, that $5 + 3 = 8$. This is one way to see how addition is "implied" by the originally stated rules.

The objects of the **integers** are the nonzero natural numbers with a "negative sign" attached to each one, zero, and the nonzero natural numbers with a "positive sign" attached to each one; that is, . . ., -3, -2, -1, 0, $+1$, $+2$, $+3$, ... The rules for the integers are a bit more complicated (see Section 3.2). There is algebraic addition, algebraic multiplication, and algebraic subtraction (*algebraic* means that we must also consider the signs of the numbers). We also have an order: -99 is less than -22, -22 is less than $+11$, and so on.

The symbol used to denote the structure of the natural numbers is $\mathbb{N}$, a special form of the letter N. The symbol used to denote the structure of the integers is $\mathbb{Z}$.

Both $\mathbb{N}$ and $\mathbb{Z}$ are very rich structures. Although their stated objects and rules are simple, the implicit objects and rules are complex and subtle. As we have already mentioned, each has an order, an addition, and a multiplication. Structure $\mathbb{Z}$ has a subtraction, but $\mathbb{N}$ has a subtraction only to a certain extent (in $\mathbb{N}$, 3 minus 5 is not defined). Both $\mathbb{N}$ and $\mathbb{Z}$ have a form of division: 20 divided by 5 is 4, whereas 23 divided by 5 is 4 with a remainder of 3. The further implications of the rules, however, quickly become complicated. For instance: the first prime number beyond 19 is, of course, 23, but what is the first prime number beyond 65537? (Don't feel bad. Nobody has yet found a quick rule to answer such a question.)

1.3 The Adjective "Discrete"

Continuing to comment on the title of the book, we have given an idea of the meaning of *mathematical structure* as a "game" consisting of "objects" and "rules." The adjective **discrete** deserves some attention. To understand first what is *not* discrete, think again of two ordinary games, tennis and darts. When the ball or the dart does not touch a line, then it is clearly "in" or "out." When it touches a line, there arise difficulties, which, unless resolved by additional rules (including possibly the verdict of an umpire), will lead to arguments. A game like craps or poker does not—barring dishonesty—leave room for such ambiguities. Neither does $\mathbb{N}$ or $\mathbb{Z}$. The reason is that, using dice or cards, two objects or two possibilities are always, according to the rules, "far apart" from each other, and thus clearly distinct. Likewise, there is no integer between 1 and 2, there are only finitely many integers between 175 and 23472, and there are only finitely many possible hands in bridge or poker. In all cases, each possibility is clearly distinct from any other. The adjective *discrete* describes this situation: any two objects are, in some sense, "far apart" from each other, and cannot be confused. An example of a mathematical structure that is *not* discrete (usually called *continuous*) is that of the real numbers, denoted $\mathbb{R}$. There are infinitely many real numbers between 1 and 2, for example, 1.1, 1.11, 1.111, 1.1111, . . .; and there are real numbers that are positive, but as close to zero as you wish, for example, 0.1, 0.01, 0.001, 0.0001, ... The complexity and subtlety of such nondiscrete structures is beyond the scope of this book. We will occasionally mention them, but we will not concentrate on them.

1.4 Usage and Mathematical Language

As a matter of usage, in some books the term *natural numbers* and the symbol $\mathbb{N}$ refer only to the numbers 1, 2, 3, 4, . . ., excluding 0. This is not wrong, but it can be confusing if you're not careful since it implies a somewhat different set of rules. If you read another text, find out which usage they follow.

The language of mathematics takes some getting used to and may seem at first to fall hard and awkwardly upon the ear. But it is nothing more than ordinary English usage, and in fact it is sometimes helpful to analyze mathematical statements as sentences. When we use the sentence "$\mathbb{Z}$ is a very rich structure," $\mathbb{Z}$ stands for the phrase "the structure of the integers" and is a *thing;* hence $\mathbb{Z}$ is a noun. Thus any sentence like "$\mathbb{Z}$ has property X" or "$\mathbb{Z}$ does not have property Y" is syntactically correct. When we use the sentence "$x < y$", x and y are both *things,* hence nouns, and the symbol $<$ stands for the predicate "is less than". In other words, "$x < y$" has the same syntax as "Jim is younger than Jane". This notion of mathematics as language will be of considerable use as we proceed.

Decimal, Binary, and Octal Arithmetic

A particularly important aspect of mathematics is the precision of its terminology. We must keep in mind that an integer existing in the infinite cosmos is one thing, but its representation by mere human beings is quite another. For instance, the integer two hundred fifty-six and the integer CCLVI and the eighth power of two are all the same integer; we simply represent it differently.

2.1 Decimal Representation of Integers

2.1.1 Expanded Notation

In Western culture, the most common and simplest representation of integers is the **positional decimal representation.** The number we just mentioned is usually written as 256. Recalling all the concepts from primary school, 6 is the number of "units," 5 the number of "tens," and 2 the number of "hundreds." A "ten" is ten units, a "hundred" is ten tens, and so on. A device to show this is the **expanded notation**

$$256 = 200 + 50 + 6.$$

Let's look at this more closely: 50 can be written as $5 \cdot 10$ (we use the raised period or dot to denote ordinary multiplication; occasionally we'll use the symbol $\times$ of elementary arithmetic or drop the symbol altogether), and 200 can be written as $2 \cdot 100$. Consequently,

$$256 = 2 \cdot 100 + 5 \cdot 10 + 6.$$

The next step is to recall that $100 = 10^2$, $10 = 10^1$, and $1 = 10^0$. We can now put some order into the expansion by writing

$$256 = 2 \cdot 10^2 + 5 \cdot 10^1 + 6 \cdot 10^0.$$

If we are dealing with an unspecified three-digit integer whose digits are a, b, and c, then the decimal representation is just abc (no multiplication intended), and we know that its value is

$$a \cdot 10^2 + b \cdot 10^1 + c \cdot 10^0.$$

Suppose now that the number of digits is also unspecified. We could, of course, write something like $abc \ldots lm$, but this would give the impression that there were exactly thirteen digits, a through m. We can avoid this by using only one letter, say a, and attaching integer subscripts to it, in the form a_0, a_1, a_2, etc. Also, the subscripts should increase from right to left, to reflect the higher powers of 10. Thus a_0 is the units digit, a_1 the tens digit, a_2 the hundreds digit, etc. Note carefully that the *value of the subscript* has nothing whatsoever to do with the *value of the digit,* only with its position in the number. The letter n will denote the highest subscript that occurs in the number. Consequently, the number of digits is $n + 1$, because we are counting from 0 to n. We can now write 256 using $n = 2$, $a_0 = 6$, $a_1 = 5$, $a_2 = a_n = 2$. Then

$$256 = a_2 \cdot 10^2 + a_1 \cdot 10^1 + a_0 \cdot 10^0.$$

To sum up:

- A nonnegative integer has a positional decimal representation $a_n a_{n-1} \ldots a_2 a_1 a_0$.
- The number of digits is $n + 1$.
- The digits are $a_0, a_1, a_2, \ldots, a_n$ (right to left).
- Each digit is a natural number less than 10.
- The integer's value is

$$a_n \cdot 10^n + a_{n-1} \cdot 10^{n-1} + \cdots + a_2 \cdot 10^2 + a_1 \cdot 10^1 + a_0 \cdot 10^0. \quad (1)$$

But even this is too cumbersome. It still suggests that there are at least five digits, and perhaps more, depending on how much ground is covered by the ellipses $\cdots$. This is not true: n may be 3, 2, 1, or even 0 for a single-digit number. Moreover, the real simplicity of the representation is still not apparent. We can solve both problems by use of the *summation notation.*

2.1.2 Summation Notation

Instead of writing an expression like formula 1 above, we can write simply

$$\sum_{k=0}^{n} a_k \cdot 10^k. \quad (2)$$

The odd looking symbol is the Greek letter *sigma* (for *s* as in summation). The letter k is called the **summation index,** and represents each of the subscripts (and exponents) from 0 through n, one at time. The expression $a_k \cdot 10^k$ is the **summation expression.** Thus, expression 2 may be translated into English as "The summation of all of the terms $a_k \cdot 10^k$, for k taking on the integer values from 0 to n inclusive.'' Expression 2 defines an integer. More accurately, it is a set of instructions for calculating an integer. The instructions implied by the summation are as follows:

1. For each integer value of k, from 0 to n inclusive, construct and evaluate $a_k \cdot 10^k$.
2. Add up (sum) all of the numbers obtained in the first step.

This grand sum is the **evaluation of the summation,** and the process is called **evaluation.**

Looking back to the example of the number 256, we had $n = 2$ (three digits), $a_0 = 6$, $a_1 = 5$, $a_2 = 2$. For summation index k taking on the values 0, 1, and 2, we have

if $k =$	then $a_k \cdot 10^k =$
0	$a_0 \cdot 10^0 = 6 \cdot 1 \quad = \quad 6$
1	$a_1 \cdot 10^1 = 5 \cdot 10 \quad = \quad 50$
2	$a_2 \cdot 10^2 = 2 \cdot 100 = \underline{200}$
	and the evaluation is 256

the value of the decimal representation $a_2a_1a_0$, which is shorthand for $\Sigma_{k=0}^{2} a_k \cdot 10^k$.

Clearly, the summation notation is a very powerful tool whose use is not limited to the decimal expansion of integers. Since we shall be using this tool frequently in this book, we point out here some of the general features of summation.

The summation index k is called a **dummy index,** because it is ''summed out'' in the process. It does not appear, therefore, in the final evaluation. Consequently, we may use any variable we wish as the summation index, provided only that we are consistent and unambiguous.

For example, consider carefully the two summations $\Sigma_{k=4}^{9} k^3$ and $\Sigma_{j=4}^{9} j^3$. They both give $4^3 + 5^3 + 6^3 + 7^3 + 8^3 + 9^3 = 1989$. It doesn't matter whether k or j is the summation variable, as long as one or the other is used consistently.

If we use an arbitrary power a instead of cubes, then $\Sigma_{b=4}^{9} b^a = 4^a + 5^a + 6^a + 7^a + 8^a + 9^a$. Here the summation index b is a dummy variable; *but the variable a is not,* since it does not appear in the index of the summation symbol but rather in the evaluation of the summation. Of course, we cannot evaluate this sum to an explicit number until a value for a is specified. Nonetheless, a survives the summation process and is hence not dummy; b does not survive and is dummy. In any summation, the summation index is always dummy.

Consider now $\Sigma_{a=4}^{9} a^a = 4^4 + 5^5 + 6^6 + 7^7 + 8^8 + 9^9 = 405{,}071{,}285$. Now a is the summation index (hence dummy) and in fact occurs more than once in the summation expression. This is perfectly valid.

Occasionally the summation index does not appear at all in the expression, as in $\Sigma_{k=1}^{6} 5$ and $\Sigma_{k=-2}^{3} 5$, both of which give the result $5 + 5 + 5 + 5 + 5 + 5 = 30$. This may require some attention. According to instruction 1 above, we must calculate the summation expression, here 5, for each value of k from 1 to 6 (or from -2 to 3). Whatever k is, 5 is 5. So, we have a 5 when $k = 1$, a 5 when $k = 2$, . . ., a 5 when $k = 6$. That is, we have six 5's. Instruction 2 tells us to add them together.

It has probably occurred to you that this expansion technique, using summations, is a terribly complicated way to write integers, which, after all, you have been writing for years with no problem. Are we not using a howitzer to kill a fly? Definitely not. The interesting and useful thing about the integers is not simply that they exist (as objects in the "game"), but that we *do things* with them (the "rules"). These rules must be proved valid for general integers, not just for specific examples such as 256.

Beware of a second meaning of the word "decimal." In a sentence like "calculate 5/7 to three decimals," the noun *decimal* refers to each of the digits *after* the decimal point. These are decimal digits, of course, but the digits before the decimal point are also decimal digits. In other words, the phrase above is a form of mathematical slang. The adjective *decimal* refers only to the fact that we are using base 10.

Now that we have a means of representing arbitrary nonnegative integers, in the next sections we can move on to the crux of the game, the algorithms of integer arithmetic.

Exercises 2.1

1. Write the following in expanded notation and as expanded summations: **(a)** 512; **(b)** 1024; **(c)** 1,000,001; **(d)** 1,000,000,000. There are some terms you may omit.
2. Write the following in expanded notation and as expanded summations: **(a)** 10101; **(b)** 20003; **(c)** 102030; **(d)** 1,000,000,005. There are some terms you may omit.
3. Do the same for four numbers of your choice.
4. Multiply 256 by 100 using the method of your choice, even a calculator. Now write 256 and 100 as expanded summations and multiply the expansions. Evaluate the resulting expansion. Do you get the same result?
5. Repeat exercise 4 for $256 \cdot 1000$, $1234 \cdot 100$, and $1234 \cdot 10000$.
6. Write the steps for multiplying $a_n a_{n-1} \ldots a_2 a_1 a_0$ by 10^k $(k \geq 0)$. Do you recognize the quick rule for multiplication by a power of 10?
7. Evaluate the following summations numerically (by hand or by calculator): **(a)** $\Sigma_{h=3}^{5} h^2$; **(b)** $\Sigma_{k=1}^{3} k^3$; **(c)** $\Sigma_{j=-5}^{7} (2j)$; **(d)** $\Sigma_{j=0}^{5} 2^j$; **(e)** $\Sigma_{h=0}^{11} 10^h$; **(f)** $\Sigma_{k=5}^{11} 10^k$; **(g)** $\Sigma_{l=0}^{3} (2^{2l})$; **(h)** $\Sigma_{k=1}^{3} k^k$.
8. Evaluate the following numerically (a calculator is advisable): **(a)** $\Sigma_{h=0}^{9} h^2$; **(b)** $\Sigma_{k=-3}^{4} k^2$; **(c)** $\Sigma_{j=0}^{3} 3^{3j}$; **(d)** $\Sigma_{h=2}^{5} h^h$.
9. Expand **(a)** $\Sigma_{x=3}^{6} x^y$; **(b)** $\Sigma_{y=3}^{6} x^y$.

10. Justify the "carry" rule in the usual algorithm for written addition of two integers a and b. [Hint: If $a = \sum_{k=0}^{n} a_k \cdot 10^k$ and if $b = \sum_{k=0}^{n} b_k \cdot 10^k$, then clearly $a + b = \sum_{k=0}^{n} (a_k + b_k) \cdot 10^k$, but the last expansion is *not* necessarily the decimal representation of $a + b$, because some of the quantities $a_k + b_k$ may be greater than 9. It is precisely this situation that is resolved by "carrying." Define c_1 as 0 and s_0 as $a_0 + b_0$ if $a_0 + b_0$ does not exceed 9, or define c_1 as 1 and s_0 as $a_0 + b_0 - 10$ if $a_0 + b_0$ exceeds 9. Then (successively when $k = 1$, $k = 2, \ldots$), define similarly c_{k+1} as 0 or 1 and s_k as $a_k + b_k + c_k$ or . . ..]

2.2 Algorithms for Decimal Arithmetic

One of the major goals of the decimal representation of integers is the efficient performance of everyday arithmetic by human beings. (Computers do these things rather differently; more about them later.) Indeed, historians often suggest, in dead earnest, that one of the reasons for the fall of the Roman Empire was that their arithmetic techniques were simply not adequate to their administrative and commercial needs. If you think that LIV + XII = LXVI is a fairly easy addition, try division! Today we use the positional decimal notation, and in primary school you were taught several techniques, or algorithms, for "doing arithmetic." Let's look at some of these in detail.

We will need, from algebra, the formulae that give the *rules of exponents:*

$$x^h \cdot x^k = x^{h+k} \qquad \text{and} \qquad (x^h)^k = x^{hk}. \tag{3}$$

2.2.1 Elementary Algorithm for Multiplication

The algorithm for written multiplication of multidigit positive integers goes thus:

1. Write the multiplicand (preferably, but not necessarily, the larger factor).
2. Beneath it, write the multiplier (the other factor), aligning the two numbers at the right.
3. Draw a horizontal bar beneath both numbers.
4. Consider the digits of the multiplier, one at a time. Multiply each of them by the multiplicand (with all the carries, etc.), and write each result *(partial product)* in a row below the bar, in such a way that the units digit of the partial product falls directly below the multiplier digit you are using.
5. When all multiplier digits have been handled, draw another horizontal bar.
6. Add the partial products, with all the carries required. This sum, written below the second bar, is the desired product.

Example The successive steps of the operation are detailed in this example of 256 × 64:

```
 256        256        256        256
× 64       × 64       × 64       × 64
----       ----       ----       ----
           1024       1024       1024
                      1536       1536
                      ----       -----
                                 16384
```

There are many things to learn from studying this procedure, besides, of course, multiplication. First, there is the nature of an algorithm. An **algorithm** is a set of clear instructions that can be executed without thinking too much, and with a minimal, but very precise, amount of previous knowledge.

In the example, we had to know beforehand how to multiply a multidigit integer by a single-digit integer; how to add multidigit numbers; what "carries" are; and how to organize the writing. But we did not have to know how to do multidigit by multidigit multiplication. That was the point of giving the algorithm. Indeed, even the previous knowledge required was probably learned via another algorithm: most of us learned how to do multidigit by single-digit multiplication in terms of single-digit by single-digit multiplication and carries, and so on. The point is that algorithms are "layered," each layer being a simple, mechanical application of still lower layers, which in turn come from still lower layers, and so on. Obviously, in the case of elementary arithmetic, the very lowest layer is the addition and multiplication tables (Figure 2.1) and, at that level, there is no substitute for rote memorization. But this applies *only* to the lowest layer, and the fact that 256 × 64 = 16384 is definitely not the sort of thing to memorize!

The second feature to be observed from the example is the need to express the algorithm in general, clear, and concise language. Even this example is somewhat deficient here. The instructions must be given in terms that can be understood in writing, or over the telephone, without further explanations, examples, or pictures. The algorithm must avoid phrases like "Take this and write it there."

Third, and less obvious, is that the algorithm has a very definite stopping point. Knowing the number of operations that are done before the stopping point allows

Figure 2.1 Decimal addition and multiplication tables.

+									
0	1	2	3	4	5	6	7	8	9
1	2	3	4	5	6	7	8	9	10
2	3	4	5	6	7	8	9	10	11
3	4	5	6	7	8	9	10	11	12
4	5	6	7	8	9	10	11	12	13
5	6	7	8	9	10	11	12	13	14
6	7	8	9	10	11	12	13	14	15
7	8	9	10	11	12	13	14	15	16
8	9	10	11	12	13	14	15	16	17
9	10	11	12	13	14	15	16	17	18

×								
1	2	3	4	5	6	7	8	9
2	4	6	8	10	12	14	16	18
3	6	9	12	15	18	21	24	27
4	8	12	16	20	24	28	32	36
5	10	15	20	25	30	35	40	45
6	12	18	24	30	36	42	48	54
7	14	21	28	35	42	49	56	63
8	16	24	32	40	48	56	64	72
9	18	27	36	45	54	63	72	81

us to measure, in some sense, the *cost* of the algorithm. Further, all results obtained are clearly identified either as temporary results no longer needed at the end (the partial products) or as final results (the product).

The fourth observation is the need to keep quite distinct the mechanics of the algorithm itself from the reasons why it works. Except at the earliest stages, there is a substantial efficiency in not letting calculation interfere with thought, nor thought with calculation. This sounds flippant, but it isn't. It is the crucial idea that makes calculators and computers possible!

Fifth, and foremost for the purposes of this book, is the consideration of the *reasons why* an algorithm works. Does the algorithm do what it is supposed to do, and no more, and does it do it in sufficient generality? This is called the **correctness** of the algorithm. Also, does the algorithm work at the lowest possible cost? This is called **efficiency.**

As an example, let's check the correctness of our multiplication algorithm for the case of general three-digit multiplicands and two-digit multipliers:

$$(a_2 \cdot 10^2 + a_1 \cdot 10 + a_0) \times (b_1 \cdot 10 + b_0).$$

Using algebra, we can multiply the first parenthesized expression by $b_1 \cdot 10$, and then by b_0, and add the two partial results. In formulae, this gives

$$(a_2 \cdot 10^2 + a_1 \cdot 10 + a_0) \times b_1 \cdot 10 +$$
$$(a_2 \cdot 10^2 + a_1 \cdot 10 + a_0) \times b_0.$$

The first line represents the product of the multiplicand by the single digit b_1, that product is then multiplied by 10. Thus, we obtain the second partial product of the algorithm, and the multiplication by 10 tells us to write it with the units digit beneath the digit b_1 or to write it "one column to the left." The second line of the formula gives the product of the multiplicand and the digit b_0, that is, the first partial product. Since addition is commutative (that is, the result does not depend on the order of the addends), the adjectives "first" and "second" refer only to the "left shifts" not to the order in which the partial products are calculated. So long as the left shifts are done correctly, the partial products may be calculated and added in any order.

A full justification of the algorithm requires an argument that is general, rather than restricted to specific numbers of digits as we have done in the example above. You can imitate the above argument in the most general case by writing the following summation, in which only the $(n + 1)$-digit multiplier b needs to be expanded:

$$a \times b = a \times \sum_{h=0}^{n} b_h \cdot 10^h = \sum_{h=0}^{n} a \cdot b_h \cdot 10^h. \tag{4}$$

In the right-hand summation, the summation index h enumerates the partial products, each of which is the multiplicand a multiplied by a single digit b_h of the multiplier b and by a power of 10. The power of 10 gives the left shift of the partial product. The partial products are then summed. This one line of summation algebra (equation 4) is the core of our proof that the algorithm is correct!

Note 1 on Theorem Proving. The power and brevity of the proof come directly from the summation representation for integers. Of course, you may have to stare at this for a while to convince yourself that it is proof positive, but it is. Note again that this is not merely a *statement* that the multiplication algorithm works generally, but a *formal, rigorous proof!* This is the way proofs of correctness are constructed: powerful, concise, and sometimes a bit obscure. But every algorithm requires a proof that it is correct. If you invent a new algorithm that is somewhat similar to an older one, you might be able to modify the older proof directly without too much fiddling. But if you have to construct a totally new proof, very likely you will have to play around with the technique, do several examples, see how things work and where the difficulties are, and only then write out the proof in full, formal generality. After all, even in the example above, we did not simply write down formula 4. But don't forget this last step. A statement of an algorithm, even with many examples, seldom constitutes a proof.

As in most proofs, the proof we just saw is based on an interpretation of a formula (formula 4 in this case). We must translate a *symbol* in the formula into its actual *meaning*. Here the multiplication by b_h denotes a partial product, that is, a multidigit by a single-digit multiplication, and the multiplication by 10^h means a left shift by h positions.

This is the first of many "Notes" on proving theorems. See the index to locate the others.

2.2.2 Elementary Algorithm for Subtraction

The usual primary school algorithm for subtraction also offers some opportunity for interesting thinking. Recall the algorithm, especially the part that mentions "borrowing," Algorithm S1, or "carrying," Algorithm S2, whichever is more familiar to you, for the subtractions

$$\begin{array}{r} 30002 \\ -\quad 35 \\ \hline 29967 \end{array} \qquad\qquad \begin{array}{r} 10000 \\ -\quad 1 \\ \hline 9999 \end{array}$$

Assuming the operation is possible, the two prevailing algorithms for subtraction go something like this:

Algorithm S1

1. Write the subtrahend beneath the minuend, units beneath units. Draw a horizontal bar beneath the subtrahend. Start with the rightmost column.
2. If the bottom digit can be subtracted from the top digit, then do the subtraction and continue at step 4.
3. If not, then *borrow;* that is, write a 1 in front of the top digit and scan the minuend, toward the left, until a nonzero digit is found. Replace this digit with the next smaller digit and all the passed zeroes (if any) with nines. Then subtract the bottom digit of the original column from the number formed from the additional 1 and the top digit. Continue at step 4.
4. Write the result beneath the bar, in the same column, and move to the next

column on the left. If there is a digit in the subtrahend, then continue at step 2. If there is not a digit, copy the remaining part of the minuend (if any) into the result, and stop. The number beneath the bar is the result of the subtraction.

Algorithm S2

1. Same as in algorithm S1.
2. The bottom digit plus how much equals the top digit? If there is a result, then continue at step 4. If there is not a result, the bottom digit plus how much equals 10 plus the top digit?
3. Write the result beneath the bar, same column, and carry 1. Then move to the next column on the left. The bottom digit (if any) plus the carry plus how much equals the top digit? If there is a result, then continue at step 4. If there is not a result, the bottom digit plus the carry plus how much equals 10 plus the top digit? Continue at step 3.
4. Same as Algorithm S1.

Check how your preferred algorithm works in the cases above. Realize the necessity to express an algorithm in general terms, and how this can create some difficulties.

In order to reach a conclusion that will be useful later, let's try to formalize and generalize the result of the second subtraction we gave as an example, 10,000 − 1. An integer consisting of all nines is a special case: all digits a_k are equal to nine, that is $a = \sum_{k=0}^{n} 9 \cdot 10^k$, where a is an $(n + 1)$-digit all-nines number. Since this is a summation, we can factor out the nines and obtain $a = 9 \times \sum_{k=0}^{n} 10^k$. Now an integer consisting only of a one followed by zeros (like 10,000) is a power of 10, something like 10^m. The second example suggests that there is a general equality of the form

$$10^m - 1 = 9 \times \sum_{k=0}^{n} 10^k. \tag{5}$$

The phrase "of the form" means that we don't yet know the relationship between m and n. Our example merely *suggests,* but *does not prove,* that $m = n + 1$. In order to find out for sure, write x for 10 in the last summation and obtain $\sum_{k=0}^{n} x^k$. Now from algebra you should recall the general formula

$$x^{n+1} - 1 = (x - 1)(x^n + x^{n-1} + \cdots + x^2 + x + 1). \tag{6}$$

If you don't recall equation 6, you can easily verify it by multiplying out the right side and cancelling terms. Now putting 10 back in for x, we have

$$10^{n+1} - 1 = (10 - 1)(10^n + 10^{n-1} + \cdots + 10^2 + 10 + 1) = 9 \times \sum_{k=0}^{n} 10^k.$$

Comparing this result with equation 5 gives $m = n + 1$, which confirms, and in fact *proves,* the informal arguments and examples we gave before.

In conclusion, for any natural number n,

$$10^{n+1} - 1 = 9 \cdot \sum_{k=0}^{n} 10^k.$$

Note 2 on Theorem Proving. For formula 6, "multiplying out and cancelling terms" really gives a *proof* of the formula. Many proofs do not involve anything more than just a brief verification, but that verification must be done *in general.*

Exercises 2.2

1. Write accurate algorithms for several written operations with integers, in whatever version is most familiar to you. Examples are addition of several addends, multidigit by single-digit multiplication, long division, etc.

2. Using algorithms with which you are familiar, find shortcuts for the following elementary operations (in base 10): **(a)** 351×101; **(b)** $382 \times 100{,}010{,}001$; **(c)** 4321×214; **(d)** 4321×114; **(e)** $20{,}000 - 2$; **(f)** $5{,}000{,}000 - 5$; **(g)** $20{,}000 - 10{,}001$; **(h)** $5{,}000{,}000 - 4{,}000{,}001$; **(i)** $1{,}000{,}000 - 10$; **(j)** $1{,}000{,}000 - 100$.

3. The following algorithm for multiplication of two positive integers is sometimes called "basic multiplication" and is used occasionally for instruction in primary schools.

(1) Write the two factors one beneath the other, aligning the two numbers on the right.

(2) Draw a horizontal bar beneath the lower number.

(3) For each digit of the multiplicand and for each digit of the multiplier calculate and write a partial product as follows:

a. The partial product is the product of the two digits.

b. The position (that is, the column number) of the units digit of the partial product is obtained as the sum of the positions of the two digits you are multiplying (remember that position 0 is the rightmost column, position 1 is the next column on the left, and so on).

c. Write the partial product in the next available new line beneath the bar, in the calculated position.

(4) Draw another horizontal bar beneath the last partial product.

(5) Add together the partial products, that is, the numbers between the two bars. Add them in column, as they are written, with carries, etc. Write the result beneath the second bar. The number just written is the result of the multiplication.

For example,

$$\begin{array}{r} 256 \\ \times\ 74 \\ \hline 24 \\ 20 \\ 8 \\ 42 \\ 35 \\ 14 \\ \hline 18944 \end{array}$$

Do other multiplications of your choice using this algorithm.

4. Prove the correctness of the above algorithm.
5. Prove the correctness of one of the two algorithms for subtraction given in Section 2.2.2.

Complements 2.2

1. There is another algorithm for multiplication of two integers called "method of division and multiplication." It is less popular than the other algorithms we have seen, but it requires no multiplication tables except that of 2.
   (1) Write the two factors as the first entries of two contiguous columns, preferably, but not necessarily, the smaller factor first.
   (2) If the last entry in the left column is 2 or more, then proceed as follows: divide that entry by 2, write the quotient as the next entry in the left column, and ignore the remainder; also, multiply by 2 the last entry of thc right column and write the result as the next entry in the right column; then repeat step 2.
   (3) When the last entry is less than 2, for each even entry in the left column, cross out the corresponding entry in the right column; then add together the remaining entries in the right column. The numbcr just calculated is the result of the multiplication.

Example 74×256:

cvcn	74	256	out
	37	512	
even	18	1024	out
	9	2048	
even	4	4096	out
even	2	8192	out
	1	16384	
		18944	

Do more multiplications of your choice using this algorithm.

2. Try to justify the correctness of the above algorithm. For a full proof, it may be better to wait until later. (See Complements 2.3.)

2.3 Binary Representation of Integers

Continuing with nonnegative integers, recall that the decimal representation for an $(n + 1)$-digit integer a is

$$a = a_n a_{n-1} \ldots a_2 a_1 a_0 = a_n \cdot 10^n + a_{n-1} \cdot 10^{n-1} + \cdots + a_2 \cdot 10^2 + a_1 \cdot 10 + a_0$$
$$= \sum_{k=0}^{n} a_k \cdot 10^k.$$

Note the role played by the number 10. Each digit represents some multiple (0 through 9) of successively increasing powers of 10. The reason for this is that there are exactly ten distinct single-digit integers, again 0 through 9. In decimal positional notation, 10 is called the **base,** or **radix,** of the system. Most probably, people got used to counting by tens because we have ten fingers (the Latin root *digit-* means finger). However, it is possible and sometimes even preferable to use a base other than 10. Any integer greater than one can serve as the base. The base cannot be one because any power of one is still one. Thus, the smallest possible base is 2.

It is quite easy to construct a **binary positional representation** for nonnegative integers using base 2. This system has several advantages, especially for use by computers, but it has one serious disadvantage: the numbers turn out to be quite long. For instance, the decimal number 266 in binary is 100001010. But since this is of no concern to a computer, and since we're discussing mathematics rather than accounting, let's take a detailed look at how such a binary system would work.

The principles of the binary system are simply these:

- First, there are only two single-digit integers, 0 and 1. (Mathematical insight coming! In each case the highest single-digit integer is just one less than the base. Is this a general property of positional notations? Yes! Why?)
- Second, each digit multiplies not a power of 10 but a power of 2.

In more precise wording, any nonnegative integer has a **binary** representation

$$b = b_m b_{m-1} \dots b_2 b_1 b_0 = \sum_{k=0}^{m} b_k \cdot 2^k, \tag{7}$$

where each digit b_k is either 0 or 1. The digits b_k are the *bi*nary digi*ts,* and are thus called **bits.**

The word "bit" has always referred to a very small value. A "bit" was also once each of the eight pieces of a breakable silver coin called a "piece of eight," and even in early American slang a dollar was composed of eight bits, so that "two bits" meant a quarter, and so on.

Binary representations are much easier to handle than decimal ones, except, of course, for their length. This is the reason that they are used inside digital computers. The circuitry is much simpler, both conceptually and technologically. If a computer were made to operate in base 10 (some were), it would have to produce and recognize, at every digit position, ten levels of voltage. Such circuits are difficult and expensive to build. A computer that operates in base 2 need only recognize "high" versus "low" voltage levels, or switches that are either "on" or "off." Of course, inside a computer, "high" voltage means several volts, not the several thousand volts of, say, a power plant. "Low" voltage means practically zero volts. Again, *discrete* simply means that we can tell them apart!

For human beings, binary numbers are easy to manipulate and illustrate a lot of good mathematics, even if they are too cumbersome for shopping lists and checkbooks. They do, however, take some getting used to, and this requires, above all, practice. Try counting in base 2. Figure 2.2 shows the first several integers; continue the table on your own for a ways to get a feel for them. Next observe the binary representations of the powers of 2 in Figure 2.3. Observe, too, the binary numbers that consist of all ones. (More about them later.) Note that even integers always end in 0 (that is, $b_0 = 0$), odd integers in 1. Can you clearly state why, using summation notation? Finally, although it's not in the table, verify for yourself that $2^{10} = 1024$ (decimal).

At all times, keep clearly in mind the distinction between an integer in the abstract and its representation in any given base. For instance, 2^{10} is whatever it is, in any base or representation. In decimal, its representation is 1024, in binary 10000000000, and in English *one thousand twenty-four.*

The number 2^{10} has a special significance for computer users because it is a reasonable power of 2 that approximates pretty well a reasonable power of 10: $2^{10} = 1024$ versus $10^3 = 1000$.
In the metric system, the prefix K (or k for *kilo-*) means one thousand, as in 1 kg = 1000 grams, 1 kHz = 1000 hertz, 1 km = 1000 meters, and even in want-ad slang $1K = $1000. In computer technology, it is convenient to count things in powers of 2, and when we count big numbers, such as the memory capacity of a computer, it's nice to have a number that is about a thousand, so that our human "feel" for decimal numbers can be used. Such a unit is $1\text{K} = 2^{10}$. You will often hear computers described in terms of having, say, 16K words (or *bytes*) of memory. The number meant here is $16 \times 1024 = 16384$, or about 16000. In fact, with computers and their peripheral gear becoming cheaper and more powerful by the day, a similar thing is happening to the prefix M: metric *mega-* means one million, as in 1 MHz = 1,000,000 hertz (cycles per second), Greek *mega* = big, computerese $1\text{M} = 2^{20}$, or roughly one million.

Figure 2.2 Counting in binary.

Decimal	Binary
0	0
1	1
2	10
3	11
4	100
5	101
6	110
7	111
8	1000
9	1001
10	1010
11	1011
12	1100
13	1101
14	1110
15	1111
16	10000

Figure 2.3 Powers of 2.

Decimal	Binary
1	1
2	10
4	100
8	1000
16	10000
32	100000
64	1000000

Exercises 2.3

1. Count in base 2 up to about 50 (decimal), or until you are quite comfortable with the system. See Figure 2.2.
2. Make a table of the powers of 2, up to at least 2^{20}, including both decimal and binary representations. See Figure 2.3.
3. Write the following as binary numerals: **(a)** $\Sigma_{h=0}^{7} 2^h$; **(b)** $\Sigma_{h=0}^{7} 2^{2h}$; **(c)** $\Sigma_{h=0}^{5} 2^{3h}$.
4. Justify (that is, prove) that the binary representation of any nonnegative integer has $b_0 = 0$ or $b_0 = 1$ according to whether the integer is even or odd, respectively. Remember, examples alone are *not* sufficient.
5. Find, quickly, a reasonable decimal approximation of 2^{30}.
6. Find a quick way to calculate the following products in binary: **(a)** 110101111 (binary) × 2 (decimal); **(b)** 111101111 (binary) × 4 (decimal); **(c)** 110110010 (binary) × 16 (decimal); **(d)** 11010110010 (binary) × 8 (decimal).
7. Generalize your method from the preceding problem to calculate $(\Sigma_{i=0}^{n} b_i 2^i) \times 2^m$.
8. Find a compact formula to calculate $\Sigma_{i=0}^{n} 2^{2i}$, $\Sigma_{i=0}^{n} 2^{3i}$. Use it to calculate $\Sigma_{i=0}^{5} 2^{2i}$ and $\Sigma_{i=0}^{2} 2^{3i}$.

Complement 2.3

1. Give a full proof of correctness of the algorithm for multiplication given in Complements 2.2. [Hint: use a binary representation of the first factor.]

2.4 Algorithms in Binary Arithmetic

We have said that any integer r greater than 1 can serve as the base (radix) for a positional representation. Only two things are required. First, we need r distinct symbols to serve as the single-digit integers. Binary has two, namely 0 and 1. Decimal has ten, 0 through 9. A base 362 counting system would need 362 distinct symbols. Second, we need complete addition and multiplication tables for all pairs of single-digit integers, and the corresponding tables for subtraction and division. These tables form the lowest "layer" for the arithmetic algorithms, and must be "memorized" (even by a computer).

Obviously, this will be much more troublesome for base 362 than base 2. Conversely, larger bases will be much more concise. In base 362, any number between 0 and 361 will require only one digit, from 362 to 131,043 (decimal; $131{,}043 = 362^2 - 1$) will require only two digits, and so on. Compare this with base 2, in which the representation of 361 (decimal) is 101101001 (binary), or nine digits. Clearly, the choice of base 10 for everyday human calculations represents a compromise between memorization and patience. Humans lack the patience to do binary arithmetic in daily life, and probably could not memorize tables for any base larger than about 20. Computers are another matter. Although it is usually too expensive to have them memorize large tables, computers, lacking a lively social life, have considerable patience.

The algorithms for arithmetic are essentially the same regardless of base; so there is a lot to learn by observing some strange slants in the usual written algorithms for arithmetic when it is done in base 2. Too, if you do work with, around, for, in, or on the same planet as a computer, you might be interested to know how *it* does arithmetic.

Let's concentrate for the moment on addition and multiplication. The tables in base 2 are exquisitely simple: $0 + 0 = 0, 0 + 1 = 1, 1 + 0 = 1, 1 + 1 = 10$ (that is, write 0, carry 1). For multiplication, $0 \cdot 0 = 0, 0 \cdot 1 = 0, 1 \cdot 0 = 0, 1 \cdot 1 = 1$. The same information appears in tabular form in Figures 2.4 and 2.5.

Figure 2.4 gives a complete tabulation. Figure 2.5 gives the same information in a more compact and more customary form. The first row and column of zeros in the multiplication table are eliminated—on the ground that those entries are understood—and then, in both tables, the headings are eliminated, because they are the same as the first entries in the rows and columns. Compare the tables for ordinary arithmetic (Figure 2.1). Note that binary multiplication does not introduce carries; only addition does.

In fact, we can simplify the process of multiplication even more. Let b be a binary number, with one or more digits. Then the multiplication of b by a single bit happens to be $b \cdot 0 = 0$ or $b \cdot 1 = b$. In other words, in binary multiplication by a single bit, you either have a zero result or a copy of the multiplicand. But you never have to modify the multiplicand. *You never really do any multiplication!* You simply take the multiplicand (1) or leave it (0).

Let's now look at the case of multidigit by multidigit multiplication in binary. Since it's so close to our previous decimal case (considerably simpler, in fact), we won't bother with a detailed explanation. Note that each "partial product" (if present) is just a copy of the multiplicand, with the appropriate "shift." The legends "take" and "leave" on the partial products are, of course, redundant: you may omit the "leave" lines. You should work through this example completely

Figure 2.4 Addition and multiplication tables in base 2.

+	0	1
0	0	1
1	1	10

×	0	1
0	0	0
1	0	1

Figure 2.5 Addition and multiplication tables in base 2, short form.

+		
	0	1
	1	10

×	
	1

to check your understanding. (The partition of the digits into groups of three is solely for clarity.)

```
           1 011 010
        ×    101 101
           ---------
           1 011 010   Take
                   0   Leave
          101 101 0    Take
        1 011 010      Take
                0      Leave
      101 101 0        Take
      ---------------
      111 111 010 010
```

Also, you should verify that, in decimal, this is simply $90 \times 45 = 4050$.

Like the multiplication algorithm, the algorithms for subtraction are practically the same in binary as in decimal. The only difference worth noting is that, in the "borrow" algorithm (Algorithm S1), whenever you borrow "across zero," you replace the zeroes with 1s (not with 9s because in binary 1 is one less than 10 and is the analogue of 9). Practice some subtractions in binary; for instance try 111111 − 101101 or 1001001 − 101101.

Now that we're experts, let's try long division! Recall the primary school method for decimal long division. Take a moment first to practice subtraction in binary and to work out (decimal) 4050 divided by 45, since that's the example we'll do in binary. The trouble with division comes in deciding how many times the divisor divides a portion of the dividend. In binary, there's no problem: It either divides once or not at all, the quotient digit being then 1 or 0, respectively. As before, you should work out the example in detail.

```
                      1 011 010
         ----------------------
 101 101 ) 111 111 010 010
           101 101
           -------
            10 010 01
             1 011 01
            ---------
                111 000
                101 101
                -------
                  1 011 01
                  1 011 01
                  --------
                        00
```

Exercises 2.4

1. Do many operations of addition, multiplication, subtraction, and division, working only in base 2. After you obtain the result, convert the data and the result to decimal, then redo the operation in decimal in any way, including calculator, to check the result. In order to convert to decimal, use formula 7 or see Section 2.5 below.

2. Compose algorithms for several written operations in base 2 (addition, multiplication, subtraction, division, etc.). Use as guidelines the corresponding elementary algorithms for base 10, but note that there are some nontrivial changes.
3. Find shortcuts for the following elementary operations in base 2: **(a)** 1011001 × 1001; **(b)** 110011 × 100010001; **(c)** 1000000 − 1; **(d)** 1000000 − 10; **(e)** 1000000 − 100.
4. Using your increased expertise, take another look at some of Exercises 2.3.

2.5 Conversion between Binary and Decimal

The conversion of the representation of an integer from base 10 to base 2, or from base 2 to base 10, requires some practice. Let a be the integer, let

$$a = a_n a_{n-1} \ldots a_2 a_1 a_0, \tag{8}$$

$$a = b_m b_{m-1} \ldots b_2 b_1 b_0 \tag{9}$$

be its representations in base 10 and 2, respectively. These give the evaluations

$$a = \sum_{k=0}^{n} a_k \cdot 10^k, \tag{10}$$

and

$$a = \sum_{h=0}^{m} b_h \cdot 2^h, \tag{11}$$

respectively.

Then the problem becomes, "Given the a_ks, find the b_hs," or vice versa. In principle, we do not need anything but formulae 10 and 11 above. In practice, however, some experience is necessary. This we shall gain through the following algorithms. Note that, in most cases, many adaptations are necessary according to the concrete environment in which you may be operating, such as hand calculation, calculator, or computer. Note also that the "easy" way is to have a calculator that performs the conversion between bases 10 and 8 and then shift by hand between bases 8 and 2 using the techniques presented in Section 2.7.1 below.

2.5.1 Conversion from Base 2 to Base 10

One algorithm for the conversion from base 2 to base 10 is as follows:

1. Write the given binary representation in a line. For practical reasons, space the digits.
2. Leave a blank line, then draw a horizontal bar.
3. Bring down the leftmost digit; that is, copy it in the same column, beneath the bar.

4. If the number just written is beneath the rightmost digit of the given binary representation, then stop. The result is the number written last under the bar.
5. Otherwise (performing the calculations in base 10) multiply by 2 the number just written, and write the result in the next column to the right, above the horizontal bar. Add the two numbers in this column, writing the result in the same column beneath the bar. Repeat from step 4.

Example Find the decimal representation of 101101 binary.

Initial layout and step 3:

```
1  0  1  1  0  1

----------------
1
```

Successive steps:

```
1  0  1  1  0  1
   2
----------------
1  2

1  0  1  1  0  1
   2  4
----------------
1  2  5

1  0  1  1  0  1
   2  4 10
----------------
1  2  5 11

1  0  1  1  0  1
   2  4 10 22
----------------
1  2  5 11 22

1  0  1  1  0  1
   2  4 10 22 44
----------------
1  2  5 11 22 45
```

The result is 45.

Note that the layout and the procedure are exactly the same as in "synthetic division." Such a procedure is called **Horner's method.** Although it looks horrible on paper, in practice it is an extremely efficient way of calculating a polynomial-like expression.

Note also that equations 10 and 11 are arithmetic *facts,* whichever way we calculate them (base 10, or 2, or 537; by hand or by calculator; or even in Roman numerals). On the other hand, formulae 8 and 9 are *representations* and use *digits,* that is, specific symbols. In our algorithm we did the calculations in base 10 and consequently we obtained a result (formula 11) in *form* 8, which is what we wanted. In other words, this algorithm (as well as the next one) is designed to make the change from base 2 to base 10, or vice versa, by doing the hard figuring in base 10.

As usual, it is not sufficient to *give* an algorithm. We must *prove* its correctness; that is, we must show that it does what it's supposed to do. In other words, are we sure the number the algorithm constructs is actually a, the same number given by formula 11?

Proof: Call h the "number just written beneath the bar." Step 3 clearly sets $h = b_m$. If $m = 0$, then we are done and there is nothing else to prove. Otherwise, the first pass of step 4 replaces h with $h \cdot 2 + b_{m-1} = b_m \cdot 2 + b_{m-1}$. The next pass (if any) replaces h with $h \cdot 2 + b_{m-2} = (b_m \cdot 2 + b_{m-1}) \cdot 2 + b_{m-2} = b_m \cdot 2^2 + b_{m-1} \cdot 2 + b_{m-2}$. And so on, until formula 11 is obtained. Q.E.D.

2.5.2 Conversion from Base 10 to Base 2

The reverse, conversion from base 10 to base 2, can be achieved by the following algorithm:

1. Write the given decimal number at the right end of a line. (The result will be written in the next line.)
2. Divide by 2 the leftmost number in the first line. Write the quotient on the left of the number and the remainder beneath the number. (Note that the remainder is 0 or 1.)
3. If the quotient is positive, then repeat step 2; otherwise, stop. The result is the number whose binary digits are written in the second line; the rightmost digit is the units digit.

Example Find the binary representation of 45. The successive steps are

```
               22 45
                   1

            11 22 45
                0  1

          5 11 22 45
            1  0  1

       2  5 11 22 45
          1  1  0  1

    1  2  5 11 22 45
       0  1  1  0  1

 0  1  2  5 11 22 45
    1  0  1  1  0  1
```

The base 2 equivalent of 45 is 101101.

We leave the proof of correctness as an exercise.

Exercises 2.5

1. Construct several of your own problems of conversion from binary to decimal, and the other way around, similar to those shown in the text. For a start, convert to binary **(a)** 1342, **(b)** 50001, **(c)** 257, **(d)** 2050; convert to decimal **(e)** 100001, **(f)** 100010, **(g)** 11110, **(h)** 101010.
2. Do again some of the exercises in problem 1 of Exercises 2.4.
3. Find a quick way to convert to decimal the following binary numbers: 10, 100, 1000, 10 . . . 0 ($n + 1$ digits), 11, 111, 1111, 11 . . . 1 ($n + 1$ digits).

Complement 2.5

1. Prove the correctness of the algorithm in Section 2.5.2 for converting from decimal to binary.

2.6 Octal and Hexadecimal Representation

Adopting the same techniques we have seen before for base 10 and base 2, we can use any integer greater than 1 as a base. We'll denote the base as r, for **radix.** Within any one representation system, r must be chosen once and for all. A single number must not have mixed bases. (There are some notable exceptions, like time and angles, but we'll not consider them now.) Then,

1. Any nonnegative number can be represented in the form $\Sigma_{k=0}^{n} c_k \cdot r^k$.
2. Each digit c_k must satisfy the inequalities

$$0 \leq c_k < r. \tag{12}$$

(Note carefully the $\leq$ and $<$).
3. The number of digits is $n + 1$.
4. The values c_k are the digits, whatever symbols you use to write them.

It isn't necessary to be rigorous about the number n. The decimal number 17 is conventionally called a two-digit number. But since 17 = 017 = 0017 = 00017 = · · ·, we could say that it has 2, 3, 4, 5, or any greater number of digits. The only restriction we place on n is that it be large enough to handle whichever integer we're trying to represent. To avoid pointless haggling, we'll simply call 17 a two-digit number, and agree that *no upper limit will ever be placed on n* (except perhaps within a computer; read on).

It is time now to establish a little rigor in our terminology. An **integer,** or a **natural number,** or a **number** in general is an abstract entity that exists quite apart from any attempt to represent it, positionally or otherwise. The representations LIV (Roman), 110110 (binary), and 54 (decimal) all are the same integer. The only "pure" representation of integers is to hold up apples (or fingers) and say "Uh, this many," but this becomes cumbersome for integers in the millions. So, we need *representations.* A **numeral** is a specific symbolic representation of a number, and in this book we talk about **numerals in base *r*.** It is a fact of life, however, that "number" is used instead of "numeral," unless there is a real danger of confusion. Using the term number for numeral is forgivable, although undesirable.

The word **digit** is another matter, and here rigorous terminology is essential. Numbers in a positional representation are composed of **digits,** each digit having two properties: its **position** and its **value.** Position is counted from right to left, with the rightmost digit occupying position 0 (since it multiplies r^0, which equals 1). Value refers to the usual notion of single-digit integer value. The positional representation combines the two, and produces the **positional value.** In the decimal number 2324, digits in positions 1 and 3 both have value 2, digit position 1 representing 20 (the positional value) and digit position 3 representing 2000. Inequality 12 simply says that each digit has a nonnegative value less than the base.

The word **digit** refers to the *digit value* and also to the *symbol* used to represent each digit. The context should decide between the two. These symbols are a function of culture and convention, not of mathematics. In theory, it doesn't matter at all what digit symbols we use, save only that we have enough of them (in fact, precisely r of them), and that they be distinguishable from each other. In practice, since our culture uses base 10, we use the symbols 0 through 9 whenever we can, so that for base 2 we use 0 and 1, for base 8 we use 0 through 7, and so on.

This last remark implies a problem regarding bases greater than 10: we need more digit symbols. For the first ten digits, we use 0 through 9 with their usual values. The most common convention is to then use letters in alphabetical order. For example, we use A for 10 (decimal), B for 11, and so on. This will cover us through base 36, far more than anybody uses in practice anyway. (If you ever have a desperate need for a base 537 system, you can choose your own digits.) Although capital letters are traditionally used, this is not universal. Calculators exist which do arithmetic in base 16 (see Section 2.7.1 below), and of necessity they use a mixture of upper and lower case letters A through F.

Two bases that are very convenient and common in connection with computers are 8 and 16. Following the tradition of using Latin word roots for adjectives (in Latin, *dec-* means 10 and *bi-* means 2), we call the base 8 system **octal** (from the Latin *oct-,* meaning 8). Base 16 is called **hexadecimal** (Greek *hex-,* 6, plus Latin *dec-,* 10). Often, hexadecimal is called simply **hex.** As we said above, octal simply uses the digits 0 through 7; hexadecimal uses 0 through 9 and A through F.

"Counting" octal or hex starts thus:

Octal	Hexadecimal	Decimal
0	0	0
1	1	1
2	2	2
3	3	3
4	4	4
5	5	5
6	6	6
7	7	7
10	8	8
11	9	9
12	A	10
13	B	11
14	C	12
15	D	13
16	E	14
17	F	15
20	10	16
21	11	17
22	12	18
⋮	⋮	⋮
36	1E	30
37	1F	31
40	20	32
41	21	33
⋮	⋮	⋮

Exercises 2.6

1. Count from 0 to 50 (decimal), representing each integer in bases 2, 8, 10, and 16.
2. Make a table of the first twelve (decimal) powers of 8 and 16 for future reference.
3. Write each of the following decimal numbers in octal form: 16, 32, 64, 128.
4. Write each of the following decimal numbers in hexadecimal form: 16, 32, 256.
5. Convert each of the following decimal numerals to binary, octal, and hex form: **(a)** 1234; **(b)** 257; **(c)** 264; **(d)** 1024; **(e)** 1025; **(f)** 1026; **(g)** 1028.
6. Write as an octal numeral **(a)** $\Sigma_{h=0}^{7} 8^h$; **(b)** $\Sigma_{h=0}^{7} 8^{2h}$; **(c)** $\Sigma_{h=0}^{5} 8^{3h}$; **(d)** $\Sigma_{h=0}^{9} 2^h$; **(e)** $\Sigma_{h=0}^{8} 2^{2h}$; **(f)** $\Sigma_{h=0}^{7} 2^{3h}$.
7. Write as an hex numeral **(a)** $\Sigma_{h=0}^{7} 16^h$; **(b)** $\Sigma_{h=0}^{7} 16^{2h}$; **(c)** $\Sigma_{h=0}^{5} 16^{3h}$; **(d)** $\Sigma_{h=0}^{9} 2^h$; **(e)** $\Sigma_{h=0}^{8} 2^{2h}$; **(f)** $\Sigma_{h=0}^{9} 2^{4h}$.

8. Convert the following to decimal. There is a quick way to do this. Do you see it?

10 octal	100 octal	1000 octal
10 hex	100 hex	1000 hex

2.7 Conversion of Bases

In Section 2.5 we saw conversions between bases 2 and 10. Conversion among other bases follows the same principles, but there are some special cases worth noting.

2.7.1 Conversions among Binary, Octal, and Hex

The conversion between binary and octal is quite simple. The bits are simply grouped by threes, and the conversion is thus done octal digit by octal digit. The correspondence of octal digits is given in Figure 2.6. Note that 8 and 9 are *not* valid octal digits.

Figure 2.6 Correspondence between octal and binary digits.

0 octal = 000 binary
1 octal = 001 binary
2 octal = 010 binary
3 octal = 011 binary
4 octal = 100 binary
5 octal = 101 binary
6 octal = 110 binary
7 octal = 111 binary

Study these two examples:

571 octal = 101 111 001 binary
17 octal = 001 111 binary (same as 1111 binary)

The spaces between the groups of three bits are solely for illustration. The reason for the three-bit groupings is that $8 = 2^3$.

Hexadecimal presents a similar situation. Since $16 = 2^4$, binary numbers are converted to hex simply by grouping the bits by fours, and using the hex digit correspondence given in Figure 2.7. Clearly, then, conversion between octal and hex entails simply regrouping the bits, and adding or dropping leading zeroes as needed.

Figure 2.7 Correspondence between hex and binary digits.

0 hex = 0000 binary
1 hex = 0001 binary
2 hex = 0010 binary
3 hex = 0011 binary
4 hex = 0100 binary
5 hex = 0101 binary
6 hex = 0110 binary
7 hex = 0111 binary
8 hex = 1000 binary
9 hex = 1001 binary
A hex = 1010 binary = 10 decimal
B hex = 1011 binary = 11 decimal
C hex = 1100 binary = 12 decimal
D hex = 1101 binary = 13 decimal
E hex = 1110 binary = 14 decimal
F hex = 1111 binary = 15 decimal

For example

577 octal = 101 111 111 binary
= 1 0111 1111 binary
= 0001 0111 1111 binary
= 17F hex

2A3B hex = 0010 1010 0011 1011 binary
= 010 101 000 111 011 binary
= 25073 octal

Conversions between any two bases that are themselves powers of 2 (2, 4, 8, 16, etc.) are accomplished by bit regrouping schemes similar to these.

2.7.2 Conversion among Other Bases

Conversions among other bases, especially to or from decimal, present a more serious problem. Here, arithmetic must be done on the number as a whole, rather than on individual digits or groups of digits. For conversions between decimal and octal or hex, there are many ways.

- One possibility is first to use the conversion schemes between binary and decimal (see Section 2.5), and then to regroup as needed for octal or hex.
- Or, those schemes can be modified to function directly in octal or hex. For hand use, you will find a table of the powers of 8 and 16 very handy (see problem 2 in Exercises 2.6).
- Working in the opposite direction, you can use a special calculator to convert decimal to octal and then convert digit-by-digit from octal to binary or hex, or the other way around.

Some calculators are able not only to convert among bases 8, 10, and 16, but actually to do arithmetic in each. For those whose dealings with computers are both intense and detailed, such a calculator might be a worthwhile investment. For the moment, however, you should practice operations such as the following, verifying all base conversions in all directions.

Example

octal		decimal		hex
5154 octal	=	2668 decimal	=	A6C hex
+ 764	=	+ 500	=	+ 1F4
6140	=	3168	=	C60

Exercises 2.7

1. Construct several of your own problems of base conversion from and to bases 2, 8, 16, and 10. In your experiments with hex, make liberal use of the digits A through F, in order to become familiar with them. Use a decimal calculator for conversion, if you wish, but not one capable of octal or hex.
2. Convert the following to decimal. There is a quick way to do these. Do you see it?

77 octal	777 octal	7777 octal
FF hex	FFF hex	FFFF hex

2.8 Arithmetic in Octal or Hex

Arithmetic done directly in octal or hex requires tables (addition, multiplication, etc.) for all pairs of digits. For example, in hex, we must know that $7 \times D = D \times 7 = 5B$ (that is, write B, carry 5). In the Exercises 2.8 you are asked to construct such tables.

For human proficiency, these tables would have to be memorized—something we do *not* recommend! More importantly, for computer implementation, such tables have to be built into the circuitry. Some advanced computers actually do their arithmetic in octal or hex, rather than in binary, since this is much faster. However, the more sophisticated circuitry makes such machines also much more expensive. Thus you will hear references to "octal computers" or "hexadecimal computers" as well as to "binary computers."

Exercises 2.8

1. For once in your lifetime, write out the full addition and multiplication tables for octal, and at least part of the tables for hex. Be alert for shortcuts; there are many. For comparison, see Figures 2.1 and 2.5.

2. Write some algorithms for written operations in base 8 or 16 (addition, multiplication, subtraction, division, etc.). Use as guidelines the corresponding elementary algorithms for base 10, but note that there are some changes.
3. Do many operations of addition, multiplication, subtraction, and some divisions. Do the arithmetic in base 8 or 16 only. After you obtain the result you may wish to convert data and result to decimal and redo the operation in any way, including calculator, to check the results.
4. Calculate quickly in octal or hex **(a)** 537 octal × 8 decimal; **(b)** 713 octal × 64 decimal; **(c)** A3F hex × 16 decimal; **(d)** 3C0 hex × 256 decimal.

2.9 Decimal Numbers and Fractions

So far, we have discussed only the representation of nonnegative integers. Recall that the symbols $\mathbb{N}$ and $\mathbb{Z}$ refer, respectively, to the structures of the natural numbers and integers. But there are other mathematical structures that you are already familiar with, and they too have their symbols. The symbol $\mathbb{Q}$ refers to the structure of the **rational numbers,** numbers that can be expressed as the quotient of two integers (with nonzero denominator). This includes proper and improper fractions, positive, zero, and negative. The symbol $\mathbb{R}$ stands for the structure of the **real numbers,** and $\mathbb{C}$ for the structure of the **complex numbers.** None of these is a discrete structure, but there are some interesting discrete phenomena hidden in $\mathbb{Q}$. Let's look at a few of them.

Even if we allow ourselves to use only finitely many digits, there are many fractions that can be expressed in base 10. For example, $1/2 = 0.5$, $1/4 = 0.25$, $7/4 = 1.75$. But some fractions, even simple ones, cannot be expressed this way with only finitely many digits: $1/3 = 0.333\ldots$ (infinitely many threes). Note carefully that 1/3 is still a rational number. Just because a number requires infinitely many digits does not mean that it's irrational. The rule is this: a number is rational if it can be expressed using an infinitely **repeating** pattern of digits at its tail, such as 1/3 (infinite repeating threes) or $1/7 = .142857\ 142857\ldots$ or $169/495 = .3\ 41\ 41\ldots$. Even 1/2 can be looked at as $.5000\ldots$, with infinitely many zeroes. Conversely, the (in)famous value π ($\pi = 3.14159265\ldots$) is irrational, not simply because it has infinitely many digits, but because those digits *do not repeat!* So is the square root of 2, or the square root of 3, or the cube root of 4. The distinction to be emphasized here is not between rational and irrational numbers, but between two classes of rational numbers—those that can be expressed using only finitely many digits (ignoring the repeating zeroes), and those that require infinitely many nonzero digits, with a string of digits repeated indefinitely. Both 1/2 and 1/3 ($1/2 = 0.5$, $1/3 = 0.333\ldots$) are rational numbers, but they differ from each other in this aspect.

2.9.1 Decimal Rationals

Base 10 numbers that can be written with finitely many digits are called **decimal rationals.** The reason is illustrated by the example 0.037 = 37 thousandths = 37/1000, which is a fraction whose denominator is a power of 10. Note that 1/5 and 3/25 are both decimal rationals because they can be written as 2/10 and 12/100, respectively. On the other hand, 1/3 and 1/7, although both fractions and hence rationals, can't be expressed as fractions *with power-of-10 denominators;* so neither is a decimal rational.

To see how decimal rationals are represented in positional notation, a little theory is needed. If x is any nonzero rational, then $1/x$ is represented as x^{-1}, $1/x^2$ as x^{-2}, and so on. In general, $1/x^n = x^{-n}$, whether n is positive, negative, or zero.

Note that if n is negative, then $-n$ is positive. Note also that the rules of exponents given in expression 3 are valid for any integer values of h and k, provided that x is not zero. Then 0.1 (= 1/10 = one tenth) can be written as 10^{-1}, $0.3 = 3 \cdot 10^{-1}$, $0.01 = 10^{-2}$, $0.005 = 5 \cdot 10^{-3}$, and so forth. Suppose a, b, c, d, and e are decimal digits. Then the number $ab.cde$ in expanded notation is $a \cdot 10 + b + c \cdot 0.1 + d \cdot 0.01 + e \cdot 0.001$, that is $a \cdot 10^1 + b \cdot 10^0 + c \cdot 10^{-1} + d \cdot 10^{-2} + e \cdot 10^{-3}$.

Even more generally, if a_k are decimal digits, then the numeral $a_n a_{n-1} \ldots a_2 a_1 a_0 . a_{-1} a_{-2} \ldots a_{-m+1} a_{-m}$ becomes, in expansion

$$a_n \cdot 10^n + a_{n-1} \cdot 10^{n-1} + \cdots + a_2 \cdot 10^2 + a_1 \cdot 10 + a_0 + \\ a_{-1} \cdot 10^{-1} + a_{-2} \cdot 10^{-2} + \cdots + a_{-m+1} \cdot 10^{-m+1} + a_{-m} \cdot 10^{-m},$$

which can be written far more succinctly as $\Sigma_{k=-m}^{n} a_k \cdot 10^k$.

Formidable as this last expression looks, it really contains *nothing new.* We have simply taken the now-familiar summation expansion and allowed the summation index k to run through negative as well as zero and positive values. Along with negative powers of 10, we have used also the corresponding negative subscripts on the digits a_k. They indicate the positions of the respective digit *after* the point (tenths, hundredths, etc.). The only thing that is even close to being new is the **decimal point,** which simply tells us which digit in the number is a_0, that is, the units digit, making it possible to distinguish the numbers 3275684 and 327.5684 .

The summation notation clarifies the discussion in the first part of the section. If there are no restrictions on n and m, (that is, if we allow any number of digits on either side of the decimal point, or even infinitely many on the right side), then we can represent, not just all of the rational numbers, but all of the real numbers. However, if either n or m (particularly m) is restricted to some maximum value, there will be certain numbers that can't be represented. We can't represent many of the reals. We can't even represent many of the rationals. If m is not allowed to exceed a given value k, then we can write only those rational numbers that can be rewritten as fractions with denominator 10^k. For example, if $m \leq 4$, then 327.5684 can be written as 3275864/10000 ($10000 = 10^4$); but 0.00001, which is clearly rational, can't be represented.

But why would n or m ever be limited? Well, for one thing, calculators and computers store numbers digit-by-digit (hence the phrase *digital computers*). Whether a computer uses base 10 or base 2, 8, or 16, no digital computer can ever store the number π exactly (nor can any notebook), since π is irrational. More to the point, computers using base 2, 8, 10, or 16 can't store the number 1/3 exactly (except by storing numerator and denominator separately and implying, rather than actually doing, the division. But that's cheating!).

When computers are used for precise calculations in science and technology, then some calculations, like "1 divided by 3", can produce small errors, called *round-off errors*. These errors occur because the computer can only store finitely many digits. Because most computers can store about eight decimal digits, consider the following three numbers, each of which has nine digits:

0.000000003 will be rounded to 0.00000000.
0.000000006 will be rounded to 0.00000001.
0.000000009 will be rounded to 0.00000001.

Now all three of these are errors; small errors perhaps, but errors nonetheless. But look at the last two in particular: 0.000000009 is greater than 0.000000006 (50% larger, actually), yet the computer will round both to 0.00000001. In other words, the computer cannot tell these two numbers apart! Now you see why our definition of discreteness is so important. In discrete structures, we must always be able to tell game objects apart, definitely and unambigously. Discrete machines like computers can *approximate* nondiscrete structures like the real and rational numbers, but cannot represent them *exactly!*

To summarize our discussion so far:

- A decimal rational is a number that can be written as a fraction whose denominator is a power of 10 (including 1, which is just 10^0).
- Any decimal rational can be written in the form $\sum_{k=-m}^{n} a_k \cdot 10^k$, where n and m are integers that are theoretically arbitrary (provided $-m \leq n$), but that may be limited in certain practical contexts.
- As usual, the a_k are decimal digits.

2.9.2 r-*nary Rationals*

All of this discussion can be easily extended to other bases, especially to the useful bases of 2, 8, and 16. All that is necessary is to substitute the chosen base for 10. Thus, we can precisely define objects called **binary rationals, octal rationals,** and **hex rationals** analogously to decimal rationals (see the exercises).

Study these examples carefully.

0.1 binary = $1/2 = 2^{-1}$	0.1 octal = $1/8 = 8^{-1}$
0.01 binary = $1/4 = 2^{-2}$	0.01 octal = $1/64 = 8^{-2}$
0.001 binary = $1/8 = 2^{-3}$	0.001 octal = $1/512 = 8^{-3}$

$$\begin{aligned} 0.1 \text{ hex} &= 1/16 = 16^{-1} \\ 0.01 \text{ hex} &= 1/256 = 16^{-2} \\ 0.001 \text{ hex} &= 1/4096 = 16^{-3} \end{aligned}$$

Furthermore, you should verify that each of these numbers is just a shade less than 1:

.11111111	binary
.77777777	octal
.99999999	decimal
.FFFFFFFF	hexadecimal

Note that the fraction one-half has the representations 1/10 = .1 in binary, 1/2 = .4 in octal, 1/2 = .5 in decimal, and 1/2 = .8 in hex.

One more of our many remarks on mathematical language. In ordinary English, there are many adjectives that are derived from numbers, for instance fourth, fifth, etc., and binary, ternary, etc. When the number is not given explicitly, such as n or r, then these adjectives become n-th or r-th or n-nary or r-nary. The same remark is valid for adjectives or nouns like quintuple, sextuple, n-tuple, r-tuple. Try to get used to such contexts.

One final point about just that—the point. It really is wrong to continue to refer to the point as a *decimal* point in bases other than 10. It's more correctly called the **radix point.** Regardless of base, it serves the same function. But most people, even mathematicians and computer scientists who should know better, still call it a decimal point in any base. You too should know better, but not insist on it. In fact, it's a *point* only in the Western Hemisphere. In Europe, the common notation for our (decimal) 20.4 is 20,4 (read "twenty decimal four" or "twenty comma four").

Exercises 2.9

1. For decimal rationals, practice conversions between fractional form and decimal form. For instance, find the decimal forms of 3/5, 7/20, and 11/125; find the fractional forms of 0.03, 10.7, and 0.125. Do more examples of your own.

2. Recall the rules for multiplying or dividing a decimal number by a power of 10. Calculate quickly in decimal 0.00375×1000, $3.07/100$, $45/10^8$, 114×10^{-11}.

3. Extend the rules used in exercise 2 to multiplication and division of a binary number by a power of 2, an octal number by a power of 8, a hex number by a power of 16.

4. Calculate the following in octal, then convert to decimal to check your answer: **(a)** 73.5 octal/64 decimal; **(b)** 73.5 octal $\times$ 64 decimal.

5. Calculate in hex, then convert to decimal to check your answer: **(a)** C5.AB hex $\times$ 256 decimal; **(b)** CD.9 hex/256 decimal.

6. State a precise definition of binary rationals, octal rationals, and hex rationals. Give the form of the summation representation for each.

7. What are the r-nary digits? What is the definition of r-nary rationals? State the form of r-nary rationals in terms of a summation representation.

8. The number one-third cannot be represented using finitely many digits in any of base 2, 8, 10, or 16; that is, it is not a binary, octal, decimal, or hexadecimal rational. Find a base r in which one-third is an r-nary rational.

9. Suppose that q and r are both powers of a same integer p, greater than 1. Prove that any r-nary rational is also a q-nary rational.

Arithmetics and Algebras

In this chapter, we shall look at some structures called **arithmetics.** Note the plural. You're probably accustomed to thinking of arithmetic in terms of one operation called addition, another called subtraction, and so on, but real life is a little more complicated.

Consider the process called *division*. What exactly is division? Suppose you had to distribute 27 feet of ribbon equally among six children. If you divide 27 by 6, you obtain $4\frac{1}{2}$ feet per child (or 4.5, if you use a calculator). Now suppose you're distributing 27 cookies among the six children. As any parent knows, the correct answer is "Four cookies per child, and hide (or eat) the remaining three to avoid arguments." Good sense, but is it good arithmetic? Finally, suppose you have to drive 27 children to the school picnic, and you can accomodate six children per car. If you divide 27 children among six children per car, you will need 5 cars. In each case, you divided 27 by 6, but in each case, you got a different answer $4\frac{1}{2}$, 4, or 5.

"Aha!" you say, "but the last two cases are not really correct. The *correct* answer is $27/6 = 4\frac{1}{2}$, but in the real world, we have to make allowances. After all, mathematics isn't *real!*" That's precisely what this book is all about. The three types of division we did above are all real, genuine, and correct. But they are different. All three are called division, but, although they are quite similar to each other, they are, in fact, three different arithmetic operations. Where they differ is in *what they are designed to accomplish*. Different goals require different tools.

Throughout this book, we shall present new objects and new operations. At times, these new ideas will have something to do with objects and operations you have learned previously. Each one, however, is different and is intended to achieve a different result. As you read, pay careful attention to what the new idea actually is, as opposed to what it *almost looks like*.

3.1 Arithmetic Modulo *m*

There are many situations in mathematics, as well as in real life, where addition behaves in a different way. For instance, consider these two examples of "addition."

- Your watch reads 5:00 A.M. After 27 hours, it reads 8:00 A.M. Does this mean that $5 + 27 = 8$?
- If you place a pencil at 45° on a protractor, and then rotate the pencil two full 360° turns, you will still read 45°. Does this mean that $45 + (2 \cdot 360) = 45$?

In both cases the answer is yes, but not according to the usual notion of addition. The addition we use is part of a new structure called the *integer arithmetic modulo m,* which we denote with the symbol $\mathbb{Z}_m$.

3.1.1 The Structure $\mathbb{Z}_m$

Definition 1 Given a positive integer m, **integer arithmetic modulo m** (or simply *mod m*) is a structure $\mathbb{Z}_m$ in which the objects are the **integers mod m** and the rules are **operations mod m.** The integer m is called the **modulus.** The integers mod m are $0, 1, \ldots, m - 1$. Note that there are m integers mod m. Addition, subtraction, and multiplication mod m are performed according to the following rule:

Perform the operation exactly as in $\mathbb{Z}$ (that is, according to the usual arithmetic). Then add or subtract m as many times as needed to make the result an integer mod m; that is, to make it nonnegative and smaller than m.

For example, suppose we let m be 5, and consider the structure $\mathbb{Z}_5$. The objects are the five integers mod 5; that is, 0, 1, 2, 3, and 4. Examples of operations are the following. What is $2 + 4$ in $\mathbb{Z}_5$? $2 + 4 = 6$ in $\mathbb{Z}$, but 6 isn't an integer mod 5. So we subtract 5 from 6 to get 1, which is now an integer mod 5. So $2 + 4 = 1$ in $\mathbb{Z}_5$. Similarly, $3 \cdot 4 = 12$ in $\mathbb{Z}$; we subtract 5 twice, and the result is 2, an integer mod 5; thus $3 \cdot 4 = 2$ in $\mathbb{Z}_5$. In both $\mathbb{Z}$ and $\mathbb{Z}_5$, $2 + 2 = 4$. But, whereas $2 + 3 = 5$ in $\mathbb{Z}$, $2 + 3 = 0$ in $\mathbb{Z}_5$. As for subtraction, $1 - 4 = -3$ in $\mathbb{Z}$; we add 5 once, and see that $1 - 4 = 2$ in $\mathbb{Z}_5$.

Operations modulo m are clearly different from the usual operations with integers, and they really should have their own symbols. Most of the time, however, we don't bother with different symbols, we just write $a + b$, $a - b$, or $a \cdot b$, or even ab. If there is a need to indicate that the operation is that in $\mathbb{Z}_m$, then we write "(mod m)" after the equation, for instance: $2 + 4 = 1 \pmod 5$, $3 \cdot 4 = 2 \pmod 5$, $1 - 4 = 2 \pmod 5$. Some authors go so far as to object to the use of an equal sign, preferring the congruence sign, $\equiv$, so that $2 + 3 \equiv 0 \pmod 5$.

3.1.2 Reduction mod m

We now look at a different way to handle $\mathbb{Z}_m$ from a practical point of view. Recall the definition of division with remainder in $\mathbb{Z}$: if n is any integer and m is a positive integer, then the division n/m is said to have **quotient** q and **remainder** r if the integers q and r satisfy the conditions $n = qm + r$, $0 \leq r < m$. For instance, 29 divided by 6 yields quotient 4 and remainder 5, since $29 = 4 \cdot 6 + 5$ and $0 \leq 5 < 6$. We might infer that -29 divided by 6 yields quotient -4 and remainder -5 (some programming languages do just that). But this is incorrect (according to our definition), because, although $-29 = (-4) \cdot 6 + (-5)$, it is not true that $0 \leq -5 < 6$. The correct result has quotient -5 and remainder 1, so that $-29 = (-5) \cdot 6 + 1$ and $0 \leq 1 < 6$.

We can now consider any integer as an integer mod m, provided that, at any time, we allow ourselves to mentally replace the integer with its remainder under ordinary division by m. This operation is called **reduction mod m.** For example, 12 reduces to 2 (mod 5), -29 reduces to 1 (mod 6), and so forth. For any integer x, the equation $x = k \pmod{m}$ simply refers to the fact that x and k reduce to the same integer mod m. Thus, $12 = 0 \pmod{6}$ and $-29 = 1 \pmod{6}$. Obviously, for any integer k mod m, we have that $k = k \pmod{m}$. In other words, we consider an integer mod m as a class of ordinary integers. We have m classes: the class of those integers that reduce to 0 mod m, the class of those integers that reduce to 1 mod m, and so on to the class of those integers that reduce to $m - 1$ mod m. The operations of addition, subtraction, and multiplication are performed exactly as in $\mathbb{Z}$, with only the additional proviso that, at any time, data, intermediate results, and final result can be reduced mod m. These operations thus yield exactly the same results as Definition 1.

The results are always the same for the two processes, so the choice is largely a matter of which process you are more comfortable with. As an example, the first process considers $\mathbb{Z}_3$ as comprising just the three objects 0, 1, and 2. The second process considers all of $\ldots, -6, -3, 0, 3, 6, \ldots$ to be "the same" as 0, all of $\ldots, -5, -2, 1, 4, \ldots$ the "same" as 1, all of $\ldots, -4, -1, 2, 5, \ldots$ the "same" as 2.

Considering the watch again, we can now simply set m equal to 24 and say that, indeed, $5 + 27 = 8 \pmod{24}$. If, like most dial watches, ours isn't capable of distinguishing A.M. from P.M., then we can use $m = 12$, from which we still get $5 + 27 = 8 \pmod{12}$. In the protractor example, we set $m = 360$, and write $45 + (2 \cdot 360) = 45 \pmod{360}$.

3.1.3 Division in $\mathbb{Z}_m$

Division in $\mathbb{Z}_m$ requires something different from the other operations. First, in this argument we refer to "even division," such as 12/3 in $\mathbb{Z}$, not to division with remainder, such as 12/5. (Note also that division by zero is never possible.) Second, the rules for division are not among the stated rules; they are implied rules in $\mathbb{Z}$ as they will be in $\mathbb{Z}_m$. Third, we should be prepared to give up the intuitive notion that the dividend is "divided up" or "shared equally," and accept the fact that division is simply a mechanical operation that is the reverse of multiplication. Think, for instance, of 12/3 as that integer which, multiplied by 3, gives 12 (that's 4). In other words, since $12 = 4 \cdot 3$, we say that $12/3 = 4$. Finally, division in $\mathbb{Z}_m$ will be defined (as a consequence of the stated rules) by trying to say that $a/b = c \pmod{m}$ whenever $a = c \cdot b \pmod{m}$. As in the case of $\mathbb{Z}$, there will be severe restrictions. Except for that, you must therefore pay close attention to the formal definition of division in $\mathbb{Z}_m$ and ignore for now your intuitive objections.

Definition 2 We define **division in** $\mathbb{Z}_m$ as follows: Given that a and b are integers mod m and b is not zero, we define $a/b = c \pmod{m}$ if there is exactly one integer $c \pmod{m}$ such that $a = c \cdot b \pmod{m}$. Otherwise a/b is undefined in $\mathbb{Z}_m$. In particular, the **multiplicative inverse of** $\boldsymbol{b}$ is defined (whenever existent) as $1/b$.

Examples $4/2 = 2 \pmod 5$ because $4 = 2 \cdot 2 \pmod 5$; $2/3 = 4 \pmod 5$ because $2 = 4 \cdot 3 \pmod 5$. But $1/2$ cannot be done in $\mathbb{Z}_4$, because none of $0 \cdot 2$, $1 \cdot 2$, $2 \cdot 2$, $3 \cdot 2$ equals $1 \pmod 4$, and there are no choices for c except 0, 1, 2, and 3. Also, $3/3$ cannot be done in $\mathbb{Z}_6$, because $1 \cdot 3 = 3 \pmod 6$, but also $3 \cdot 3 = 3 \pmod 6$, $5 \cdot 3 = 3 \pmod 6$, and the results would be 1, 3, *and* 5.

There is a process for division that is practical for small numbers and theoretically possible in general. In order to calculate (when possible) a/b, (1) reduce a and b mod m [*optional*]; (2) in $\mathbb{Z}$ try, for each value of c from 0 to $m - 1$, if $a - c \cdot b$ is a multiple of m. If so, then $a = c \cdot b \pmod{m}$ and c is a possible result. If only one possible result exists, then it is *the* result. If no result or more than one possible result exists, then $a/b \pmod{m}$ is undefined.

Division in $\mathbb{Z}_m$ always works in at least these special cases, for any modulus m ($m > 1$), and for any integer $a \pmod{m}$:

$$\begin{aligned} a/1 &= a \pmod{m}; && \\ a/(m-1) &= m - a && \text{if } a \neq 0, \\ a/(m-1) &= 0 && \text{if } a = 0; \\ a/b &= a \cdot (1/b) && \text{if } 1/b \text{ is defined.} \end{aligned}$$

In other cases, division may or may not be defined, depending on a, b, and m. Try several more examples of these cases in order to confirm your understanding.

Exercises 3.1

1. Reduce each of the integers -7, -2, -1, 0, 1, 2, 7 for each of the following moduli: **(a)** 3; **(b)** 5; **(c)** 6.

2. Practice addition and multiplication mod m for several values of m, such as $m = 2, 3, 5$, and 6. Also practice division, remembering that some cases won't work.

3. **(a)** In $\mathbb{Z}_5$, calculate (and reduce mod 5) each of the following:
(1) $2 + 4$, **(2)** $3 - 4$, **(3)** 4×2, **(4)** $0 - 2$, **(5)** $4/3$, **(6)** $2/3$,
(7) $4 - 2 \times 3$;
(b) In $\mathbb{Z}_3$, calculate (and reduce mod 3) each of the following:
(1) 1^2, **(2)** 2^2, **(3)** 1^3, **(4)** 2^3, **(5)** $1/2$, **(6)** $2/1$, **(7)** $2/2$, **(8)** $1 - 2/1$,
(9) $2 - 1/2$;
(c) In $\mathbb{Z}_4$, calculate (and reduce mod 4) each of the following:
(1) 1^2, **(2)** 1^3, **(3)** 1^4, **(4)** 2^2; **(5)** 2^3, **(6)** 2^4, **(7)** 3^2, **(8)** 3^3, **(9)** 3^4;
(d) In $\mathbb{Z}_5$, calculate (and reduce mod 5) each of x^2, x^3, x^4, x^5 for each $x = 1, 2, 3, 4$.

4. Construct the entire addition and multiplication tables for $\mathbb{Z}_2$, $\mathbb{Z}_3$, $\mathbb{Z}_5$, and $\mathbb{Z}_6$. Use Figure 2.1 as an example of layout, but note a substantial difference: in that table (and in similar tables for other bases) we provide sums and products of single-digit numbers only. In this exercise we want the tables for *every* integer mod m. The task is possible (at least when m is small) because there are only finitely many integers mod m.

5. Prove that for any integers x and y, for any natural number k, and for any modulus m ($m > 0$), if x reduces to u and if y reduces to v, then
(1) $x + y$ and $u + v$ reduce to the same integer mod m;
(2) xy and uv reduce to the same integer mod m;
(3) x^k and u^k reduce to the same integer mod m.
Can the last statement be extended to cases in which $k < 0$? Why or why not? Note that this proof is necessary to give sense to the process in Section 3.1.2.

3.2 Formal Arithmetic Properties

Whichever way the structure $\mathbb{Z}_m$ is presented, it inherits from $\mathbb{Z}$ several arithmetic properties, though not all. We list here the more important arithmetic properties, with comments about their validity in $\mathbb{Z}$ or $\mathbb{Z}_m$.

These properties go under several names; we give the more common ones, but you should expect (and become used to) changes of phrasing according to normal English usage. Also, keep in mind that the word "arithmetic" when used as an adjective is pronounced a-rith-MET-ic.

- **Associative property of addition.** For every x, every y, and every z, $(x + y) + z = x + (y + z)$. We say that "addition is associative," or that "+ has the associative property," or "associativity is valid for +," or the like.
- **Associative property of multiplication.** For every x, every y, and every z, $(xy)z = x(yz)$.

- **Commutative property of addition.** For every x and every y, $x + y = y + x$.
- **Commutative property of multiplication.** For every x and every y, $xy = yx$.
- **Distributive property of multiplication over addition.** For every x, every y, and every z, $(x + y)z = xz + yz$. We say that "multiplication distributes over addition," or "$\cdot$ distributes over $+$". Thus, addition and multiplication are formally related to each other.
- **Existence of an additive identity.** For every x, $x + 0 = x$ and $0 + x = x$. We describe this by saying that 0 is the **additive identity.**
- **Existence of a multiplicative identity.** For every x, $x \cdot 1 = x$ and $1 \cdot x = x$. That is, 1 is the **multiplicative identity.**
- **Existence of a multiplicative annihilator.** For every x, $x \cdot 0 = 0$ and $0 \cdot x = 0$. That is, 0 serves not only as the additive identity but also as the **multiplicative annihilator.**
- **Existence of an additive inverse.** For every x, there is a y such that $x + y = 0$. We denote this object y as $-x$ and call it the **additive inverse of** $\boldsymbol{x}$.
- **Existence of the inverse operation of addition.** For every choice of x and y, there is an object z such that $z + y = x$. We denote this object z as $x - y$, and call it the **difference** of x and y. Thus subtraction is the inverse of addition, and it follows that $-y = 0 - y$ and $x - y = x + (-y)$.
- **Existence of powers.** For every x, we set $x^2 := x \cdot x$, $x^3 := x^2x$, and in general, for any *natural* number k, $x^{k+1} := x^kx$. The expression x^k is called the **power** of x with **exponent** k. We complete the definition by setting $x^1 := x$; and $x^0 := 1$, the multiplicative identity. It can be proved that (with these definitions) the rules of exponents (that is, formulae 3 in Section 2.2) are valid for all nonnegative values of h and k.

We use here the special symbol $:=$ to mean "is defined to be." Equality $(=)$ is a statement of purported truth which must be justified. Definition $(:=)$ is just an arbitrary assignment, a form of shorthand. Once a symbol is defined, we can deal with it just like any other mathematical object. To see the difference between $=$ and $:=$ compare the following two statements: $3^2 := 3 \cdot 3$ and $3^2 \cdot 3^2 = 3^4$. The first is an arbitrary shorthand, the second is a resulting fact.

The properties listed so far are valid in $\mathbb{Z}$ as well as in $\mathbb{Z}_m$.

- **Existence of a multiplicative inverse.** If $x \neq 0$, and if there is a single object y such that $y \cdot x = 1$, then we say that y is the **multiplicative inverse of** $\boldsymbol{x}$. Here there is a difference: in $\mathbb{Z}$, only 1 and -1 have multiplicative inverses. In $\mathbb{Z}_m$, 1 and -1 still have multiplicative inverses, but there may be other objects as well that have multiplicative inverses. In particular, if m is a prime number, then every nonzero object in $\mathbb{Z}_m$ has a multiplicative inverse (see Exercises 3.2).
- **Existence of divisors of zero.** There is also a second difference: if we know that (in $\mathbb{Z}$) $xy = 0$, then it must be that $x = 0$ or $y = 0$ or both. This is not always true in $\mathbb{Z}_m$, since we have seen that $2 \cdot 3 = 0 \pmod 6$, but $2 \neq 0 \pmod 6$ and $3 \neq 0 \pmod 6$. Such "factors" are called **divisors of zero.** The property that $xy = 0$ implies $x = 0$ or $y = 0$ is true in $\mathbb{Z}_m$ whenever m is a prime integer (see the exercises).

- **Trichotomy of order.** For every x and every y in $\mathbb{Z}$, one and only one of the following is true: $x < y$, $x = y$, $y < x$. This **trichotomy axiom** holds for the rational and real numbers also. We say also that "$<$ is trichotomic," or the like.
- **Transitivity of order.** In $\mathbb{Z}$, if $x < y$ and if $y < z$, then $x < z$. This is known as the **transitive property** of $<$. We say that "$<$ is transitive."
- **Consistency of order with addition and multiplication.** In $\mathbb{Z}$, if $0 < x$ and if $0 < y$, then $0 < x + y$ and $0 < xy$. We say that "sums and products of positive integers are positive," where "positive integer" means an integer x such that $0 < x$.
- **Uniformity of order under addition.** In $\mathbb{Z}$, if $x < y$ and if z is any integer, then $x + z < y + z$. We say that "order is uniform under addition."

As soon as we try to combine order properties with arithmetic operations, we run into trouble in $\mathbb{Z}_m$. For instance, two order properties of $\mathbb{Z}$ that *fail* in $\mathbb{Z}_m$ are consistency with addition and uniformity. Suppose we insist that the integers mod m be ordered in the usual sequence $0 < 1 < 2 < \cdots < m - 1$. Then uniformity quickly fails because $(m - 1) + 1 = 0$: if we take $x = 0$, $y = m - 1$, and $z = 1$, then $x + z = 1$, $y + z = 0$, $x < y$ (that is, $0 < m - 1$), and it is *not* true that $x + z < y + z$ (that is, that $1 < 0$). The consistency property we have already seen to fail in some examples in $\mathbb{Z}_6$ (such examples of failure are called **counterexamples**). If we had a notion of $0 < 1$, $1 < 2$, $2 < 3$, $\cdots$ in $\mathbb{Z}_6$, then we would have $0 < 2$ and $0 < 3$, but we have seen that $2 \cdot 3 = 0$ in contradiction of consistency.

"But," you say, "we invented a new addition and multiplication for $\mathbb{Z}_m$; why not invent a new order as well? Can't we simply list them in a different order, so that the order properties will hold in that new order?" Yes, we *can* list them in different order, but no, the order properties will *not* hold. To see this, suppose we take $\mathbb{Z}_3$ and invent some new ordering. There aren't many choices; all we can do is select one of these six orders:

$$0 < 1 < 2, \quad 0 < 2 < 1, \quad 1 < 0 < 2, \quad 1 < 2 < 0, \quad 2 < 0 < 1, \quad 2 < 1 < 0.$$

Transitivity would give, respectively,

$$0 < 2, \quad 0 < 1, \quad 1 < 2, \quad 1 < 0, \quad 2 < 1, \quad 2 < 0.$$

Adding 1 to both sides and invoking uniformity would give, respectively,

$$1 < 0, \quad 1 < 2, \quad 2 < 0, \quad 2 < 1, \quad 0 < 2, \quad 0 < 1,$$

each of which contradicts the tentative choice because of the axiom of trichotomy. In short, the definitions of addition and multiplication in $\mathbb{Z}_m$ dictate that we cannot possibly satisfy all the order properties.

Note 3 on Theorem Proving. Note that a proof of the type just given is called a "proof by exhaustion" (*exhaustion of cases,* to be precise), because it proves the desired property by giving a separate proof in each of all possible cases. This technique for proving theorems is convenient when the number of possibilities is rather small, so that we can afford a separate proof in each case.

You might ask why we give these "Notes on Theorem Proving" in bits and pieces. Would it not be better to present a complete set of rules by which you'd be able to construct any proof? It would, of course, but such a set of rules does not exist (at least not today). There are, however, a few techniques that help in the process. You'll learn them a few at a time in these Notes (see the index to find them all), and you will so build up the experience necessary to construct your own proofs.

Exercises 3.2

1. State whether each of the following equalities is true or not:
   (a) In $\mathbb{Z}_2$: **(1)** $1 + 1 = 1$, **(2)** $1 + 1 = 0$, **(3)** $1 \cdot 1 = 1$, **(4)** $1 \cdot 1 = 0$, **(5)** $1^2 = 1$.
   (b) In $\mathbb{Z}_3$: **(1)** $2 + 1 = 0$, **(2)** $2 + 1 = 2$, **(3)** $2 + 2 = 1$, **(4)** $1 - 2 = 2$, **(5)** $2 \cdot 2 = 1$, **(6)** $2 \cdot 1 = 2$, **(7)** $2^2 = 1$, **(8)** $1^2 = -2$, **(9)** $1/2 = 2$, **(10)** $2/1 = 2$.
   (c) In $\mathbb{Z}_5$: **(1)** $3 + 4 = 2$, **(2)** $3 + 4 = 1$, **(3)** $3 - 4 = 4$, **(4)** $2 - 4 = 7$, **(5)** $-3 = 2$, **(6)** $2 \cdot 2 = 4$, **(7)** $2 \cdot 3 = 1$, **(8)** $2^2 = 1$, **(9)** $3^2 = -1$, **(10)** $3^3 = 2$, **(11)** $3^3 = 4$, **(12)** $1/3 = 2$, **(13)** $1/2 = 3$.
2. Solve for x:
   (a) In $\mathbb{Z}_2$: **(1)** $x + 1 = 0$, **(2)** $x - 1 = 0$.
   (b) In $\mathbb{Z}_3$: **(1)** $2x + 1 = 0$, **(2)** $2x - 1 = 0$, **(3)** $1 - x = 2$.
   (c) In $\mathbb{Z}_5$: **(1)** $3x = 1$, **(2)** $3x = 4$, **(3)** $3x + 1 = 0$, **(4)** $2x = -4$, **(5)** $4x = 2$.
3. Solve simultaneously each of the following pairs of equations for x and y:
   (a) In $\mathbb{Z}_2$: **(1)** $x + y = 0, x = 1$, **(2)** $x + y = 0, x = 0$.
   (b) In $\mathbb{Z}_3$: **(1)** $x + y = 0, x - y = 1$, **(2)** $x + 2y = 1, x = 2$, **(3)** $x + 2y = 2, y = 0$.
   (c) In $\mathbb{Z}_5$: **(1)** $x + 4y = 0, x + 2y = 1$, **(2)** $x = y, x + 3y = 1$, **(3)** $x + 4y = 0, x + y = 2$.
4. Using the multiplication table for $\mathbb{Z}_5$, note the presence of a one in each row. What is the significance of these ones? Observe the multiplication table for $\mathbb{Z}_6$. Are there any zeroes? What do they indicate? Are there any ones? What do they indicate?
5. In $\mathbb{Z}$, the powers x^k $(k > 0)$ are all different from each other unless $x = 0$ or $x = \pm 1$. Is the same true in $\mathbb{Z}_m$ (when $m > 3$)? Why or why not?
6. In $\mathbb{Z}_5$, make a table of powers x^k for each integer x mod 5 and for each natural number k up to 10. Don't forget to reduce mod 5. Do you notice anything special? Find a shortcut to calculate x^k for large k, and calculate x^{437} for each x. Then, for each x, calculate $\sum_{i=0}^{393} x^i$.
7. Prove that in $\mathbb{Z}_m$, when m is prime, **(a)** if $x \neq 0$, then x has a multiplicative inverse; **(b)** $xy = 0$ only if $x = 0$ or $y = 0$. This proof may offer some difficulty; see Section 3.1.3 for a hint.

3.3 A Practical Example: ISBN

A practical application of the structure $\mathbb{Z}_{11}$ is the International Standard Book Number (ISBN). See Figure 3.1 for the description of the ISBN system given in the User's Manual of the ISBN Agency. Its practical goal is to expedite book orders, acquisitions, and deliveries. The ISBN for this book, 0-15-517683-8, is printed at the bottom of the back cover near the spine.

Assume that the digits of the ISBN are (left to right) $d_{10}d_9 \ldots d_2d_1$. The check digit will then be d_1. Note that the stated rule on the check digit (see Recommen-

Figure 3.1 Description of the ISBN. Excerpted—by permission—from "The ISBN System, User's Manual" by the *International ISBN Agency,* Berlin, 1975.

- The purpose of this International Standard is to coordinate and standardize internationally the use of book numbers so that an ISBN identifies one title, or edition of a title, from one specific publisher, and is specific to that title or edition.
- Every ISBN consists of ten digits and whenever it is printed it is preceded by the letters ISBN.
- The ten-digit number is divided into four parts of variable length, each part when printed being separated by a hyphen.
- The four parts are as follows:
  1. **Group identifier.** This part identifies the national, geographic or other similar grouping of publishers.
  2. **Publisher identifier.** This part identifies a particular publisher within a group.
  3. **Title identifier.** This part identifies a particular title or edition of a title published by a particular publisher.
  4. **Check digit.** This is a single digit at the end of the ISBN, which provides an automatic check of the correctness of the ISBN.
- Publishers with a large output of books are assigned a short publisher identifier; publishers with a small output of books are assigned a longer publisher identifier. The title identifier is assigned by the publisher from within the range of numbers assigned to him.
- The "check digit" is computed as the result of a calculation on the other nine digits. According to the ISO Recommendation No. 2108 "the check digit is calculated on a modulus 11 with weights 10-2, using X in lieu of 10 when ten would occur as check digit." This means that each of the first nine digits of the ISBN (i.e., excluding the check digit itself) is multiplied by a number ranging from 10 to 2 and that the sum of the products thus obtained, plus the check digit, must be divisible, without remainder, by 11.

Sample of publisher and title identifiers, assuming a group identifier of one digit only:

Publisher Identifier	Title Identifier
0–7	00–19
80–94	200–699
950–997	7,000–8,499
9980–99899	85,000–89,999
99900–999999	900,000–949,999
	9,500,000–9,999,999

dation No. 2108 in the description) can be restated by saying that d_1 must be such that

$$\sum_{i=1}^{10} i \cdot d_i = 0 \pmod{11}. \tag{1}$$

Note also that each of d_2 through d_{10} ranges from 0 to 9, and d_1 may range from 0 to 10, with the agreement that—in this context—10 is spelled X.

Example The first nine digits of the ISBN of their User's Manual are 3-88053-002- . The check digit should be calculated so that $10 \cdot 3 + 9 \cdot 8 + 8 \cdot 8 + 7 \cdot 0 + 6 \cdot 5 + 5 \cdot 3 + 4 \cdot 0 + 3 \cdot 0 + 2 \cdot 2 + d_1 = 0 \pmod{11}$; that is (remember reduction mod 11), $8 + 6 + 9 + 0 + 8 + 4 + 0 + 0 + 4 + d_1 = 0 \pmod{11}$, $6 + d_1 = 0 \pmod{11}$, which gives $d_1 = 5 \pmod{11}$.

Exercises 3.3

1. Look up the ISBN of this book and of other books you may have around. Check whether it is a correct ISBN, that is, whether equation 1 is satisfied. (There are some books—very few—that bear an incorrect ISBN. Nobody's perfect!)
2. According to the rules, multiplying by 10, 9, . . ., 2, 1 left to right (and adding) should give a multiple of 11. Show that if you multiply by 1, 2, . . ., 9, 10 (still left to right) and add, then you obtain an equally good check.
3. Show that a suitable formula for calculating the check digit when the other nine digits are given is $d_1 = -\sum_{i=2}^{10} i \cdot d_i \pmod{11}$.
4. Give yourself any nine decimal digits as the first nine digits of a fictitious ISBN. Calculate the check digit using either the exercise above or the given rule.
5. Experience shows that the most common errors that occur in processing orders by means of ISBN's are (a) mistyping one digit, and (b) interchanging the positions of two digits. Start from an actual ISBN. Simulate both types of mistakes, one at a time, and then check whether the result is a valid ISBN.
6. Prove, in general, that a single mistake of type (a) or (b) above is always detected by the ISBN system.
7. Starting with an actual ISBN, find several cases in which other types of errors go undetected, for instance: mistyping two digits, or interchanging two digits and mistyping one. Note that although some of these cases go undetected, most are detected.

3.4 New Types of Algebra

You may have noticed, as we progress, that the distinction between arithmetic and algebra begins to blur, or even to fade away. One of the primary reasons for this is that we are not content to view arithmetic simply as specific calculations, such as addition, in a specific setting, such as the integers. We have examined several structures, each with an *addition-like* operation, a *multiplication-like* operation, and so on. We have tried to compare these different operations and see what they have in common and—more importantly—how they differ.

3.4.1 Summary of Formal Properties

Let's summarize some of the things we've observed, or that you may have observed yourself in structures we haven't considered in detail. Refer to Section 3.2.

- **Addition** and **multiplication.** In all of the structures we've mentioned ($\mathbb{N}$, the natural numbers; $\mathbb{Z}$, the integers; $\mathbb{Q}$, the rationals; $\mathbb{R}$, the reals; $\mathbb{C}$, the complex numbers; $\mathbb{Z}_m$, the integers mod m) for any two objects x and y, the objects $x + y$ and xy always exist in the structure. Furthermore, addition and multiplication are associative and commutative; each has an identity; and the additive identity is also the multiplicative annihilator. We have always represented the additive identity as 0, although the connotation of zero being "nothing" has not always been true. The multiplicative identity has always been denoted by 1, although again the connotation of "one" might not always hold. (A tricky but unimportant exception is $\mathbb{Z}_1$, in which 0 is also the multiplicative identity, since 0 is the only object in the structure!)
- **Additive inverse.** In every one of those structures except $\mathbb{N}$, there is an additive inverse, $-x$, for each x. Once again, the connotation of the "$-$" may mislead. In $\mathbb{Z}_5$, 2 and 3 are each other's additive inverses, that is, $2 = -3$ and $3 = -2 \pmod 5$.
- **Subtraction.** In all of those structures, if y has an additive inverse $-y$, then the subtraction $x - y$ is defined for any x and any y. Note that $x - y$ is always defined in $\mathbb{Z}$, $\mathbb{Q}$, $\mathbb{C}$, $\mathbb{R}$, and $\mathbb{Z}_m$, but only sometimes in $\mathbb{N}$.
- **Multiplicative inverse.** In $\mathbb{Q}$, $\mathbb{R}$, and $\mathbb{C}$, and in $\mathbb{Z}_m$ when m is prime, every x other than 0 has a multiplicative inverse; in $\mathbb{Z}$, only 1 and -1 do.
- **Division.** In all of the structures we've considered, if y has a multiplicative inverse, then the division x/y is defined. Thus, division is always defined (when the divisor is not zero) in $\mathbb{Q}$, $\mathbb{R}$, and $\mathbb{C}$, and in $\mathbb{Z}_m$ when m is prime, but division will fail in some cases in $\mathbb{Z}_m$ (m not prime), in $\mathbb{N}$, and in $\mathbb{Z}$.
- **Order.** All of $\mathbb{N}$, $\mathbb{Z}$, $\mathbb{Q}$, and $\mathbb{R}$ have an ordering that is both consistent with their arithmetic and satisfies the trichotomy axiom and transitivity. Structure $\mathbb{Z}_m$ has no such ordering. Parenthetically, neither has $\mathbb{C}$. In $\mathbb{C}$, the very notions of $<$ and $>$ have no meaning.

If we were to ask, "What is the single most important thing that all of these structures have in common?" the answer we would most likely get is, "They all have to do with numbers. That is, they are all counting systems of some fashion." That's true, but it's not the most important thing. The most important aspect that these structures have in common is that *their arithmetics are so similar in form.* There are many structures, with very different concrete meanings, which nonetheless have the same *form* as those we've studied. Many have nothing whatever to do with counting, measurement, or numbers. Later in this book, we'll study several of them. For now, just to show you that there is more to mathematical structures than just counting systems, let's look at a simplified version of a *non-numerical* discrete structure: two-terminal switching algebra.

3.4.2 Switching Algebra

Imagine water flowing through a pipe, or electricity flowing through a wire, or, for that matter, a path through a garden. At various points, we have objects (valves, switches, gates) that either allow passage ("on") or halt passage completely ("off"). At any moment, each object is completely on or completely off. Now imagine that, instead of a single pipe, wire, or path, we have an interconnected network of them, but still with only one starting point and one exit point. Interspersed in the network are various of the objects, some on, the rest off. Our question: Is it possible to start at the starting point and "flow" to the exit point? That is, given the on/off state of each object, is the network as a whole on or off?

For simplicity, let's limit our analogy to electricity in a network of wires and switches. Consider Figure 3.2, which shows the only two ways in which two switches can be combined.

Let's call the switches x and y. In Figure 3.2a the two switches are wired in **parallel;** electricity can flow through either switch, so the entire network is on if either one switch alone is on or if both are on; if both switches are off, then the network is off. Now look at Figure 3.2b, where the switches are wired in **series.** Since electricity must flow through both switches, the network is on only if *both* switches are on; otherwise it is off.

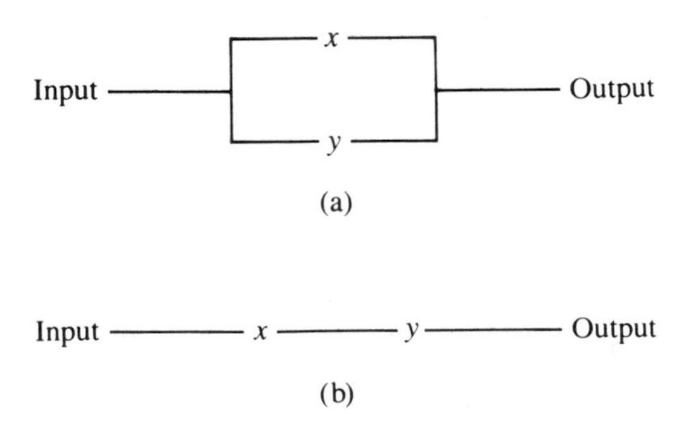

Figure 3.2 Two switches combined (a) in parallel and (b) in series.

Now let's have some fun. Let's denote by 0 a switch that is off, and by 1 a switch that's on. Consequently, $x = 0$ means that switch x is off, and $x = 1$ means that switch x is on. Furthermore, let's use the symbol $\vee$ to mean "wired in parallel with," so that $x \vee y$ means "the network consisting of switch x wired in parallel with switch y." Similarly, let the symbol $\wedge$ mean "wired in series with," so that $x \wedge y$ denotes switches x and y in series. You should carefully confirm that the states of the resulting networks are represented in Figure 3.3.

There is no uniform way to read $\vee$ and $\wedge$ in switching algebra; "parallel" and "series" are possibilities in this instance. More acceptable readings are "or" and "and," respectively, as we'll see later.

These tables look similar to addition and multiplication tables we used before and should be read accordingly. If x and y are given concretely, look in the table for the intersection of the row headed by the value of x and the column headed by the value of y; the entry in that position is the value of $x \vee y$ or $x \wedge y$, respectively. If we analyze those tables as we did those for $\mathbb{N}$ and $\mathbb{Z}$, we can make the following observations:

- Operations $\vee$ and $\wedge$ are both commutative.
- Both can readily be extended to more than two switches by verifying (from the tables) that $(x \vee y) \vee z = x \vee (y \vee z)$ and $(x \wedge y) \wedge z = x \wedge (y \wedge z)$ for every choice of x, y, and z. That is, both operations are associative. See Figures 3.4a and b.
- Operation $\vee$ has an identity, 0, which is also an annihilator for operation $\wedge$. (See Section 3.2)
- Operation $\wedge$ has an identity, 1.
- Object 0 is its own inverse under operation $\vee$, but 1 has no inverse. Similarly, object 1 is its own inverse under $\wedge$, but 0 has none. So for each $\vee$ and $\wedge$, one object has an inverse, one does not.
- Again from the tables, it is easy to verify that *each* of the two operations distributes over the other. This is remarkably different from ordinary arithmetic, in which $(xy) + z$ is *not* the same as $(x + z)(y + z)$. However, $(x \wedge y) \vee z$ *is* the same as $(x \vee z) \wedge (y \vee z)$.

Since $x \vee 0 = x$ for all x, the similarity between $\vee$ and addition is striking. Similarly, $x \wedge 1 = x$ for all x, so $\wedge$ looks remarkably like multiplication. Adding force to the argument is the fact that $x \wedge 0 = 0$ for all x.

$\mathbb{N}$, $\mathbb{Z}$, $\mathbb{Q}$, $\mathbb{R}$, $\mathbb{C}$, and $\mathbb{Z}_m$ all have similar properties for their additions and multiplications. In fact, some authors actually use the terms "addition" and "multiplication" (and even the symbols $+$ and $\cdot$) for $\vee$ and $\wedge$, in spite of the fact that

$\vee$	0	1
0	0	1
1	1	1

$\wedge$	0	1
0	0	0
1	0	1

Figure 3.3 Tables of operations $\vee$ and $\wedge$.

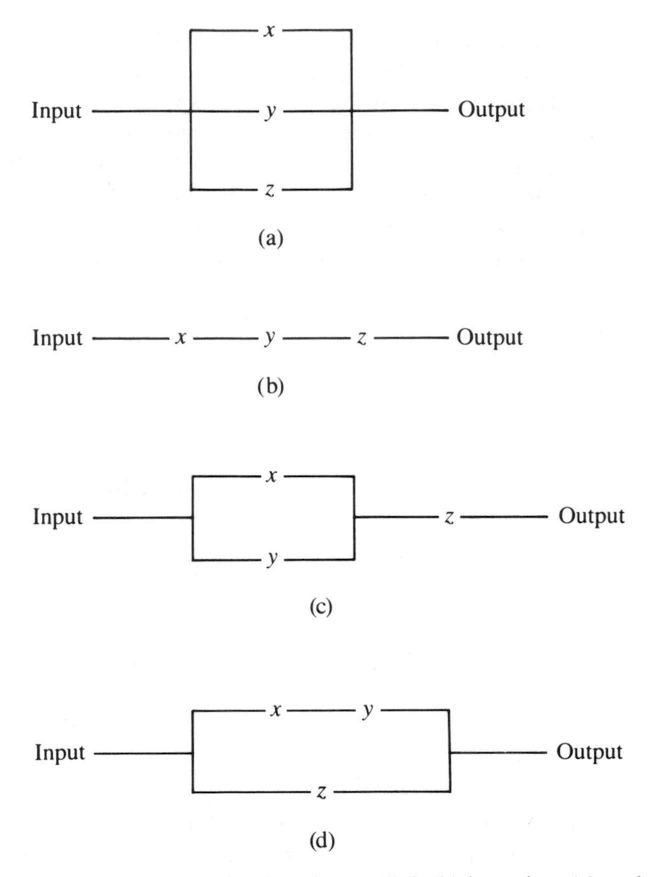

Figure 3.4 Three switches combined (a) in parallel, (b) in series, (c) and (d) mixed.

$\vee$ and $\wedge$ are clearly not numerical operations. This usage is all right, provided care is taken not to change the substance of the tables in Figure 3.3 and provided we keep always in mind that addition or multiplication in, say, $\mathbb{Z}$, is not the same *calculation* as addition or multiplication in switching algebra—although their *form* and *properties* are markedly similar. It is these forms and properties that we are interested in, and that we shall study in detail as we progress.

In conclusion, we have an *algebra,* the **two-terminal switching algebra:** the *objects* are the states "off" and "on" (or 0 and 1, respectively), and the *rules* are the operations $\vee$ and $\wedge$ as given by the tables we produced. Some of the *implied* rules of this algebra are those we've seen, such as commutativity, associativity, distributivity, and identity properties.

Note 4 on Theorem Proving. The properties above (commutativity, associativity, and so on) provide more examples of proofs that are simple and direct verifications, but that may require us to organize the order of the steps and to patiently split the general statement into several cases. Let's investigate in detail the proof of some of the statements above. First of all, because the identity properties are simpler than the others we may want to prove those first. In order to prove the identity property for 0 (that is, $x \vee 0 = x$ for every x), keep in mind that, in the present context, "every" really means "either of 0 or 1." So all we have to do is prove that $0 \vee 0 = 0$ and $1 \vee 0 = 1$ (exhaustion of cases), and both results come directly from the tables. Second, let's see how to prove a "longer" property, such as

$$(x \vee y) \vee z = x \vee (y \vee z). \tag{2}$$

We could again proceed by exhaustion of cases, considering separately all possibilities for x, y, and z. A slightly shorter way would be to use the identity property we just proved and see what happens to equation 2 when x, y, or z is 0. The formula is verified immediately in each of the three cases: $(0 \vee y) \vee z = y \vee z$ and $0 \vee (y \vee z) = y \vee z$, $(x \vee 0) \vee z = x \vee z$ and $x \vee (0 \vee z) = x \vee z$, $(x \vee y) \vee 0 = x \vee y$ and $x \vee (y \vee 0) = x \vee y$. The remaining case, namely $(1 \vee 1) \vee 1 = 1 \vee (1 \vee 1)$, is verified directly from the tables. **Q.E.D.**

The abbreviation Q.E.D. used here and earlier abbreviates the Latin *quod erat demonstrandum*, which means "that which was to be proved." It is a convenient and customary way to mark the end of a proof.

Exercises 3.4

1. Let a, b, c, d, and e be switches, each either on (1) or off (0). Draw a switching network corresponding to each of the following formulae: **(a)** $a \vee b \vee c \vee d$, **(b)** $a \wedge b \wedge c \wedge d$, **(c)** $((a \vee b) \wedge c) \vee d$, **(d)** $a \vee ((b \wedge c) \vee d)$, **(e)** $(a \wedge b) \vee (c \wedge d)$, **(f)** $((a \vee b) \wedge (c \vee d)) \wedge e$.
2. Write formulae for the switching networks in Figure 3.4.
3. Evaluate $(a \vee a)$ and $(a \wedge a)$ for each a. Since the operation $\wedge$ is similar to multiplication, we can set $a^2 := a \wedge a$. (Compare *existence of powers* in Section 3.2.) Give consistent definitions for a^0, a^1, and a^k $(k \geq 1)$. Evaluate each of these expressions.

Complement 3.4

1. In ordinary arithmetic, a symbol like $2a$ can be interpreted in two ways: as the product $2 \times a$ or as the sum $a + a$; the results are the same, of course. Similarly, $3a$ is $3 \times a$ or $a + a + a$, and so on. But if we have a structure like switching algebra that has an addition-like operation (here $\vee$), and if a is an object in that structure, then the first interpretation may fail. In switching algebra, we can consider $a \wedge a$, $0 \wedge a$, $1 \wedge a$, but not $2 \wedge a$, because 2 is not an object in our algebra. The second interpretation still holds, however: $2a := a \vee a$, $3a := a \vee a \vee a$, and so on. Also, $1a$ is defined quite naturally as just a. What about $0a$? Give a formal definition of ka when k is a natural number and a is an object in

a structure that has an addition-like operation. As a guideline, use *existence of powers* in Section 3.2. Distinguish the cases $k = 0, k = 1, k > 1$. Consider the possibility $k < 0$.

In switching algebra, calculate $0a$, $1a$, $2a$, $3a$ for every value of a (that is, for each of the two values of a). Do the same in $\mathbb{Z}_2$, $\mathbb{Z}_3$, $\mathbb{Z}_5$, and (for several values of a) in $\mathbb{Z}$.

3.5 Arithmetic and Geometric Series

In this section, we look at two special types of summations that arise commonly in many areas of mathematics, but particularly in discrete mathematics. We look at these now not only to confirm your understanding of summations but also because these two techniques, *arithmetic and geometric series,* will be needed later in the book. Thus, as you read this section, you will want to make note not only of the general techniques used but also of the specific results.

Suppose we have a list of numbers (integer or real, positive, negative, or zero) $a_0, a_1, a_2, \ldots, a_n$, a total of $n + 1$ numbers, which are indexed from 0 to n to indicate that they occur in this order. Such a list is called a *sequence* or *progression*. An individual number is called an **element** or **term** of the sequence and is denoted by a_i. We could have an infinite sequence; in that case, we usually write $a_0, a_1, a_2, \ldots$. For instance, the sequence of natural numbers is written as $0, 1, 2, \ldots$ (infinite), and the sequence of multiples of 3 between 1 and 100 is $3, 6, 9, \ldots, 99$ (finite). For now, we are going to concern ourselves only with finite sequences. The series could start with a_1 instead of a_0 if you wish, so long as you're consistent. For our purposes, it's more convenient to start with a_0.

Now suppose we take some finite sequence and sum the elements. That is, we wish to evaluate $\Sigma_{i=0}^{n} a_i$. Such a summation is called a **series summation** or simply a **series.** Of course, this summation presents no theoretical difficulties; we have all the numbers of the sequence, and we need only add them up. However, there are two particular cases of sequences and their series summations where even this addition step can be shortened. To see how, let's look at the original sequences.

3.5.1 Arithmetic Sequences

In the first special case, suppose our given sequence is such that each element in the sequence differs from the preceding one by some constant, which we denote by h. Such a sequence is called an *arithmetic sequence,* and h is the *arithmetic shift*. Clearly, the first element a_0 must be specified explicitly. But now our sequence looks like this:

$$
\begin{aligned}
a_0 &= a_0 \\
a_1 &= a_0 + h \\
a_2 &= a_1 + h \\
&\vdots \\
a_i &= a_{i-1} + h \\
&\vdots
\end{aligned}
$$

(Both of the examples we gave above are arithmetic sequences. For the natural numbers, $a_0 = 0$ and $h = 1$; and, for the positive multiples of 3, $a_0 = 3$ and $h = 3$.) This means that the entire sequence can be written as follows when a_0 and h are given:

$$
\begin{aligned}
a_0 &= a_0 \\
a_1 &= a_0 + h \\
a_2 &= a_1 + h = (a_0 + h) + h = a_0 + 2h \\
a_3 &= a_2 + h = (a_0 + 2h) + h = a_0 + 3h \\
&\vdots \\
a_i &= a_{i-1} + h = (a_0 + (i - 1)h) + h = a_0 + ih \\
&\vdots
\end{aligned}
$$

In other words, the entire sequence is completely determined by just two numbers, a_0 and h. Any element a_i is simply $a_0 + ih$. Since the entire sequence depends only on a_0 and h, the series $\sum_{i=0}^{n} a_i$ must also depend only on a_0, h, and, of course, n.

Is there a short way to evaluate the series to take advantage of these observations? Yes, there is. To see how, we first note that a summation is just a fancy notation for addition and that therefore the following manipulation of terms is perfectly legitimate:

$$
\begin{aligned}
&a_0 + (a_0 + h) + (a_0 + 2h) + \cdots + (a_0 + nh) \\
&\qquad = (a_0 + a_0 + \cdots + a_0) + (0 + h + 2h + \cdots + nh),
\end{aligned}
$$

where each of the ellipses on the right indicates that we are adding $n + 1$ terms within each pair of parentheses. We can factor a_0 from the first sum and h from the second. More importantly, we can write the whole thing in summation notation as

$$
\begin{aligned}
\sum_{i=0}^{n} a_i &= \sum_{i=0}^{n} (a_0 + ih) \\
&= \left(\sum_{i=0}^{n} a_0\right) + \left(\sum_{i=0}^{n} ih\right) \\
&= a_0\left(\sum_{i=0}^{n} 1\right) + h\left(\sum_{i=0}^{n} i\right) \\
&= a_0(n + 1) + h(0 + 1 + 2 + \cdots + n).
\end{aligned}
$$

(Stop! Do you see why $\Sigma_{i=0}^{n} 1 = n + 1$?) Look closely at the last expression in parentheses on the right. This is simply the series for the natural numbers, the sum of 0 through n. If we knew this number, the rest would be a few multiplications and additions. *Every arithmetic series reduces to the special case of the natural number series!* All we have to do now is to find a short way to evaluate $0 + 1 + 2 + \cdots + n$, that is, $\Sigma_{i=0}^{n} i$. In fact, since the first term, 0, adds nothing to the series, all we have to evaluate is $\Sigma_{i=1}^{n} i = 1 + 2 + \cdots + n$. Let's see how we can do that quickly.

There is a marvelously simple trick to evaluating this sum. For convenience, let's define the symbol S_n to be the sum of the first n positive natural numbers, that is,

$$S_n := \sum_{i=1}^{n} i = 1 + 2 + \cdots + n,$$

which is the same as

$$S_n := 1 + 2 + 3 + \cdots + (n - 2) + (n - 1) + n,$$

and also the same as

$$S_n := n + (n - 1) + (n - 2) + \cdots + 3 + 2 + 1,$$

because addition is commutative. Suppose we add the two sums together, term by term, from left to right. We obtain

$$\begin{aligned} S_n + S_n = {} & [1 + n] + [2 + (n - 1)] + [3 + (n - 2)] + \cdots \\ & + [(n - 2) + 3] + [(n - 1) + 2] + [n + 1]. \end{aligned}$$

The value of each of the n terms in brackets [ ] on the right is $n + 1$. Thus we can write $2S_n = n(n + 1)$, or simply $S_n = n(n + 1)/2$. For instance, then, $1 + 2 + 3 + \cdots + 100 = 100 \cdot 101/2 = 5050$.

We can summarize our results as follows:

Definition 3 A **finite arithmetic sequence** is a sequence $a_0, a_1, a_2, \ldots, a_n$, having **initial term** $\boldsymbol{a_0}$ and **arithmetic shift** $\boldsymbol{h}$, such that, for all positive i, $a_i := a_{i-1} + h$.

Theorem 4 The corresponding **finite arithmetic series** is given by

$$\sum_{i=0}^{n} a_i = a_0(n + 1) + hn(n + 1)/2 = (n + 1)(a_0 + hn/2). \tag{3}$$

In particular, $\Sigma_{i=1}^{n} i = n(n + 1)/2$.

For an example, we return to the multiples of 3 between 1 and 100. Here $a_0 = 3$, $h = 3$, $n = 32$; there is a total of 33 terms, a_0 through a_{32}; and $a_i = 3(i + 1)$. Thus $\Sigma_{i=0}^{32} 3(i + 1) = (32 + 1)(3 + 3 \cdot 32/2) = 1683$.

3.5.2 Geometric Sequences

The second special sequence we want to look at is the **geometric sequence.** Here again, the first term is given explicitly, and each term after the first is defined in terms of its immediate predecessor. But here, the new term is obtained by *multiplying* the preceding term by the constant **geometric shift,** which we again call h. That is, once a_0 has been given, each successive term a_i is obtained as $a_i := a_{i-1}h$. (We use the symbols a_i and h here too, not to confuse you, but simply to emphasize that these symbols are arbitrary designations. It's the *ideas* we want you to understand and remember, not specific symbols.) Once again, any given element of the sequence can be written as (the term a_0 is given)

$$\begin{aligned} a_0 &= a_0 \\ a_1 &= a_0 h \\ a_2 &= a_1 h = (a_0 h)h = a_0 h^2 \\ &\vdots \\ a_i &= a_{i-1}h = a_0 h^i \\ &\vdots \end{aligned}$$

Again we see that the entire sequence depends only an a_0 and h. The corresponding *geometric series* is then

$$\sum_{i=0}^{n} a_i = a_0\left(\sum_{i=0}^{n} h^i\right) \tag{4}$$

and it too depends only an a_0, h, and n. Note the way that a_0 was factored out of the last summation. You should write this out for yourself to verify that this is legitimate.

Now we have a crucial difference between arithmetic and geometric series. Every arithmetic series reduces to the special case of the natural number series $1 + 2 + 3 + \cdots + n$, but in equation 8, after we factor out a_0, we have to deal with the value h that appears in the special geometric series

$$\sum_{i=0}^{n} h^i. \tag{5}$$

In particular, if $h = 1$, then all the terms in the sequence are just 1, and the series summation is $\sum_{i=0}^{n} 1^i = n + 1$. On the other hand, if $h \neq 1$, then we already saw a formula that gives us the result, formula 6 in Chapter 2. We reproduce it here, with an h in place of x:

$$h^{n+1} - 1 = (h - 1)(h^n + h^{n-1} + \cdots + h^2 + h + 1). \tag{6}$$

Let's again denote by S_n the summation in equation 5 (it's *my* symbol; I can redefine it however I want to!):

$$S_n := \sum_{i=0}^{n} h^i = 1 + h + h^2 + \cdots + h^n. \tag{7}$$

If we multiply S_n by $h - 1$, we obtain the right-hand side of equation 6. Therefore $h^{n+1} - 1 = (h - 1)S_n$ and

$$S_n = (h^{n+1} - 1)/(h - 1). \tag{8}$$

Let's summarize this result (including equations 4, 7, and 8) in the form of a definition and a statement.

Definition 5 A **finite geometric sequence** is a sequence $a_0, a_1, a_2, \ldots a_n$, with **initial term** a_0 and **geometric shift** h such that, for all positive i, $a_i := a_{i-1}h$.

Theorem 6 The corresponding **finite geometric series** is given by

$$\sum_{i=0}^{n} a_i = a_0(n + 1) \qquad \text{if } h = 1; \tag{9}$$

$$\sum_{i=0}^{n} a_i = a_0(h^{n+1} - 1)/(h - 1) \quad \text{if } h \neq 1. \tag{10}$$

In particular, $\Sigma_{i=0}^{n} h^i = (h^{n+1} - 1)/(h - 1)$ if $h \neq 1$.

By way of example, let's consider a sequence of powers of 10, say 1, 10, 100, . . . 1,000,000. Here $a_0 = 1$, $h = 10$, and $n = 6$. Thus we seek $\Sigma_{i=0}^{6} 10^i$. Since $h \neq 1$, equation 10 yields $\Sigma_{i=0}^{6} 10^i = (10^7 - 1)/9 = 1{,}111{,}111$ (which we've seen somewhere before).

Exercises 3.5

1. Prove that equation 3 for the sum of an arithmetic series can be rewritten in the form $\Sigma_{i=0}^{n} a_i = (n + 1)(a_0 + a_n)/2$ (in words, the number of terms multiplied by the arithmetic average of first and last term).
2. Calculate **(a)** $\Sigma_{k=200}^{300} (6k)$; **(b)** $\Sigma_{h=20}^{30} (11h)$; **(c)** $\Sigma_{k=22}^{32} (3k + 1)$; **(d)** $\Sigma_{j=21}^{31} 3^j$.
3. Calculate the sum of all **(a)** the integers from 1 to 299; **(b)** the integers from 300 to 399; **(c)** the even integers from 2 to 98; **(d)** the even integers from 200 to 266; **(e)** the multiples of 3 from 3 to 99; **(f)** the powers of 2 from 2 to 1024; **(g)** the powers of 2 from 256 to 16384; **(h)** the powers of 3 from 3 to 2187; **(i)** the powers of 3 from 81 to 19683.
4. Calculate the sum of **(a)** the first 21 positive even integers; **(b)** the first 15 positive odd integers; **(c)** the first 7 positive multiples of 4; **(d)** the first 11 positive multiples of 11.
5. Calculate the sum of the integers between 1000 and 10,000,000 inclusive.
6. Calculate the sum of the integers between 10^m and 10^n inclusive, where $0 \leq m \leq n$.
7. Calculate the sum of the powers of 10 between 1000 and 1,000,000 inclusive.
8. Calculate the sum of the powers of 10 between 10^m and 10^n inclusive, where $0 \leq m \leq n$.
9. Calculate **(a)** $\Sigma_{i=0}^{n} x^{2i}$; **(b)** $\Sigma_{i=0}^{n} x^{3i}$; **(c)** $\Sigma_{i=m}^{n} x^i$, for $m \leq n$; **(d)** $\Sigma_{i=m}^{n} x^{2i}$, for

$m \le n$. Note that the results will be simple expressions involving x, n, and possibly m, but not i, which is a dummy variable in each summation. Test your expressions with specific values for x, n, and m.

10. Using the new tools, redo problem 2 of Exercises 2.7.

3.6 The Binomial Theorem

From the study of standard algebra, you should be familiar with the formulae

$$(x + y)^2 = x^2 + 2xy + y^2,$$
$$(x + y)^3 = x^3 + 3x^2y + 3xy^2 + y^3.$$

Even if you don't immediately recognize (or don't quite remember) them, you can reconstruct them by multiplying out. Or they can easily be derived from the now-familiar definitions

$$(x + y)^0 := 1, \quad (x + y)^1 := x + y, \quad \text{and} \quad (x + y)^{k+1} = (x + y)^k(x + y), \qquad (11)$$

when k is any natural number (see *existence of powers* in Section 3.2). The examples above are the cases in which $k + 1 = 2$ and 3. For larger exponents, we could simply continue the long multiplication, calculating each new, larger polynomial in turn. But there is an easier way. Moreover, the easier way is a technique that is of considerable importance besides the expansion of polynomials. Suppose we state our results so far in a table, omitting the plus signs. For each exponent k, we understand that we are to add the several terms horizontally in the appropriate row.

$k = 0$	1			
$k = 1$	x	y		
$k = 2$	x^2	$2xy$	y^2	
$k = 3$	x^3	$3x^2y$	$3xy^2$	y^3
⋮	⋮			

The rows are indexed by k ($k = 0$ for row 0, $k = 1$ for row 1, etc.) The pattern of exponents in the literal part is quite readily apparent: In each row, there are $k + 1$ terms, which we will index with the index i ($i = 0, 1, 2, \ldots, k$); then, counting from left to right in row k, term i has literal part $x^{k-i}y^i$ (recall that x^3 is also x^3y^0). You should immediately try to extend the table by a few rows, in only the literal parts, to see this pattern. But what about the coefficients? Since the literal parts are simple to supply later, let's rewrite the table listing only the coefficients. We get

$k = 0$	1					
$k = 1$	1	1				
$k = 2$	1	2	1			
$k = 3$	1	3	3	1		
$k = 4$	1	4	6	4	1	
$k = 5$	1	5	10	10	5	1
⋮	⋮					

We've given you a few more rows for comparison. Is there a simpler way to generate the coefficients than multiplying out polynomials? Yes, there is, and it is given formally by the next definition and the next theorem. But let's see it informally first. The power $(x + y)^k$ is a polynomial whose coefficients are provided by row k of the table above (Pascal's triangle) and whose literal parts are provided by row k of the table further above (we've already stated how this is constructed). In order to construct successive rows of Pascal's triangle, we need only three facts: (1) each successive row will have one more entry than the preceding one; (2) the first and last entries are 1; (3) adding two consecutive entries of any known row gives as a result the entry in the next row that falls beneath the second of the two entries we've added.

In order to prove the theorem, we have to consider a *general* polynomial multiplication as in equation 11. First, we write the general formula

$$(x + y)^k = C_{k,0}x^k + C_{k,1}x^{k-1}y + C_{k,2}x^{k-2}y^2 + \cdots + C_{k,k-1}xy^{k-1} + C_{k,k}y^k$$
$$= \sum_{i=0}^{k} C_{k,i}x^{k-i}y^i. \tag{12}$$

Note very carefully that the notation $C_{k,i}$ refers only to the coefficients within row k; each row has its own complete set of coefficients: $C_{k,0}, C_{k,1}, \ldots, C_{k,k}$. Now calculate $(x + y)^{k+1}$ according to equations 11 and 12. To use formula 11 we must expand the product of formula 12 (as written) by $(x + y)$. What will be the coefficient of term i in row $k + 1$, i.e., what is $C_{k+1,i}$? The key rests in seeing that the needed literal part is $x^{k+1-i}y^i$. When we multiply formula 12 by x we obtain the term $\cdots + C_{k,i}x^{k-i}y^ix + \cdots$. When we multiply it by y, we obtain the term $\cdots + C_{k,i-1}x^{k+1-i}y^{i-1}y + \cdots$. *Only these two terms* contribute to term i of row $k + 1$.

Now we want to compare this result with what we would obtain if we applied formula 12 to the expression $(x + y)^{k+1}$. We must rewrite formula 12 as it appears when k is replaced by $k + 1$:

$$(x + y)^{k+1} = C_{k+1,0}x^{k+1} + C_{k+1,1}x^ky + C_{k+1,2}x^{k-1}y^2 + \cdots$$
$$+ C_{k+1,k}xy^k + C_{k+1,k+1}y^{k+1}$$
$$= \sum_{i=0}^{k+1} C_{k+1,i}x^{k+1-i}y^i.$$

The comparison gives

$$C_{k+1,i}x^{k+1-i}y^i = (C_{k,i}x^{k-i}y^i)x + (C_{k,i-1}x^{k-(i-1)}y^{i-1})y.$$

Since the literal parts can be cancelled from each side, we have that

$$C_{k+1,i} = C_{k,i} + C_{k,i-1}. \tag{13}$$

But we have a problem. This formula doesn't work when $i = 0$ (since there is no such thing as $C_{k,-1}$), nor when $i = k + 1$ (since there is no such term as $C_{k,k+1}$). These difficulties can be handled with the easy observation that *within any row, the first and last coefficients are always 1.* In formula 12, for every k: $C_{k,0} = 1$, $C_{k,k} = 1$.

Formula 13 provides a way to generate each successive row of the coefficient table, given only the preceding row. Obviously, the table could be expanded indefinitely, to show higher and higher values of k. Expressing all of our observations (including formula 13) in words, we first define *Pascal's triangle*—as a table—and then use it to state a form of the *binomial theorem.*

Definition 7 ***Pascal's triangle*** is the table constructed as follows:

(1) The rows are indexed by k ($k = 0, 1, 2, \ldots$).
(2) Row k has $k + 1$ terms, indexed by i ($i = 0, 1, 2, \ldots k$).
(3) Row 0 contains the single entry 1.
(4) The entry with index i in row k is denoted $C_{k,i}$.
(5) Within row k, the first and last entries, $C_{k,0}$ and $C_{k,k}$ are both 1.
(6) The sum of any two consecutive entries in any row other than row 0 is the entry that appears in the next row beneath the second of the two entries. (Formula 13 says the same thing, but is more concise.)

Theorem 8: ***Binomial theorem.*** For any natural number, k, the expansion of $(x + y)^k$ is a polynomial whose terms, indexed by i ($i = 0, 1, 2, \ldots k$), are monomials that have as their literal parts the expressions $x^{k-i}y^i$ and as their coefficients the entries $C_{k,i}$ from Pascal's triangle. Equivalently,

$$(x + y)^k = \sum_{i=0}^{k} C_{k,i}x^{k-i}y^i.$$

Pascal's Triangle is named for the seventeenth century mathematician Blaise Pascal, who studied it. Pascal's triangle has many crucial implications beyond polynomial expansion, and we shall see much more of it later in the book. In fact, you should regard Pascal's triangle, not as an artifact of polynomial expansion, but as an independent mathematical object interesting in its own right. The fact that it can be used for polynomial expansions is a happy coincidence.

Exercises 3.6

1. Expand $(x + y)^9$; $(a + b)^5$; $(t + 1)^6$; $(u - v)^8$; and $(u^2 + 1)^9$.

2. Each sequence below contains a few consecutive elements of a row of Pascal's triangle; use only *addition* to construct as many elements as possible in the next rows of Pascal's triangle. **(a)** . . . 969 3876 11628 27132 50388 75582 . . .; **(b)** . . . 3003 5005 6435 . . .; **(c)** . . . 816 3060 8568 18564 . . ..

3. Prove that within row k the sum of the terms of Pascal's triangle is 2^k. [Hint: $x := 1$ and $y := 1$.]

4. Prove that within any row other than row 0, if the terms are assigned alternating positive and negative signs, then they will sum to 0. [Hint: $x := 1$, $y := -1$.]

5. Using the addition rule in Pascal's triangle (see equation 13), and still using ones at the edges of the table, do the additions in $\mathbb{Z}_5$. That is, construct Pascal's triangle mod 5. If you are patient enough (to row 10 or, better, row 16) you'll discover a strange pattern.

6. Do the same in $\mathbb{Z}_3$. You'll need less patience, but a sharper ability for pattern recognition.

7. Do the same in the switching algebra, using $\vee$ instead of $+$, and observe that nearly nothing happens.

Selections, Permutations and Combinations

A *string* is a sequence of symbols placed beside each other with no gaps. The most common example of a string over an alphabet is a word in ordinary language. In an English word, the letters are selected from the English alphabet and are placed in a definite order. If different letters are selected, or even if the same letters are selected, but arranged in a different order, the result is a different word. An example is provided by the words "depart" and "parted." It is possible to select the same letter more than once, but once again, the order determines whether two words are the same; the words "add" and "dad" both contain one *a* and two *d*s, but are clearly different.

Even this simplistic definition, however, needs to be more detailed. Consider the words "us", "Us", and "US". Are they all the same word? The first two probably *mean* the same thing (as in "all of us"), but even these two would not be considered the same by a typesetter. The third would probably be interpreted by most Americans as the abbreviation for "United States". So the *distinctness* of these words (that is, whether they are different words) may be open to some ambiguity. Moreover, consider the word "twenty-five". When we specified the English alphabet, would you have included the hyphen as a letter? Should it be included at all? This question plays no end of havoc with computerized mailing lists. Here again, the choice of alphabet is open to interpretation. This chapter will clarify such ambiguities and investigate some specific types of strings.

4.1 Strings and Words over an Alphabet

There are some further complications in discussing words as strings. Not all languages have the same alphabet. Russian uses an alphabet vaguely similar to English, but the Chinese "alphabet" is completely different. (Those who view

English as the ''natural'' alphabet would do well to remember that native speakers of Cantonese, the major Chinese language, outnumber English speakers by nearly three to two! Also, a single symbol of written Chinese is not really a letter in the Western meaning, but the collection of such symbols can still be considered as an ''alphabet.'') Even the notion of a ''word'' can be tricky; in studying diction, nonsense words can be a powerful tool, and even the familiar Christmas carol ''Deck the Halls'' contains the nonsense words ''fa-la-la,'' and so on. Finally, a ''language'' can mean something other than a natural (human) language; it could mean a computer programming language, or even the language of mathematics. Consequently, when we consider ideas like these, we need more precise definitions. In this section, we shall, still rather informally, look at some of them.

Definition 1 A **finite alphabet** consists of a finite number of distinct symbols, called just **symbols,** or **letters,** or **characters.**

The upper-case English alphabet is a finite alphabet, with or without the customary inclusion of special characters like the hyphen, period, and comma. We may also want to include the space, or blank, as a character in its own right; ''by law'' and ''bylaw'' are not the same string. The digits of a counting system provide another example of an alphabet; another alphabet may comprise the same digits and again special characters like a radix point and positive and negative signs.

Another question is the ''validity'' of a string. In our most obvious example, that of strings from the English alphabet, we have made no distinction of validity between SORT and SQRT. Both are four-letter strings by our definition, but the latter is a nonsense word. If by ''valid'' we mean simply that a string has four letter, then both of these are valid. If, however, we interpret validity to mean that the string is an actual English word—one that has meaning—then ''SORT'' is valid, but ''SQRT'' is not. In most computer languages, however, ''SQRT'' has a meaning—''the square root of''. Mathematics cannot determine *meaning*. It can, however, determine *form,* and often that's enough for our purposes. For us, validity means that a string has a particular form, according to some clearly stated rule. Any string that satisfies the statement of form will be called a *word.* Note that this distinction between *string* and *word* is not universally accepted; different books may use the terms differently.

Suppose we take the alphabet of the ten decimal digits 0123456789, and construct all strings of nine digits. Any such string will be a valid word in the sense that it's a nine-digit natural number. If that's all we want to know, fine. But suppose we insert two hyphens and write the string in the form 123-45-6789. Now it looks like a Social Security number. (The length is still 9 since only nine digits are chosen. For our purposes here, the hyphens are solely for readability. Of course, our typesetter or IRS officer could argue that now we have a word of length 11 from the alphabet 0123456789-, hyphen included. But we'll say that, for us, the length is still 9.) The question is whether any nine-digit number *could* be a valid Social Security number. As it happens, the answer is no. The Social Security

Administration does not issue numbers having three initial zeroes. Thus, we can immediately conclude that 000456789 *is* a valid nine-digit integer, but *could not possibly* be a Social Security number, even if hyphens were added. Finally, if we have the number 123-45-6789, which *is* a valid SSN, we ask whether any individual actually *has* this number. Form alone cannot determine this; this is part of *meaning*. Remember, mathematics can determine *validity (form)* but not *meaning*.

Definition 2 Given a finite alphabet, a **string** is a sequence of symbols from that alphabet, possibly with repetitions. The **length** of the string is the number of symbols that comprise it, repetitions included. A **word** is a string whose form satisfies some clearly stated criteria of validity.

In the next several sections, we're going to consider forming strings from finite alphabets under various types of formal rules of validity. In each case, we are going to ask three questions, in this order:

1. What are the formal rules of validity in each case? Namely, are certain symbols allowed in certain positions? Are repeated symbols allowed in any given word? Are the valid words all of a fixed, given length? If not, what length restrictions apply?
2. Subject to these rules, how many possible distinct valid words are there?
3. How might we go about making a list of all such valid words, so that each word appears in the list once and only once?

Briefly, the three questions are *What?, How many?,* and *How can we get them all?* As an example to point out the different nature of these three questions consider the alphabet of the ten digits, 1234567890.

1. What? All strings of length 5 (that is, all five-digit decimal integers, leading zeroes included);
2. How many? 100,000 (the numbers from 00000 to 99999 are as many as the numbers from 1 to 100,000; that is, 100,000);
3. How can we get them all? (a) start from 00000; (b) add 1 (and keep the leading zeroes, if any); (c) if you've got 99999, then stop; otherwise, resume from step (b).

In each case, the finite alphabet will be given. As usual, we assume only that each symbol in the alphabet is distinct from the others, and that the size of the alphabet is known and finite. As for the valid words, sometimes they will be of fixed length (like the SSN example), sometimes not (like an ordinary language); sometimes repetitions will be allowed (SSN or ordinary language), sometimes they will not (a shuffled deck of cards, ready to be dealt, in a one-deck card game); sometimes the order in which the symbols appear in a word will be essential (digit representation of an integer or shuffled deck of cards), sometimes it will not (a hand in bridge). But in every case, these three questions will be asked, in the order shown.

Exercises 4.1

1. Write all possible strings fitting the criteria given: **(a)** 2-character strings from alphabet abc; **(b)** 1-character strings from alphabet abc; **(c)** 1-character strings from alphabet ab; **(d)** 2-character strings from alphabet ab; **(e)** 3-character strings from alphabet ab; **(f)** 1-character strings from the alphabet that comprises the letter a only, **(g)** 2-character strings from the alphabet that comprises the letter a only.
2. Write all the 3-character strings from the alphabet consisting only of the symbols 0 and 1. Describe this list in terminology that does not contain the symbols 0 and 1 themselves.
3. Write several 4-character strings from the alphabet 0123456789ABCDEF. Describe with an appropriate phrase what the full list would be.
4. Suppose you have a finite alphabet with at least one symbol. How many strings of length 0 are there? Be careful! [Hint: In how many ways can you proceed to take nothing out of an alphabet?]
5. Find the number of all **(a)** 3-character strings from the alphabet that contains no characters, and which is called the empty alphabet; **(b)** 2-character strings from the empty alphabet; **(c)** 1-character strings from the empty alphabet; **(d)** 0-character strings from the empty alphabet. [Hint: In how many ways can you proceed to take one, two, or three symbols from an empty alphabet? Nothing from an empty alphabet?]
6. An arithmetical formula in a very elementary arithmetic is a word from the alphabet containing the symbols $0123456789+-\cdot/()$ (decimal points, positive and negative signs for numbers, and ordinary sentence punctuation are not allowed.) For example, $9 + (5 \cdot 4)$ is an arithmetical formula. Is every string from such an alphabet an arithmetical formula? Why or why not? Can you prescribe formal rules of validity for arithmetical formulae? The formal rules may present some difficulty. Be careful and clear. Try to construct formulae according to the rules you stated (not according to your experience), and then check them against your experience. Do they "work"? If you have difficulty, eliminate ( ) from the alphabet and try the problem again.
7. Repeat problem 6 for $01\vee\wedge()$ (referring to switching algebra).

4.2 Selections

To begin a detailed discussion of finite alphabets, let's first look at the alphabet itself. As we've said before, there are really only two things we insist upon for a finite alphabet: that it contain only finitely many symbols and that each of the symbols be distinct from any others. We will use the letter n, a natural number, to indicate the size of the alphabet, the number of symbols in it. For the upper-case English alphabet, $n = 26$ (although we've seen that for some purposes this may not be enough). For the alphabet of decimal digits 0123456789, $n = 10$. We could even have, as a very special case, an alphabet containing no characters, that is $n = 0$. Such an alphabet is said to be **empty.** If $n > 0$, the alphabet is said to be **nonempty.** In the discussion below, we are going to develop formulae that depend upon the value of n. Unless we specifically say that the alphabet must be nonempty, these formulae will hold for any value of n, including $n = 0$.

Let's look first at the simplest type of validity rule, the *selection.* (There's that word "rule" again. Remember the game?) In a selection, finitely many characters are chosen (selected) from the alphabet and placed in a specific order in the word. The same character may be selected as many times as desired. No other restrictions apply. Thus, every string is a word; that is, every string is considered valid. In fact, some authors don't even bother using the term *selection*; they simply talk about all possible strings. We want to be a little more specific because later we shall consider more restrictive cases. The only thing to remember about selections is that two words are considered the same only if they contain *exactly the same symbols in exactly the same order.*

There's nothing to stop us from talking about selections from the English alphabet, but of course these selections will not all be meaningful English words. The best example of a selection process is one you've been dealing comfortably with for years. If we define our alphabet to be the decimal digits 0123456789 ($n = 10$), then any selection from this alphabet is simply a numeral that represents a natural number, leading zeroes included. In forming the integer 1004, we selected, in order, a 1, two 0s, and a 4. Note that zero was selected twice; there are no restrictions on repetition. Note also that the order of the digits is important: the number 1400 also has a 1, two 0s, and a 4, but 1400 is not the same number as 1004. If we define our alphabet as comprising just the symbols 0 and 1, then any selection is a *binary* natural number, and vice versa. For the alphabet 01234567, we have strings that are the *octal* natural numbers, and for the alphabet 0123456789ABCDEF we have the *hexadecimal* natural numbers. The most important use for the idea of selections is in positional systems like these.

So far, we haven't put any limits on the size of the selection word. Suppose we consider as valid only words of precisely length k. (We begin here to use k, a natural number, for word length.) Such a word is called a *k-selection.* Obviously, in the case of all natural numbers in a counting system, no limits are possible, since a natural number can be arbitrarily large. We can, however, talk about k-digit natural numbers. Here we have a minor problem—whether or not to allow leading zeroes. For string-forming purposes, we'll allow leading zeroes. That is, 005 is a valid 3-digit number, 05 is a valid 2-digit number, and 5 is a valid 1-digit number. This complication won't cause any irreparable harm. The case of varying k will be discussed a little later. Before we go on to consider the second and the third question from the last section, let's formally restate the definition of a selection.

Definition 3 **Selection** is a process in which words are formed from a finite alphabet with no restrictions on the form of the words. Consequently, in a selection, (1) repeated symbols are allowed and (2) two words are distinct if they have different lengths, or if they contain different symbols, or if they contain the same symbols in different order. A word formed in this way is also called a **selection.** A ***k*-selection** is a selection of length k.

Example The 2-selections from alphabet ABCD are: AA, AB, AC, AD, BA, BB, BC, BD, CA, CB, CC, CD, DA, DB, DC, DD.

The term *selection* thus serves two purposes. First, it refers to a process, that is, a set of rules. Remembering the notion of a game, the objects are now the symbols of the alphabet, the rules are the selection rules, and the implied objects are the selections. In this sense, selection is a process that applies to any one given alphabet. Once an alphabet has been chosen, we also use the term *selection* to mean a particular word constructed via the selection process on that alphabet. This distinction between *the* selection process and *a* selection word is minor, and it would be pedantic to make much of it. The definition above answers the first question of the previous section: *What is the process?*.

Now let's consider the second question: *How many?* Given an alphabet of size n, how many k-selections can we form? We'll use the symbol $S(n, k)$ to denote their number. That is, we set

Definition 4 $S(n, k) :=$ number of k-selections from an alphabet of size n.

Let's go back to the decimal example and try to count the number of four-digit decimal numbers. In this case, $n = 10$ and $k = 4$. Since every selection word is a string, and every string is a sequence, we'll start by writing the selection as a sequence $a_1a_2a_3a_4$, where a_i are decimal digits. (Again, you could number right-to-left if you like, or number right-to-left, 0 to 3, as we did for positional-base representations. The effect is the same, and our way is a little clearer in the present context.) How many such selections are there? Our technique to count them is the following. If we know how many possibilities there are for a_1, how many for a_2, how many for a_3, how many for a_4 *individually,* then the total number of possibilities is simply the product of these four numbers. Here $n = 10$, so that each a_i could be any of ten digits. Thus for each of ten choices for a_1, there are ten choices for a_2; for each of ten a_1s *and* ten a_2s, there are ten possible a_3, and so on. Multiplying all of these numbers together, we find that the total number of selections is just $10 \cdot 10 \cdot 10 \cdot 10 = 10^4$. The significance of the number $S(10, 4) = 10^4$ becomes clear when you look at the process in general. The process is easily generalized and yields the following theorem, which is the answer to the question of how many, for the case of the selections.

Theorem 5 The number of distinct k-selections from an alphabet of size n is n^k. That is, $S(n, k) = n^k$.

Proof: As for the proof in general, remember that the number of possible selections is simply the product of the numbers of possible choices for each of the elements in the sequence, at least when the numbers involved are positive. Since each element can be any of n alphabet characters, and since there are k elements in the selection word sequence, the product is n^k.

Now let's consider the special cases in which $n = 0$ or $k = 0$. Suppose first that the alphabet is empty, that is, $n = 0$, but $k > 0$. How many k-selections are there? Any attempt to select a character from the alphabet will fail, regardless of how many times (k) we try. Since no selections can be formed, our answer should be 0. But $0^k = 0$ for any positive k, so our formula holds even when $n = 0$ and $k > 0$.

Now let's suppose that the alphabet is nonempty ($n > 0$) but $k = 0$. That is, the process consists of ignoring the alphabet entirely, and not even attempting to form selections. How many possible ways are there to do this? This sounds like an absurd question, reminiscent of medieval arguing about the number of angels dancing on the head of a pin. In fact, though, it's important for compatibility with later cases that we make the following observation: there is precisely one way to do nothing, and that is—do nothing. In other words, for any alphabet, there is one way to take nothing from it; there is exactly *one* 0-selection. Does our formula reflect this? For any positive n, $n^0 := 1$, so our formula holds even when $n > 0$ and $k = 0$.

The only remaining question is what to do if $n = 0$ *and* $k = 0$. That is, how many ways are there to ignore an alphabet which is empty anyway? Since we said that there is *one* way to ignore anything (an empty alphabet, in this case) our answer ought to be 1, and in fact $0^0 := 1$ by the usual definition in $\mathbb{N}$. So far, our formula n^k contains within it three subtle special cases ($n = 0, k > 0$; $n > 0, k = 0$; $n = 0, k = 0$) and works beautifully in each case. **Q.E.D.**

Beware: $0^0 := 1$ in $\mathbb{N}$; $x^0 := 1$ for any positive x in $\mathbb{N}$, $\mathbb{Z}$, $\mathbb{Q}$, $\mathbb{R}$; but 0^0 is a very ambiguous symbol in almost any structure other than $\mathbb{N}$.

At the other extreme, what happens when either n or k or both are allowed to become infinite? (We are not going to need these cases, but thinking about them is good exercise.) First of all, we'll need a symbol for an "infinite number". The usual symbol is ∞, but we'll use α, the first letter of the Greek alphabet, to emphasize that *in the present context,* with caution, and as far as possible, we intend to see that "an infinite number" can be treated like a counting number.

The case of an infinite alphabet ($n = \alpha$) is the easiest to deal with (even though the stress here is on *finite* alphabets): if k is a natural number and $k \geq 1$, then even the first character will have infinitely many choices, and $\alpha^k := \alpha$. When $k = 0$ we set (as usual) $\alpha^0 := 1$.

Now suppose that n is finite, but k is allowed to become infinite, $k = \alpha$. If $n = 0$, we want an answer of 0, as before. Sure enough, we set $0^\alpha := 0$. If $n = 1$, we want an answer of 1, since all we can do is to select that one character repeatedly. Here again, $1^\alpha := 1$. Finally, if $n > 1$, n^α is, again, an infinite number. This last case is precisely what happens when we consider all of the natural numbers. The alphabet, 0123456789 in decimal, for example, is finite, but the number of digits in a number is unrestricted, so the number of selections is infinite. We may want to stress this fact by writing *every* natural number with infinitely many leading zeroes, followed by as many significant digits as necessary: 3 = . . . 00003, 2000 = . . . 0002000, 10^{50} = . . . 000100 . . . 00 (infinitely many zeroes before the 1, fifty zeroes after the 1). Thus our formula works even better than promised: it works even in the infinite cases.

Expressing natural numbers in a positional-base representation is the most obvious example of a selection process. This fact gives an important clue to answering the third question from Section 4.1: how might all possible k-selections be enumerated? By **enumeration,** we mean a list of all possible k-selections, such that each k-selection appears once and only once in the list. An important remark is that, for an alphabet of size n, any k-selection from that alphabet is simply a k-digit natural number expressed in base n. "But wait!" you say, "Digits must have values from 0 through $n - 1$. The symbols in any arbitrary alphabet aren't necessarily digits." Aren't they? The symbols in an alphabet can have any meaning we care to assign them, provided only that we're consistent. The only requirements for the n digits of a positional system in base n are that they be distinct and that they be assigned distinct numerical values from 0 to $n - 1$. All we have to do is to assign one symbol of the alphabet as 0, another as 1, another as 2, and so on, until the last remaining symbol is assigned the value $n - 1$. Which symbol gets which assignment doesn't matter.

For example, in the 26-letter upper-case English alphabet, the most natural assignment would be A $\mapsto$ 0, B $\mapsto$ 1, C $\mapsto$ 2, . . ., Z $\mapsto$ 25. (From now on, we shall use the symbol $\mapsto$: B $\mapsto$ 1 means "to B we assign 1.") Then, suppose we want to enumerate all 3-selections. We know before we start that the number of words will be $S(26, 3) = 26^3 = 17576$. To enumerate them, all we need to do is to count from 0 to 17575, and convert each natural number to this new base 26 system. The first and last few such numbers are shown here.

AAA = 0 (decimal)
AAB = 1
AAC = 2
.
.
.
NOV = November (calendar)
= 9173 (decimal)
NOW = The present; this very instant (dictionary)
= National Organization for Women (current affairs)
= Negotiable Order of Withdrawal (banking)
= 9174 (decimal)
.
.
.
ZZX = 17573
ZZY = 17574
ZZZ = 17575

Note again that the highest natural number here is $26^3 - 1 = 17575$. It isn't even necessary to index the positions from right to left, but we've done that here so that you can see these selections as numbers.

To sum up, a selection is the most general type of string over a finite alphabet. No restriction is made about using any symbol multiple times, and the only way that two selections are considered the same is that they contain exactly the same symbols in exactly the same order. The number of possible distinct k-selections from an alphabet of size n is $S(n, k) = n^k$. To enumerate all the k-selections, simply represent in base n the natural numbers from 0 to $n^k - 1$, interpreting the symbols of the alphabet as n-nary digits.

Exercises 4.2

1. Calculate $S(4, 2)$; $S(4, 3)$; $S(5, 2)$; $S(5, 3)$.
2. Write all the six-digit binary numerals, leading zeroes included. Convert each to octal form.
3. Without looking back to previous chapters, **(a)** how many 28-digit octal numerals are there? **(b)** How many 19-digit hex numerals? **(c)** How many 527-digit base-254 numeral? (Just write the formulae; don't multiply them out!)
4. Let the alphabet be ABC. Write all 2-selections from the given alphabet, once each. Now, calculate $S(3, 2)$ and check whether you have written all 2-selections.
5. Repeat question 4 for **(a)** the 3-selections from the alphabet AB; **(b)** the 4-selections from the alphabet comprising the symbol A only; **(c)** the 0-selections from the alphabet ABC; **(d)** the 0-selections from the alphabet AB; **(e)** the 4-selections from the empty alphabet; **(f)** the 1-selections from the empty alphabet; **(g)** the 0-selections from the empty alphabet.
6. In problems 4 and 5 above, you may have noticed that, although they are simple, you still need an organized way to enumerate selections. Using the algorithm given at the end of this section (or any other efficient and correct algorithm) start enumerating the 3-selections from the alphabet STUVWXYZ, or the 4-selections from the alphabet FT. Stop when you really understand how the algorithm works. You may also want to do similar problems on a computer. Calculate by hand or calculator, in advance, how much output you'll get; do not waste paper!

4.3 Restricted Strings

The rest of this chapter is devoted to cases in which validity is restricted to certain forms of strings. We'll consider two classes of such *restricted strings* in this section because they'll reinforce your understanding of strings. In the first, suppose that a certain position in a string is restricted to only certain symbols. For instance, how many strings out of the English alphabet have three letters, of which the second *must be* P, Q, R, or S? Nothing new here; we again simply multiply the number of choices for each position. Since positions 1 and 3 are unrestricted, each has 26 possibilities. Position 2 has four possibilities. The answer is thus $26 \cdot 4 \cdot 26 = 2704$. A second interesting variation is one in which certain positions in the string must match, whatever letter is actually used. For example, how many four letter words are there in which positions 2 and 4 are the same? Multiplying individual choices, we get $26 \cdot 26 \cdot 26 \cdot 1 = 26^3 = 17576$. Why only one choice for position 4? Because it must match position 2; once the second letter is chosen, whatever it is, that letter (one choice) must also occupy position 4. Put another way, since the second and fourth letters must match, they have 26 choices *between them.*

Obviously there are many ways to restrict the validity of strings; these are merely two examples, toys that help you understand the problem. But there are other types of restrictions that arise commonly in practical problems. Probably the most important of these occurs when we are forbidden to use the same letter more than once. It is to this problem that we next turn our ruthless gaze.

Exercises 4.3

1. In English, there are 5 vowels (A, E, I, O, and U), 20 consonants, and Y, which we're going to treat as a special case, neither consonant nor vowel. How many five-letter strings are there in which positions 1 and 3 are vowels, 2 and 4 are consonants, and 5 is a Y? (Example EVERY).

2. How many nine-letter strings can be formed from the English alphabet, if the first three letters are to be selected from A through L, the next three contain no L, and the last three are equal to each other?

3. How many five-letter words can be formed from the alphabet ABCDE under each of the following restrictions: **(a)** first and last letter are to be different from each other; **(b)** the words contain exactly one C; **(c)** the words contain no repeated letters?

4. How many eight-digit decimal numerals are possible if leading zeroes are not allowed and the last three digits are the same?

5. How many eight-digit octal numerals are possible if leading zeroes aren't allowed and the last three digits are the same?

6. How many 19-letter strings are there in which the first letter is C, the fifth is H, and the twelfth and fourteenth letters are the same?

7. A three-word sentence has the pattern subject/verb/object. Subject and object are to be selected from a list of 2000 given nouns, the verb from a list of 1500 given verbs. **(a)** How many such sentences can be formed? **(b)** How many can be formed under the condition that the object be different from the subject? **(c)** How many can be formed if the object has to be different from the verb, given that 200 words appear in both lists?

8. In the computer language FORTRAN, the name of a variable is made up of from one to six symbols, the first of which is a letter from the English alphabet and the others (if any) are letters or decimal digits. How many possible names can be formed?

9. Find the number of four-digit numerals (leading zeroes included) for which no two consecutive digits are equal to each other.

10. A telephone number for national direct dialing contains a three-digit area code, a three-digit exchange code, and a four-digit station number, for a total of ten digits. (We ignore the possible prefix 1, which is not really part of the number, and we ignore the emergency 911.) The first digit of the area code is other than 0 and 1; the second digit of the area code is 0 or 1. The first and second digits of the exchange are other than 0 and 1; the third digit of the exchange code is other than 0. Is any ten-digit number a valid telephone number? Why or why not? How many ten-digit numbers are there? How many possible direct dialing numbers are there?

11. (a) As we already mentioned, a Social Security number is a 9-selection from the alphabet 0123456789 in which the first three digits are other than 000. How many SSNs are there? Again, don't multiply out. **(b)** In a slightly different form, a SSN is an 11-selection from the alphabet 0123456789- (hyphen included) such that the fourth and the seventh position must contain a hyphen, the other positions must contain digits, and the first three positions are other than 000. How many SSNs are there in this new sense? Compare with the previous answer.

4.4 Permutations

There is one string-validity rule that is encountered far more often than any other, both in theoretical and applied mathematics. This is *nonrepetition,* the requirement that no symbol from the alphabet occur more than once in any one word. In fact, we've already made use of this restriction in the last section, although we didn't explicitly state it. Recall that we said that it is possible to number the n symbols of a finite alphabet from 0 through $n - 1$, so that each symbol gets one and only one assignment. This is the same as placing the symbols of the alphabet in a specific sequence, that is, constructing an n-selection with no repetitions. How many ways are there to do this? In other words, given n symbols, how many possible *orderings* of those symbols are there?

The idea that the symbols of an alphabet might have an order is precisely the idea behind nonrepetition. If the alphabet is ordered, there is a unique first symbol, a unique second symbol, and so on. Most of the alphabets you deal with have an implied order, so that 0 comes before 1 (digits), and A comes before B (ordinary letters). But is this order inherent or simply an arbitrary convention? In the case of digit values, it's inherent; each digit value is exactly 1 greater than its predecessor. In other alphabets, though, it's simply a convention. If you had been taught the English alphabet as ZYX . . . CBA as a child, it would seem today to be perfectly natural. In the Greek alphabet, the equivalent of Z is the sixth letter zeta (ζ); for a Greek, it would sound ridiculous to hear that ζ is the last letter, because omega (ω) is. Whenever we deal with alphabets *other than digit values,* all notions of order are just arbitrary convention. Since our aim is to deal with an arbitrary alphabet, it's up to us—if convenient—to specify any order that alphabet is to have.

4.4.1 Permutations of the Alphabet

In summary,

Definition 6 **Permutation** of an alphabet of size n is the process of constructing those n-selections that contain no repetitions. Each of these selections is called a **permutation of the alphabet** or an **ordering** of the alphabet. Equivalently, a permutation of the alphabet is a string that comprises all the symbols of the alphabet, exactly once each. The number of such permutations is denoted $P(n)$.

For example, the permutations of the alphabet ABCD are ABCD, ABDC, ACBD, ACDB, ADBC, ADCB, BACD, BADC, BCAD, BCDA, BDAC, BDCA, CABD, CADB, CBAD, CBDA, CDAB, CDBA, DABC, DACB, DBAC, DBCA, DCAB, DCBA.

The string-formation process we're discussing is called "permutation," and, as before, the term "permutation" also refers to any word thus formed. To calculate $P(n)$, we use the same technique we used for selections. We can represent any permutation as a sequence, calculate the number of possible choices for each position, and multiply the numbers calculated. For example, suppose we wanted to find $P(26)$, the number of permutations of the English alphabet. As before, we write a permutation as a sequence $a_1a_2a_3 \ldots a_{25}a_{26}$, and, as before, the number of choices for a_1 is 26. But what of a_2? Whatever letter we chose as a_1 is now unavailable, so that a_2 can be any of only 25 choices. It doesn't matter *which* of the 26 letters was chosen as a_1; whichever it was, there are only 25 left. Similarly, for a_3, there remain only 24 choices, once a_1 and a_2 have been chosen. Continuing in this way, there are two choices for a_{25} and only one choice (the only remaining letter) for a_{26}. Consequently, $P(26) = 26 \cdot 25 \cdot 24 \cdot \ \ldots \ \cdot 2 \cdot 1 = 403{,}291{,}461{,}126{,}605{,}635{,}584{,}000{,}000$, or about $4.03 \cdot 10^{26}$. An essential ingredient in this example was the product of the integers 1 to 26. Such a construct is important in many respects and deserves a special definition.

Definition 7 For any natural number n, the **factorial of n** (or **n factorial**) is denoted $n!$ and is defined thus:

$n! :=$ product of the first n positive integers if $n > 1$;
$1! := 1$;
$0! := 1$.

Consequently, for every natural number n, $(n + 1)! = (n + 1) \cdot n!$

The factorial function is one of the fastest growing functions in common use, with 7! already over a thousand, 10! over a million, and 13! or 14! beyond the integer capacity of most computers. Even among specially designed scientific computers and calculators, it's rare to find a machine that can deal with 70!, which is greater than 10^{100}.

Theorem 8 The number of permutations of an alphabet of size n is $n!$ In other words, $P(n) = n!$

Proof: The process we used in the example is easily generalized. When $n>1$, each permutation of the alphabet is a sequence $a_1a_2a_3 \ldots a_{n-1}a_n$, and the number of choices for a_1 is n. Whatever letter we chose as a_1 is now unavailable, so that a_2 can be any of only $n - 1$ choices. Similarly, for a_3, there remain only $n - 2$ choices, once a_1 and a_2 have been chosen. Continuing in this way, there are two choices for a_{n-1} (the next-to-last letter of the word), and only one choice (the only remaining letter) for a_n. Consequently, $P(n) = n \cdot (n - 1) \cdot (n - 2) \cdot \ldots \cdot 2 \cdot 1$, which is precisely $n!$

When $n = 1$, we have to "order" the only symbol of the alphabet, and there is clearly only one way to do that. That is, $P(1) = 1$. Since $1! := 1$, the formula is still valid.

When $n = 0$, we have the usual "ignoring" situation: we do not take any symbol out of anywhere, and there is just one way to do that. That is, $P(0) = 1$. Since $0! := 1$, our formula is valid again. Q.E.D.

4.4.2 k-Permutations

Let the size of the alphabet be still n. We may want to keep the nonrepetition rule for forming words, but we may also want to accept words that contain any number of letters. Such a process is still called *permutation,* but a resulting word is now called a *k-permutation,* where k is, as usual, the word length.

Definition 9 **Permutation** is the process of constructing selections without repetitions from a given alphabet. For any assigned length k, any k-selection without repetitions is called a ***k*-permutation** from the given alphabet. The number of k-permutations from an alphabet of size n is denoted $P(n, k)$.

In other words, the only rule of validity for words in the permutation process is that no letter be repeated in the same word. Consequently, in a k-permutation, for each given k, repeated symbols are not allowed and two words are distinct if they contain different symbols or if they contain the same symbols in different order. As an example, the 3-permutations from the alphabet abcd are

abc abd acb acd adb adc
bac bad bca bcd bda bdc
cab cad cba cbd cda cdb
dab dac dba dbc dca dcb

Note that the permutations of the entire alphabet (which we saw previously) are the n-permutations; we are in the special case in which $k = n$. Consequently, $P(n, n) = P(n) = n!$ When k is any natural number, we'll have to recalculate the number of k-permutations, but the basic ideas will be the same as before. The result is

Theorem 10 If $k > 0$, then the number of k-permutations from an alphabet of size n is the product of k consecutive integers the largest of which is n. In other words, if $k > 0$, then $P(n, k) = n(n-1)(n-2)\cdots(n-k+1)$. If $k = 0$, then $P(n, 0) = 1$. If $k > n$, then $P(n, k) = 0$. If $k < 0$, then we set $P(n, k) := 0$.

Proof: When $0 < k \leq n$, the process of calculating $P(n, k)$ is nearly the same as for $P(n)$. A k-permutation is a sequence $a_1a_2a_3 \ldots a_k$. The number of choices for a_1 is n. Again, a_2 can be any of only $n-1$ choices. For a_3 there remain only $n-2$ choices, once a_1 and a_2 have been chosen. Continuing in this way, we have k numbers to multiply, $n, n-1, n-2, \ldots$. We have so proved the first part of the theorem. For the second part, the only new question is to decide what the last factor is. We proceed as follows:

$$
\begin{aligned}
\text{1st factor} &= n &&= n-(1-1)\\
\text{2nd factor} &= n-1 &&= n-(2-1)\\
\text{3rd factor} &= n-2 &&= n-(3-1)\\
\text{4th factor} &= n-3 &&= n-(4-1)\\
&\vdots\\
k\text{th factor} & &&= n-(k-1)
\end{aligned}
$$

The last factor is $n-(k-1)$, the same as $n-k+1$, as we wanted to show. This argument includes also the case in which $k = 1$, when the only "factor" is n.

It's time again to consider special cases. Does the formula for $P(n, k)$ properly describe what to do with empty alphabets and empty strings? Does it apply to the case when $k > n$? If $k > n$, then we must have $P(n, k) = 0$ because we have to select k distinct elements from n elements, and that's impossible. Our formula reflects that, because the product of more than n consecutive integers from n down includes zero and is therefore zero. The second part of the formula has to be adjusted. To avoid misunderstandings, we set $P(n, k) := 0$ when $k > n$ or when $k < 0$. When $k = 0$, we have to select nothing, whether from an empty alphabet or from a nonempty alphabet does not matter. In either case, there is one way, and $P(n, 0) = 1$. **Q.E.D.**

Example The number of 5-permutations from the upper-case English alphabet is therefore $P(26, 5) = 26 \cdot 25 \cdot 24 \cdot 23 \cdot 22 = 7{,}893{,}600$.

The formula above for $P(n, k)$ is convenient for hand calculation. There is however another formula that is easier to memorize, very inefficient for hand calculation, but useful on a calculator that has the factorial function. Note that some calculators have the function $P(\ ,\ )$ directly.

Theorem 11 If $0 \le k \le n$, then $P(n, k) = n!/(n - k)!$

Proof: If $0 < k < n$, then the numerator is the product of all integers from 1 to n; the denominator is the product of all the integers from 1 to $(n - k)$, and they all cancel out, leaving $[(n - (n - k)]$—that is, k—consecutive factors, the highest being n. So, the formula checks. If $k = 0$ or if $k = n$, then the expressions reduce to $P(n, 0)$ or $P(n, n)$ on the left and to $n!/n!$ or $n!/0!$ on the right. Since $0! := 1$, both sides are equal to each other and the formula checks again. Q.E.D.

4.4.3 Enumeration

We have just seen that a permutation is a string over a finite alphabet with the restriction that no letter be repeated in the same string. The number of possible distinct k-permutations from an alphabet of size n is $P(n, k) = n(n - 1) \cdots (n - k + 1)$.

Finally, let's turn our attention to enumerating, or listing, all k-permutations. The easiest way, at first glance, is to enumerate all k-selections, as described in the last section, and then eliminate any that contain repeated symbols. But this method, although correct, would be very wasteful implemented on a computer. Such exhaustive, "brute force" techniques would achieve the result, but at exorbitant cost. Fortunately, the k-selection algorithm can be modified to produce k-permutations very efficiently. The modification is quite straightforward once you understand how the algorithm actually *does* work as opposed to how you *want* it to work, but it is not simple.

As usual, we have first to work out a plan of attack, to see what we want to accomplish, where the difficulty is, and how to bypass it. We said that the basic idea is to modify the algorithm for enumerating selections, which is, basically, "add 1." In a decimal example, this implies the following: If the right-most position is not 9, then just add 1 to it; if the right-most position is 9, then "carry," that is, bypass the 9s, replacing them with the lowest possible value, 0, and add 1 at the next position that does not contain a 9 (unless, of course, we are "out of range," in which case we stop). The new algorithm will have to do the same, with one exception: whenever we "add 1" (at any position) or write "the smallest possible values," we must do it without creating repetitions. As an example ($n = 10$, $k = 4$), take the 4-permutation 1098. What's the next 4-permutation? Not 1099 since that contains a repetition, not 1100 (repetitions, again), nor any of 1101 to 1199, not 1200, not 1201, not 1202, but 1203. We've increased by 1 (twice) the third position from the right and, in each of the remaining two positions, we've placed the lowest possible values that do not introduce repetitions (0 and 3).

At this point, we can try to write an algorithm. Assuming that $0 < k \le n$, an algorithm to enumerate the k-permutations of the n objects 0 to $n - 1$ (digit values) is the following:

1. Start from the k-permutation $0, 1, \ldots, k-1$.
2. Write down the k-permutation just found.
3. Let p be the right-most position in the k-permutation.
4. Let h be the value of the digit in position p. If h is $n - 1$, then continue at step 6. Otherwise, from among the values $h + 1$ to $n - 1$ (inclusive) select the smallest one that does not equal any of the digits (if any) in the positions on the left of position p. If such a value does not exist, then again continue at step 6. If it does, place it in position p and continue at step 5.
5. If position p is the right-most position, then continue at step 2. If not, from the sequence 0 to $n - 1$, select the lowest possible values (as many as the number of positions on the right of p) that do not coincide with any of the values now at position p and on the left of p (if any). Place these values, in order, in the positions on the right of p. Continue at step 2.
6. If p is not the left-most position, then move to the position left of p, rename this position as p, and continue at step 4. If p is the left-most position, then stop: all k-permutations have been enumerated.

Example Let's apply our algorithm to obtain an enumeration of the 3-permutations from alphabet 0123. Now $n = 4$ and $k = 3$.

Steps 1 and **2** start from string 012. **Step 3** makes p the right-most position of the string. Counting positions right to left, 0 to 2, the value of p is 0. **Step 4** sets $h = 2$, the value of the digit in position p. Since $h < n - 1$, the only available choice is 3, which is placed in position p to obtain string 013. Step 4 then carries the process to step 5, which in turn leads back to steps 2 and 3. **Step 3.** $p = 0$. **Step 4.** The value of the digit in position p is $h = 3$, which equals $n - 1$. This carries the process to step 6. **Step 6** then changes the value of p to 1. Since 1 is not the left-most position, the process continues at step 4. **Step 4.** The value of the digit in position $p = 1$ is $h = 1$. Since $h < n - 1$, the values 2 and 3 are tested and 2 is selected. The string is now 02? and the process continues at step 5. **Step 5.** Since p is not the right-most position, the values 0, 1, 2, and 3 are tested. Values 0 and 2 are excluded because they coincide with values in the first part of the string. Between the remaining values 1 and 3, one value must be selected, and the lower is chosen and placed in position 0, to obtain string 021.

You can see now how the algorithm works, but we have not nearly exhausted the 3-permutations. To save space and to avoid tedious repetitions, we merely list below all 3-permutations in the order they are produced by the algorithm. You can work through some of the details of the process. It may be worthwhile to see, for instance, how the algorithm proceeds after having written 032 or 321 at step 2. The enumeration obtained is

012 013 021 023 031 032 102 103 120 123 130 132
201 203 210 213 230 231 301 302 310 312 320 321.

Exercises 4.4

1. Calculate $n!$ for each value of n from 0 to 10. If you use a calculator, calculate also $11!$, $12!$, . . ., until your calculator declines to provide *an exact integer value*.
2. Calculate $P(3, 1)$, $P(3, 2)$, $P(3, 3)$, $P(4, 1)$, $P(4, 2)$, $P(4, 3)$, $P(4, 4)$.
3. Calculate $P(n, 1)$ for any positive n. Calculate $P(1, 1)$.
4. Find the number of all possible **(a)** 6-digit octal numerals with no repeated digits; **(b)** 3-digit decimal numerals with no repeated digits; **(c)** 3-digit hex numerals with no repeated digits; and **(d)** 3-letter words from the English alphabet (nonsense words included) with no repeated letters.
5. A word obtained from another word by changing the order of its letters is called an *anagram* of the original word. Including nonsense words and the original word, how many anagrams does each of the following words have? **(a)** CAR; **(b)** CARD; **(c)** CARDS; **(d)** SCARED. List the anagrams of **(e)** CAR; and **(f)** CARD.
6. The arrow in the figure below is to be painted in three different colors (background, arrowhead, and stem). Seven colors are available. How many color schemes are there? Enumerate them in some organized fashion.

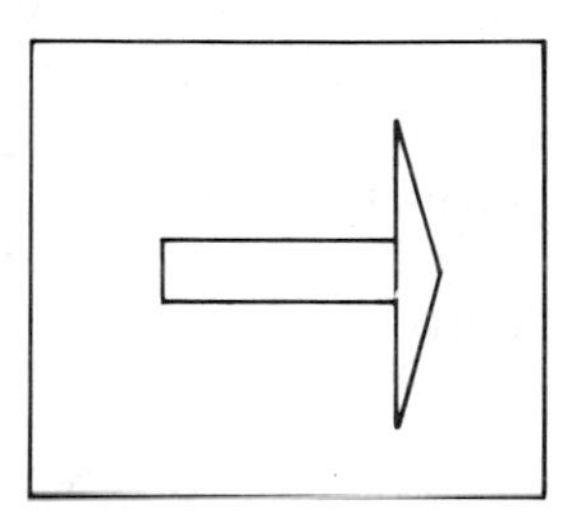

7. Prove the correctness of the algorithm for enumerating k-permutations (given in the text). Make sure to show that each of the given constructions is possible and that at the "stop" all k-permutations have been listed, once each.

4.5 Combinations

In both selections and permutations, the order in which the symbols are selected from the alphabet is crucial in distinguishing one word from another. If our alphabet is, say, abcd, then the word "bad" is different from the word "dab", even though each contains an a, a b, and a d. In permutations, we have the added restriction of nonrepetition of symbols. In both cases, however, we are very careful to say that the order in which symbols are selected plays the primary role in distinguishing words, and therefore also in counting and enumerating all possible words.

There is, however, a kind of word-formation rule in which order doesn't matter at all. The easiest way to demonstrate this is by an example you've probably seen before, in some form or another: Every morning the nine justices of the U. S. Supreme Court enter a meeting room and shake hands. Each justice shakes the hand of each of the other justices. How many handshakes actually occur? To analyze this in our own context, let's first take our alphabet to consist of nine justices. Since only two people can shake hands at one time, our permissible "words" each consist of two justices. Moreover, no justice shakes his or her own hand, so that we have the restriction of nonrepetition; no justice is "selected" twice for the same handshake. So, at first glance, we follow the previous technique and say that each of 9 justices shakes each of $9 - 1 = 8$ other hands, so that there occur a total of $9 \cdot 8 = 72$ handshakes. But that's not right! Suppose Justice Blackmun shakes Justice O'Connor's hand. Let's write this word as Blackmun-O'Connor. Is there any difference between the words Blackmun-O'Connor and O'Connor-Blackmun? That is, does it matter which one of the two shakes the other's hand? Of course not. The only thing that we want to know is which two justices participate in any one handshake and also how many such distinct pairs there are. In other words, here the order of selection of the two justices *doesn't matter*. The words Blackmun-O'Connor and O'Connor-Blackmun are "the same." Our first guess of 72 handshakes has included every handshake twice; the correct answer is one-half of this, or 36.

Another example is a hand in a card game like bridge. A hand contains 13 cards, but the order in which they are dealt is irrelevant: you can reorder them in your hand as you wish, without altering the substance of the game.

Consider a "smaller" example, chosen to clarify an important point—the case of the 3-permutations from the alphabet abcd (see after Definition 9). We list them here in a different layout:

abc	abd	acd	bcd
acb	adb	adc	bdc
bac	bad	cad	cbd
bca	bda	cda	cdb
cab	dab	dac	dbc
cba	dba	dca	dcb

We've placed in the same column those 3-permutations that all comprise the same letters, although in different orders. So, if order does not matter, we shall consider all the words in one column as "the same."

In general, we are going to consider a process called *combination*. Suppose that we have a class of many k-permutations, each containing *exactly the same symbols*, just in different orders. As k-permutations, they're considered distinct, but the whole class constitutes just one combination. As before, we use n to denote the size of the alphabet ($n = 9$ in the Supreme Court example, $n = 52$ in the bridge example, $n = 4$ in the abcd example), and k to denote the size of the combination ($k = 2$ for the Supreme Court, $k = 13$ for bridge, and $k = 3$ for abcd). Because of the requirement of nonrepetition, we must have, as before, the condition that $0 \leq k \leq n$. We summarize this as

Definition 12 ***Combination*** is a process in which k-permutations from an alphabet are assigned to classes in such a way that each class consists of all k-permutations that contain the same letters, in all possible orders. Each such class is called a **k-combination.** The number of k-combinations from an alphabet of size n is denoted $C(n, k)$.

Informally, k-combinations are "words" of length k such that (1) repetitions are not allowed and (2) order is irrelevant in the sense that two words are considered "equivalent" if they comprise the same symbols, possibly in different orders. Usually, a k-combination is denoted by giving any one of the k-permutations in it. For instance Blackmun-O'Connor (which refers also to O'Connor-Blackmun), or abc (which refers also, as a combination, to acb, bac, . . .). Such a k-permutation is called a *representative* of the k-combination. The expression $C(n, k)$ is usually read as "the number of combinations of n symbols (or objects), taken k at a time" or, colloquially, "n choose k." We shall look next at evaluating $C(n, k)$, that is, at counting combinations. The general principles involved will be made clearer by considering first the specific case of the abcd example above. There are a total of 24 3-permutations in abcd, because $P(4, 3) = 4 \cdot 3 \cdot 2 = 24$. Each column contains those 3-permutations that are obtained from each other by permuting the three letters; each column contains therefore six permutations *of those three letters,* because $3! = 3 \cdot 2 \cdot 1 = 6$. The number of columns is therefore $24/6 = 4$. This is a devious way to count four columns, but it will work in general, when we cannot just write everyting down and then count.

Considering now the general $C(n, k)$, let n and k be given, and suppose $0 \leq k \leq n$. We have defined a class as consisting of those k-permutations that consist of exactly the same k symbols, just in different orders. Comparing two different classes, we note two things: first, any two different classes involve different symbols; second, each class contains the same number of permutations, namely $k!$ "Whoa, there!" you say. "Where did *that* come from?" Look back. We said that each class contains *all* k-permutations of *its* k symbols. Consider for the moment the alphabet made up of only *those* k symbols. The k-permutations in that class are *the* permutations of the new alphabet, and their number is therefore $P(k)$, which is $k!$ for all classes. Next we look at all of the classes taken together. Every possible k-permutation of the original alphabet is in there somewhere, in one (and only one) of the classes.

Recall now a piece of very easy arithmetic. If you have x classes with y students in each, then you have xy students all together. Apply this to our case: we have x classes (x is unknown) and each contains $k!$ k-permutations ($y = k!$); so the total number of k-permutations is xy. (Read that sentence again; it's subtle and important.) We recall that the total number of k-permutations is $P(n, k)$, so $xy = P(n, k)$, $x = P(n, k)/y$, and $x = P(n, k)/k!$ But x is the number of classes, that is, the number of k-combinations, $C(n, k)$. We have so proved that $C(n, k) = P(n, k)/k!$ Using theorems 10 and 11, we can rewrite this in either form $n(n - 1)(n - 2) \ldots (n - k + 1)/k!$, $n!/(k!(n - k)!)$. With the addition of a few remarks, we have so proved the following theorem.

Theorem 13 If $0 \le k \le n$, then the number of k-combinations of n objects is $C(n, k) = P(n, k)/k! = n!/(k!(n - k)!)$. If $k > 0$, then another form is $C(n, k) = n(n - 1)(n - 2) \cdots (n - k + 1)/k!$ As usual, we take care of unusual cases separately: If $k < 0$ or if $k > n$, then $C(n, k) := 0$.

Also, in every case, $C(n, k) = C(n, n - k)$.

Proof: We have already proved the first part. The final statement is obtained thus: by the first part of the theorem, $C(n, h) = n!/(h!(n - h)!)$. We can replace h with $(n - k)$, then $C(n, n - k) = n!/((n - k)!(n - (n - k))!) = n!/((n - k)!k!)$, which is the same as $C(n, k)$. (We'll see later another much nicer proof.) Q.E.D.

Let's review two examples to confirm your understanding. First, in the Supreme Court example above, we have $P(9, 2) = 72$ 2-permutations of the 9 justices. But there are two 2-permutations per class (handshake): Blackmun-O'Connor and O'Connor-Blackmun together make up a single class; Blackmun-Scalia and Scalia-Blackmun, another, and so on. Since $2! = 2$, we have $P(9, 2)/2! = 36$ classes, hence $C(9, 2) = 36$. This calculation is conducted especially for the present learning stage. In a concrete problem-solving situation, the calculation would be simply $C(9, 2) = P(9, 2)/2! = 9 \cdot 8/2 = 9 \cdot 4 = 36$.

In a larger example, the one on the alphabet abcd, we have already conducted the "learning" calculation. The direct calculation would be $C(4, 3) = P(4, 3)/3! = 4 \cdot 3 \cdot 2/(3 \cdot 2 \cdot 1) = 4$.

Note a few calculational tricks. The second formula in theorem 13, $C(n, k) = n(n - 1)(n - 2) \cdots (n - k + 1)/k!$, says, practically: lay down $1 \cdot 2 \cdot 3 \cdot \ldots \cdot k$ in the denominator; on the top, start with $n(n - 1)(n - 2) \cdots$ and write as many factors as at the bottom. This is the most efficient formula for hand calculation when k does not exceed one-half of n. If k is larger, then calculate $C(n, n - k)$ instead. For example, $C(26, 21) = C(26, 5) = 26 \cdot 25 \cdot 24 \cdot 23 \cdot 22/(1 \cdot 2 \cdot 3 \cdot 4 \cdot 5)$. Note also that $C(n, k)$ is always an integer, so all factors in the denominator must cancel; do that before multiplying.

With a calculator, the situation is different. If the calculator has the factorial function, then formula $n!/[k!(n - k)!]$ is much more efficient to use, provided the calculator can handle a number as big as $n!$. Some calculators even have the function $C(\ ,\ )$ directly.

Finally, lets look at the enumeration of k-combinations. We'll use as our tool the listing given above for the 3-combinations from the alphabet abcd ($n = 4$, $k = 3$). The four 3-combinations are, abc abd acd bcd. Remember that the only thing we care about in a combination is which symbols participate; the order of the symbols is irrelevant. Each k-combination can then be *completely characterized* (that is, we can get all the information we need) simply by noting which symbols are used. In any k-combination, each symbol from the alphabet either occurs or does not occur. That suggests that a scheme using binary arithmetic might be useful. In order to tell which of the n symbols of the alphabet we take, we mark those symbols with ones and the others with zeroes. Let the alphabet (of size n) be conventionally ordered in any one way. Then the marks 0 and 1 give a sequence of zeroes and ones, that is, a binary n-digit number. In this example, the situation is as follows:

a	b	c	d	
1	1	1	0	for abc (or acb or bac . . .)
1	1	0	1	for abd (or adb . . .)
1	0	1	1	for acd (or adc . . .)
0	1	1	1	for bcd (or bdc . . .)

In general, for any n and k, imagine the symbols of the alphabet are arbitrarily but consistently ordered. For any n-digit binary number, each symbol of the alphabet is automatically assigned a digit of the binary number, according to the given order of the alphabet and the customary order of the digits in the number. That is, binary digit b_{n-1} will refer to the first symbol of the alphabet, binary digit b_{n-2} to the second symbol, . . ., binary digit b_1 to the next-to-last symbol, binary digit b_0 to the last symbol:

1st symbol	2nd symbol	3rd symbol . . .	$(n-1)$st symbol	nth symbol
b_{n-1}	b_{n-2}	b_{n-3} . . .	b_1	b_0

When the b_i are given concretely, we can form the k-combination by taking those symbols of the alphabet for which the corresponding b_i is 1. The process works also in reverse.

This procedure gives an inefficient but correct way to list the k-combinations from an alphabet of size n. We simply list all of the n-digit binary numbers, 2^n in all, from 0 (all zeroes) to $2^n - 1$ (all ones), in the usual arithmetic order. From this list, we locate all entries that have exactly k ones. Listing the corresponding alphabet symbols for each such number completes the process. In the exercises you'll be asked to modify this algorithm to an efficient one. *Note carefully that we are not listing permutations, but simply symbols for k-combinations,* that is, for classes of k-permutations. Within each class, the order of symbols is irrelevant.

The technique of using binary numbers to characterize combinations yields a powerful and unexpected benefit. (Mathematicians call this kind of luck serendipity.) Each binary number completely characterizes a combination, and each combination can be completely characterized by a binary number. Thus, if we ask for the total number of combinations, including all possible values for k, that number must be the same as the number of n-digit binary numbers. Thus, we have proved

Theorem 14 If n is a natural number, then $\Sigma_{k=0}^{n} C(n, k) = 2^n$.

This is but one of the many interesting properties of the **combinatoric function** $C(n, k)$. This function is so multi-faceted that it deserves some special study on its own. In the next section we'll examine some of its properties.

To sum up, a combination is a class of strings over a finite alphabet, each class being defined as comprising all permutations that contain the same letters. The number of possible distinct k-combinations from an alphabet of size n is $C(n, k) = n(n-1) \cdots (n-k+1)/k!$

Exercises 4.5

1. Make a table of the combinatoric function $C(n, k)$ for n from 0 to 6, and for all permissible values of k. List n vertically, increasing from top to bottom, and k horizontally, increasing from left to right. Save this table; you'll need it for the next section.

2. Compute $C(24, 0)$, $C(24, 1)$, $C(24, 2)$, $C(24, 3)$, $C(100, 0)$, $C(100, 1)$ $C(100, 2)$, $C(100, 3)$.

3. Compute $C(999, 0)$; $C(999, 1)$; $C(999, 2)$; $C(999, 997)$; $C(999, 998)$; $C(999, 999)$. Be efficient!

4. In a chess tournament, each player is to play against each of the other players. How many matches are necessary if there are 20 players?

5. Cardboard disks are to be painted with different colors on the two sides. The available colors are R, O, Y, G, B, I, V. How many different color patterns are possible?

6. A six-member committee is to be chosen from an assembly of 25 persons. How many different choices are possible?

7. **(a)** Two committees are to be chosen from an assembly of 25 persons. One committee must comprise three persons, the other five. No one is to serve in both committees. How many different choices are possible? **(b)** Two six-member committees are to be chosen from an assembly of 25 persons. No one is to serve in both committees. How many different choices are possible?

8. Consider the figure below, which is supposed to represent a transparency. The two outer rings have different colors, selected from among ten available colors. The three 120-degree slices have different colors, selected from twenty available colors (different from the first ten). How many different coloring patterns are there? Justify your answer. Express the result as an arithmetic formula, but do not perform the final multiplications and divisions.

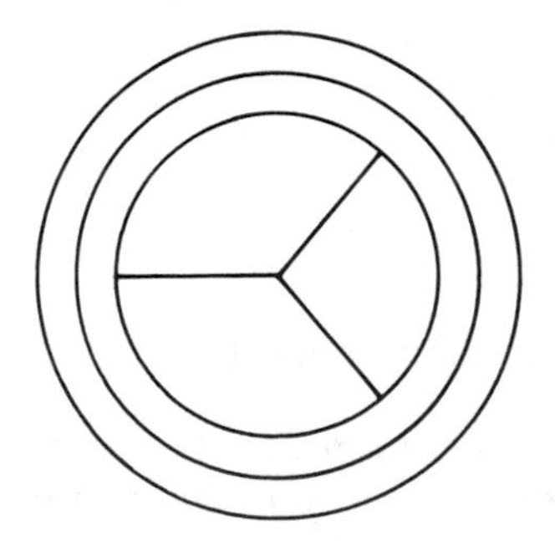

9. An *n-gon* is a polygon having n sides and n vertices. A *diagonal* is a line connecting two nonadjacent vertices. How many diagonals does an n-gon have? Draw the cases in which $n = 3$, $n = 4$, $n = 5$, and $n = 6$ to check your answers visually.

10. Construct a correct and efficient algorithm to enumerate the k-combinations, for any given k and n. Use the binary scheme mentioned in the text, but "cut the waste." [Hint: start from 1 . . . 10 . . . 0 (the binary number consisting of k ones followed by $n - k$ zeroes) and "push" the ones to the right, reluctantly. The crucial point is to locate the right-most pair 10 (if any)]. Do not limit yourself to any one computer language. Apply your algorithm to a few concrete values for n and k, and check that it does what it's supposed to do. (If not, back to the drawing board!) Alternatively, run and check the algorithm on a computer.

11. Find the number of each of the following. Calculate completely. **(a)** All 23-combinations of 26 distinct objects; **(b)** all 4-permutations of 6 distinct objects; **(c)** all 3-selections of 7 distinct objects; **(d)** all 28-combinations of 29 distinct objects; **(e)** all 6-permutations of 6 distinct objects; **(f)** all 3-selections of 3 distinct objects.

12. Write in a systematic fashion without duplications **(a)** all 2-permutations from alphabet ABCDE; **(b)** all 5-selections from alphabet 01; **(c)** one representative from each of all 3-combinations from alphabet 12345; **(d)** all 1-permutations from alphabet ABCDE; **(e)** all 3-selections from alphabet 01; **(f)** one representative from each of all 5-combinations from alphabet 12345.

13. There are two kinds of domino tiles: those with one of the numbers 0 to 6 repeated, for instance

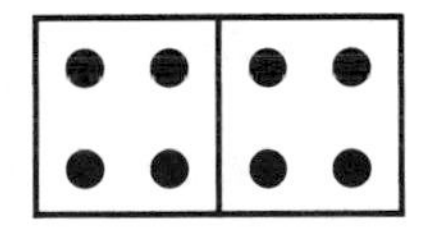

and those with two different numbers, 0 to 6, for instance

Find the number of different tiles of each kind. Find the total number. Calculate completely.

4.6 The Combinatoric Function $C(n, k)$

The combinatoric function $C(n, k)$, which we developed in the previous section, has many interesting and unexpected properties. In this last section, we look at some of the more surprising ones. In particular, we'll see that $C(n, k)$ is not really only a combinatoric function at all, but something much more general. Before we get ahead of ourselves, however, let's review the more mundane features of

$C(n, k)$. First, there is the matter of language and notation. The combinatoric function is also often called the *combination function,* although neither name is really universal. The notation $C(n, k)$ is probably the best one, since it's consistent in form with the selection function $S(n, k)$ and with the permutation function $P(n, k)$. Another common notation is

$$\binom{n}{k},$$

read "n above k." Actually, this is probably a little more common than $C(n, k)$, but is also a little dangerous for beginners to use, because it's constantly being confused with the parenthesized fraction (n/k). A third type of notation is now all but obsolete: the ${}_nC_r$ notation (as in ${}_5C_2 = 10$), and its cousins ${}_nP_r$ and ${}_nS_r$. In fact, the combinatoric function was, for a while, known as the NCR function and was constantly being confused with the company that makes cash registers. We suggest the notation $C(n, k)$ and will use it throughout this book, but you should recognize the others when you see them. Then there's the matter of language. People often refer to $C(n, k)$ as "the number of combinations of n things taken k at a time," or simply "n things taken k at a time," or even the colloquial "n choose k." All are acceptable, if a bit vague. Just remember that laziness is all right, so long as it's rigorous and painstaking laziness.

4.6.1 Properties of the Combinatoric Function

Now let's look at the mathematical properties of $C(n, k)$. In the first exercise of Exercises 4.5 you constructed a partial table of $C(n, k)$. Did you notice any peculiarities about this table? First, of course, if you constructed the table properly, any value of $C(n, k)$ for which $k < 0$ or $k > n$ is zero, and you needn't bother including them at all. All other values are positive. In fact, all values for which $k = 0$ or $k = n$ are 1. But you've probably noticed something else, a feature we mentioned briefly in the last section. Each row—that is, the values of $C(n, k)$ for a fixed n and all permissible k—has a sort of mid-point symmetry. For instance, row $n = 4$ has entries 1, 4, 6, 4, 1, and row $n = 5$ has 1, 5, 10, 10, 5, 1. This is the meaning of the last property of theorem 13. This *combinatoric symmetry* is really not as surprising as it looks. In English, "taking k things out of n" is the same as "leaving $n - k$ things out of n." But the words "taking" and "leaving" are simply imagery, and could easily be exchanged. So there are exactly as many ways to "take k things out of n" as there are to "take (the other) $n - k$ things." This is the "nicer" proof we mentioned in theorem 13 about the symmetry property.

Speaking of rows in the table of $C(n, k)$ did you notice what happens if you sum the entries in each row of your table? Do it now, if you haven't already. What you will find is a reiteration of theorem 14, namely that $\Sigma_{k=0}^{n} C(n, k) = 2^n$.

By far the most important observation, however, is one you might have missed. With the exception of the first row ($n = 0$, consisting of only the entry 1), each row of the table can be constructed from the previous row, without ever invoking factorials at all! If you add any two consecutive entries in any row, the result is the entry in the next row beneath the second of the entries you added (see Figure 4.1). The first and the last entries of each row can be considered exceptions, but they are always one and, if you consider the nonexistent entries as zero, then these entries are not exceptions at all. Stop here, and verify your own table. In formulae, our property is

$$C(n + 1, k) = C(n, k - 1) + C(n, k). \tag{1}$$

What this means is that, in evaluating $C(n, k)$ for any n and k, we never really have to use the factorial function, multiplication, or division directly.

As usual, many verifications do not necessarily constitute a proof. We have to prove equation 1 in general. One way is the following: Consider an alphabet of size n. Then consider another alphabet, of size $n + 1$, obtained from the previous one by adjoining one extra symbol. Any one k-combination from the second alphabet either contains the extra symbol or it doesn't. Consequently, the number of k-combinations from the second alphabet = $C(n + 1, k)$ = number of k-combinations that *do not* contain the extra symbol + number of k-combinations that *do* contain the extra symbol. Let's consider the two addends separately. The combinations in the first addend are really combinations from the first alphabet; their number is therefore $C(n, k)$. Those that contribute to the second addend all contain the extra symbol; if you ignore it, you have a $(k - 1)$-combination from the first alphabet again; the number of such combinations is $C(n, k - 1)$. Going back to the addition, we have $C(n + 1, k) = C(n, k) + C(n, k - 1)$, which proves formula 1. Check on your own what happens for "small" values of n and k.

Moreover, we've seen this exact same feature before, in a very different context. Look back at the discussion of Pascal's triangle in Section 3.6. There we

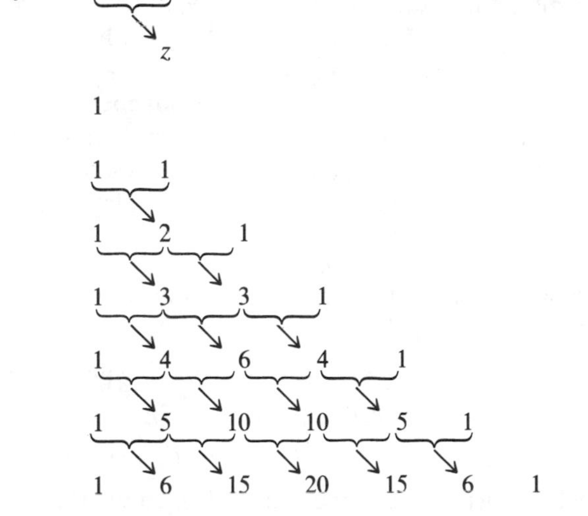

Figure 4.1 Recursive property of $C(\ ,\ )$.

were discussing the coefficients for expanding $(x + y)^n$. We used then the notation $C_{n,k}$. If you compare your table with Pascal's triangle, you'll see that they're identical (at least for the part you can reasonably write; see Exercises 4.6). The expression $C_{n,k}$, which is the coefficient of $x^k y^{n-k}$ and of $x^{n-k} y^k$, is simply the same as $C(n, k)$. This use of $C(n, k)$ far predates the study of alphabets and combinations, and the original name for $C(n, k)$ is the *binomial coefficient function*. But today, $C(n, k)$ is most often used in the context of combinations, so the original name can be a bit confusing. An observation we made before bears repeating. If we expand $(1 + 1)^n$ using the binomial theorem, we end up simply summing the binomial coefficients for n. Since $(1 + 1)^n = 2^n$, we have the result we expected. See theorem 14 in this section and Exercises 3.6.

4.6.2 Recursion

In mathematics, this sort of surprise interconnection is more common than you might suppose. Two mathematicians, working on totally different problems, and possibly centuries apart, discover that the same tool is the key to both solutions. The mathematics of combinations was worked out by Leonhard Euler in the late 1700s. Imagine his excitement (and confusion) when he found that the crucial factors were the binomial coefficients discovered by Blaise Pascal nearly 150 years earlier! What was the connection? In hindsight, of course, you could look at the binomial expansion from just the right slant, and see it as a combination problem (see Complements 4.6). But the real moral of the story is that *$C(n, k)$ is a basic mathematical tool, independent of any particular application.* It is a binomial coefficient, a combinatoric counter, and, long after Pascal and Euler were dead, it became the foundation for the science of probability. Is it any wonder that $C(n, k)$ has no universal name or notation?

This multiplicity of guises of $C(n, k)$ has a more immediate implication for us. The notion that $C(n, k)$ can be built up with each row depending only on the last one is called **recursion,** and the fundamental statement $C(n + 1, k) = C(n, k - 1) + C(n, k)$ in equation 1 is called a **recurrence relation.** We can use the recurrence relation to define $C(n, k)$ from the very beginning, as if we had never heard of binomial coefficients, combinations, or factorials. Study the following theorem carefully to see that it does, in fact, completely characterize $C(n, k)$; nothing else is needed.

Theorem 15 For any natural number n and any integer k, the function $C(n, k)$ is defined recursively by

$$\begin{aligned} C(n, k) &:= 0 \quad \text{if} \quad k < 0 \quad \text{or} \quad k > n, \\ C(n, k) &:= 1 \quad \text{if} \quad k = 0 \quad \text{or} \quad k = n; \\ C(n + 1, k) &:= C(n, k - 1) + C(n, k) \quad \text{if} \quad 0 < k \leq n. \end{aligned}$$

There are three things to notice about this recursive definition. First, it is complete; nothing else is needed. Second, it is recursive, the values for $n + 1$ are calculated in terms of those for n. Third, although the definition is recursive, *recursion by itself is never sufficient;* the first two lines of the formula are called **initial conditions,** and serve to get the recursion going. In Pascal's triangle, the first row, and the first and last entry in each row, must be given explicitly. Only then can recursion supply the rest.

More specifically, the first formula in theorem 15 tells where the nonessential parts of the construction are. The second formula tells you that the beginning and the end of each line is a 1; this is the part that gets the recursion going. The actual recursion comes from the third formula; applying the formula to row $n = 1$ fills in row $n + 1 = 2$, applying the formula again to row $n = 2$ fills in row $n + 1 = 3$, and so on, indefinitely.

It has probably occurred to you that the factorial function $n!$ itself could also be defined recursively. Very true. In theorem form, it becomes

Theorem 16 For any natural number n, the factorial function $n!$ is defined recursively by

$$\begin{aligned} n! &:= 1 \quad \text{if} \quad n = 0; \\ (n+1)! &:= (n!)(n+1) \quad \text{if} \quad n \geq 0. \end{aligned}$$

Once again, the recursive definition completely characterizes the function and consists of an initial condition and a recurrence relation.

Exercises 4.6

1. Use the recursive definition to construct a table of $n!$ when $n = 0, 1, 2, \ldots, 10$.
2. Use the recursive definition to construct again your table of $C(n, k)$ when $n = 0, 1, 2, \ldots, 10$ and for all permissible value of k.
3. Calculate completely each of the following:

(a) $\binom{3}{0} + \binom{3}{1} + \binom{3}{2} + \binom{3}{3}$,

(b) $\binom{4}{0} + \binom{4}{1} + \binom{4}{2} + \binom{4}{3} + \binom{4}{4}$,

(c) $\binom{5}{0} + \binom{5}{1} + \binom{5}{2} + \binom{5}{3} + \binom{5}{4} + \binom{5}{5}$. Be efficient.

4. Use the recurrence properties of $C(n, k)$ and $C_{n,k}$ to verify that Pascal's triangle and the table of the combinatoric function are actually the same. (Work *in general!*)

Complements 4.6

1. Another proof of the coincidence of $C_{n,k}$ and $C(n, k)$ is based on a combinatorial interpretation of the expansion of $(x + y)^n$, which is the product $(x + y)(x + y)$

$\cdots (x + y)$ (n factors). In order to get the monomial $x^{n-k}y^k$ you must select y in k of the factors and x in the others. Work out a general proof, when $n > 1$.

2. Formula 1 can also be proved in a strictly algebraic form. Express each of the two addends as a fraction in the form given by theorem 13, and then "add fractions." Conduct the calculation carefully. [Hint: you can find the answer in Section 9.2.2.]

3. Another example of a recursive process is provided by the **Stirling numbers of the second kind,** $S[n, k]$. They are defined as follows, for each natural number n and each integer k:

$$\begin{aligned} S[n, k] &:= 0 \quad \text{if} \quad k \le 0 \quad \text{or if} \quad k > n; \\ S[n, n] &:= 1 \quad \text{if} \quad n > 0; \\ S[n+1, k] &:= k\, S[n, k] + S[n, k-1] \quad \text{if} \quad 0 < k \le n. \end{aligned}$$

As usual, construct a few lines of the table. The stress of this exercise is to see how a table can be constructed recursively, without knowing "what the thing is about." We'll see later (in Section 5.7) that these numbers have a concrete practical meaning.

4.7 Summary

We have considered strings from an alphabet (Section 4.1) and several processes to choose *words* from among all possible *strings:* the words are those strings that satisfy some specifically given criteria. Among the many criteria for choosing strings, we have concentrated on two: whether repetitions are allowed or not; and whether the order in which elements appear in a string is relevant or not. These two criteria provide four processes: (1) *selections,* in which repetitions are allowed and order counts (Section 4.2); (2) *permutations,* in which no repetitions are allowed and order counts (Section 4.4); (3) *combinations,* in which no repetitions are allowed and order does not count (Section 4.5); and (4) *combinations with repetitions,* or *multisets* (not included here; see Section 7.2.4), in which repetitions are allowed and order does not count.

For each of the four types of strings, the essential questions are (1) *what* are they, (2) *how many* are there, and (3) *how can we enumerate* them?

The basic formulae that answer question (2) are: The number of *selections* of k objects out of n objects is $S(n, k) = n^k$. The number of *permutations* of k objects out of n objects is $P(n, k) = n(n-1) \cdots (n-k+1) = n!/k!$. The number of *combinations* of k objects out of n objects is $C(n, k) = n(n-1) \cdots (n-k+1)/k!$ $= n!/(k!(n-k!))$.

The last formula provides a function that is important on its own, the *combinatoric function* $C(n, k)$ (Section 4.6), which can also be constructed by *recursion* (Section 4.6.2).

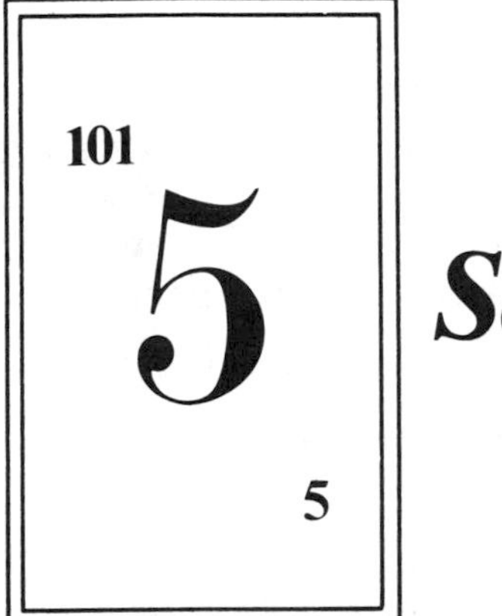

5 Set Theory

Saint Augustine once said of time, "If no one asks me what is time, I know. But if someone asks me 'What is time?' I don't know." If Augustine were alive today, he would probably say the same thing about the notion of a set. Intuitively, the words "set," "collection," "family," and "aggregate," mean about the same thing. There is no difficulty, in simple situations, in doing some operations on sets: comparing or combining two sets, taking the common part of two sets, and so on. The goal of set theory is to formalize these intuitive concepts, so that we can extend good-sense reasoning to more complex situations. Set theory is based on a handful of *axioms,* basic statements from which we can deduce the properties of sets. They are the "rules of the game." Remember Chapter 1? A game comprises objects, rules, and—out of necessity—vocabulary and implied rules.

Axioms are not the same thing as definitions since definitions are simply arbitrary conventions, the "vocabulary." Nor are axioms quite the same thing as theorems; theorems are the "implied rules" and require proof. Axioms are the statements on which the theory is based. Usually they are a rigorous clarification of common sense. As we go through them, you should set aside your initial reaction of "Of course! How could anyone doubt this?" and ask rather, "Is it really impossible to prove this?" Unfortunately, it is. But, what if an axiom is not true? Go back to a nonmathematical game, such as basketball. Among the rules, you'll find that the ball must have a certain shape, size, and weight. What if your ball doesn't meet those standards? Simple: you are not playing basketball; perhaps you are playing some other game. The same is true for a mathematical theory or structure: if, in a specific situation, an axiom of that theory or structure is not true, then you are not studying that theory or structure.

The axioms dealt with in this chapter are known as the axiom of extension, the axiom of specification, the axiom of pairing, the axiom of existence, the axiom of unions, the axiom of infinity, and the axiom of powers. Their statements will be given as the need arises; see Sections 5.1.4 to 5.1.6 for the first four and Sections 5.3.1, 5.6, and 5.8 for the others, respectively. The goal of this chapter is *not* to establish a logically rigorous foundation of set theory. Rather, we intend to show how a few basic principles can be combined to produce operational rules that are useful in applications, both at the theoretical level and in the applied sciences.

5.1 Elements and Sets

Continuing with our analogy to familiar games, recall tic-tac-toe. At the beginning there is only the grid and no objects. The objects (the X's and the O's) are produced by playing the game, that is, by applying rules. Similarly, in set theory, we start from a fundamental concept *(belonging)* and use the axioms. This will produce the objects, that is, the sets and their elements.

5.1.1 Examples of Sets

Before we go on, let's look at some examples of sets and their elements. You already know some sets, such as the English alphabet, the Greek alphabet, the set of the natural numbers, the set of the integers. Then the following statements are true:

X is an element of the English alphabet,
α is an element of the Greek alphabet,
-5 is not an element of the set of the natural numbers,
-5 is an element of the set of the integers.

5.1.2 Belonging

The most basic concept of set theory is that of *belonging*. The statement

$$x \in \mathrm{A}$$

stands for "x **belongs to** A", or "x is a member of the set A", or "x is an **element** of the set A", or "x is **contained in** the set A", or, simply, "x is in A". The symbol $\in$ is a special form of the Greek letter ϵ (epsilon = short *e*) and is a graphic shorthand for these phrases. The phrases themselves are simply variant of the verb "is". In fact, the statement "$x \in A$" is a *proposition.* It has a subject (x), a verb *(is)*, and a predicate adjective *(an element of A)*. Moreover, whenever we consider a specific object x and a specific set A, the statement "$x \in A$" is either true or false. The negation of that statement is written "$x \notin A$" and is read "x is not an element of the set A".

The examples of Section 5.1.1 may be rewritten as the following statements:

X $\in$ English alphabet,
$\alpha \in$ Greek alphabet,
$-5 \notin$ set of the natural numbers,
$-5 \in$ set of the integers.

There is a crucial implication in the statement "$x \in A$". We are to assume that A is a set, but nothing has been said about x. That is, x itself may or may not be a set in its own right. Consider, for example, "the graduating class of 1988 is an element of the set of graduating classes of this century". But the graduating class of 1988 is itself a set of graduating seniors. In other words, there's absolutely nothing wrong with speaking of a set that contains, as its elements, other sets. But be careful! The distinction between sets and elements is relative, but the two terms are *not synonymous!* Suppose Jim is a graduating senior in 1988. Then "Jim is an element of the graduating class of 1988". But we said before that "the graduating class of 1988 is an element in the set of graduating classes of this century". Does it therefore follow that "Jim is an element in the set of graduating classes of this century"? No, it does not! Jim is a single student, not a graduating class. This distinction would be valid even if Jim were the only senior in his class. In formulae, if $x \in A$ and if $A \in B$, it is *not* necessarily true that $x \in B$. This restriction on the notion of elements of a set is subtle and difficult at first, but have patience. For the moment, just remember that elements and sets are not the same thing, although a set may be an element of another set. See also *Russell's paradox* in Complements 5.2.

5.1.3 Description, Enumeration, and Specification

We've seen examples of sets you already know: the English alphabet, the Greek alphabet, the set of the natural numbers, the set of the integers. Other examples are the set of the binary digits, that is, the set that comprises the elements 0 and 1, and the set of the positive integers, that is, the set of those integers that are greater than 0. In this last case we've done something new. Let's analyze it.

A set can be given in three ways: by *description,* by *enumeration,* and by *specification.* We used **description** when we said "English alphabet", "Greek alphabet", "natural numbers", or "integers". These are phrases that *describe* sets. They refer to a previous definition. Sometimes a symbol is a sufficient description. Consider the symbol $\mathbb{Z}$ used in Chapters 1 and 2. The structure $\mathbb{Z}$ of the integers is more than just the set of integers. It includes also the operations, or rules, on the integers, such as addition, multiplication, and order. However, it's customary (although an abuse of language) to use the symbol $\mathbb{Z}$ to mean also just the set of the integers. The statement $5 \in \mathbb{Z}$ is syntactically correct (it's also true). Similar remarks apply to the structures $\mathbb{N}$, $\mathbb{Q}$, and $\mathbb{Z}_m$.

A set is given by **enumeration** when we list its elements individually. In ordinary writing, they are usually separated by commas and enclosed in braces; for instance, {0, 1, 2, 3, 4, 5, 6, 7, 8, 9}. You may say "But that's the set of the decimal digits!" In that case, you have given a definition: Set of the decimal digits := {0, 1, 2, 3, 4, 5, 6, 7, 8, 9}. The phrase "A := {a, b, c}" is read as "A is defined to be a set containing all of, and only, the elements a, b, and c''. We could, of course, define the alphabets by enumeration. If we're very careful with our use of ellipses, we could even say things like: English alphabet := {A, B, . . ., Z}, Greek alphabet := {α, β, . . ., ω}, Decimal digits := {0, 1, . . ., 9}.

But how does one enumerate the set $\mathbb{N}$ of natural numbers? There are infinitely many elements of the set; enumeration is, shall we say, time-consuming. Of course, we could use open-ended ellipses, as $\mathbb{N} := \{0, 1, 2, \ldots\}$. But what, for example, are we to make of $C := \{12, 178, 642, 5319, \ldots\}$? Even worse, what do you mean by $\{1, 2, 4, 7, \ldots\}$? You may think thus: start from 1, add 1, get 2; add 2, get 4; add 3, get 7; add 4, get 11; etc. Then you meant $\{1, 2, 4, 7, 11, 16, 22, 29, 37, \ldots\}$. But you might think thus: start from 1 and 2; add them, add 1, get 4; add the last two, 2 and 4, add 1, get 7; add the last two, 4 and 7, add 1, get 12; etc. Then you meant $\{1, 2, 4, 7, 12, 20, 33, 54, \ldots\}$. The point is that ellipses in an enumeration are nearly always ambiguous, unless the context gives additional, unambiguous suggestions.

Specification is a technique that defines a new set in terms of one already defined. It works like this.

Definition 1 If A is a known set and if S is a proposition about an unspecified element of A, then the notation

$$\{x \in A \mid S\}$$

refers to the set of precisely those elements of A for which the proposition S is true. Proposition S is the **specification.**

A shorthand form such as $\{x \mid S\}$ is sometimes used, but *only when* the known set A is clearly understood from context.

Chapter 6 contains a detailed discussion of the meaning of *proposition*. For now, consider it as a condition (of the form ''x must be of some specified type'', or ''x must *not* satisfy some criterion''). That is, a proposition expresses a sort of a test that x must satisfy to be in the new set. A *proposition* is a statement that, for each element x of the given set, is right or wrong.

Some examples illustrate the method:

$\{x \in \mathbb{Z} \mid x \text{ is even}\}$,
$\{x \in \text{graduating class of 1988} \mid x \in \text{football team}\}$,
$\{x \in \mathbb{N} \mid x \text{ is not a prime}\}$,
$\{x \in A \mid x = x\}$,
$\{x \in A \mid x \neq x\}$.

The last two examples present special cases, which we'll examine in a moment.

5.1.4 Equality of Sets

The first axiom of set theory gives the definition of equality between two sets.

Axiom of extension. Two sets are equal if and only if they contain the same elements.

This is a typical example of a situation that seems trivial, but, on second thought, you'll see that it's less trivial than it sounds. Anybody would accept the fact that $\{a, b\} = \{a, b\}$. But what about $\{b, a\}$, $\{a, a, b\}$, and $\{a, b, a, b, b\}$? By the axiom of extension, they are all the same set. In other words, repetitions of elements and the order in which the elements are mentioned are irrelevant in the definition of a set. In this context, we are ready to accept a looser interpretation of the term "enumeration." When we give a set by enumeration, we allow repetitions, which do not make any difference anyhow (in a *list,* repetitions are excluded). A translation of these remarks into more formal terms is the following:

Definition 2 Given that A and B are sets, the statement $A = B$ means: (1) For every x in A, $x \in B$; *and* for every x in B, $x \in A$. Or, equivalently, (2) if $x \notin B$, then $x \notin A$; *and* if $x \notin A$, then $x \notin B$. The negation of the statement $A = B$ is $A \neq B$. It means, in an equivalent form, that there exists an x such that $x \in A$ and $x \notin B$, *or* there exists an x such that $x \in B$ and $x \notin A$.

Note that this type of argument will turn out to be very important in solving problems in set theory. See also *Degeneracy* in Complements 5.2.

5.1.5 Remarks

A question that we should have asked when we used description, enumeration, and specification is "Can we do that?" In other words, if we use one of these methods to give a set, are we respecting the rules of the game? For *description,* the answer is simple: A description relies on a previous definition, which was up to us, provided that at that time we respected the rules of the game. For *specification*, we have an axiom (that is, a rule of the game) that states precisely what we need.

Axiom of specification. If A is a given set and if S is a proposition that, when applied to each element of A, is either true or false, then there exists a set composed of precisely those elements of A for which S is true.

For enumeration, the answer is a little more involved. In the case of "small" sets, we have another rule of the game.

Axiom of pairing. If a and b are given objects, then there exists a set whose elements are precisely a and b.

Such a set is denoted $\{a, b\}$, and we have solved the problem in at least one case. In the special case in which $a = b$, equality of sets (which we have just seen) tells us that the set $\{a, a\}$ is the same as the set that comprises the only element a. For such set, we use the symbol $\{a\}$, that is, we define $\{a\} := \{a, a\} =$ the **singleton** of a, and we solve the problem again. For larger (and smaller) sets, we need additional arguments, and we'll return to the problem later.

5.1.6 The Empty Set

Whatever the objects of a game are considered to represent, whether numbers, letters, planets, or something else, we always hold that for any conceivable object x, it's always true that $x = x$. It is equally true that the statement $x \neq x$ is always false. Returning to two examples we considered at the end of Section 5.1.3, when A is a given set, we can form the sets $\{x \in A \mid x = x\}$ and $\{x \in A \mid x \neq x\}$. Since $x = x$ always, the first set is simply A. Similarly, because it's *never* true that $x \neq x$, there is *no* element that satisfies the specification for the second set. Therefore, for any set A, the set $\{x \in A \mid x \neq x\}$ has no elements. It is called an **empty set.**

If we had started from another set B, the set $\{x \in B \mid x \neq x\}$ would again be a set with no elements. Would it be the same as $\{x \in A \mid x \neq x\}$? In other words, is it true that the set that doesn't contain any apples is equal to the set that doesn't contain any oranges? By the axiom of extension, the answer is affirmative (see part (2) of Definition 2; see also Degeneracy in Complements 5.2). So we know that all empty sets are equal to each other, no matter where they came from. That is, the set of apples that doesn't contain any apples is exactly the same as the set of oranges that doesn't contain any oranges.

But is there an empty set? Another rule of the game provides the answer.

Axiom of existence. There exists a set.

If we start from any one of the existing sets (and we know now that some do exist), then we know how to construct the only empty set. So we have finally a positive result (rather meager, to tell the truth):

Theorem 3 There exists a unique set with no elements.

Definition 4 The only set with no elements is called the **empty set,** or the **null set,** or the **vacuous set** and is denoted $\varnothing$ or $\{\ \}$.

The moment we have the set $\varnothing$, some of the properties we saw before (which ones, specifically?) prove the existence of many sets, for example $\{\varnothing\}$, $\{\varnothing, \{\varnothing\}\}$, $\{\varnothing\}$, $\{\varnothing, \{\varnothing\}\}\}$, $\{\{\varnothing\}\}$, $\{\{\{\varnothing\}\}\}$. Are these sets all different from each other? Why? Can you define more sets using similar techniques? In this context, you should take a few moments to define, in words, the meaning of the notations A, $\{A\}$, $\{\{A\}\}$, $\{\{\{A\}\}\}$, ... For example set $\{\{A\}\}$ is the set whose only element is the set whose only element is A. What, for example, is meant by $C := \{\{A\}, \{A, \{A\}\}\}$?

5.1.7 Cardinality of a Set

In keeping with the study of finite mathematics, we have to have a way of denoting the number of elements in a set. This number is called the **cardinality** of the set and is denoted by $\#A$ or $|A|$. Example: if $A := \{4, 5, 6\}$, then $\#A = 3$, or, equivalently, $\#\{4, 5, 6\} = 3$. Be very careful about notation: A is a set, but $\#A$ is a number. Specifically, $\#A$ is either a natural number (for **finite** sets; that is, those that can be enumerated in the usual sense), or some "infinity" (for infinite sets such as $\mathbb{N}$ and $\mathbb{Z}$). What about the case in which $\#A = 0$? Such a set contains no elements at all and is therefore an empty set.

An accurate study of cardinalities is fascinating, but quite difficult. We'll limit ourselves to cardinalities of finite sets; these cardinalities are ordinary natural numbers. That's the reason why natural numbers are also called "counting numbers."

Exercises 5.1

1. For each of the following sets, enumerate its elements, and state the set's cardinality. **(a)** $\{x \in \mathbb{N} \mid x < 5\}$; **(b)** $\{x \in \mathbb{N} \mid x < 5$ and x is divisible by $2\}$; **(c)** $\{x \in \mathbb{Z} \mid |x| < 3\}$ ($|x|$ denotes absolute value in $\mathbb{Z}$); **(d)** $\{x \in \mathbb{Z}_5 \mid x^2 = 1 \pmod 5\}$; **(e)** $\{x \in$ English alphabet $\mid$ the upper case symbol for x does not include straight line segments$\}$; **(f)** $\{x \in \{0, 1, 2, 3\} \mid x \neq 0$ and $x \neq 1$ and $x \neq 3\}$; **(g)** $\{x \in \{0, 1, 2, 3\} \mid x = 4\}$; $\{x \in \mathbb{N} \mid x$ is an entry of line 4 of Pascal's triangle and $x < 11\}$; **(h)** $\{x \in \{a, b\} \mid x \in \{b, c\}\}$ (a, b, c are distinct); **(i)** $\{x \in \{a, \{a\}\} \mid x \neq a\}$; **(j)** $\{x \in \{a\} \mid x \in \varnothing\}$.
2. Give more examples of your own. Present examples of sets from the classroom or from real life by description, enumeration, or specification.
3. For each of the following pairs of sets, state the cardinality of each set, and state whether the two sets are equal to each other. **(a)** $\{0, 1, 2, 3, 4\}$, $\{x \in \mathbb{N} \mid x < 5\}$; **(b)** $\{-2, -1, 0, 1, 2\}$, $\{x \in \mathbb{Z} \mid x^2 < 5\}$; **(c)** {A, E, I, O, U}, $\{x \in$ English alphabet $\mid x$ is a vowel$\}$; **(d)** The set of the 3-selections from set $\{0, 1\}$, $\{000, 001, 010, 100, 111\}$; **(e)** $\{x \in \mathbb{N} \mid x < 11\}$, $\{x \in \mathbb{N} \mid x$ is an entry of rows 0 to 5 of Pascal's triangle$\}$.
4. Answer the questions after Definition 4 in Section 5.1.6.

5.2 Set Inclusion

Now that we've discussed the notion of an object being an element of a set, we turn to the comparison of two sets. Let's consider first all possible sets that can be constructed using the objects 1, 2, and 3. There is, of course, set $\{1, 2, 3\}$, which is made up of all of them, and set { } (the empty set $\varnothing$), which contains none. Then there are those sets that comprise two of the given objects: {2, 3}, {1, 3}, {1, 2}. Finally, there are the singletons, {1}, {2}, and {3}. All of these sets have one property in common: their elements are elements of set {1, 2, 3}. This is the basic idea of being a *subset* of set {1, 2, 3}. The same idea applies to the comparison of (for instance) set {1} with set {1, 2}: "all" elements of the former are also elements of the latter.

It may be instructive to represent with a diagram the fact that "all elements of a set __ are also elements of set __". This is what an arrow denotes in Figure 5.1.

Suppose A and B are two sets.

Definition 5 The set A is called a **subset** of the set B, or is said to be **included** in B, if every element of A is also an element of B. We denote this by $A \subseteq B$. In formulae, $A \subseteq B$ if for all x in A, $x \in B$. Set B is sometimes called a **superset** of A.

In the previous example, for instance, $\{1, 2\} \subseteq \{1, 2, 3\}$, $\{1\} \subseteq \{1, 2\}$, $\{1\} \subseteq \{1, 2, 3\}$, and even $\{1, 2, 3\} \subseteq \{1, 2, 3\}$.

Look closely at the language of the definition, especially at the formula. The formula has a very specific form, namely "For all x in A, $S(x)$," where A is a set, x is any element of A, and $S(x)$ is a proposition (a true/false assertion) about x. Such a formula can be read, "For every element x in set A, the statement $S(x)$ is true." Here, $S(x)$ is the proposition that states $x \in B$. (See also Degeneracy in Complements 5.2). You should become accustomed to thinking, speaking, and writing in this kind of language.

The negation of the statement $A \subseteq B$ is $A \not\subseteq B$. The statement $A \not\subseteq B$ means that there is at least one element of A that is not in B. There may exist many such; perhaps *no* element of A is in B. All it takes is one. Thus, in formulae,

Definition 6 $A \not\subseteq B$ means that there exists an x in A such that $x \notin B$.

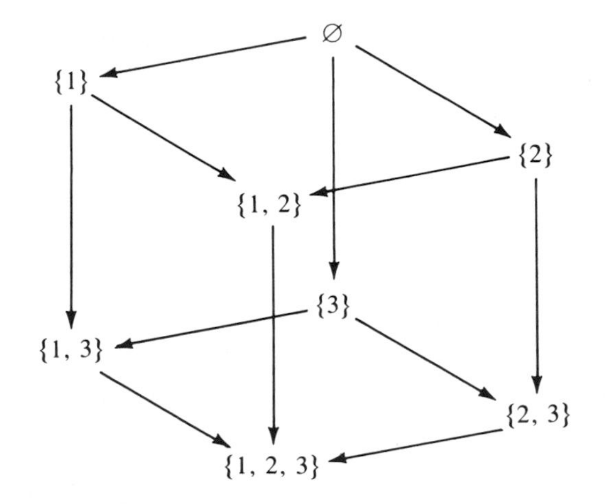

Figure 5.1 All subsets of {1, 2, 3}.

This is again a mathematical proposition of the general form "There exists x in A such that $S(x)$." You should become used to this form, too.

The statement $A \subseteq B$ assumes that both A and B are sets. In particular, the statements $A \in B$ and $A \subseteq B$ have very different meanings. For instance, the set of even natural numbers is a subset (*not* an element) of the set of natural numbers; 4 is an element (*not* a subset) of the set of even natural numbers. Very specifically, contrast these two statements: (1) If $A \in B$ and $B \in C$, then $A \in C$. (*Wrong!* In the example of the graduating class we have seen that this statement is false, in general.) (2) If $A \subseteq B$ and $B \subseteq C$, then $A \subseteq C$. (*Right!* This is always true.) In fact, this last result is so important that we state and prove it as a theorem.

Theorem 7 Let A, B, and C be sets. If $A \subseteq B$ and if $B \subseteq C$, then $A \subseteq C$.

Proof: We must show that for all x in A, $x \in C$. Suppose that x is any element of A. Since $A \subseteq B$, this implies that $x \in B$. Since $B \subseteq C$, the fact that $x \in B$ implies that $x \subset C$. Thus, for all x in A, $x \in C$. Hence $A \subseteq C$. Q.E.D.

The fact that $A \subseteq B$ does not exclude the possibility that A and B are exactly the same set. After all, for all x in A, it is true that $x \in A$. Thus, *every set is a subset of itself*. That is, for any set A, the statement $A \subseteq A$ is always true. It's also true that, for any set A, $\varnothing \subseteq A$. (See Degeneracy in Complements 5.2 below.)

If, in set inclusion, we want to exclude equality, there is a new symbol. The statement $A \subset B$ means that A is a subset of B but A is not the same as B, that is, B includes all of A and more besides.

Definition 8 $A \subset B$, means $A \subseteq B$ and $A \neq B$. Equivalently, $A \subset B$ means that for all x in A, $x \in B$, but there exists an x in B such that $x \notin A$. Set A is then called a **proper subset** of B.

The defining propositions are complex; go back and read them again. In particular, the word "but" is just another way of saying "and". For example, consider the sets $A := \{4, 5, 6\}$ and $B := \{4, 5, 6, 7, 8\}$. Clearly A is a subset of B, but A is a proper subset, because there exists at least one element of B that is not in A (in fact there are two, 7 and 8). Some books use the symbol $\subset$ to mean either $\subset$ or $\subseteq$. As long as you realize that there are two ideas at issue, that notation is not too confusing.

The concept of subset is related to set equality. You've probably already noticed (or should have) that set equality $A = B$ is simply the double requirement $A \subseteq B$ and $B \subseteq A$. Set equality is a straightforward notion, but it does have one subtle twist, a trap you must be careful to avoid. Suppose we define A as a set, and further define $B := \{A\}$. Is A equal to B? In other words, is A equal to $\{A\}$? In general, no. Set B is a set containing one element: the set A. Consequently, if A contains no elements or more than one element, then $A \neq \{A\}$. Even if A contains exactly one element, then the equality above is true only if that single element is A itself; and this happens only for a *very* strange set. We cannot logically exclude this possibility, but for all practical purposes you may assume that it won't happen. In particular, $\varnothing \neq \{\varnothing\}$, as we saw at the end of Section 5.1.6.

Analogies with Arithmetic. There are some analogies between set theory and arithmetic, which we shall point out as we encounter them. A few have to do with set inclusion and inequality of numbers. The strongest analogies in this area are the following. For every choice of a, b, c,

in $\mathbb{N}$	in set theory
if $a \leq b$ and $b \leq c$, then $a \leq c$	if $a \subseteq b$ and $b \subseteq c$, then $a \subseteq c$
if $a \leq b$ and $b \leq a$, then $a = b$	if $a \subseteq b$ and $b \subseteq a$, then $a = b$
$0 \leq a$	$\varnothing \subseteq a$

Exercises 5.2

1. For each of the following sets, state its cardinality, and list all of its subsets. **(a)** {a, b}; **(b)** {a, b, c}; **(c)** {∅, {a}, {a,{a}}}; **(d)** {∅, {∅}}.
2. The following sets are subsets of the upper-case English alphabet. For each given pair, state whether the two sets are equal or not: **(a)** {A, B}, {B, A}; **(b)** {A, B, A}, {B, A, B}; **(c)** {A, B, C}, {C, C, A}; **(d)** {A, B, C}, {C, C, B, A}.
3. The following sets are subsets of the uppercase English alphabet. For each of the following pairs, state whether the first set is a subset of the second set: **(a)** {A, B}, {A, B, C}; **(b)** {A, B}, {C, A}; **(c)** {A, B, A}, {A, B}; **(d)** { }, {A}; **(e)** { }, { }; **(f)** {A, B}, {A, A, A, A}.
4. For each of the following pairs of sets, state the cardinality of each, and state which of the following statements (if any) is true:
   (i) First set = second set.
   (ii) First set $\subset$ second set.
   (iii) Second set $\subset$ first set.
   (a) $\{2, 3, 4\}$, $\{x \in \mathbb{N} \mid x < 5\}$; **(b)** $\{-3, -2, -1, 1, 2, 3\}$, $\{x \in \mathbb{Z} \mid x^2 < 5\}$; **(c)** {A, E, I, O, U}, $\{x \in$ English alphabet $\mid x$ is not a vowel$\}$. **(d)** The set of 4-selections from {0, 1}, {binary representations of x with no leading zeroes $\mid x \in \mathbb{N}$ and $x < 16$ decimal}; **(e)** The set of 4-selections from {0, 1}, {binary representation of x possibly with leading zeroes $\mid x \in \mathbb{N}$ and $x < 16$ decimal}; **(f)** {00, 01, 10, 11}, {binary representation of x with no leading zeroes $\mid x \in \{1\}$}. **(g)** {00, 01, 10, 11}, {binary representation of x possibly with leading zeroes $\mid x \in \{1\}$}, **(h)** $\{x \in \mathbb{N} \mid x < 11\}$, $\{x \in \mathbb{N} \mid x$ is an entry of row 5 of Pascal's triangle$\}$; **(i)** $\{x \in \mathbb{N} \mid x$ is an entry in column 2 of Pascal's triangle$\}$, $\{\Sigma_{h=1}^{k} h \mid k \in \mathbb{N}$ and $k > 0\}$.
5. Prove that the sets {∅}, {∅, {∅}}, {{∅}, {∅, {∅}}}, {{∅}}, {{{∅}}} are all different from each other. See Sections 5.1.6 and 5.2. Define more sets, using similar techniques, and prove that they differ from each other and from the sets above.
6. Prove that $A = B$ if and only if $A \subseteq B$ and $B \subseteq A$.

Complements 5.2

1. **Degeneracy.** Throughout this book, we stress the importance of looking at mathematical language in the same way that we look at ordinary language. Mathematics is not a separate language, but rather takes the insight and imagination of ordinary discourse and adds rigor, clarity, and precision. Nowhere is this more pronounced that in the area of mathematical definitions, and in particular, in a situation known as a *degeneracy*. Many definitions in mathematics have roughly the same grammatical form. In set theory, for example, it looks something like this: "$P(A)$ is defined to be true if, for all x in A, $S(x)$ is true," where $P(A)$ is some property of a set A, for instance a statement to be defined, and $S(x)$ is some proposition regarding an unspecified element x of A. For instance, consider the definition of a subset: $A \subseteq B$ if, for all x in A, $x \in B$. Here $P(A)$ is the statement being defined, namely that $A \subseteq B$, and $S(x)$ is the proposition $x \in B$. The key phrase in the definition is "for all x in A." But now a question arises. What happens if A is empty, so that there *is no* x in A, and thus no $S(x)$ to even consider? That is, the propositions $S(x)$ are **degenerate,** and yield no information. Is the statement $P(A)$ then true?

   For another example in ordinary language, consider the statement, "If the moon is square, then the air temperature here is 25,000 degrees." Each component is clearly false, but is the composite statement true? Put in the conditional: "If the moon were square, then the air temperature here would be 25,000 degrees." Perhaps that sounds a little "truer." How do we define it, true or false? There is no universally correct answer to this, because definitions are by nature arbitrary. In formulating definitions, the most important consideration is that the definitions say exactly and consistently what we want them to say, and that no ambiguities remain. Degeneracy is one important tool for removing such ambiguity. We'll study it more deeply when we take up the study of logic.

   As it happens, it's much more convenient to define the statements above about $A \subseteq B$ (when $A = \varnothing$), or about the moon, as being *degenerately true*. In Definition 2 (set equality) we have taken care of this degeneracy by giving the definition in two forms, which are equivalent to each other thanks to the degeneracy convention. For Definition 5 (set inclusion), we need to invoke the degeneracy convention in order to apply it to the case in which the first set is empty. The result is, as already mentioned, that $\varnothing \subseteq B$ whenever B is a set.

2. **Russell's paradox.** If you're having trouble keeping clear the ideas of a set's *elements* versus its *subsets,* you're in excellent company, at least historically. The developers of set theory in the nineteenth century were bedeviled by the inherent ambiguity of whether a set could contain another set as an element. In particular, the stumbling block was this: if a set can contain sets as elements, can it contain *itself* as an element? That is, can there exist a set A such that $A \in A$? First of all, this idea conjures images of an infinitely retreating chain of a set containing itself, which in turn contains itself, which in turn contains itself,

and so on forever in both directions. In fact, the real problem is much more fundamental, much simpler, and much more disturbing.

Lord Bertrand Russell (1872–1970) showed that such a situation would lead to a *paradox,* a logical statement that is true but appears to be false. Russell's argument, which is a model of elegant simplicity, runs as follows: Suppose it were possible to have a set satisfying the condition $A \in A$. Suppose also we have a collection of sets, that is, a set V whose elements are sets. Some sets in V might thus contain themselves as elements, others might not. Let's collect into a set T those which do not: $T := \{A \in V \mid A \notin A\}$.

The question now is, is T an element of T? Suppose $T \in T$. Then the specification is not satisfied by T, and $T \notin T$. That is, if $T \in T$, then $T \notin T$. Confused? Let's try going the other way. Suppose $T \notin T$. Then, since it doesn't contain itself, T qualifies for membership in T, so $T \notin T$ implies that $T \in T$. In either case, the statement $T \in T$ is both true and false. More precisely, if it's false, then it's true; if it's true, then it's false. In one word, it's a paradox. Russell's "answer" to this fatal paradox is simple. The set T cannot be an element of V, as we have tacitly assumed in our reasoning. In other words, no matter how big a collection of sets is, it leaves out some set. Restated once more, "the set whose elements are all the sets" *does not exist.*

Russell's paradox has come into the popular mathematical literature by way of a clever (if a bit dated) analogy: the "barber paradox." Imagine a town sealed off from the world, which no one either enters or leaves. Every man in town must either shave himself or be shaved by the barber, who is also a man. No man who is able to shave himself may use the services of the barber. The question is: who shaves the barber? Playing with this problem a little bit should convince you that it's simply a restatement of the paradox we saw before.

Still another form of the same paradox is the following: A catalog is a list of items, for instance the books of a library or the items sold by a mail-order firm. The catalog itself may be one of the items: some small libraries used to have their catalog in book form, as opposed to the usual card catalog form, and some firms sell their own catalog. So, it may happen that a catalog lists itself. A scholar decided to make a list of those catalogs that do not list themselves. Should the scholar's catalog include itself as an item?

5.3 Union

There are three most basic operations on sets: union, intersection, and complement. In this section, we derive the first of them.

5.3.1 Union of Two Sets

The intuitive idea of the union of two sets can best be illustrated by example. Suppose we give $A := \{\mathrm{u, v, w}\}$, $B := \{\mathrm{x, y, z}\}$, $C := \{\mathrm{p, q, r, s}\}$, $D := \{\mathrm{r, s, t}\}$. Then the **union** of A and B is denoted by $A \cup B$ and is defined as $A \cup B :=$ {u, v, w, x, y, z} and similarly the union of C and D is $C \cup D :=$ {p, q, r, s, t}. The symbol $\cup$ for **union** is similar to the letter U, making it easier to remember. Note that this type of union is a **binary operator** on sets, in the same sense that addition (+) is a binary operator on numbers. All this means is that both $\cup$ and + require two **arguments**; compare the syntax of the set expression $A \cup B$ with the arithmetic expression $a + b$. For sets A and B, the union $A \cup B$ is itself a set. Thus, we've defined a new set in terms of old ones.

Be especially careful to understand exactly what the set $A \cup B$ contains as elements. The set $A \cup B$ always includes sets A and B as *subsets,* but might not (and usually does not; see Degeneracy in Complements 5.2) contain the *sets A* and *B* as *elements*. It contains the elements from the sets A and B. You've already seen that this distinction between sets and elements is important. Moreover, look at the example $C \cup D$ above. Both C and D individually contain the elements r and s. In $C \cup D$, these elements are listed, but only once. If we looked only at $C \cup D$, we couldn't tell which elements came from C and which from D. Some come from both. Even worse, if we knew C and $C \cup D$, we still wouldn't know exactly what elements were in D.

In general, for any two sets A and B, we'll try this tentative definition:

Definition 9 Given sets A and B, the **union** of A and B is denoted $A \cup B$ and is defined as the set $A \cup B := \{x \mid x \in A \text{ or } x \in B\}$.

This definition is slightly improper, in that we haven't specified a "known set" for x (see *specification* in Section 5.1.3). Thus, our definition above relies to a certain extent on intuition and doesn't rigorously conform to correct mathematical syntax. Another rule of the game comes to the rescue:

Axiom of unions. If C is any collection of sets, then there exists a set V such that, for every X in C, $X \subseteq V$.

In the case of two sets A and B, we have $C := \{A, B\}$, and we know the existence of a set V such that $A \subseteq V$ and $B \subseteq V$. Then the definition of union can be rephrased correctly.

Definition 10 Given sets A and B, the **union** of A and B is denoted $A \cup B$ and is defined as the set $A \cup B := \{x \in V \mid x \in A \text{ or } x \in B\}$, where V is a set such that $A \subseteq V$ and $B \subseteq V$.

What if we had used another set, say W, in place of V? Would we have obtained the same set $A \cup B$? Yes! (See Complements 5.3.)

In ordinary language, "or" means "one or the other, but not both". Very likely, that's the way you would understand it if you were told "do your homework or watch television." In mathematical language, by convention, "or" always means "one or the other, or both".

5.3.2 Formal Properties of Union

Like addition, union has certain properties that are very similar in form to those of addition in $\mathbb{N}$, and one that is glaringly different. We'll list some of the properties; you find the exception. One should also prove these properties on the ground of the definitions above and the properties we have already proved.

Let A, B, C be any sets.

- **Commutativity.** $A \cup B = B \cup A$. (*Example.* Suppose $A := \{1, 2\}$ and $B := \{2, 3\}$. Then each of $\{1, 2\} \cup \{2, 3\}$ and $\{2, 3\} \cup \{1, 2\}$ is simply the set $\{1, 2, 3\}$.)
- **Associativity.** $(A \cup B) \cup C = A \cup (B \cup C)$. (*Example.* Furthermore, suppose $C := \{3, 4\}$. Then the sets $(\{1, 2\} \cup \{2, 3\}) \cup \{3, 4\}$ and $\{1, 2\} \cup (\{2, 3\} \cup \{3, 4\})$ are, respectively, $\{1, 2, 3\} \cup \{3, 4\}$ and $\{1, 2\} \cup \{2, 3, 4\}$, and both reduce to $\{1, 2, 3, 4\}$.)
- **Idempotency.** $A \cup A = A$. (*Example.* $\{1, 2\} \cup \{1, 2\} = \{1, 2\}$.)
- **Identity.** $A \cup \varnothing = A$. (*Example.* $\{1, 2\} \cup \{\ \} = \{1, 2\}$.)
- $A \subseteq A \cup B$ and $B \subseteq A \cup B$. (*Example.* $\{1, 2\} \subseteq \{1, 2, 3\}$ and $\{2, 3\} \subseteq \{1, 2, 3\}$.)

We can now clarify the question of whether we can use enumeration to present a finite set. We have already solved the problem for the cases $\{\ \}$, $\{a\}$, and $\{a, b\}$. The other cases can be handled (but not with total ease) by setting $\{a, b, c\} := \{a, b\} \cup \{c\}$, $\{a, b, c, d\} := \{a, b, c\} \cup \{d\}$, and so on. Give some thought to the problem. See also the Axiom of Infinity in Complements 5.6.

Analogies with Arithmetic. There are other enlightening analogies between formal properties of set theory and formal properties in arithmetic. We have already noted that both the union of two sets and addition of two numbers are binary operations: $a + b$ has two arguments and so does $a \cup b$. For every choice of a, b, c:

in $\mathbb{N}$	in set theory
$a + b = b + a$	$a \cup b = b \cup a$
$(a + b) + c = a + (b + c)$	$(a \cup b) \cup c = a \cup (b \cup c)$
$a + 0 = a$	$a \cup \varnothing = a$
$a \leq a + b,\ b \leq a + b$	$a \subseteq a \cup b,\ b \subseteq a \cup b$

In $\mathbb{Z}$, for each a there is a number, ^{-}a, such that $a + (^{-}a) = 0$. In $\mathbb{N}$, such property fails (except when $a = 0$) and so does the analogue in set theory. Unless $a = \varnothing$, there is no set x such that $a \cup x = \varnothing$. In set theory, for every a, $a \cup a = a$. This has *no* analogue in $\mathbb{N}$ or $\mathbb{Z}$: $a + a \neq a$, unless $a = 0$.

5.3.3 Union of Several Sets

Because of associativity, expressions like $A \cup B \cup C$, $A \cup B \cup C \cup D$, $A \cup B \cup C \cup D \cup E$, and so on, require no parentheses. But when we deal with large numbers of sets, the notation gets cumbersome. We can use for set union a technique very similar to summation notation (see Section 2.1). Study the following two examples carefully. Suppose we have fifty sets, $A_1, A_2, \ldots, A_{50}$, and we want to write their union. We could write $A_1 \cup A_2 \cup \ldots \cup A_{50}$, but a far more intelligent thing to write would be $\bigcup_{i=1}^{50} A_i$. This is almost exactly like summation notation in arithmetic. The variable i is a dummy index, and A_i is the **union expression,** similar to a summation expression. Just as in summation, the idea is to "evaluate" each of the A_i (if needed), then form their union. Sometimes we don't even want to index the sets with a numeric dummy variable. Suppose that S is a collection of sets, that is, a set whose elements are sets. The union of all these sets could be written as $\bigcup_{A \in S} A$ even though the individual member sets of S don't have names, subscripted or not. The specification that A (the unnamed set) is in the collection S replaces the role of the subscript. Here the term A is serving as the dummy variable and also as the union expression. Its range is "for all A in S."

Some examples will help clarify these notions. Work through each one for yourself.

Examples (1) Given $A_i := \{i - 1, i\}$,

$$\bigcup_{j=1}^{4} A_j = \{0, 1\} \cup \{1, 2\} \cup \{2, 3\} \cup \{3, 4\},$$

which happens to be the same as $\{0, 1, 2, 3, 4\}$.

(2) Given $S := \{\{0\}, \{1\}, \{2\}, \{3\}\}$,

$$\bigcup_{A \in S} A = \{0, 1, 2, 3\}.$$

(3) Given $S := \{\{0, 1\}, \{1, 2, 3\}, \{3, 4, 5, 6\}\}$

$$\bigcup_{A \in S} A = \{0, 1, 2, 3, 4, 5, 6\}.$$

(4) $\bigcup_{j=3}^{5} \bigcup_{i=1}^{j} \{i - 1, i\} = \bigcup_{j=3}^{5} \{0, 1, \ldots, j\} = \{0, 1, 2, 3\} \cup \{0, 1, 2, 3, 4\} \cup \{0, 1, 2, 3, 4, 5\} = \{0, 1, 2, 3, 4, 5\}$.

(5) $\bigcup_{x \in \text{alphabet}} \{x\} = \text{alphabet}$.

A special case arises when S is a **singleton,** that is, when $\#S = 1$. Let Q be the only element in S (Q is a set). Then

$$\left(\bigcup_{x \in S} x\right) = \left(\bigcup_{x \in \{Q\}} x\right) = \left(\bigcup_{x = Q} x\right) = Q \quad (\textit{not } \{Q\}\text{. Why?})$$

The degenerate case arises when S itself is empty: $\bigcup_{A \in \varnothing} A = \varnothing$.

All these are consequences of the general definition below, which relies on the axiom of unions.

Definition 11 Suppose that S is a collection of sets. The union of all these sets is denoted $\bigcup_{A \in S} A$ and is defined as $\bigcup_{A \in S} A := \{x \in V \mid \text{there exists an } A \text{ in } S \text{ such that } x \in A\}$, where V is a set such that, for every A in S, $A \subseteq V$.

Exercises 5.3

1. The following sets are subsets of the uppercase English alphabet. For each of them, enumerate its elements and state the set's cardinality. **(a)** {A, B} ∪ {B, C}; **(b)** {A, B, C} ∪ {B, C, A}; **(c)** {A, B} ∪ {B, C} ∪ {C, D}; **(d)** {A, B, C} ∪ { }.
2. For each of the following pairs of sets, enumerate—once each—the elements of the union of the two sets. **(a)** $\{x \in \mathbb{N} \mid x < 7\}$, $\{x \in \mathbb{N} \mid x \text{ is an entry of row 4 of Pascal's triangle}\}$; **(b)** $\{x \in \mathbb{Z} \mid x^2 < 7\}$, $\{x \in \mathbb{N} \mid x < 7\}$; **(c)** $\{x \in$ English alphabet $\mid x$ is a vowel$\}$, $\{x \in$ English alphabet $\mid x$ precedes F in an alphabetical order$\}$; **(d)** {00, 01, 10, 11}, {3-digit binary representations of $x \mid x \in \mathbb{N}$ and $x < 5\}$; **(e)** $\{x^2 \mid x \in \mathbb{N} \text{ and } x < 5\}$; $\{\sum_{h=0}^{k} (2h+1) \mid k \in \mathbb{N} \text{ and } k < 5\}$; **(f)** $\{x \in \mathbb{Z} \mid x^2 < 1\}$, $\{x \in \mathbb{N} \mid x^2 < 0\}$.
3. Enumerate the elements of each of the following sets, and state the set's cardinality. **(a)** {1, 2, 3, 4} ∪ {2, 3, 4, 5}; **(b)** {1, 2, 3, 4} ∪ {5, 6, 7, 8}; **(c)** $\{x \in \mathbb{Z} \mid |x| < 5\} \cup \{x \in \mathbb{Z} \mid 4 < x < 9\}$; **(d)** $\{x \in \mathbb{Z} \mid |x| < 5\} \cup \{x \in \mathbb{Z} \mid |x - 2| < 5\}$; **(e)** $\bigcup_{j=1}^{3} \{x \in \mathbb{Z} \mid j < x < j + 3\}$; **(f)** $\bigcup_{j=1}^{3} \{j^2\}$; **(g)** $\bigcup_{j=-531}^{531} \{j - j\}$; **(h)** $\bigcup_{j=1}^{10} \{j, j + 2, j + 4\}$.

Complements 5.3

1. Prove that if A, B, C are sets, then $A \cup (B \cup C) = (A \cup B) \cup C = \bigcup_{X \in \{A,B,C\}} X$.
2. Prove the formal properties of union given in Section 5.3.2.
3. Prove that the union, as defined in Definition 10 or Definition 11, is independent of the choice of set V.

5.4 Intersection

Now let's do something different. Instead of combining sets, let's take their common part. This leads to *intersection,* the second of the three most basic operations on sets.

5.4.1 Intersection of Two Sets

Definition 12 Given sets A and B, the **intersection** of A and B is written $A \cap B$ and is defined by $A \cap B := \{x \in A \cup B \mid x \in A \text{ and } x \in B\}$.

This definition of intersection is perfectly proper, since we now have a known set, $A \cup B$, and a correct specification. As with union, $A \cap B$ is now a set in its own right. Compare the definitions of intersection and union. In particular, notice the use of the connective *and* versus *or*. This is the main point here: union required membership in A *or* B; intersection requires membership in A *and* B. If we look back to the beginning of Section 5.3.1, to sets A, B, C, and D, we quickly verify that $C \cap D = \{\text{r}, \text{s}\}$ and $A \cap B = \emptyset$. The second example leads to a definition.

Definition 13 Sets A and B are said to be **disjoint** when their intersection is empty.

Like union, the operation of intersection of two sets is a binary operation. Before we list its formal properties, let's look at two alternative definitions of intersection. "Wait a minute!" you say. "If we give three definitions of the same thing, don't we have to *prove* that all three define exactly the same thing?" Yes. So we'll have to state and prove two properties and then see how they could be used as definitions. In fact, we're going to leave the proofs up to you, although we'll give you some hints.

Theorem 14 Given sets A and B, $A \cap B = \{x \in A \mid x \in B\}$.

Proof: You must show that these two sets are equal. Recall that set equality is a bidirectional subset relationship. Define $C := \{x \in A \mid x \in B\}$ and prove that $C \subseteq A \cap B$ and $A \cap B \subseteq C$. If you need more guidance, see the proof of Theorem 16 below.

Theorem 15 Given sets A and B, $A \cap B = \{x \in B \mid x \in A\}$.

Proof: The proof here involves the same idea as that for theorem 14. All we have to do to use these as definitions is to change equality ($=$) to definition ($:=$) in either of the last two statements and use that statement as a definition, and change $:=$ to $=$ in Definition 12 and consider its statement as a theorem. Which one is chosen in any given situation is a matter of convenience. You should remember all three of them.

Another useful property is

Theorem 16 If $A \subseteq B$, then $A \cap B = A$.

Proof: We have to show two facts: $A \cap B \subseteq A$ and $A \subseteq A \cap B$. To show that $A \cap B \subseteq A$, let x be an element of $A \cap B$ and prove that $x \in A$. By the definition of intersection, the fact that $x \in (A \cap B)$ implies that $x \in A$. To show that $A \subseteq A \cap B$, let x be an element of A and show that $x \in (A \cap B)$. Since $A \subseteq B$, the fact that $x \in A$ implies that $x \in B$. Since $x \in A$ and $x \in B$, the definition of intersection implies that $x \in A \cap B$.

Since $A \cap B \subseteq A$ and $A \subseteq A \cap B$, we have proved that $A \cap B = A$. Q.E.D.

5.4.2 Formal Properties of Intersection

Like union, intersection has certain properties you should study. These properties are very similar in form to those of union as well as to those of multiplication in $\mathbb{N}$. We'll list some of the properties; you find the exceptions in the analogies with union and with arithmetic. Then prove these properties using the definitions above and the properties we have already proved. For any sets A, B, C,

- **Commutativity.** $A \cap B = B \cap A$.
- **Associativity.** $(A \cap B) \cap C = A \cap (B \cap C)$.
- **Idempotency.** $A \cap A = A$.
- **Annihilation.** $A \cap \varnothing = \varnothing$.
- **Distributivity.** $A \cap (B \cup C) = (A \cap B) \cup (A \cap C)$ and $A \cup (B \cap C) = (A \cup B) \cap (A \cup C)$.
- $A \cap B \subseteq A$ and $A \cap B \subseteq B$.

Let's compare the last two properties with those for union. Notice that the subset relations are reversed. Combining these relations with the analogues for union, we can now see that $A \cap B \subseteq A \subseteq A \cup B$ and $A \cap B \subseteq B \subseteq A \cup B$.

Venn Diagrams. A pictorial way to recall (or guess) set-theoretical properties is provided by **Venn diagrams.** We visualize a set as if it were a region of the plane, the elements being points of the plane. Then the intersection of two sets is the region common to two given regions, etc. The process is best illustrated by Figure 5.2: parts b, c, and d represent $A \cap B \cap C$, $A \cap (B \cup C)$, and $(A \cap B) \cup C$, respectively, where A, B, C are sets as given in part a. Keep in mind, however, that Venn diagrams are an excellent way to understand how things go, to have a hint to the solution of a problem, or to find a strategy for the proof of a theorem, but they *do not* provide definitions, solutions, or proofs. Make some Venn diagrams to verify the formal properties listed above for intersection and those for union in the previous section.

Analogies with Arithmetic. Comparison of intersection of two sets with multiplication of two numbers points up more analogies between formal properties of set theory and formal properties in arithmetic. First, both are binary operations. Second, for every choice of a, b, and c,

in $\mathbb{N}$	in set theory
$a \cdot b = b \cdot a$	$a \cap b = b \cap a$
$(a \cdot b) \cdot c = a \cdot (b \cdot c)$	$(a \cap b) \cap c = a \cap (b \cap c)$
$a \cdot 0 = 0$	$a \cap \varnothing = \varnothing$
$a \cdot (b + c) = (a \cdot b) + (a \cdot c)$	$a \cap (b \cup c) = (a \cap b) \cup (a \cap c)$.

Two properties are missing from the analogy: (1) In $\mathbb{N}$, the other distributive property is not valid. In general, $a + (b \cdot c)$ is *not* the same as $(a + b) \cdot (a + c)$, but $a \cup (b \cap c)$ *is* the same as $(a \cup b) \cap (a \cup c)$. (2) In $\mathbb{N}$, $a \cdot 1 = a$ for every a. In set theory, is there a special set u such that $a \cap u = a$ for every a? In full generality, no. There is, however, a way to remedy the situation as we shall see below.

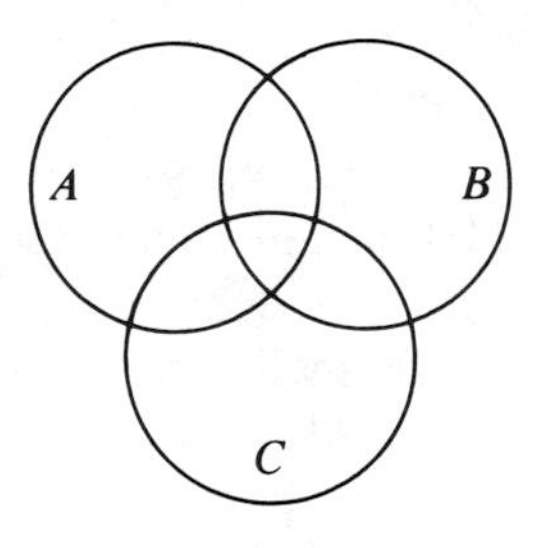

(a) Sets A, B, C

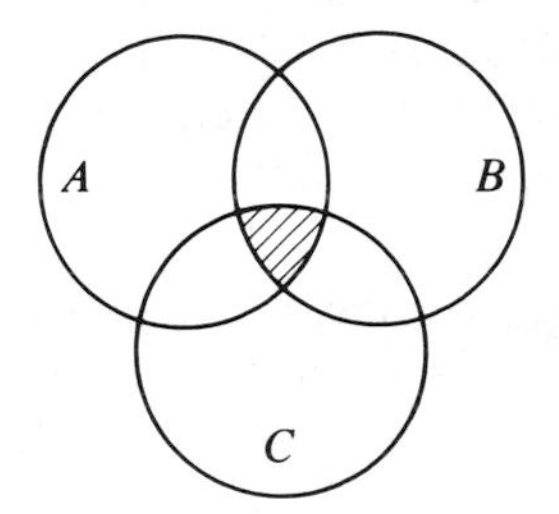

(b) $A \cap B \cap C$

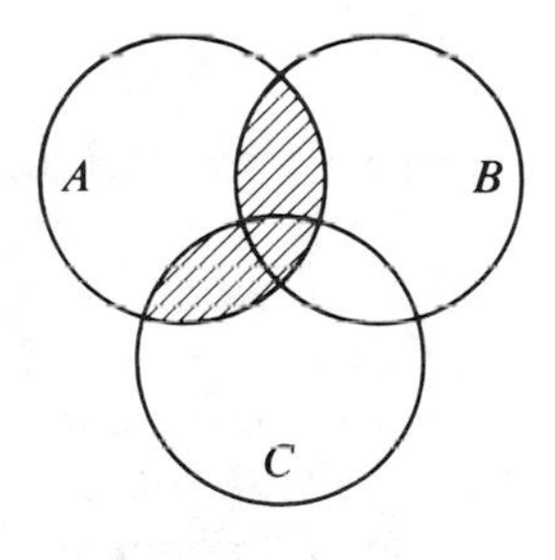

(c) $A \cap (B \cup C)$

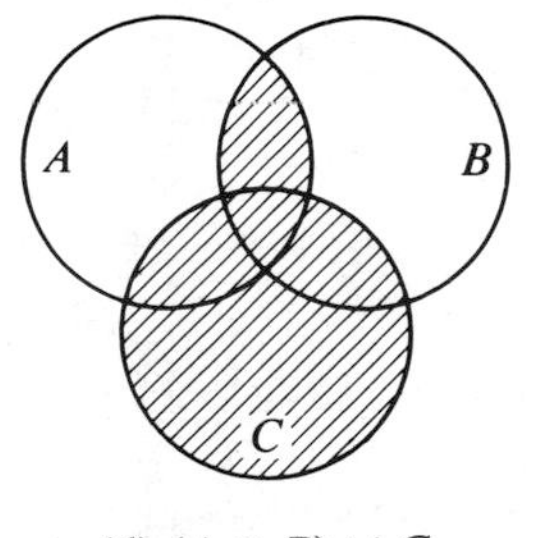

(d) $(A \cap B) \cup C$

Figure 5.2 Venn diagrams.

5.4.3 Universal Set

Some old books on set theory start out by describing something called "the universal set," which is somewhat defined as containing everything. This idea of the universal set as some cosmic repository for "everything that exists" is not really tenable (see "Russell's paradox" in Complements 5.2). We'll introduce a much more mundane and formally defensible definition for a universal set U. A universal set is a specific set, not just an idea. In any given setting, a **universal set** U is a set that includes as subsets all the sets we are considering in that given context. In other words, in that setting we are restricting our considerations to subsets of the given set U. No cosmic implications are intended. A universal set plays several roles and resolves several doubts. A modest example is the identity property of intersection, which was missing above; we give it here. For a more important role, read on.

Analogy with Arithmetic. If we restrict our considerations to the subsets of any given universal set u, then the missing analogy between multiplication and intersection reappears.

For every choice of a,

in $\mathbb{N}$	in set theory
$a \cdot 1 = a$	$a \cap u = a.$

5.4.4 Intersection of Several Sets

A case in which a universal set plays an important role is the intersection of several sets. As in the case of union (see Section 5.3.3), the associative property of intersection allows us to skip parentheses in expressions like $A \cap B \cap C$, $A \cap B \cap C \cap D$, and $A \cap B \cap C \cap D \cap E$. But when we have many sets, we use the same technique as in the case of union, employing symbols such as $\bigcap_{i=0}^{n} A_i$ and $\bigcap_{A \in S} A$. Let's formally define the second one, which is actually the most general case.

Definition 17 Suppose that S is a nonempty collection of sets. The intersection of all these sets is denoted $\bigcap_{A \in S} A$ and is a defined as

$$\bigcap_{A \in S} A := \left\{ x \in \bigcup_{A \in S} A \mid \text{for every } A \text{ in } S, x \in A \right\}.$$

Note the condition that $S \neq \varnothing$, which we did not have in the case of the union; degeneracy took care of that then. In the case of intersection, however, degeneracy would not help, unless we assign a universal set U as above, we assume that each element of S is a subset of U, and we complete the definition just given by defining $\bigcap_{A \in \varnothing} A := U$.

Examples Work these out carefully and thoroughly.

(1) $\bigcap_{i=3}^{5} \{i-2, i-1, i\} = \{1, 2, 3\} \cap \{2, 3, 4\} \cap \{3, 4, 5\} = \{3\}$.

(2) Given $S := \{\{0, 1, 2, 3\}, \{1, 2, 3, 4\}, \{2, 3, 4, 5\}\}$, $\bigcap_{A \in S} A = \{2, 3\}$.

(3) $\bigcap_{i=3}^{5} \bigcup_{j=0}^{i} \{j\} = \bigcap_{i=3}^{5} \{0, 1, \ldots, i\} = \{0, 1, 2, 3\} \cap \{0, 1, 2, 3, 4\} \cap \{0, 1, 2, 3, 4, 5\} = \{0, 1, 2, 3\}$.

Exercises 5.4

1. The following sets are subsets of the uppercase English alphabet. For each of them, enumerate its elements, and state the set's cardinality. **(a)** {A, B} ∩ {B, C}; **(b)** {A, B, C} ∩ {B, C, A}; **(c)** {A, B, C, D, E} ∩ {B, C, D, E, F} ∩ {C, D, E, F, G}; **(d)** {A, B, C} ∩ {B, C, D, E, F}.

2. Enumerate the elements of each of the following sets, and state the set's cardinality. **(a)** $\{1, 2, 3, 4\} \cap \{2, 3, 4, 5\}$; **(b)** $\{1, 2, 3, 4\} \cap \{5, 6, 7, 8\}$; **(c)** $\{x \in \mathbb{Z} \mid x < 5\} \cap \{x \in \mathbb{Z} \mid x > 0\}$; **(d)** $\{x \in \mathbb{Z} \mid |x| < 5\} \cap \{x \in \mathbb{Z} \mid |x - 2| < 5\}$; **(e)** ({a, b, c} ∪ {b, c, d}) ∩ {a, c, e}; **(f)** {a, b, c} ∪ ({b, c, d} ∩ {a, c, e}); **(g)** $\{x \in \mathbb{N} \mid x \text{ is even}\} \cap \{x \in \mathbb{N} \mid x \text{ is prime}\}$.

3. For each of the following pairs of sets, enumerate (once each) the elements of the intersection of the two sets. **(a)** $\{x \in \mathbb{N} \mid x < 7\}$, $\{x \in \mathbb{N} \mid x$ is an entry of row 4 of Pascal's triangle$\}$; **(b)** $\{x \in \mathbb{Z} \mid x^2 < 7\}$, $\{x \in \mathbb{N} \mid x < 7\}$; **(c)** $\{x \in$ English alphabet $\mid x$ is a vowel$\}$, $\{x \in$ English alphabet $\mid x$ precedes F in an alphabetical order$\}$; **(d)** {00, 01, 10, 11}, {2-digit binary representations of $x \mid x \in \mathbb{N}$ and $x < 4\}$; **(e)** $\{x^2 \mid x \in \mathbb{N}$ and $x < 5\}$, $\{\sum_{h=0}^{k} (2h + 1) \mid k \in \mathbb{N}$ and $k < 5\}$; **(f)** $\{x \in \mathbb{Z} \mid x^2 < 1\}$, $\{x \in \mathbb{N} \mid x^2 < 0\}$.

4. The definition of the symbol $^+$ is as follows:

Definition 18 Given that A is a set, the **successor** of A is denoted A^+ and is defined by $A^+ := A \cup \{A\}$.

Be sure to understand the syntax of the following formulae. Then construct a few more examples of your own. **(a)** $\{a, b\}^+ = \{a, b, \{a, b\}\}$; **(b)** $\varnothing^+ = \varnothing \cup \{\varnothing\} = \{\varnothing\}$, **(c)** $\{\varnothing\}^+ = \{\varnothing\} \cup \{\{\varnothing\}\} = \{\varnothing, \{\varnothing\}\}$; **(d)** $\{\varnothing, \{\varnothing\}\}^+ = \{\varnothing, \{\varnothing\}\} \cup \{\{\varnothing, \{\varnothing\}\}\} = \{\varnothing, \{\varnothing\}, \{\varnothing, \{\varnothing\}\}\}$. Reduce each of the following expressions to a single symbol: **(e)** $(x \cup y) \cap \varnothing$; **(f)** $(x \cap x) \cup \varnothing$; **(g)** $x \cap (x^+)$; **(h)** $x \cup (x^+)$.

5. Prove, as indicated in the text, that the three definitions of intersection are equivalent. See Theorems 14 and 15.

Complements 5.4

1. Prove the formal properties of intersection given in Section 5.4.2.
2. Prove the two properties stated in Theorem 19. Show how they can be used to give two alternative definitions of a subset that are equivalent to Definition 5.

Theorem 19 (1) $A \subseteq B$ if and only if $A \cup B = B$. (2) $A \subseteq B$ if and only if $A \cap B = A$.

5.5 Complementation

Finally, armed with the operations of union and intersection, and with the null and universal set, we can define set complement. This is the third and last of the basic operations on sets.

Definition 20 Given a set A, subset of a given universal set U, the **complement set** of A is denoted by A' and is defined as $A' := \{x \in U \mid x \notin A\}$.
We also say, briefly, that A' is the **complement** of A.

Note that some books use different symbols such as $C(A)$, an overhead bar $\bar{A}$ instead of a prime $'$, etc. Also, we don't have to write U as the "known set"; it is assumed when we don't write anything else, provided, of course, that it is clearly given, explicitly or by context.

Examples If $U = \{0, 1, 2, 3, 4, 5, 6, 7, 8, 9\}$, then $\{0, 2, 4, 6, 8\}' = \{1, 3, 5, 7, 9\}$, $\{0, 1, 2, 3, 4\}' = \{5, 6, 7, 8, 9\}$, $\varnothing' = \{0, 1, 2, 3, 4, 5, 6, 7, 8, 9\}$, $\{0, 1, 2, 3, 4, 5, 6, 7, 8, 9\}' = \varnothing$. If $U = \mathbb{N}$ and if $A = \{x \in \mathbb{N} \mid x < 5\}$, then $A' = \{x \in \mathbb{N} \mid x \geq 5\}$. Confirm your understanding of the term complement by observing the Venn diagrams in Figure 5.3: parts b, c, and d represent $A \cup B$, $A \cap B$, and $(A \cup B)'$, respectively, where the universal set U and sets A, B are given in part a.

Exercises 5.5

1. Given that $U = \mathbb{N}$, find the complement of each of the following sets; present it by description, enumeration, or specification. **(a)** $\{x \in \mathbb{N} \mid x > 3\}$; **(b)** $\{x \in \mathbb{N} \mid x > 0\}$; **(c)** $\{x \in \mathbb{N} \mid x \text{ is even}\}$; **(d)** $\{x \in \mathbb{N} \mid 3 < x < 5\}$.
2. Given that $U = \mathbb{N}$, find the complement of each of the sets: **(a)** $\{x \in \mathbb{N} \mid x = x^2\}$; **(b)** $\{x \in \mathbb{N} \mid x = |x|\}$; **(c)** $\{x \in \mathbb{N} \mid x^2 > x\}$.

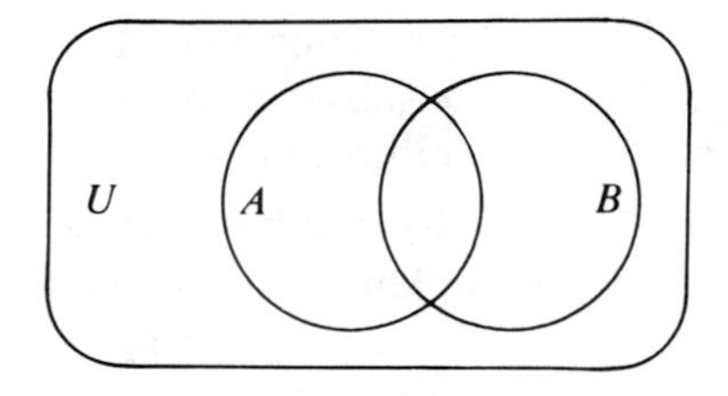

(a) Sets A, B and universal set U

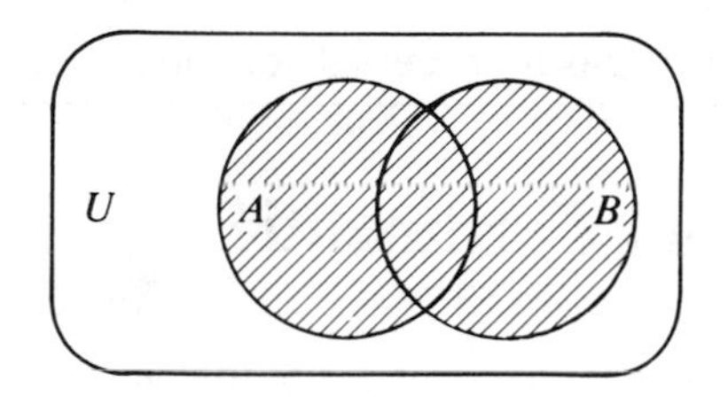

(b) $A \cup B$

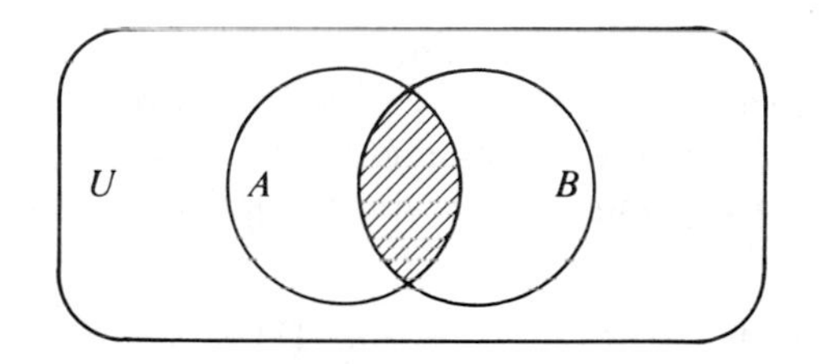

(c) $A \cap B$

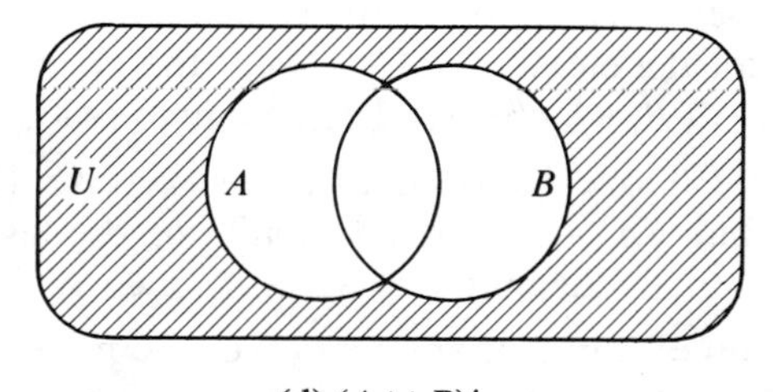

(d) $(A \cup B)'$

Figure 5.3 Venn diagrams

3. Give more examples of your own. Try to think of sets (classroom examples or from real life) and a subset of each; present the complement by description, enumeration, or specification.

4. For each of the following pairs of sets, enumerate (without repetition) the elements of the complement of the second set within the first set U, considered as the universal set. **(a)** $U :=$ English alphabet, $\{x \in U \mid x \text{ is a consonant}\}$; **(b)** $U :=$ English alphabet, $\{x \in U \mid x \text{ follows G in alphabetical order}\}$; **(c)** $U :=$ set of 3-selections from $\{0, 1\}$, $\{000, 111\}$; **(d)** $U := \{\text{3-selections from } \{0, 1\}\}$, $\{x \in U \mid \text{selection } x \text{ contains precisely two 1s}\}$. **(e)** U as in part c, $\{x \in U \mid \text{selection } x \text{ contains at least one 0}\}$; **(f)** $U := \{x \in \mathbb{N} \mid x \text{ is an entry of Pascal's triangle}\}$, $\{x \in U \mid x > 10\}$.

5. Given that U is a universal set and $A \subseteq U$, prove that $A \cup A' = U$, $A \cap A' = \varnothing$.

5.6 Finite Sets and Cardinalities

We have said in Section 5.1.7 that a finite set is a set whose elements can be enumerated in the usual sense. We also said that we are going to study cardinalities of finite sets only. Recall that the cardinality of set A is denoted $\#A$. Sets $\{4, 5, 6\}$, $\{x \in \mathbb{N} \mid x \geq 1 \text{ and } x \leq 256\}$, and the uppercase English alphabet are all examples of finite sets, with cardinalities 3, 256, and 26, respectively; $\varnothing$ is a finite set with cardinality 0.

At the intuitive level, proving the statements we are considering in this section is quite easy. If, however, you want to conduct the proof on rigorous set-theoretical grounds, the way is long and hard. We are not going to take that way, of course, but we are going to give you an indication of how it can be done (see Complements 5.6).

Theorem 21 Let A and B be finite sets. (1) If $A \subseteq B$, then $\#A \leq \#B$. (2) If $A \subset B$, then $\#A < \#B$. (3) If $A = \varnothing$, then $\#A = 0$; and if $\#A = 0$, then $A = \varnothing$. (4) If A and B are disjoint, that is, if $A \cap B = \varnothing$, then $\#(A \cup B) = \#A + \#B$. This property is the set-theoretical form of the first grade maxim that "adding is putting together." (5) In any case, $\#(A \cup B) = \#A + \#B - \#(A \cap B)$. This property is a simple form of the *principle of inclusion and exclusion*. Basically, count the elements of A, count the elements of B, and exclude the elements of $A \cap B$, which have been counted twice.

Proof: Prove Theorem 21 as an exercise. [Hints: For the first claim, suppose $A \subseteq B$ but $\#A > \#B$. What then? For the last, consider $A \cup B$ as the union of three mutually disjoint sets: $\{x \in A \mid x \notin B\}$, $\{x \in B \mid x \notin A\}$, $A \cap B$.]

As an example of application of the principle of inclusion and exclusion consider the following: In a box there are 18 math books and 25 hardcover books. Of these books, 8 are hardcover math books. How many books are there in the box? Result: $\#\{\text{math}\} + \#\{\text{hard}\} - \#(\{\text{math}\} \cap \{\text{hard}\}) = 18 + 25 - 8 = 35$.

Exercise 5.6

1. Find a general formula for $\#\{x \in \mathbb{Z} \mid x \geq m \text{ and } x \leq n\}$, where m and n are any two given integers. Be careful about the case in which $m > n$.
2. Given that $\#X = k > 1$, find a compact formula for $\#\bigcup_{n=0}^{9} \{n\text{-selections from } X\}$.
3. In an airplane there are 50 passengers; 30 of them are American; 20 of the passengers are British; 10 of the passengers are journalists; 40 of the passengers are engineers; 6 of the passengers are American journalists. How many are British engineers?
4. A statistical report on a factory contains the following data: The payroll consists of 900 employees, 800 of whom participate in the company's pension plan. Of the 900 employees, 200 have a seniority of five years or more, and 80 of these participate in the company's pension plan. There must be an error in the data. Why?
5. Prove Theorem 21 in the text (properties of cardinalities of sets).

Complements 5.6

1. We mentioned in the text that cardinalities may present some difficulties, especially in the case of infinite sets. But there are also some twists for finite sets. For example, what do you really mean by 3 or "three"? Not three apples or three fingers, just "three". Don't feel bad if you do not have a clear-cut answer. *After* humankind had learned very well how to count, it still took millennia to answer this question. One way out is the following: Define $0 := \varnothing$ (this is just a symbol; nothing has been done). Define $1 := 0^+$. Recall the definition of successor of a set; in this case, $0^+ = \{0\}$. Define $2 := 1^+ = \{0, 1\}$, $3 := 2^+ = \{0, 1, 2\}$, $4 := 3^+ = \{0, 1, 2, 3\}$, and so on. Can we really go on forever? The answer is given by another rule of the game.

Axiom of infinity. There exists a set W with the following properties: (1) $\varnothing \in W$; (2) if $y \in W$, then $y^+ \in W$.

Using the technique outlined above, we can construct the set $\{0, 1, 2, 3, 4, 5, \ldots\}$, that is, the set of the natural numbers, in a strictly set-theoretical form. The usual symbol for it is ω. After that, we can introduce (still in a set-theoretical way) order, addition, multiplication, and all the rest. As we said, the way is long and tortuous, but it's worth thinking about. For all practical purposes, if you understand the structure $\mathbb{N}$, then you have all you need in the present context.

5.7 Partitions

The intuitive idea of *partition of a set* is simple: just split the set into nonoverlapping nonempty subsets. When you cut or break a pie, not necessarily into the usual radial slices, each particle of the pie belongs to precisely one of the pieces. For example, consider all possible ways to split the set {1, 2, 3, 4}. One way is to split it into four pieces, each piece containing just one element: {1}{2}{3}{4}. Another extreme way is to "split it into one piece": {1, 2, 3, 4}. Or, we can have two pieces, each containing two elements or one containing one element and the other containing three elements: {1, 2}{3, 4}, {1, 3}{2, 4}, {1, 4}{2, 3}; {1}{2, 3, 4}, {2}{1, 3, 4}, {3}{1, 2, 4}, {4}{1, 2, 3}. Finally, we can have three pieces, two of which necessarily contain one element each and the third one two elements: {1}{2}{3, 4}, {1}{3}{2, 4}, {1}{4}{2, 3}, {2}{3}{1, 4}, {2}{4}{1, 3}, {3}{4}{1, 2}.

The following formal definition gives precision to the ideas we just presented:

Definition 22 Let A be a given set. A **partition** of A is a collection T of subsets of A such that: (1) For every C in T, $C \neq \varnothing$. (2) For every x in A, there exists a C in T such that $x \in C$. (3) For every C in T and for every D in T, either $C = D$ or C and D are disjoint.

Briefly, a partition is a collection of nonempty, mutually disjoint subsets whose union is the given set. Each element of T is called a **part** or a **block** of the partition. Let $C(A)$ denote the set whose elements are all the partitions of A. Let $C_k(A)$ be the set whose elements are those partitions of A made up of exactly k parts. Clearly, there are no partitions if $A = \varnothing$. By degeneracy, $C_0(A) = \varnothing$ for every A; if $A = \varnothing$, then $C(A) = \varnothing$ and $C_k(A) = \varnothing$ for every natural number k.

Examples Set {0, 1, 2, 3, 4, 5, 6, 7, 8, 9} can be partitioned into the parts {0, 1, 2}, {3, 5, 7, 9}, {4, 6, 8}; then a partition is {{0, 1, 2}, {3, 5, 7, 9}, {4, 6, 8}}. Other partitions of the same set are, for instance, {{0, 2, 4, 6, 8}, {1, 3, 5, 7, 9}}; {{0}, {1, 2, 3, 4, 5, 6, 7, 8, 9}}; {{0}, {1}, {2}, {3}, {4}, {5}, {6}, {7}, {8}, {9}}; {{0, 1, 2, 3, 4, 5, 6, 7, 8, 9}}. As you can see, there are a great many of them. Note especially the last two, which are somewhat extreme cases.

Let A be the uppercase English alphabet, written in any one given font, provided only that the letters I and O are written just like this. Let $\alpha := \{x \in A \mid$ the symbol for x is made of straight line segments only$\}$, and let $\beta := \{x \in A \mid x \notin \alpha\}$. Verify that $\{\alpha, \beta\}$ is a partition of A. Why were we so particular about I and O?

Let A be a given, nonempty alphabet of size n. Let Q be the set of all k-permutations of elements of A. Partition Q by placing in a same part those k-permutations that contain the same elements (see Chapter 4). In other words, the parts are the k-combinations as classes of k-permutations. We have also seen that $\#Q = P(n, k)$ and that the cardinality of each part is $k!$ Then the number of parts is $P(n, k)/k!$ This is what we called $C(n, k)$.

These preceding examples and those in the exercises should show how partitions are important in practical problems, such as problems of "classification."

Exercises 5.7

1. List all those partitions of $\{0, 1, 2, 3, 4, 5\}$ for which $\{2, 3, 4, 5\}$ is one of the parts.
2. Find all the partitions of each of the following sets. Assume that a, b, c are distinct elements. **(a)** $\{a, b\}$; **(b)** $\{a, b, c\}$; **(c)** $\{a\}$. Do this problem also on the computer, if you wish. Consider also sets with four or five elements.
3. Consider the set of all 5-digit binary strings. Partition this set according to the number of ones that appear in each string. How many elements are in each block? Generalize your results to n-digit binary strings.
4. Describe in correct set-theoretical terms the partitions indicated below in nearly ordinary language: **(a)** split the integers into even and odd integers; **(b)** split the binary numerals into those that comprise an even or an odd number of ones; **(c)** partition the natural numbers according to the fact that they have or do not have proper divisors; **(d)** partition the set of the second degree polynomials with real coefficients according to the number of distinct real roots. Find more examples of your own.
5. Verify that any nonempty set A admits the partition $\{A\}$ and the partition $\{\{x\} \mid x \in A\}$. When are these two partitions the same?
6. Let T be a partition of a nonempty set V. Let W be a nonempty subset of V. Prove that $\{X \cap W \mid (X \in T) \wedge (X \cap W \neq \varnothing)\}$ is a partition of W.
7. Let U be a set of cardinality 2 or more, even infinite. Let A be any proper, nonempty subset of U. Verify that $\{A, \{x \in U \mid x \notin A\}\}$ is a partition of U. Why the condition $\#U > 1$?

Complements 5.7

1. **Stirling numbers.** Let A be a finite, nonempty set and define $n := \#A$. Clearly, A admits precisely one partition into one part, which is A itself; the partition is then $\{A\}$. If $n = 1$, then this is the only partition of A. The question we want to address is, given the natural number k, what is the number of partitions of A into exactly k parts? Let $S[n, k]$ denote such number. When $k = 0$ or $k > n$ the result is clearly 0. When $n = 1$ or $k = 1$, we have already seen that the result is 1. Assuming that $S[n, k]$ is known for a certain value of n and for every value of k, calculate $S[n + 1, k]$. [Hint: Mark any one element of A, say z; some partitions admit set $\{z\}$ as one of the parts, some do not. Consider the two collections separately. If you ignore z, the first kind of partitions give partitions of set $\{x \in A \mid x \neq z\}$ into $k - 1$ parts. The second kind give partitions of

set $\{x \in A \mid x \neq z\}$ into k parts, and z can be adjoined to each one of the parts, one at a time. Count all possibilities together. The result is in Complements 4.6.] Construct $S[n, k]$ for all values of n up to 5 or 6.

2. Apply the result above to calculate the number of partitions of each of the following sets into each possible numbers of parts **(a)** $\{a\}$; **(b)** $\{a, b\}$; **(c)** $\{a, b, c\}$. Do this problem also on the computer, if you wish, for slightly larger sets. You may also want to generate all partitions, or all partitions with a given number of parts.

3. Let $\{S_i \mid 1 \leq i \leq n\}$ be a partition of set A ($n > 0$). Let $\{T_i \mid 1 \leq i \leq m\}$ be a partition of S_1. Prove that $\{S_i \mid 2 \leq i \leq n\} \cup \{T_i \mid 1 \leq i \leq m\}$ is a partition of A.

4. Let S and T be partitions of a same set A. Show that $\{X \cap Y \mid X \in S, Y \in T, X \cap Y \neq \varnothing\}$ is a partition of A. It is called the *cross-partition* obtained from S and T. [Beware: the number of elements in the cross-partition is not always $(\#S) \cdot (\#T)$.] For example, consider this case: A class comprises Audrey, Bruce, Charles, David, Elmer, Frank, Gina, and Helene. Audrey is a senior; Bruce, Charles, David, and Elmer are juniors; Frank and Gina are sophomores; Helene is a freshman. The partition of that class according to seniority is therefore {{Audrey}, {Bruce, Charles, David, Elmer}, {Frank, Gina}, {Helene}}. The partition male/female is {{Audrey, Gina, Helene}, {Bruce, Charles, David, Elmer, Frank}}. The cross-partition is {{Audrey}, {Bruce, Charles, David, Elmer}, {Frank}, {Gina}, {Helene}}. Note that we might have had eight blocks, but there are only five, in this case.

5. Find the cross-partition of $\{0, 1, 2, 3, 4, 5, 6, 7, 8, 9\}$ obtained from the partitions $\{\{0, 2, 4, 6, 8\}, \{1, 3, 5, 7, 9\}\}$, $\{\{0, 1, 2, 3, 4\}, \{5, 6, 7, 8, 9\}\}$.

6. Let A be a finite set. Let X be any partition of A. Recall the definitions of $C(A)$, $C_k(A)$. Verify **(a)** $\#C(A) = \sum_{k \in \mathbb{N}} \#C_k(A)$; **(b)** $\#C(A) = \sum_{k=1}^{\#A} \#C_k(A)$; **(c)** $\#A = \sum_{Y \in X} \#Y$; **(d)** $\{C_k(A) \mid 1 \leq k \leq \#A\}$ is a partition of $C(A)$.

5.8 The Power Set

We have discussed the idea of subsets of a given set. In this section we shall collect all those subsets together and give this collection a name. As an example, see Figure 5.1, where we enumerated all subsets of set $\{1, 2, 3\}$. If we give, by enumeration, a set whose elements are all these sets, we obtain $\{\varnothing, \{1\}, \{2\}, \{3\}, \{2, 3\}, \{1, 3\}, \{1, 2\}, \{1, 2, 3\}\}$. This is, for this example, the collection we are after.

5.8.1 Definition of Power Set

Definition 23 For a set A, the **power set** of A is denoted $\mathcal{P}A$ and is defined as the set whose elements are the subsets of A.

Examples (1) If $A := \{a, b\}$, then $\{a\}$, $\{b\}$, $\{a, b\}$ and $\varnothing$ are all subsets of A. In fact, they are the only subsets of A. We assume, of course, that $a \neq b$. Note that $\{a, b\}$ is the same as A. Thus, $\mathcal{P}\{a, b\} := \{\{a\}, \{b\}, \{a, b\}, \varnothing\}$.
(2) If $B := \{x\}$, then $\mathcal{P}B := \{\varnothing, \{x\}\} = \{\varnothing, B\}$.
(3) $\mathcal{P}\varnothing := \{\varnothing\}$.

The power set has a few wrinkles we should note.

1. Since $\varnothing$ is a subset of any set A, it is always true that $\varnothing \in \mathcal{P}A$. Thus, a power set can never be empty, since it contains, as an element, the set $\varnothing$.
2. Since any set A is a subset of itself, $A \in \mathcal{P}A$. When $A = \varnothing$, this and the last comment are the same. For a nonempty set A, the set $\mathcal{P}A$ must have at least two elements: $\varnothing$ and A itself.
3. The statements $X \in \mathcal{P}A$ and $X \subseteq A$ are equivalent. The statements $x \in A$, $\{x\} \subseteq A$, and $\{x\} \in \mathcal{P}A$ are equivalent. The statement $\{A\} = \mathcal{P}A$ is equivalent to $A = \varnothing$.

Look carefully at the syntax *and* at the meaning of these statements. Then the usual question arises: can we do it? That is, does the power set exist as a set on its own? The answer is yes, because of another rule of the game.

Axiom of powers. Given that A is any set, there exists a set whose elements are the subsets of A.

5.8.2 Cardinality of the Power Set

Suppose A is a finite set. What is the cardinality of $\mathcal{P}A$? Pay close attention to the following theorem and proof. Not only is the theorem of great importance in its own right, but the proof illustrates one of the most useful ideas of mathematical thinking.

Theorem 24 Suppose A is a finite set and define $n := \#A$. There are exactly $C(n, k)$ distinct sets B such that $B \subseteq A$ and $\#B = k$. Furthermore, $\#\mathcal{P}A = 2^n = 2^{\#A}$.

For the symbol $C(n, k)$ see Section 4.6. The symbol $2^{\text{natural number}}$ has the usual arithmetic meaning of power, even it the "natural number" has a strange form in this case.

Proof: We consider the set A as an alphabet of size n, just as in Chapter 4. Consider any subset B of A, where $\#B = k$. Set B is a set of elements (symbols) chosen from A (the alphabet), such that the order of selection is irrelevant and repetitions are not allowed, because we want just a subset. In the terminology of string formation, B is simply a k-combination from A. Theorem 4.13 gives the number of k-combinations of n objects: there are exactly $C(n, k)$ such combinations. Hence there are $C(n, k)$ subsets B. In addition, if k is unrestricted, then we simply consider all combinations (subsets), and from the same theorem, there are 2^n subsets of A. Q.E.D.

The fact that $\#\mathcal{P}A = 2^{\#A}$ explains the name "power set." You must commit this result to memory since we will use it often. In the proof, of course, we could have gone to basics and characterized any subset of A as a binary integer between 0 and $2^n - 1$, and gone on just as in theorem 4.14, or near the end of Section 4.5. But why bother? We've already done that, and here we need only recall that work. Mathematicians go to great and careful pains to be lazy; in this sense, "lazy" really means "efficient." Now you see that the whole idea of strings and combinations, not to mention the combinatoric function $C(n, k)$, are not isolated notions, but general mathematical tools with many applications!

The fact that a combination is simply a subset of the alphabet means that we can begin now to write combinations as we should—as subsets. If our alphabet is $A := \{a, b, c\}$, then there are $C(3, 2) = 3$ possible distinct 2-combinations: $\{a, b\}$, $\{a, c\}$, and $\{b, c\}$. The *set* of all possible distinct 2-combinations is $\{\{a, b\}, \{a, c\}, \{b, c\}\}$. We could also call $\{a, b\}$ an **unordered pair** of elements from A. Similarly, any k-combination, that is, any subset of cardinality k, can also be called an **unordered k-tuple** from A. It exists, of course, only if $0 \leq k \leq n$. In our example, the set A itself is an unordered 3-tuple (or triple) of elements of A. The power set of A has cardinality $2^3 = 8$ and is $\mathcal{P}A = \{\varnothing, \{a\}, \{b\}, \{c\}, \{a, b\}, \{b, c\}, \{a, c\}, \{a, b, c\}\}$. In this sense, the empty set corresponds to the *null* combination, or 0-combination.

Exercises 5.8

1. Write explicitly (enumerate) each of the following sets, and state its cardinality. Verify that the cardinality formulae given in the text are correct in these cases. **(a)** $\mathcal{P}\{a, b, c\}$; **(b)** $\mathcal{P}\varnothing, \mathcal{P}\mathcal{P}\varnothing, \mathcal{P}\mathcal{P}\mathcal{P}\varnothing, \ldots$; **(c)** $\mathcal{P}\{a\}, \mathcal{P}\mathcal{P}\{a\}, \mathcal{P}\mathcal{P}\mathcal{P}\{a\}, \ldots$. Things grow enormously after a very few steps. This is described, in the colorful jargon of mathematics, as "combinatorial explosion." Stop when you understand what's going on. But be very careful not to confuse $\{a\}$, $\{\{a\}\}$, $\{\{\{a\}\}\}$, $\{a, \{a\}\}$, etc.

   Do this problem on the computer as well, if you wish. Be sure to handle output properly.

2. For each of the following pairs of sets A, L, find the cardinality of L. **(a)** $A = \{0, 1, 2, 3, 4\}$, $L = \{X \in \mathcal{P}A \mid \#X = 3\}$; **(b)** $A = \{0, 1, 2, 3, 4\}$, $L = \{X \in \mathcal{P}A \mid \#X > 3\}$; **(c)** $A = \{0, 1, 2, 3, 4\}$, $L = \{X \in \mathcal{P}A \mid \#X < 3\}$; **(d)** $A = \{0, 1, 2\}$, $L = \{X \in \mathcal{P}\mathcal{P}A \mid \#X = 2\}$; **(e)** $A = \{0, 1, 2\}$, $L = \{X \in \mathcal{P}\mathcal{P}A \mid \#X < 2\}$; **(f)** $A = \{0, 1, 2, 3\}$, $L = \{X \in \mathcal{P}A \mid X \cap \{0, 3\} = \varnothing\}$; **(g)** $A = \{0, 1, 2, 3\}$, $L = \{X \in \mathcal{P}A \mid X \cap \{0, 3\} \neq \varnothing\}$; **(h)** $A = \{0, 1, 2, 3\}$, $L = \{X \in \mathcal{P}A \mid X \cap \{0, 3\} = \{0\}\}$.

3. Given $A := \{1, 2\}$, $B := \{1, 2, 3\}$, calculate: **(a)** $\#(A \cup B)$, **(b)** $\#(A \cap B)$; **(c)** $\#\mathcal{P}A$; **(d)** $\#\mathcal{P}\mathcal{P}A$; **(e)** $\#\mathcal{P}B$; **(f)** $\#\mathcal{P}\mathcal{P}B$.

Complements 5.8

1. Prove that the following definition of intersection of sets is equivalent to Definition 17 and includes the degenerate case. If U is any given universal set and if $S \subseteq \mathcal{P}U$, then $\bigcap_{A \in S} A := \{x \in U \mid \text{for every } A \text{ in } S, x \in A\}$.
2. Let A be a set. Let X be any partition of A. Recall the definitions of $C(A)$, $C_k(A)$ from Section 5.7. Verify **(a)** $X \subseteq \mathcal{P}A$; **(b)** $X \in \mathcal{P}\mathcal{P}A$; **(c)** $X \in C(A)$; **(d)** $X \in C_{\#X}(A)$; **(e)** $C(A) \subseteq \mathcal{P}\mathcal{P}A$; **(f)** $C(A) \in \mathcal{P}\mathcal{P}\mathcal{P}A$.

5.9 Cartesian Product Sets

Now let's consider cases where we *do* care about the order in which elements (symbols) are taken from a set A (alphabet). When $k = 2$, we are considering the 2-selections from A, such as (x, y), where x and y are elements of A. Such a 2-selection is called an **ordered pair** from A, or a 2-term **sequence** of elements of A. Note the use of parentheses (,) rather than braces { , }. This indicates that we picked elements x and y in A, and that x has a privilege (it's the first one of them). In general, $(x, y) \neq (y, x)$. The only time that $(x, y) = (y, x)$ is when $x = y$.

5.9.1 k-*fold Cartesian Products*

After these preliminaries, we have all the ingredients to proceed swiftly and introduce a few new concepts.

Definition 25 Given a set A and a natural number k, the **k-fold Cartesian product** of A is denoted A^k and is defined to be the set of all k-selections from A (see Section 4.2). In particular,

$$A^2 := \{(x, y) \mid x \in A \text{ and } y \in A\};$$
$$A^1 := \{(x) \mid x \in A\};$$
$$A^0 := \{(\ )\} = \{\Lambda\}.$$

The symbol ( ) denotes the **null string,** also denoted Λ. These sets are sometimes referred to as **cross-products,** and are also denoted by $A \times A$ (when $k = 2$), $A \times A \times A$ (when $k = 3$), and so on. A k-selection is also called an **ordered k-tuple.** (See also Complements 5.9.)

Furthermore, we must not confuse A^1 with A. The *element* x is not the same as the *string* (x). The former is an element of A, the latter is an element of A^1. The name and notation for Cartesian products intentionally mimic the familiar coordinate systems in analytical geometry. The name comes from the inventor of analytical geometry, the French mathematician René Descartes (or des Cartes). The real number line is usually symbolized as $\mathbb{R}^1$, the *xy*-plane as $\mathbb{R}^2$ or $\mathbb{R} \times \mathbb{R}$, and so on. In the plane, for instance, we know that the point (3.5, 4.7) is interpreted as $x = 3.5$, $y = 4.7$ and that it is not the same as the point (4.7, 3.5). On the line, (2.7) is a point (an element of $\mathbb{R}^1$), and 2.7 is a number (an element of $\mathbb{R}$). This reinforces the fact that, in general, $A^1 \neq A$.

5.9.2 Cartesian Product

It's quite easy, and desirable, to extend the notion of Cartesian product to cases involving different sets. A simple example is the following. Suppose two quizzes (quiz 1 and quiz 2) are administered to three students, say *a, b,* and *c*. Then each quiz paper is an ordered pair (quiz *n*, student *x*), that is, $(1, a)$, $(1, b)$, $(1, c)$, $(2, a)$, $(2, b)$, $(2, c)$. We can extend this idea to the general case.

Definition 26 Given that A and B are sets, the **Cartesian product** of A and B is denoted $A \times B$ and is defined as $A \times B := \{(x, y) \mid x \in A, y \in B\}$.

(Actually, we should use a symbol like $\{(x, y) \in (A \cup B)^2 \mid x \in A, y \in B\}$ in order to respect the syntax of the specification. But there is no real danger of confusion here.)

For example, $\{u, v\} \times \{0, 1, 2\} = \{(u, 0), (u, 1), (u, 2), (v, 0), (v, 1), (v, 2)\}$. Remember, though, that these products don't commute: in general $A \times B \neq B \times A$, unless $A = B$. In another form, $(u, v) = (x, y)$ if and only if $u = x$ and $v = y$. But, with that warning, we can intuitively define sets such as $A \times (B \times B)$, $(B \times A) \times B$, $A \times (B \times (C \times D))$, and so on. Furthermore, we can even define sets such as $(A \times B)^2$. For the empty set, we need only note that, for any set A, $A \times \varnothing = \varnothing$ and $\varnothing \times A = \varnothing$.

When there is no danger of ambiguity, we use an expedient to write a little bit less than we should. Instead of (x, y), we write just xy. With this convention, we have $\{u, v\} \times \{0, 1, 2\} = \{u0, u1, u2, v0, v1, v2\}$.

5.9.3 Cardinalities of Cartesian Products

The fact that Cartesian products really contain selections tells us immediately how to calculate their cardinalities.

Theorem 27 Given finite sets A and B and a natural number k,

(1) $\#(A^k) = (\#A)^k$ if $A \neq \varnothing$ or if $k > 0$;
(2) $\#(\varnothing^0) = 1$;
(3) $\#(A \times B) = (\#A) \cdot (\#B)$ (1).

Proof: Before we start the proof, let's clarify some terminology and notation. The elements of a k-fold Cartesian product are selections. The elements of a Cartesian product are restricted selections. Be sure to understand the syntax of the formulae above: A is a set, $\#A$ is a natural number, (natural number)$^{\text{natural number}}$ is a natural number, set$^{\text{natural number}}$ is a set. The symbol $\times$ denotes Cartesian product, and the centered dot denotes ordinary multiplication.

For the first two parts of the theorem, the results are given in Section 4.2. For part (3), we can use the following similar argument. Assume first that $m := \#A > 0$ and $n := \#B > 0$. A pair (x, y) is a restricted selection from set $A \cup B$, the restrictions being that $x \in A$ and $y \in B$. These restrictions give m choices for the first position and n choices for the second position. There are therefore $m \cdot n$ possible choices and our conclusion is valid. If $m = 0$ or if $n = 0$ there is no possibility to construct a pair (x, y), and our formula is valid again because $m \cdot n = 0$. Q.E.D.

Equation (1) translates in set-theoretical terms the etymological meaning of *m times n*. For instance, when $m = \#A = 2$ and $n = \#B = 3$, think of the array

$$\begin{matrix} * & * & * \\ * & * & * \end{matrix}$$

A similar remark is that the Cartesian product $A \times B$ can be visualized as the set of the boxes of a double-entry table whose rows are labelled by the elements of A and whose columns are labelled by the elements of B. For instance, if $A = \{1, 2\}$ and $B = \{a, b, c\}$, the table is

	a	b	c
1			
2			

Returning to the example of two quizzes administered to three students, the empty boxes in the table represent the quiz papers. They can be filled with the answers, with the grades, or with something else (but this adds to the concepts of Cartesian product).

Formula (1) is the set-theoretical form of a property that is called the *rule of product.*

Rule of Product If an action A1 can be performed in m different ways, and if a second action A2 can be performed independently in n different ways, then the combined action (A1 and A2) can be performed in mn different ways. In our case, action A1 is the choice of one element among the m elements of set A, action A2 is the choice of one element among the n elements of set B, and the combined action is therefore the choice of *one* element from set $A \times B$. Note that we've used the rule of product on several occasions when counting selections or permutations.

Exercises 5.9

1. Enumerate each of the following sets, and state its cardinality. Verify that the cardinality formulae given in the text are correct in these cases. **(a)** $\{a, b, c\} \times \{A, B\}$; **(b)** $\{a, b, c\} \times \{A\}$; **(c)** $\{a, b, c\} \times \varnothing$; **(d)** $\mathcal{P}(\{a\} \times \{b\})$; **(e)** $(\mathcal{P}\{a\}) \times (\mathcal{P}\{b\})$; **(f)** $\mathcal{P}(\varnothing \times \varnothing)$; **(g)** $(\mathcal{P}\varnothing) \times (\mathcal{P}\varnothing)$; **(h)** $\mathcal{P}(\{0, 1\}^2)$.

2. Given that $A := \{0, 1\}$, find the cardinality of each of the following sets: **(a)** $A^3 \cup A^4$; **(b)** $A^3 \cap A^4$; **(c)** $\bigcup_{k=0}^{3} A^k$; **(d)** $\bigcup_{k=1}^{3} A^k$; **(e)** $A \times$ (English alphabet); **(f)** $A^3 \times A^4$; **(g)** $A^0 \times A^1$; **(h)** $\mathcal{P}A^0$; **(i)** $\mathcal{P}(A^0 \times A^1)$; **(j)** $A^2 \times \mathcal{P}\{0, 1, 2\}$; **(k)** $\mathcal{P}(A^2 \times \{1\})$; **(l)** $(\mathcal{P}(A^2)) \times A$.

3. Given $A := \{1, 2, 3\}$, $B := \{a, b\}$ $(a \neq b)$, $C := \{t \in \mathbb{Z} \mid 1 \leq t \leq 375\}$. Give each of the following sets by enumeration: **(a)** $A \times A$; **(b)** $A \times B$; **(c)** $\mathcal{P}B$. Calculate each of the following cardinalities (don't multiply out): **(d)** $\#\mathcal{P}C$; **(e)** $\#\mathcal{P}(A \times B)$, **(f)** $\#(C \times (C \times (C \times C)))$.

4. If A is the set of decimal digits (or binary digits or octal digits), how can you characterize $A \times A$ in strictly arithmetical terms? What of A^k when $k \in \mathbb{N}$?

5. Prove that $\#((A \times B) \times C) = \#(A \times (B \times C))$ for any finite sets A, B, C.

6. Prove the following theorems. For any sets A, B, C, and D: **(a)** $(A \cup B) \times C = (A \times C) \cup (B \times C)$; **(b)** $(A \cap B) \times C = (A \times C) \cap (B \times C)$; **(c)** if $A \subseteq C$ and $B \subseteq D$, then $A \times B \subseteq C \times D$.

   Don't forget to check the case in which any of A, B, C, or D is empty!

7. Let S be a partition of set A and T be a partition of set B. Show that $\{V \times W \mid V \in S, W \in T\}$ is a partition of $A \times B$.

Complements 5.9

1. In Definition 25 we defined a Cartesian product in terms of k-selections. Does this definition respect the rules of set theory? The substance of the definition is correct and, for all practical purposes, we don't need anything more. But the form leaves something to be desired. Let's see how we can use the axioms of set theory to define $A \times A$ (the other cases present no major problems). We need first to define a 2-selection from A, that is, the object (x, y) when $x \in A$ and $y \in A$. We must state that we want the two elements x and y and that x must come first. In other words, we must give the set $\{x, y\}$ and the set $\{x\}$. We have then given two sets; consequently, we can also give a set whose elements are those two sets. In other words, we have defined $(x, y) := \{\{x, y\}, \{x\}\}$. Now let x and y be arbitrary elements of A; we need a set that contains as elements all of (x, y). To have this set, observe that (1) each of $\{x, y\}$ and $\{x\}$, being a subset of A, is an element of $\mathcal{P}A$; (2) since $\{x, y\}$ and $\{x\}$ are elements of $\mathcal{P}A$, the

set $\{\{x, y\}, \{x\}\}$ is a subset of $\mathcal{P}A$, that is, an element of $\mathcal{P}\mathcal{P}A$. This is the set we want. Then the definition of $A \times A$ is $\{(x, y) \in \mathcal{P}\mathcal{P}A \mid x \in A \text{ and } y \in A\}$. (This is a lot of work just to "set things straight." As we said before, all you need in practice is to note the difference between *set* $\{x, y\}$ and *sequence* (x, y), and to observe that $(x, y) = (u, v)$ if and only if $x = u$ and $y = v$.)

5.10 Summary

Theoretically, we based set theory on the statement "object x is (or is not) an element of set A" (notations: $x \in A$ or $x \notin A$) and on the axioms of set theory listed at the beginning of the chapter. From a practical point of view, the essential operations on sets are:

- **union** of two sets: $A \cup B$, the set which contains those elements that belong to A plus those that belong to B including (only once) those that belong to both A and B (Section 5.3).
- **intersection** of two sets: $A \cap B$, the set which contains those elements that belong both to A and B (Section 5.4).
- **complement** A' of a set A in a universal set U: A' is the set which contains those elements of U that do not belong to A; it is assumed that A is a *subset* of U, that is, that every element of A is also an element of U (Sections 5.4.3 and 5.5).

Consequently, we have several notions and properties:

- **methods to present a set:** description, enumeration, specification (Section 5.1.3).
- **the empty** set $\varnothing$: the set with no elements (Section 5.1.6).
- **set inclusion** $A \subseteq B$; that is, $A \cap B = A$ or, equivalently, every element of A is also in B (Section 5.2).
- **equality of sets,** $A = B$: A and B have the same elements or, equivalently, $A \subseteq B$ and $B \subseteq A$ (Section 5.1.4).
- **the power set of A,** $\mathcal{P}A$: the set comprising as elements the subsets of A (Section 5.8).
- **The k-fold Cartesian product of set A,** A^k: its elements are the k-term sequences of elements of A (Section 5.9).
- **the Cartesian product of sets A and B,** $A \times B$: its elements are the *ordered pairs* (x, y), with x in A and y in B (Section 5.9).
- **the cardinality of set A,** $\#A$: the number of elements of set A (Sections 5.1.7 and 5.6).
- **a partition of a set A:** a "slicing" of A into nonoverlapping parts (Section 5.7).

Some important cardinality properties in the finite cases are $\#(A^k) = (\#A)^k$; $\#(A \times B) = (\#A) \cdot (\#B)$; $\#\mathcal{P}A = 2^{\#A}$; and $\#(A \cup B) = (\#A) + (\#B) - \#(A \cap B)$.

The most important properties of operations on sets are the *ten formal properties* (see a complete list in Section 6.2.1); their usual names are commutativity of union, commutativity of intersection, associativity of union, associativity of intersection, distributivity of intersection over union, distributivity of union over intersection, identity for union, identity for intersection, complement property of union and complement property of intersection.

Structures and Logic

In Chapter 5 we acquired the most important tools in set theory, such as union, intersection, and Cartesian products. In this chapter we shall use those tools to develop several **structures.** Some of them will be elaborations of ideas we have encountered before, for example, strings (Section 5.9) and switching algebra (Section 3.4); some will be new, such as algebra of sets or Boolean algebras. In other chapters we'll study more structures, with emphasis always on structures that have applications, some of which we shall mention. Note that at this point we have already reached a rather sophisticated level in mathematical language. We are therefore able to use a compact presentation of the new material. Throughout, be sure you understand the meaning of concepts and statements that are presented (in full or in part) by means of symbolic language.

6.1 The Algebra of Strings

This section elaborates on the mechanism of Cartesian product sets and proceeds in a direction that will be important, for instance, for the theory of languages in computer science.

Let A be a given finite nonempty alphabet. With the notion of the k-fold Cartesian product A^k firmly in hand, we can now do operations on the selections from A. Recall from Sections 4.2 and 5.9 the definition of *k-selection* (or *string*), that is, a sequence of k elements from A. To make things a little easier, we're going to start subscripting the elements of A in a string. Instead of writing $A \times A = \{(x, y) \mid x \in A, y \in A\}$, we prefer to write $A \times A = \{(x_1, x_2) \mid x_1 \in A, x_2 \in A\}$, when $k = 2$. In general, we can speak of an ordered k-tuple as $(x_1, x_2, \ldots, x_k)$, where $x_1 \in A$, $x_2 \in A$, $\ldots$, $x_k \in A$. We can even economize the notation further by writing $(x_i)_{i=1,2,\ldots,k}$ or even $(x_i)_{i=1}^{k}$. Stare at that for a moment; it's not as forbidding as it looks.

Suppose we have two products A^r and A^s. What is the relationship between these two and the single product A^{r+s}? Following the lead of exponentiation in $\mathbb{N}$ and $\mathbb{Z}$, we might be tempted to say $A^{r+s} = A^r \times A^s$. Although this statement is correct in substance, it is not correct in form and, in this case, an error in form may have bad practical implications. For instance, suppose $A :=$ {a, b, c, d, e, f, g}. Then $u :=$ (a, b, c) $\in A^3$, $v :=$ (d, e, f, g) $\in A^4$, and $w :=$ (a, b, c, d, e, f, g) $\in A^7$. Isn't w simply the ordered pair of tuples (u, v)? No! $(u, v) =$ ((a, b, c), (d, e, f, g)) in $A^3 \times A^4$, which is very much like w, but not quite the same thing. The practical meaning is that one of the two interpretations keeps track of the point at which the two strings were joined together, and the other interpretation does not. Another example is provided by ordinary English: the string **her** is in A^3 (A being the 26-letter English alphabet), the string **ring** is in A^4, the string **herring** is in A^7 and is clearly different from the string **her ring.** Is the string **her ring** in A^7? Certainly not. (It can be variously interpreted as an element of $A^3 \times A^4$, or $A^3 \times \{\text{space}\} \times A^4$, or $(A \cup \{\text{space}\})^8$, but not of A^7.) The point is this: not only do ordered tuples not commute under Cartesian product, they do not even associate. In general, $((u, v), t) \neq (u, (v, t))$, and neither is equal to (u, v, t). Those extra parentheses are important, and can't be thrown away at will.

6.1.1 Concatenation of Strings

There is an operation, however, that does "throw parentheses away," and does what we really have in mind now. That operation is called *concatenation* of strings and is usually denoted with a colon. Keep in mind, however, that in different contexts many other symbols are used. To concatenate two strings literally means to write one string after the other string to form a third string. This idea is so straightforward that words make it confusing, so let's give a definition through examples.

Definition 1 Examples of **concatenation** of strings: (1) (a, b):(c, d, e) := (a, b, c, d, e). (2) If $x :=$ (a, b) and $y :=$ (c, d) then $x{:}y =$ (a, b):(c, d) := (a, b, c, d).

(A formal definition would need some mildly nasty notations like: if $u = (x_i)_{i=1}^r$ and if $v = (y_i)_{i=1}^s$, then $u{:}v := (z_k)_{k=1}^{r+s}$ where $z_k = x_k$ if $1 \leq k \leq r$, $z_k = y_{k-r}$ if $r < k \leq r + s$.)

We shall need to know some conventions about symbols. One is that if x is a string, then $x^0 := \Lambda$, $x^1 := x$, $x^2 := x{:}x$, $x^3 := x{:}x{:}x$, . . . (recall that Λ is the null string). In particular, (a, b, c)^3 := (a, b, c, a, b, c, a, b, c). Another we have already seen: when there is no danger of confusion, a string like (a, b, c) is denoted simply abc.

Concatenation of strings is not commutative, because $u{:}v$ is seldom the same string as $v{:}u$. Concatenation is associative, however; that is, $(u{:}v){:}t = u{:}(v{:}t)$. The null string Λ is the identity for concatenation: $u{:}\Lambda = u$ and $\Lambda{:}u = u$ for any string u.

6.1.2 Operations on Strings

There are several derived operations on strings and sets of strings: concatenation of two or more sets of strings, the *free semigroup* on an alphabet, and *iteration* on a set of strings. Such operations are important in many theoretical and practical applications, such as abstract algebra and computer science.

6.1.2.1 Concatenation of Two Languages

When we have two languages (that is, two sets of strings) we can concatenate each string in the first set with each string in the second set. For instance, if $L :=$ {ab, ac} and $M :=$ {ax, xy}, we obtain the set {abax, abxy, acax, acxy}. This process is similar to Cartesian product, but not quite the same: study the example $L :=$ {xy, x}, $M :=$ {yz, z}; we'd obtain the strings xyyz, xyz, xyz, xz, and their set would be {xyyz, xyz, xz} with three elements, not four.

Definition 2 A **language** over alphabet A is defined as any set of strings whose symbols are in A. Let L and M be two languages. Then the **concatenation** of the two languages L and M is denoted $L : M$ and is defined by $(L : M) := \{u{:}v \mid u \in L, v \in M\}$.

Example Note the role of the null string Λ in this operation. If $L =$ {ab, a} and if $M = \{\Lambda$, b}, then $L{:}M =$ {ab, a, abb}. Note also that we're not really concatenating the sets themselves, but the strings within the sets. With this distinction clearly in mind, we can then easily see that $A^r{:}A^s = A^{r+s}$, which is what we had in mind before. Note also that the operation of concatenation of sets of strings is associative, and we can therefore consider operations like $L{:}M{:}P$, or $L{:}L{:}L$, and so on.

6.1.2.2 The Free Semigroup on an Alphabet

If A is a finite nonempty alphabet, then the set A^* is defined to be the set of all finite strings formed from the elements of A; that is, A^* is the set of all finite selections from A. For instance, if $A :=$ {a, b, c}, then some elements of A^* are the strings a, b, c, aa, ab, ac, ba, bb, bc, ca, cb, cc, aaa, aab, aac, aba, abb, abc, . . . and, of course, the null string Λ. Note that the concatenation of any strings in A^* is again a string in A^*. In our example, ab:bc = abbc, which is in A^*.

Like $\mathbb{N}$ and $\mathbb{Z}$, A^* is a structure; it has objects (strings) and rules (concatenation, which is associative and has an identity). As such it has the rather forbidding name *free semigroup* of A, a name we'll explain later.

Definition 3 For any nonempty finite set A, the **free semigroup** of A, denoted by A^*, is defined as the structure made up of the set $A^* := \bigcup_{k\in\mathbb{N}} A^k$ and the operation of concatenation of elements of A^*, with associativity and with identity Λ. When $A = \varnothing$, $\varnothing^* := \{\Lambda\}$ and concatenation is null.

In this context, there are several forms of shorthand in the notation. They may be used when there is no ambiguity (but ambiguity is by no means uncommon): $xyz \ldots$ stands for $(x, y, z, \ldots)$ when $x, y, z, \ldots$ are elements of A; $x^2, x^3, \ldots$ stand for $xx, xxx, \ldots$ when x is an element of A; $uv, u^2, u^3, \ldots$ stand for $u{:}v, u{:}u, u{:}u{:}u, \ldots$ when u and v are elements of A^*.

As a consequence, a language over A is simply a subset of A^*.

Note that A^* is **closed** under concatenation; that is, the concatenation of any two elements of A^* is in A^*. Since A^* contains all possible selections from A, of arbitrarily great length, A^* is infinite, even though each A^k is finite.

Examples Let D be the set of the digits of a positional-base arithmetic, for example binary ($D := \{0, 1\}$) or decimal ($D := \{0, 1, 2, \ldots, 9\}$). Then D^1 consists of all one-digit numbers, D^2 of all two-digit numbers, and so on. Since k, the number of digits in a given number, is not limited, D^* contains representations of all natural numbers and is infinite. This is a typical case in which, following the daily practice, we write these strings without parentheses, as in 256 rather than (2, 5, 6). But note carefully that $D^* \neq \mathbb{N}$. Recall that a distinction must be made between abstract natural numbers in $\mathbb{N}$ and specific representations in D^*. The strings 1, 01, 001, 0001, . . . all represent the same natural number, 1, but are different strings in D^*. Specifically, $1 \in D^1$, $01 \in D^2$, $001 \in D^3$, and so on. Thus, D^* contains not only representations of all natural numbers, but contains infinitely many representations for each. In addition D^* contains Λ, the null sequence, which is the only element of D^0. There is no corresponding element of $\mathbb{N}$. Don't confuse the zero-sequence Λ with the one-sequence 0.

Let A be the set of instructions that some computer is capable of performing. At this level, we consider each instruction as a single element, although an instruction is usually a string of symbols. A *program* is a sequence of elements of A. What does it mean to execute a program on this computer? A given program might execute different ways given different input data, and we want now to characterize, not the program itself, but a given execution of that program. Such an execution consists of a string of instructions from A, some instructions possibly occurring many times, others possibly not occurring at all. Any execution is thus characterized by a string in A^*. Before the execution begins, we start from the string Λ. Whenever an instruction is executed (if any), we adjoin to our string that instruction as an extra term. Suppose the execution halts after k instructions, either correctly or in error. The execution is then ultimately a string in A^k, containing all the instructions performed, in order. Programmers often speak of a program having an "infinite loop," but that's really unrealistic. No execution string is infinitely long, but it can (in theory) be finite but arbitrarily long.

6.1.2.3 Iteration

When L is a given set of strings, we may want to concatenate with each other any finite number of strings from L, in all possible ways. For instance, let $L := \{\text{a}, \text{xy}\}$. Concatenating strings of L with each other, in all possible ways, any number of times, gives infinitely many strings, some of which are a, xy, aa, axy, xya, xyxy, aaa, aaxy, axya, axyxy, and, of course, the null string Λ. This construction is formalized as follows.

Definition 4 Let L be a language. The **iteration** of L, or **Kleene star** of L, is denoted $L^\star$ and is defined by $L^\star := \bigcup_{k\in\mathbb{N}} L^{\langle k\rangle}$, where the symbol $L^{\langle k\rangle}$ is defined temporarily as follows: when $k > 1$, $L^{\langle k\rangle}$ denotes the concatenations of k languages equal to L (for instance, $L^{\langle 2\rangle} = L{:}L$, $L^{\langle 3\rangle} = L{:}L{:}L$), and $L^{\langle 1\rangle} := L$, $L^{\langle 0\rangle} := \{\Lambda\}$.

The Kleene star is named for the American mathematician Stephen Cole Kleene (CLEAN), born in 1909.

Note that a language of A is any subset of A^*, that is, any element of $\mathcal{P}A^*$, the power set of A^*. Therefore, the set of all languages on A is $\mathcal{P}A^*$. We can concatenate two languages or iterate a language (and obtain a language in either case). Also, we can take the union or the intersection of two languages, as subsets of A^* (and again obtain a language). In other words, the set $\mathcal{P}A^*$ of all languages on A is closed under the unary operation of iteration and the binary operations of concatenation, union, and intersection.

Example If $A := \{x, y\}$ and if $L := \{x, yy\} \subseteq A^*$, it is understood that the elements of L are (x) and (y, y). Then several elements of $L^\star$ are: Λ, x, xx, xxx, . . ., yy, yyyy, . . ., xyy, xxyy, xxxyy, . . ., xyyx, xxxyyxyyyyx, ... All the elements of $L^\star$ are the finite strings of none, one, or more x's and, in any positions, none, one, or more *pairs* of consecutive y's.

In many books, the same form of star or asterisk is used for the Kleene star and for the free semigroup on an alphabet. A distinction is really necessary because the free semigroup strings together *elements* of A as symbols, whatever they may be, while $\star$ concatenates together elements of L, which must be *strings*. When L is a language, we can do either, but with different results. In a computer language, for instance, alphabetic and special symbols are strung together to form variable names, operational symbols, and expressions. Such names or symbols or expressions are concatenated together to form statements; and statements are strung together to form programs.

Also, in the case in which A is an ordinary alphabet and L is the set of those strings that are the words of an ordinary language, what we here have called "language" could be called "vocabulary".

Exercises 6.1

1. Write without repetition all strings of lengths 1, 2, and 3 from the alphabet {R, S, *}. Use some systematic order. [Recall that you may write abc rather than (a,b,c).]

2. Given the alphabet $K :=$ {f, g, h}, $K^* :=$ {finite strings over K}, $L :=$ {fg, gf} $\subseteq K^*$, and $M := \{\Lambda, \text{f}, \text{g}\} \subseteq K^*$, write by enumeration each of the following sets: **(a)** K^2; **(b)** $L \cup M$; **(c)** $L \cap M$; **(d)** $L{:}M$; **(e)** $L{:}L$; **(f)** $M{:}M$.

3. When alphabet A is defined by $A :=$ {a, b}, write the following sets explicitly: **(a)** $\bigcup_{i=0}^{2} A^i$; **(b)** $\bigcup_{i=0}^{3} A^i$; **(c)** $\bigcup_{i=1}^{3} A^i$; **(d)** $\bigcup_{i=2}^{3} A^i$.

4. When alphabet A is defined by $A :=$ {x}, write the following sets explicitly: **(a)** $\bigcup_{i=0}^{2} A^i$; **(b)** $\bigcup_{i=0}^{3} A^i$; **(c)** $\bigcup_{i=1}^{3} A^i$; **(d)** $\bigcup_{i=2}^{3} A^i$.

5. Given that $A := \{0, 1, 2\}$, find the cardinality of each of the following subsets of A^*: **(a)** $\bigcup_{h=0}^{2} A^h$; **(b)** $A^2 \cup A^3$; **(c)** $A^2{:}A^3$; **(d)** $A^2 \cap A^3$; **(e)** $\{\Lambda\}{:}A^3$; **(f)** $A^0{:}A^3$; **(g)** $\varnothing{:}A^4$.

6. Given that $A := \{0, 1, 2\}$ and $B := \{2, 3\}$, find the cardinality of each of the following subsets of $(A \cup B)^*$: **(a)** $A^2 \cap B^2$; **(b)** $A^2 \cup B^2$; **(c)** $A^2{:}B^3$; **(d)** $(\bigcup_{h=0}^{2} A^h){:}(\bigcup_{k=0}^{2} B^k)$; **(e)** $(\bigcup_{h=0}^{2} A^h) \cap (\bigcup_{k=0}^{2} B^k)$.

7. The given alphabet is $K :=$ {x, y}. Given are also $A :=$ {y, xy} $\subseteq K^*$, $B :=$ {xx} $\subseteq K^*$. **(a)** Write an 8-element subset of K^*. **(b)** Write an 11-element subset of $A^\star$. **(c)** Write an 11-element subset of $(A{:}B)^\star$. Give, by enumeration, each of the following: **(d)** $A \cap B$; **(e)** $A \cup B$; **(f)** $A{:}B$; **(g)** $A{:}A$; **(h)** $B{:}B$; **(i)** $B{:}A$.

8. If the cardinality of A is n ($\#A = n$), and if $k \geq 0$, find an explicit, compact formula for $\#(A^k)$ and one for $\#(\bigcup_{i=0}^{k} A^i)$. You may want to recall Section 3.5. Test your formula on the exercises above.

9. Suppose $\#A = n > 0, k > 0$. Construct an algorithm to systematically produce all elements of A^k. [Hint: See the positional-base example in Section 4.2.] Apply the algorithm to the cases in which $n = 5, k = 2$; $n = 2, k = 5$. Work this problem also on the computer, if you wish, with several choices for n and k.

10. Suppose $A :=$ {a, b}, $L := \bigcup_{i=1}^{3} A^i$. Write explicitly **(a)** a few elements of $L \times L$; **(b)** a few elements of $L \times L \times L$; and **(c)** several elements of $L^\star$, for instance those of length not exceeding 3, or 4, or 5. Do this problem also on the computer, if you wish. You may take an alphabet with more letters and longer strings. Watch out for a "combinatorial explosion." Calculate in advance how many strings you'll get.

6.2 The Algebra of Sets

We have developed set theory starting from a handful of *axioms*, but its real power lies in the fact that, once acquired, set theory can be presented in a more practical and more compact form, as a formal structure. In fact, at the operational level, a fortunate circumstance occurs. If we limit consideration to the subsets of a given universal set U, then we have proved, on the ground of the axioms, a few formal properties of union, intersection, and complementation (see Section 6.2.1). The rest follow automatically by using the *syntax* (that is, the *form*) of those formal properties. We need not refer back to the definitions of subset, union, etc. any more. The exact procedure for establishing (that is, *proving*) these other properties might not be obvious, but the fact that it can be done at all is remarkable enough. In this section, we'll show examples of how this works.

6.2.1 Ten Basic Formal Properties

Let U be a given nonempty set, and define $A := \mathcal{P}U$ (the power of set of U). We note first that A is closed under union, intersection, and complementation. That is, for any choice of x and y in A, $x \cup y$, $x \cap y$, x' are also in A. Recall that $x' := \{z \in U \mid z \notin x\}$. We have proved—or indicated how to prove—the following ten formal properties of set theory:

1a. Commutativity of union: $x \cup y = y \cup x$.
1b. Commutativity of intersection: $x \cap y = y \cap x$.
2a. Associativity of union: $(x \cup y) \cup z = x \cup (y \cup z)$.
2b. Associativity of intersection: $(x \cap y) \cap z = x \cap (y \cap z)$.
3a. Distributivity of intersection over union: $x \cap (y \cup z) = (x \cap y) \cup (x \cap z)$.
3b. Distributivity of union over intersection: $x \cup (y \cap z) = (x \cup y) \cap (x \cup z)$.
4a. Identity for union: $\varnothing \cup x = x$, $x \cup \varnothing = x$.
4b. Identity for intersection: $U \cap x = x$, $x \cap U = x$.
5a. Complement property of union: $x \cup x' = U$.
5b. Complement property of intersection: $x \cap x' = \varnothing$.

For the proofs, see Section 5.3.2 and Complements 5.4. We've listed these properties in pairs because the two properties in each pair are so similar (more on that later). With each pair we've given the name by which those properties are commonly known. These properties and their names should be memorized.

6.2.2 Derived Properties

But what of all the other properties we derived before, such as idempotency and annihilators? Really, we didn't *have* to prove them while we were doing set theory; we can prove them as theorems (actually as pairs of theorems). The only "tools" we want to use for the proofs are the ten formal properties above and, later, theorems proved from these properties. Each step is justified by property number or theorem reference. We give only a few proofs, the others are left as exercises. We've also given some of the theorems' names, so you can see not only their form, but also their significance.

Theorem 5 ***Idempotency.*** For every x in A
(a) $x \cup x = x$;
(b) $x \cap x = x$.

Proof:

part (a)

$x \cup x$	$= (x \cup x) \cap U$	(by property 4b)
	$= (x \cup x) \cap (x \cup x')$	(5a)
	$= x \cup (x \cap x')$	(3b)
	$= x \cup \varnothing$	(5b)
	$= x$	(4a)
$\therefore\ x \cup x$	$= x$	Q.E.D.

part (b)

$x \cap x$	$= (x \cap x) \cup \varnothing$	(4a)
	$= (x \cap x) \cup (x \cap x')$	(5b)
	$= x \cap (x \cup x')$	(3a)
	$= x \cap U$	(5a)
	$= x$	(4b)
$\therefore\ x \cap x$	$= x$	Q.E.D.

If you compare the sequence of steps in parts (a) and (b) of the proof, you'll understand how these proofs occur in pairs. Also note how the roles of $\varnothing$ and U are here reversed.

Theorem 6 ***Uniqueness of identities.*** For any given element b in A
(a) if $x \cup \mathrm{b} = x$ for all x, then $\mathrm{b} = \varnothing$;
(b) if $x \cap \mathrm{b} = x$ for all x, then $\mathrm{b} = U$.

Proof:

part (a)	$\varnothing \cup \mathrm{b} = \varnothing$	(substitute $x = \varnothing$)
but	$\varnothing \cup \mathrm{b} = \mathrm{b}$	(4a)
$\therefore$	$\mathrm{b} = \varnothing$	Q.E.D.

part (b). Exercise.

Thus, not only are $\varnothing$ and U identities for $\cup$ and $\cap$, but they are the only identities.

Theorem 7 ***Annihilators.*** For every x in A
(a) $x \cup U = U$;
(b) $x \cap \varnothing = \varnothing$.

Proof:

part (a)

$x \cup U$	$= x \cup (x \cup x')$	(by property 5a)
	$= (x \cup x) \cup x'$	(2a)
	$= x \cup x'$	(Theorem 5)
	$= U$	(5a)
$\therefore\ x \cup U$	$= U$	Q.E.D.

part (b). Exercise.

Theorem 8 ***Uniqueness of complement.*** For any x in A, if y is such that $x \cup y = U$ and $x \cap y = \varnothing$, then $y = x'$.

Proof:

$$
\begin{aligned}
x' &= x' \cap U && \text{(by property 4b)}\\
&= x' \cap (x \cup y) && \text{(hypothesis)}\\
&= (x' \cap x) \cup (x' \cap y) && \text{(3a)}\\
&= \varnothing \cup (x' \cap y) && \text{(1b, 5b)}\\
&= x' \cap y && \text{(4a)}\\
&= (x' \cap y) \cup \varnothing && \text{(4a)}\\
&= (x' \cap y) \cup (x \cap y) && \text{(hypothesis)}\\
&= (x' \cup x) \cap y && \text{(3a, 1a)}\\
&= U \cap y && \text{(1a, 5a)}\\
&= y && \text{(4b)}\\
\therefore\quad x' &= y && \text{Q.E.D.}
\end{aligned}
$$

The term *hypothesis* simply refers to the "if" statement given in the theorem. In this case the hypotheses were that $x \cup y = U$ and $x \cap y = \varnothing$.

Here we see that the complement of any set is unique; that is, that any set can have only one complement. This is useful enough in its own right, but it has an unexpected additional benefit. We can use Theorem 8 to establish the equality of two sets simply by showing that both are the complement of a third set.

Theorem 9 ***Double complement.*** For any w in A, $(w')' = w$.

Proof: If we apply properties (5a, 1a) and (5b, 1b) to w, we obtain $w' \cup w = U$, $w' \cap w = \varnothing$. Now apply properties 5a and 5b to w' and obtain $w' \cup (w')' = U$, $w' \cap (w')' = \varnothing$. Clearly $(w')'$ is the complement of w'. But by Theorem 8, w is also the complement of w'; hence the two must be equal, $w = (w')'$. Q.E.D.

Theorem 10 ***Duality of identities.*** (a) $\varnothing' = U$; (b) $U' = \varnothing$.

Proof: Exercise.

We now consider the first set theory result which is new to us, in that we haven't considered it before. The theorems of this pair are among the most important in all of set theory; they are known collectively as *DeMorgan's laws.*

Theorem 11 ***DeMorgan's laws.*** For every x in A and every y in A

(a) $(x \cup y)' = x' \cap y'$;

(b) $(x \cap y)' = x' \cup y'$.

Proof: For part (a), we will show that $(x \cup y) \cup (x' \cap y') = U$ and $(x \cup y) \cap (x' \cap y') = \varnothing$, and then apply Theorem 8 to conclude that $(x' \cap y') = (x \cup y)'$.

$$
\begin{aligned}
(x \cup y) \cup (x' \cap y') &= ((x \cup y) \cup x') \cap ((x \cup y) \cup y') && \text{(by property 3b)}\\
&= ((x \cup x') \cup y) \cap (x \cup (y \cup y')) && \text{(2a, 1a)}\\
&= (U \cup y) \cap (x \cup U) && \text{(5a)}\\
&= U \cap U && \text{(1a, Theorem 7(a))}\\
&= U && \text{(Theorem 5(b))}
\end{aligned}
$$

$$
\begin{aligned}
(x \cup y) \cap (x' \cap y') &= (x \cap (x' \cap y')) \cup (y \cap (x' \cap y')) && \text{(3a)}\\
&= ((x \cap x') \cap y') \cup (x' \cap (y \cap y')) && \text{(2b, 1b)}\\
&= (\varnothing \cap y') \cup (x' \cap \varnothing) && \text{(5b)}\\
&= \varnothing \cup \varnothing && \text{(1b, Theorem 7(b))}\\
&= \varnothing && \text{(Theorem 5(a))}
\end{aligned}
$$

By Theorem 8, $(x' \cap y') = (x \cup y)'$. Part (b) is left as an exercise. Q.E.D.

The double negation property and DeMorgan's laws are of crucial importance because of their syntax. They allow us to calculate the complement, not merely of a single set, but of an entire set expression. They are the set-theoretic analogues of negating an arithmetic expression, as in $^-(a + b) = {}^-a + {}^-b$.

Another new result follows.

Theorem 12 If x and y are in A, then $x \cap y = x$ if and only if $x \cup y = y$.

Proof: We leave the proof as an exercise. Be sure to prove the theorem using only the ten basic formal properties and any of the theorems already proved. Start by showing that $(x \cup y) \cup y' = U$ and $(x \cup y) \cap y' = \varnothing$. Then use Theorem 8.

One final remark: Among the properties in Section 6.2.1 we have somewhat lost set inclusion. But we can recover it (that is, we can redefine it formally) by means of Theorem 5.19 of Complements 5.4. Try it. If you have doubts, see the definition of order in a Boolean algebra given in Section 6.3.2.

Note 5 on Theorem Proving. Don't make too much of the distinction between formal proof and informal proof. The distinction doesn't depend on the form in which you write the proof; it depends on the substance of the proof. If you provide all the details and the reasons for each step, then the proof is formal. If you provide only a reasonably complete guideline on how to conduct the proof and leave some details to the reader, then the proof is informal. If your guideline is too flimsy and you leave out many details, then it's not a proof, just a set of hints.

Note some vocabulary connected with an *if/then* statement. The statement in the *if* part is the *hypothesis* (or *premise*), the statement in the *then* part is the *conclusion* (or *thesis*). When conducting the proof, the hypothesis is to be accepted as a given fact. (See also Degeneracy in Complements 6.5.) In general, in a statement of the type "if p, then q", we use many different phrasings for the same concept: "q only if p", "p is a sufficient condition for q", "q is a necessary condition for p", and so on.

When we are proving several properties in a row, each property we have already proved can be used in proving the next properties.

6.2.3 Duality

Surely by now you've grown tired of seeing (and proving) set-theoretical results in pairs. You've probably noticed (or should have) that these pairings follow a definite pattern. It's time for us to exploit that pattern in a systematic way.

Definition 13 Suppose S is any set-theoretical expression or theorem about subsets of a universal set U. The **dual** of S is the same expression or theorem, except that

- all occurrences of $\cup$ are replaced by $\cap$,
- all occurrences of $\cap$ are replaced by $\cup$,
- all occurrences of U are replaced by $\varnothing$, and
- all occurrences of $\varnothing$ are replaced by U, but
- the complement notations are not altered.

Take a moment now to go back and verify that all of the paired theorems, *especially the formal properties,* are in fact duals of each other. Theorem 8 stands alone because its two hypotheses are each other's duals; Theorem 8 is **self-dual,** and so is Theorem 12. We can cut our future set-theoretic workload almost exactly in half by establishing the following theorem:

Theorem 14 ***Principle of duality.*** Let S be any set-theoretic theorem about subsets of a universal set U. If S is true, then its dual is likewise true; if S is false, then its dual is likewise false.

Proof: If S is true, its proof will ultimately involve some sequence of steps involving the ten formal properties of set theory. Since each formal property has a dual, the proof of the dual of S will consist of a sequence of precisely the duals of the steps in the proof of S. Hence the dual of S will also be true.

If S is false, the dual of S must also be false. If the dual of S were true, then its dual (the dual of the dual, or the original S) would also be true. But S is false, hence its dual must also be false. Q.E.D.

Note 6 on Theorem Proving. Theorem 14 is called a **metatheorem,** a theorem about theorems. Metatheorems play a central role in the formal study of languages, especially computer programming languages. The reason we've been so insistent that mathematics be looked at as a language is so that we can make use of metatheorems. In subsequent chapters, we shall make considerable use of this powerful tool.

Summary. For any set U, we have three operations under which $\mathcal{P}U$ is closed: union, $\cup$, intersection, $\cap$, and complementation, $'$. We also have ten formal properties and all of their consequent theorems (most of which, obviously, haven't been detailed). This structure is called the **algebra of the subsets of U,** or more briefly, the **set algebra of U.** With the algebra of set theory firmly in tow, we now move on to explore others.

Exercises 6.2

1. Complete the proofs of the theorems in the text. Don't simply refer to Theorem 14; that's logically correct, but this exercise asks you to do the proofs independently. Looking up to the proof of part (a) while proving part (b) is advisable, though.

2. When $U := \{a\}$, construct $A := \mathcal{P}U$. List the results of $x \cup y$, $x \cap y$, and x' for all possible x, y in A.

3. Other operations can readily be defined on A, the power set of any given universal set U. For example, **(a)** *difference:* $x - y := x \cap (y')$; **(b)** *symmetric difference:* $x + y := (x \cup y) - (x \cap y)$; **(c)** *Sheffer stroke:* $x \mid y := (x \cap y)'$.
Given that $U := \{a, b, c, d, e, f\}$; $x := \{a, b, c, d\}$; and $y := \{c, d, e\}$, calculate **(d)** $x - y$; **(e)** $x + y$; **(f)** $x \mid y$.

Complements 6.2

1. Use Venn diagrams to verify intuitively that all ten formal properties of set theory are true.

2. Prove formally **(a)** $x \cap (x \cup y) = x$; **(b)** $x \cup (x \cap y) = x$; **(c)** $x \cap y = x$ if and only if $x \cup y = y$ (theorem 12); **(d)** if, for some w, $x \cap w = y \cap w$ and $x \cap w' = y \cap w'$, then $x = y$.

3. For each of the following construct a Venn diagram (see Figures 5.2 and 5.3), then prove formally: **(a)** $(x \cup y) - z = (x - z) \cup (y - z)$; **(b)** $x - (y \cup z) = (x - y) \cap (x - z)$; **(c)** $x + y = (x \cup y) \cap (x' \cup y')$; **(d)** $x' = (x \mid x)$; **(e)** $x \cup y = ((x \mid x) \mid (y \mid y))$; **(f)** $x \cap y = ((x \mid y) \mid (x \mid y))$.

6.3 Boolean Algebras

Boolean algebras are new structures which we will construct "by abstraction." They present one of the most typical cases in which abstraction is put to work constructively in two different directions. We observe first that some concrete structures (like switching algebra in Section 3.4 and algebra of sets in Section 6.2) have a few formal properties in common, properties that have a power of their own, independent of their meaning in one or the other structure.

The second step is to reverse the logical situation. We "postulate" those few properties; that is, we consider a structure whose objects are totally unspecified but are required only to satisfy those few formal properties. In other words, we use those properties as the axioms, or the "rules of the game," of our new structure. The third step is to draw from the axioms as many derived properties as we can and want to. We are then sure that all of these properties are valid in any concrete structure for which the axioms are valid. The fourth and final constructive step is to apply those derived properties to any concrete situation in which the axioms are valid. At this level, we'll have the *applications* of that abstract structure.

In this section, the process goes to work as follows: We have already observed that the ten basic formal properties of set theory (Section 6.2.1) show promise to be useful in other circumstances. We now define a Boolean algebra as any set that contains any elements, of any kind, provided only that those ten formal properties are satisfied. We shall next develop formally the derived properties of Boolean algebras on the ground of the axioms only. Later on, we'll be able to apply these results to several other structures. See, for instance, Sections 6.4 (logic) and 6.6 (switching networks) and Chapter 14 (applications of Boolean algebras).

6.3.1 Definition of Boolean Algebra

Definition 15 Let A be any set containing at least two elements. Let $\vee$ and $\wedge$ denote two binary operations on A (that is, two given rules), each of which, starting from any two elements x and y of A, produces an element of A as a result. Such results are denoted $x \vee y$ and $x \wedge y$, respectively. Let $'$ be a unary operation, that is, a given rule that, starting from any element x of A, produces an element of A, denoted x' or $\sim x$. Let two special, distinct elements of A be denoted 0 and 1, respectively. Let the data above satisfy the following ten axioms for every choice of x, y, and z in A:

1a. Commutativity of $\vee$: $x \vee y = y \vee x$.
1b. Commutativity of $\wedge$: $x \wedge y = y \wedge x$.
2a. Associativity of $\vee$: $(x \vee y) \vee z = x \vee (y \vee z)$.
2b. Associativity of $\wedge$: $(x \wedge y) \wedge z = x \wedge (y \wedge z)$
3a. Distributivity of $\wedge$ **over** $\vee$: $x \wedge (y \vee z) = (x \wedge y) \vee (x \wedge z)$.
3b. Distributivity of $\vee$ **over** $\wedge$: $x \vee (y \wedge z) = (x \vee y) \wedge (x \vee z)$.
4a. Identity for $\vee$: $0 \vee x = x$, $x \vee 0 = x$.
4b. Identity for $\wedge$: $1 \wedge x = x$, $x \wedge 1 = x$.
5a. Complement property of $\vee$: $x \vee x' = 1$.
5b. Complement property of $\wedge$: $x \wedge x' = 0$.

Then the sextuple $(A, \vee, \wedge, ', 0, 1)$ is called a **Boolean algebra.**

In analogy to logic, $x \vee y$ is usually read "x or y", $x \wedge y$ "x and y", x' "not x"; in analogy to set theory, x' is also read "complement of x". In analogy to arithmetic, 0 and 1 are read "zero" and "one", respectively. But recall that these are analogies: no concrete connotation is attached at this stage to any of the ingredients of a Boolean algebra. In applications, 0 or 1 will be states of electronic circuits (like "off" and "on"), or special propositions (like a "true" one or a "false" one), or special sets (like $\varnothing$ or a universal set), and so on.

Note also that the ten axioms (they are axioms now) are exactly the ten properties of Section 6.2.1 (they were properties then) with the following changes:

Replacements $\cup \mapsto \vee$, $\cap \mapsto \wedge$, complementation $\mapsto$ $'$, $\varnothing \mapsto 0$, $U \mapsto 1$.

This shows immediately that, if any nonempty set U is given, then the set $\mathcal{P}U$, with the operations of union, intersection, and complementation, and with the special elements $\varnothing$ and U, is a Boolean algebra.

6.3.2 Derived Properties of Boolean Algebras

We can save much work now. Observe that in Section 6.2.2 we took great pain in conducting proofs on the ground of the ten basic formal properties *only,* without reference to the then concrete meanings in set theory. Consequently, those proofs can be reproduced in any abstract Boolean algebra, subject only to the replacements listed near the end of Section 6.3.1. Below we give you a sample; do the other substitutions as exercises. As a partial conclusion, the following theorems and definitions are valid in every Boolean algebra.

Theorem 16 ***Idempotency.*** For every x in A
(a) $x \vee x = x$;
(b) $x \wedge x = x$.

Proof:

part (a)

$$
\begin{aligned}
x \vee x &= (x \vee x) \wedge 1 && \text{(by axiom 4b of Definition 15)}\\
&= (x \vee x) \wedge (x \vee x') && \text{(5a)}\\
&= x \vee (x \wedge x') && \text{(3b)}\\
&= x \vee 0 && \text{(5b)}\\
&= x && \text{(4a)}\\
\therefore\ x \vee x &= x && \text{Q.E.D.}
\end{aligned}
$$

part (b)

$$
\begin{aligned}
x \wedge x &= (x \wedge x) \vee 0 && \text{(4a)}\\
&= (x \wedge x) \vee (x \wedge x') && \text{(5b)}\\
&= x \wedge (x \vee x') && \text{(3a)}\\
&= x \wedge 1 && \text{(5a)}\\
&= x && \text{(4b)}\\
\therefore\ x \wedge x &= x && \text{Q.E.D.}
\end{aligned}
$$

Theorem 17 ***Uniqueness of identities.*** For any given b in A,
(a) If $x \vee b = x$ for all x, then $b = 0$;
(b) If $x \wedge b = x$ for all x, then $b = 1$.

Theorem 18 ***Annihilators.*** For every x in A,
(a) $x \vee 1 = 1$;
(b) $x \wedge 0 = 0$.

Theorem 19 ***Uniqueness of complement.*** For any element x of A, if y is such that $x \vee y = 1$ and $x \wedge y = 0$, then $y = x'$.

Theorem 20 ***Double complement.*** For any element w, $(w')' = w$.

Theorem 21 ***Duality of identities.***
(a) $0' = 1$;
(b) $1' = 0$.

Theorem 22 ***DeMorgan's laws.*** For every x and every y in A,
(a) $(x \vee y)' = x' \wedge y'$;
(b) $(x \wedge y)' = x' \vee y'$.

Theorem 23 For every x and every y in A, $x \wedge y = x$ if and only if $x \vee y = y$.

Definition 24 Suppose S is any expression or theorem in Boolean algebra. The *dual* of S is the same expression or theorem, except that

- all occurrences of $\vee$ are replaced by $\wedge$,
- all occurrences of $\wedge$ are replaced by $\vee$,
- all occurrences of 1 are replaced by 0,
- all occurrences of 0 are replaced by 1,
- the complement notations are not altered.

Theorem 25 ***Principle of duality.*** Let S be any theorem of a Boolean algebra. If S is true, then its dual is likewise true; if S is false, then its dual is likewise false.

Theorem 26 For every x and every y in A,
(a) $x \wedge (x \vee y) = x$;
(b) $x \vee (x \wedge y) = x$;
(c) if, for some w, $x \wedge w = y \wedge w$ and $x \wedge w' = y \wedge w'$, then $x = y$.

Also, we can use the axioms to establish an *order* in A:

Definition 27 The statement $x \leq y$ means $x \vee y = y$ or, equivalently, $x \wedge y = x$. (For the equivalence of the two parts, see Theorem 23.)

6.3.3 Elementary Boolean Algebra

The simplest examples of Boolean algebra occur when set A has precisely two elements. We'll call such a Boolean algebra an *elementary* Boolean algebra. Since 0 and 1 must be distinct elements of A, the only choice we have is to decide which of the two elements of A is 0 and which is 1. In abstract, either choice is possible. In concrete cases, there may be reasons to prefer one to the other; for instance, if $U := \{x\}$ and $A := \mathcal{P}U = \{\varnothing, \{x\}\}$, then there is a good reason to select $\varnothing$ to be the 0 of our Boolean algebra. After the choice of 0 and 1 (and if we want a Boolean algebra), we have no more choices. The properties of a Boolean algebra dictate all of the operations:

1. $0 \vee 0 = 0,\ 0 \vee 1 = 1,\ 1 \vee 0 = 1$ by axioms 1a and 4a.
2. $0 \wedge 1 = 0,\ 1 \wedge 0 = 0,\ 1 \wedge 1 = 1$ by 1b and 4b.
3. $1 \vee 1 = 1,\ 0 \wedge 0 = 0$ by idempotency.
4. $0' = 1,\ 1' = 0$ by duality of the identities.

The entire structure is then uniquely determined. Verifying that it is actually a Boolean algebra is quite immediate. We leave this verification as an exercise.

6.3.4 Algebra of Sets

Comparing Sections 6.2 and 6.3.1 shows immediately (as we have already remarked) that another example of Boolean algebra is provided by the algebra of the subsets of a given nonempty universal set. Note that this Boolean algebra is very general because the universal set is an arbitrary set, finite or infinite. The only restriction is that it be nonempty.

Exercises 6.3

1. Verify that the construction of an elementary Boolean algebra given in Section 6.3.3 actually satisfies the axioms in Section 6.3.1 and therefore provides a Boolean algebra. In particular, verify the associativity, distributivity, and complement properties.
2. Prove the theorems in Section 6.3.2, that is, the derived properties of a Boolean algebra. Use Section 6.2.2 as a guideline.
3. Let OR, AND, NOT be the operations of the elementary Boolean algebra we defined in Section 6.3.3 and denoted there $\vee$, $\wedge$, $'$, respectively. Verify that the following is a Boolean algebra:

   $A := \{00, 01, 10, 11\}$;
   $\vee$ is defined as BITWISE OR, that is $((i, j) \vee (k, l)) := (i \text{ OR } k, j \text{ OR } l)$;
   $\wedge$ is defined as BITWISE AND;
   $'$ is defined as BITWISE NOT;
   0 is 00, 1 is 11.

4. Prove that the analogous structure is a Boolean algebra when $A := \{0, 1\}^k$, for any given positive integer k and, in particular, when $k = 3$.
5. Given that $A := \{0, 1\}^3$ (i.e., $A := \{000, 001, 010, \ldots, 111\}$) and that $\vee$, $\wedge$, and $\sim$ are bitwise operations (BITWISE OR, AND, and NOT, respectively), we have a Boolean algebra. Calculate explicitly each of the following: **(a)** $011 \vee 010$; **(b)** ~ 011; (c) $(\sim 101) \wedge 011$; **(d)** $\sim(101 \wedge 011)$; **(e)** $010 \wedge 111$; **(f)** $010 \vee 000$.
6. In any Boolean algebra (not necessarily one containing two elements only) verify that the following equalities are valid for every choice of x, y, and z: **(a)** $x \wedge (y \vee x) = x$; **(b)** $\sim((\sim x) \wedge \sim(y \wedge \sim z)) = (x \vee y) \wedge \sim(z \wedge \sim x)$.
7. In any Boolean algebra, prove **(a)** if $x \le y$ and if $y \le z$, then $x \le z$; **(b)** if $x \le y$ and if $y \le x$, then $x = y$; **(c)** $x \le x$.

Complements 6.3

1. In any Boolean algebra, define more operations in analogy to problem 3 of Exercises 6.2.
2. [This may be difficult; we'll see it again in Chapter 14.] Let the set A of a Boolean algebra have more than two elements. Then there is an element w other than 0 and 1. **(a)** Define $X := \{u \in A \mid u \wedge w' = 0\}$, $Y := \{v \in A \mid v \wedge w = 0\}$. **(b)** Let z be any element of A and set $z_1 := z \wedge w$, $z_2 := z \wedge w'$. **(c)** Prove that $z =$

$z_1 \vee z_2$, $z_1 \in X$, $z_2 \in Y$. **(d)** Prove also that, if $z = r \vee s$, if $r \wedge w' = 0$, and if $s \wedge w = 0$, then $r = z_1$ and $s = z_2$. That is, every element z of A has a unique representation as above in terms of z_1 and z_2. **(e)** Prove that X is a Boolean algebra with the same $\vee$ and $\wedge$ operations as on A (call them OR_1, AND_1, respectively), the same 0, but a different "1" (which one?) and a different complementation (call it NOT_1). Prove the same for Y and call OR_2, AND_2, NOT_2 the operations in Y. **(f)** Furthermore, prove that if $z = z_1 \vee z_2$ as above and if $t = t_1 \vee t_2$ in a similar manner, then $z \vee t$ (in A) $= (z_1 \text{ OR}_1\ t_1) \vee (z_2 \text{ OR}_2\ z_2)$, $z \wedge t$ (in A) $= (z_1 \text{ AND}_1\ t_1) \vee (z_2 \text{ AND}_2\ t_2)$, z' (in A) $= (\text{NOT}_1\ z_1) \vee (\text{NOT}_2\ z_2)$.

In other words, the given Boolean algebra can be obtained by combining two "smaller" Boolean algebras with a technique similar to the one we used in problem 3 of Exercises 6.3. The technical name for this "combining" is **direct product.**

6.4 Elementary Logic and Truth Values

The purpose of this section is to provide an introduction to what is variously called *elementary formal logic* or the *algebra of propositions* or *propositional calculus*. A more rigorous treatment will be considered in Chapter 14. For the moment, let's take a look at the basic ideas forming the study of logic.

6.4.1 Propositions and Logical Operators

In mathematics and logic, the term *sentence* or *proposition* has a much narrower meaning than in ordinary language. "Have you read the title of this section?" or "Read the title of this section!" are sentences that we will *not* accept as propositions. Replies like "I have" or "I did" are fragmentary sentences which—allowing for the ambiguity—we will accept. Clear-cut statements like "x is less than y" are acceptable, provided x and y are numbers; "a transistor is less than a mosquito" is unacceptable because it's meaningless gibberish. The defining feature of a mathematical or logical proposition is that it's restricted to being either *true* or *false*. No other interpretation is intended. We don't require that all propositions be true; "33 is less than 15" is an acceptable proposition, even though it's patently false. The line between informal statement and formal proposition is not always clear. Some propositions are controversial, in that they contain elements of both truth and falsity, as for example, "Rutherford B. Hayes was a great president". We will limit ourselves to those propositions whose truth or falsity can be *unambiguously decided once all the relevant facts are known*. What remains is still a considerable class of propositions. Even if a proposition is unambiguous, we may still not know whether it's true or false. For integers N and M, "$N \geq M$" is an acceptable proposition, but without knowing what N and M are, we have no idea whether "$N \geq M$" is true or false. But it must be one or the other! It cannot be neither, and it cannot be both. The fact that a proposition is true or false is called the **truth value** of the proposition. For brevity, we'll denote the value of true by the letter T, and false by F. Sometimes, in other settings, we'll see alternative notations for true/false, such as 1/0, Y/N, yes/no, on/off, up/down. We'll show how some of these other notations arise, and why they're useful.

Sometimes we have two propositions, p and q, whose truth values are unknown, but we do know that they *always have the same value;* that is, both are true or both are false. For example, "n is even" and "n is a multiple of 2" are either both true or both false. We shouldn't denote this by $p = q$, since p and q aren't the *same* statement; they simply have the same truth value. So, we'll use the **equivalence** symbol, $p \Leftrightarrow q$. A more rigorous definition of equivalence will be given later.

6.4.2 The Logical Operators AND, OR, and NOT

Propositions can be combined to form new propositions, called **compound propositions.** This is done by means of the **logical operators** or **logical connectives.** The three most common logical operators are AND, OR, and NOT.

For example, the proposition "$3 < 7$ and 4 is even" is true because both of its elementary proposition (or **arguments,** or **components**) are individually true. "The moon is square and Chicago is in Illinois" is false, since one of its arguments is false. "The moon is square or Chicago is in Illinois" is true. The negation of "every integer is even" is "not every integer is even"; the former is false, the latter is true, of course, but both are clear-cut statements. The negation of "$x = 7$" (true or false that it may be) is, in ordinary arithmetic, "$x \neq 7$", but we'll often write "NOT ($x = 7$)" in logical jargon. The proposition "$x < 7$ or $x = 7$" is usually written "$x \leq 7$".

Both AND and OR are *binary* operators, since each requires two arguments. Negation is a *unary* logical operator, in that it takes only one argument.

If p and q are propositions, then the propositions "p OR q", "p AND q", and "NOT p" are themselves propositions, having a syntax similar to $A \cup B$, $A \cap B$, or A' in set theory and $x + y$, $x \cdot y$, or ^-x in arithmetic. Formally,

Definition 28 Suppose p and q are propositions. The **disjunction** of p, q (that is, the proposition "p OR q") is denoted by $p \vee q$ and is defined to be true if either of p and q is individually true, or if both are; otherwise it's false. The **conjunction** of p, q (that is, the proposition "p AND q") is denoted by $p \wedge q$ and is defined to be true if each of p and q is individually true; otherwise it's false. The **negation** of p (that is, the proposition "NOT p" is denoted by $\sim p$ or p' or $\neg p$ and is a proposition that is defined as being true if p is false, and false if p is true.

Some minor but confusing discrepancies exist between logical language and ordinary language. Most are unimportant, and some are even useful. Note carefully our use of the logical operator OR. The statement p OR q is shorthand for "p is true or q is true or both are true". This use of OR is called the *inclusive* OR. In nonmathematical language, the *exclusive* OR is more common: "one of p or q, but not both". Lawyers use the phrase *and/or* to mean the inclusive or. We will never use that phrase; when we mean to use the exclusive OR, we'll specifically say so. "The food and service are good" becomes in logical terms, "The food is good and the service is good". The word *but* is simply another way of expressing the logical AND: "The food is good but the service is bad" is equivalent to "The food is good and the service is bad". Logical language sometimes produces awkward sentence constructions: "Not 'the moon is square' " is better rendered as "The moon is not square". Somewhat more subtle is the negation of statements containing *all/none* or *everyone/no one* phrases. The negation of "Everyone in this class wears glasses" is, grammatically and logically, "Not everyone in this class wears glasses". But the statements "Everyone in this class wears glasses" and "No one in this class wears glasses" are *not* negations of each other, either grammatically or logically. They're simply different statements. Although these two propositions cannot both be true, they *could be*—and probably are—both false. We'll see this in more detail in Section 6.5.3.

We're also going to use the letters T and F as propositions themselves. That is, T is a conventional proposition that always has the value *true,* and F is a conventional proposition that always has the value *false.* Used in this way, T and F are called **self-defining propositions.**

In summary, the binary operations $\vee$, $\wedge$ and the unary operation $\sim$, when applied to propositions, produce propositions.

Exercises 6.4

1. In each of the following cases, the propositions p, q are given. State the truth values of (1) p, (2) q, (3) $\sim p$, (4) $\sim q$, (5) $(p \wedge q)$, (6) $(p \vee q)$, (7) $\sim(p \wedge q)$, and (8) $\sim(p \vee q)$. **(a)** p: Chicago is in Illinois; q: The moon is a satellite. **(b)** p: Chicago is in Illinois; q: The moon is square. **(c)** p: Chicago is in Montana; q: The moon is a satellite. **(d)** p: Chicago is in Montana; q: The moon is square.

2. Split each of the following common mathematical abbreviations into its elementary argument propositions: **(a)** $x \leq 5$; **(b)** $y \geq 7$; **(c)** $5 < x < 9$; **(d)** $5 \leq x < 9$; **(e)** $5 \leq x \leq 9$; **(f)** $x \not\subset A$; **(g)** $A \cap B \subseteq A \subseteq A \cup B$.

3. State the truth value of each of the following proposition: **(a)** $3 < 7$; **(b)** $3 \leq 7$; **(c)** $7 < 7$; **(d)** $7 \leq 7$; **(e)** $3 = 7$; **(f)** $7 = 7$; **(g)** $A \subseteq A$; **(h)** $A \in \{A\}$; **(i)** $\varnothing \subseteq \{\varnothing\}$; **(j)** $\varnothing \in \{\varnothing\}$; **(k)** $\{\varnothing\} \in \{\varnothing\}$; **(l)** $\{\varnothing\} = \{\{\varnothing\}\}$; **(m)** $x^2 - y^2 = (x - y) \cdot (x + y)$; **(n)** $x^2 + 2xy + y^2 = (x + y)^2$.

4. Express the truth value of each of the following in terms of the truth value of p. Each of these instances has a standard mathematical name. What is it? **(a)** $p \wedge p$; **(b)** $p \vee p$; **(c)** $p \wedge T$; **(d)** $p \wedge F$; **(e)** $p \vee T$; **(f)** $p \vee F$.

5. State the truth value of each of $\sim T$, $\sim F$, $p \vee \sim p$, $p \wedge \sim p$.

6.5 Algebra of Propositions

In this section we are finally going to put to use results acquired before. After a few preliminary remarks, we will be able to apply to the algebra of propositions all the results we have seen in Boolean algebra (and, potentially, others as well; see Chapter 14).

6.5.1 Formal Rules

From a strictly logical point of view, the interest in a simple or compound proposition is not in its phrasing, but in its *truth value*. The set of all possible truth values is very simple: $\{F, T\}$. For reasons that will become clear in a moment, we'll use $\{0, 1\}$ instead. The set of the truth values of any propositions is closed under operations $\vee$, $\wedge$, $\sim$. There is therefore the possibility that we may compare the present situation with Boolean algebra. We have a set with two special elements, to be compared with 0 and 1 of a Boolean algebra; we have two binary operations and a unary one, to be compared with $\vee$, $\wedge$, and $'$ of a Boolean algebra; and we have proved—or we can easily verify—that the ten axioms of a Boolean algebra hold. The only exception seems to be that, at times, $\Leftrightarrow$ is replaced by the symbol $=$. But this is perfectly correct: we are referring only to truth values and, if $p \Leftrightarrow q$, then (truth value of p) $=$ (truth value of q), and conversely. Consequently, if we consider propositions as symbols that stand for their truth values, then the algebra of propositions is a Boolean algebra for which the set A has two elements. We know therefore, without further proofs, that all the properties of Boolean algebra are valid for the algebra of propositions.

All we have to do at this point is to get practice by observing some of them. Note, however, that these observations are not trivial: a formal statement or expression of Boolean algebra, when translated in terms of logical operators, may provide some new insight in the meaning of a compound proposition. Typical examples are double negation and DeMorgan's laws. Expressed in terms of ordinary language, they become: (1) "it is not true that it is not true that . . ." is equivalent to ". . ."; (2) "it is not true that p and q" is equivalent to "p is not true or q is not true"; and (3) "it is not true that p or q" is equivalent to "p is not true and q is not true".

Formally,

Theorem 29 **Double negation.** For any proposition p: $(\sim(\sim p)) \Leftrightarrow p$.
DeMorgan's laws. For any propositions p, q: $(\sim(p \wedge q)) \Leftrightarrow ((\sim p) \vee (\sim q))$, $(\sim(p \vee q)) \Leftrightarrow ((\sim p) \wedge (\sim q))$.

We have already remarked that we need no proofs. The statements above are direct applications of properties of Boolean algebra.

6.5.2 Truth Tables

Compound propositions involve one or two arguments, which may be simple or compound propositions and may have arguments on their own. Ultimately, a compound statement may involve any number of arguments. Dealing with propositions that have three or four or a dozen arguments would quickly exhaust even the most vivid intuition, so we must have some systematic tools for doing so. One tool we have seen, Boolean algebra. Another is the graphical representation known as the **truth table.** In a truth table, the truth value of a compound proposition is given for every possible selection of truth values of all of its arguments. For the three compound propositions we've introduced so far, $\sim p$, $p \vee q$, and $p \wedge q$, we have the tables in Figure 6.1

p	$\sim p$
T	F
F	T

p	q	$p \vee q$
T	T	T
T	F	T
F	T	T
F	F	F

p	q	$p \wedge q$
T	T	T
T	F	F
F	T	F
F	F	F

Figure 6.1 Truth tables of NOT, OR, AND.

The table is read in the usual way. For instance, the third line of the second and third table states, if p is false and q is true, then $p \vee q$ is true and $p \wedge q$ is false. Before proceeding with the applications of truth tables, notice how they're constructed. The table for unary operation $\sim$ has only two rows; the other two have four rows each. On the left side of the table, we systematically list every possible selection of truth values for the arguments. Chapter 4 tells us how many such selections there are and how to enumerate them. If there are n arguments, and since $\#\{T, F\} = 2$, we must have 2^n rows in the truth table. In order to enumerate them, we list the binary integers from 0 to $2^n - 1$, identifying 0 as F and 1 as T. The order of the rows is unimportant, although traditionally they're listed in descending binary order from top to bottom. For any proposition involving three independent arguments, p, q, and r, the left side of the truth table would look like the table below, excluding the parenthetical parts, or replacing, for instance, TTT with 111, TTF with 110, and so on.

p	q	r		
T	T	T	(binary 111 = decimal 7)	
T	T	F	(binary 110 = decimal 6)	
T	F	T	(binary 101 = decimal 5)	
T	F	F	(binary 100 = decimal 4)	
F	T	T	(binary 011 = decimal 3)	
F	T	F	(binary 010 = decimal 2)	
F	F	T	(binary 001 = decimal 1)	
F	F	F	(binary 000 = decimal 0)	

Here is an example of "calculation" with truth tables. We'll establish again one of DeMorgan's laws.

Theorem 30 If p and q are any propositions, then $(\sim(p \wedge q)) \Leftrightarrow ((\sim p) \vee (\sim q))$.

Proof: Proof is via the truth table; see the table below. Calculate the conclusion of the theorem (that is, the "then" part, which is a compound proposition) for each selection of the truth values of the arguments. In order to facilitate the calculation, calculate the compounds that are successively needed to obtain the final expression; reserve a column for each of them. The solid column of *T*s at the right is proof that the final expression is always true, that is, that the two expressions are indeed equivalent, as claimed. Q.E.D.

p	q	$p \wedge q$	$\sim(p \wedge q)$	$\sim p$	$\sim q$	$\sim p \vee \sim q$	$(\sim(p \wedge q)) \Leftrightarrow ((\sim p) \vee (\sim q))$
T	*T*	*T*	*F*	*F*	*F*	*F*	*T*
T	*F*	*F*	*T*	*F*	*T*	*T*	*T*
F	*T*	*F*	*T*	*T*	*F*	*T*	*T*
F	*F*	*F*	*T*	*T*	*T*	*T*	*T*

Stop here and make sure that you see where all of these truth values came from. The values for p and q were supplied as described above; the rest were calculated. Each new column is the result of one single operation $\sim$, $\vee$, or $\wedge$. Read on only when you're comfortable with the mechanics of this process. Note also that, when you have experience, you can bypass a few columns.

Truth tables can also be used to define new logical operations on propositions or new Boolean operations in Boolean algebra. We've already used truth tables to define (or redefine) operators $\sim$, $\vee$, and $\wedge$. We can likewise use tables to define, for instance, the **exclusive OR:**

p	q	p XOR q
T	*T*	*F*
T	*F*	*T*
F	*T*	*T*
F	*F*	*F*

Note that the verbal description "one or the other, but not both" would give the following definition in formula: p XOR q means $(p \vee q) \wedge \sim(p \wedge q)$. In Exercises 6.5 you'll be asked to prove its equivalence with the truth table above.

Another operator is one we've used consistently throughout both set theory and logic. This is **implication,** the proposition "If p is true, then q is true". Implication is usually denoted as $p \Rightarrow q$. Its truth table is

p	q	$p \Rightarrow q$
T	*T*	*T*
T	*F*	*F*
F	*T*	*T*
F	*F*	*T*

Note that implication is degenerately true if p is false; see Degeneracy in Complements 6.5. Note also that the verbal definition—including the degeneracy convention—gives the following definition in formula: $p \Rightarrow q$ means $(\sim p) \vee (p \wedge q)$. In the exercises you'll be asked to prove its equivalence with the truth table above. In English, implication is read in several ways. We may say "If p, then q", or "q if p", or "p implies q". Less obviously, we may also say "p only if q", although you may have to stare at the sentence for a while to see why this holds.

The two operators $p \Rightarrow q$ and $q \Rightarrow p$ are *not* equivalent. The latter statement is called the **converse** of the former. However, if we have *both* $p \Rightarrow q$ *and* $q \Rightarrow p$, then we must have that $p \Leftrightarrow q$ (which explains the notation). You'll be asked in the exercises to prove this using a truth table proof. Looking above, though, we see that $p \Leftrightarrow q$ could be read as "p implies q, and conversely", or "p if q, *and* p only if q", or briefly, "p if and only if q". In fact, the phrase *if and only if* is so important that it's sometimes abbreviated *iff*, as in "p iff q", a dangerous practice, and one that often results in accusations of poor spelling. Still, the notation is useful in that it reinforces that equivalence is simply bidirectional implication.

Two operators that *are* equivalent are $p \Rightarrow q$ and $(\sim q) \Rightarrow (\sim p)$. The latter statement is called the **contrapositive** statement of the former. This deserves a theorem, the proof of which we leave as an exercise.

Theorem 31 Let p and q be propositions. Then $p \Rightarrow q$ if and only if $(\sim q) \Rightarrow (\sim p)$.

Note 7 on Theorem Proving. Theorem 31 is often a powerful tool in proving an if/then property. Proving that $p \Rightarrow q$ is equivalent to proving that $(\sim q) \Rightarrow (\sim p)$. This is the essence of a "proof by **contradiction.**" In practice, a proof by contradiction proceeds more or less thus. In order to prove that p implies q: (1) Deny the conclusion; that is, use $\sim q$ as a temporary hypothesis. (2) On the ground of this temporary hypothesis, prove the negation of the original hypothesis; that is, prove $\sim p$ (or any contradictory statement, like $x \neq x$, or $1 < 0$). (3) Then conclude "if q were false, then we would obtain that p is false (or that $x = x$ is false, etc.), which would be a contradiction. Consequently, q cannot be false and must therefore be true."

6.5.3 Logic Quantifiers

In Complements 5.2 we pointed out the importance of propositions of the type "for every __" and "there exists a __ such that __". It's time to formalize such constructs in the context of algebra of propositions. We also noted that the negation of such a construct may offer some trouble, even in ordinary situations (see, for instance, Section 6.4.2). Our formalization will take care of that too.

Definition 32 Let S be a given set. Let $P(x)$ be a proposition that involves an unspecified element x of S and that, for each specific choice of x in S, is either true or false. Then the statement $\forall x \in S \mid P(x)$ is usually read "for all x in S, statement P is true" and is defined to be equivalent to the statement $\{x \in S \mid P(x)\} = S$. The statement $\exists x \in S \mid P(x)$ is usually read "there exists x in S such that statement P is true" and is defined to be equivalent to the statement $\{x \in S \mid P(x)\} \neq \varnothing$. The symbols $\forall$ and $\exists$ are the **quantifiers.**

Our phrase "there exists x in S" is shorthand for "there exists *at least one* x in S." Perhaps many elements of S satisfy P; perhaps all do. All it takes is one.

Examples $(\forall x \in \mathbb{Z} \mid x - x = 0)$ is true; $(\forall x \in \mathbb{Z} \mid x^2 \geq 0)$ is true; $(\exists x \in \mathbb{Z} \mid x^2 < 0)$ is false; $(\forall x \in \mathbb{Z} \mid (\exists t \in \mathbb{Z} \mid x + t = 0))$ is true; $(\exists t \in \mathbb{Z} \mid (\forall x \in \mathbb{Z} \mid x + t = 0))$ is false.

As usual, the set S may be omitted from the quantified statement, but *only when* it's clearly given by the context. Parentheses may be necessary for clarity. In the last example above, the following would be correct, but perhaps confusing: $\exists t \in \mathbb{Z} \mid \forall x \in \mathbb{Z} \mid x + t = 0$.

Theorem 33 (a) $(\sim(\forall x \in S \mid P(x))) \Leftrightarrow (\exists x \in S \mid \sim P(x))$;
(b) $(\sim(\exists x \in S \mid P(x))) \Leftrightarrow (\forall x \in S \mid \sim P(x))$.

In words, to negate a quantified statement, change the quantifier and negate the core statement.

Proof: Exercise. [Hint: use the definition and set theory. Note that the use of quantifiers combines algebra of propositions and algebra of sets.]

Examples Consider the two false statements from the examples above; their negations must be true. The negations are, in several equivalent forms each: (1) $\sim(\exists x \in \mathbb{Z} \mid x^2 < 0)$, that is, $\forall x \in \mathbb{Z} \mid \sim(x^2 < 0)$, or $\forall x \in \mathbb{Z} \mid x^2 \geq 0$. (2) $\sim(\exists t \in \mathbb{Z} \mid (\forall x \in \mathbb{Z} \mid x + t = 0))$, that is, $\forall t \in \mathbb{Z} \mid \sim(\forall x \in \mathbb{Z} \mid x + t = 0)$, that is, $\forall t \in \mathbb{Z} \mid (\exists x \in \mathbb{Z} \mid x + t \neq 0)$. Note also the importance of the order in which multiple quantifiers appear in an expression. We'll see more examples in the exercises.

Exercises 6.5

1. Using algebra of sets and Boolean algebra as guidelines, express several of the derived properties in terms of algebra of propositions. In the text we gave an example in terms of DeMorgan's laws and double negation. Find some others.
2. Negate the propositions that appear in Exercises 6.4. When appropriate, find equivalent formulae for the result. When possible, state the truth value of the result.
3. In terms of logical operators, find ordinary language equivalents of the ten basic and several derived properties of Boolean algebras. See Section 6.3.
4. Construct the truth table for each of the following propositions: **(a)** $(\sim p) \wedge q$; **(b)** $\sim(p \wedge q)$; **(c)** $(p \wedge q) \vee r$; **(d)** $p \wedge (q \vee r)$; **(e)** $p \wedge q \wedge r$; **(f)** $p \vee q \vee r$; **(g)** $p \wedge q \wedge \sim(p \wedge q)$; **(h)** $p \vee q \vee \sim(p \wedge r)$; **(i)** $p \wedge (q \vee r) \wedge \sim q$; **(j)** $(p \wedge q) \vee (q \wedge r) \vee (r \wedge p)$; **(k)** $(p \wedge q \wedge r) \vee ((\sim p) \wedge (\sim q) \wedge (\sim r))$.
5. For XOR and for $\Rightarrow$ prove that the two definitions given in the text, truth table and formula, are equivalent to each other.

6. Show that $\Rightarrow$ is the analogue, in algebra of propositions, of $\subseteq$ in set theory and of $\leq$ in Boolean algebra (see Definition 27).

7. In set theory, what is the syntactic analogue for the proposition $p \Leftrightarrow q$?

8. Prove each of the following using truth tables or any other method but this time do *not* use results from Boolean algebra. (This is proof of the fact that algebra of propositions satisfies the axioms of a Boolean algebra.) If any of these propositions has a standard mathematical name, state it. What are these ten propositions collectively known as?

$p \vee q \Leftrightarrow q \vee p$
$(p \vee q) \vee r \Leftrightarrow p \vee (q \vee r)$
$p \wedge (q \vee r) \Leftrightarrow (p \wedge q) \vee (p \wedge r)$
$p \vee F \Leftrightarrow p$
$p \vee (\sim p) \Leftrightarrow T$

$p \wedge q \Leftrightarrow q \wedge p$
$(p \wedge q) \wedge r \Leftrightarrow p \wedge (q \wedge r)$
$p \vee (q \wedge r) \Leftrightarrow (p \vee q) \wedge (p \vee r)$
$p \wedge T \Leftrightarrow p$
$p \wedge (\sim p) \Leftrightarrow F$

9. The following are to be proved. You may use either truth table proofs or Boolean algebra formal proofs using the results of exercise 8. If you invoke the appropriate derived properties of Boolean algebra, then the proof will be really short. If any of these propositions has a mathematical name, state it.

$p \vee p \Leftrightarrow p$
If $(p \vee q) \Leftrightarrow p$ for all p, then $q \Leftrightarrow F$
$(p \vee T) \Leftrightarrow T$
If $(p \vee q) \Leftrightarrow T$ and if $(p \wedge q) \Leftrightarrow F$, then $q \Leftrightarrow \sim p$
$\sim(\sim p) \Leftrightarrow p$
$\sim F \Leftrightarrow T$

$p \wedge p \Leftrightarrow p$
If $(p \wedge q) \Leftrightarrow p$ for all p, then $q \Leftrightarrow T$
$(p \wedge F) \Leftrightarrow F$
$\sim T \Leftrightarrow F$

10. Prove $((p \Rightarrow q) \wedge (q \Rightarrow p))$ iff $(p \Leftrightarrow q)$.

11. Prove Theorem 31 (about the contrapositive of a statement).

12. For each of the propositions below, (1) write a formula of propositional algebra equivalent to the given proposition, (2) state the truth value of the given proposition and justify your statement. **(a)** For each natural number h there exists a natural number k such that $k < h$; **(b)** For each natural number h there exists a natural number k such that $k \leq h$; **(c)** For some natural number h every natural number k satisfies the condition $h < k$; **(d)** For some natural number h every natural number k satisfies the condition $h \leq k$.

13. In a Boolean algebra on set A, consider each of the following statements. Is it true? Why? Express the given statement as a formula containing quantifiers. **(a)** For each p there exists a q such that $p \wedge q = p$. **(b)** There exists a p such that for every q we have $p \wedge q = p$. **(c)** There exists a p such that for every q we have $p \wedge q = 0$.

14. Prove Theorem 33 (the rules for negating quantified statements).

15. Verify the examples in Section 6.5.3.

16. Negate each of the following statements. Find some equivalent formula for each negation. Then express the result in ordinary language, in a suitable, equivalent form. Give the truth value of each of the given statements and of its negation. **(a)** $\forall x \in \mathbb{N} \mid \exists y \in \mathbb{N} \mid x = y^2$; **(b)** $\forall x \in \mathbb{N} \mid \exists y \in \mathbb{N} \mid x \cdot y = 1$; **(c)** $\forall x \in \mathbb{N} \mid \forall y \in \mathbb{N} \mid x < y$; **(d)** $\exists x \in \mathbb{N} \mid \forall y \in \mathbb{N} \mid x + y = x$; **(e)** $\exists x \in \{0, 1\} \mid \exists y \in \{0, 1\} \mid \exists z \in \{0, 1\} \mid ((x \neq y) \wedge (y \neq z) \wedge (z \neq x))$.

Given $T := \{0, 1, 2, 3\}$; **(f)** $\forall B \in \mathcal{P}T \mid \#B = 3$; **(g)** $\forall B \in \mathcal{P}T \mid \#B > 4$; **(h)** $\forall B \in \mathcal{P}T \mid \exists x \in T \mid B \cup \{x\} = B$.

17. Give the truth value of each of the following statements. For each of the false statements, construct its negation, in a suitable, equivalent form. Do more examples of your own. For instance, express some of the properties of set theory, Boolean algebra, etc. **(a)** $\forall x \in \mathbb{N} \mid \forall y \in \mathbb{N} \mid (\exists t \in \mathbb{N} \mid (x + t = y) \vee (y + t = x))$; **(b)** $\forall x \in \mathbb{N} \mid \exists t \in \mathbb{N} \mid x + t = x$; **(c)** $\exists t \in \mathbb{N} \mid \forall x \in \mathbb{N} \mid x + t = x$; **(d)** $\forall x \in \mathbb{N} \mid \forall t \in \mathbb{N} \mid x + t = x$; **(e)** $\exists x \in \mathbb{N} \mid \exists t \in \mathbb{N} \mid x + t = x$.

Given $T := \{0, 1, 2, 3\}$, **(f)** $\exists B \in \mathcal{P}T \mid \forall C \in \mathcal{P}T \mid B \cap C = \emptyset$; **(g)** $\forall B \in \mathcal{P}T \mid \forall C \in \mathcal{P}T \mid B \cap C = \emptyset$; **(h)** $\forall B \in \mathcal{P}T \mid \exists C \in \mathcal{P}T \mid B \cap C = \emptyset$; **(i)** $\exists B \in \mathcal{P}T \mid \exists C \in \mathcal{P}T \mid B \cap C = \emptyset$.

18. Prove the following in general: **(a)** if $((T \neq \emptyset) \wedge (\forall x \in S \mid \forall y \in T \mid P(x, y)))$, then $(\forall x \in S \mid \exists y \in T \mid P(x, y))$. **(b)** $(\forall x \in S \mid \forall y \in T \mid P(x, y))$ is equivalent to $(\forall (x, y) \in S \times T \mid P(x, y))$, which in turn is equivalent to $(\forall y \in T \mid \forall x \in S \mid P(x, y))$.

State and prove similar properties when some $\forall$ is changed to an $\exists$.

19. Prove that if S is finite and nonempty, then **(a)** $(\forall x \in S \mid P(x)) \Leftrightarrow \bigwedge_{x \in S} P(x)$; **(b)** $(\exists\, x \in S \mid P(x)) \Leftrightarrow \bigvee_{x \in S} P(x)$. How would you handle the degenerate case in which S is empty?

Complements 6.5

1. We've defined the *Sheffer stroke* operator by $p \mid q :\Leftrightarrow \sim(p \wedge q)$. Draw the truth table for $p \mid q$. Prove, using a method of your choice, that **(a)** $p \mid p \Leftrightarrow \sim p$; **(b)** $((p \mid q) \mid (p \mid q)) \Leftrightarrow (p \wedge q)$; **(c)** $((p \mid p) \mid (q \mid q)) \Leftrightarrow (p \vee q)$.

2. How many possible distinct unary logical operators are there? That is, how many truth tables are there of the form below. Enumerate them, and associate with each one a logical meaning.

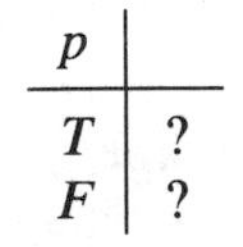

p	
T	?
F	?

3. How many possible distinct binary logical operators are there? Enumerate them by truth table, and associate with each one a logical meaning. Recognize those we have already seen. For each, try to find a formula in terms of the operators $\sim$, $\vee$, $\wedge$; see the definition of Sheffer stroke as an example.

4. If a logical operator has n arguments ($n \geq 0$), how many rows will its defining truth table have? How many possible distinct n-ary logical operators are there?

5. Degeneracy. Students of logic have traditionally been put through a good deal of misery at the hands of the implication operator $\Rightarrow$. Its truth table is

p	q	$p \Rightarrow q$
T	T	T
T	F	F
F	T	T
F	F	T

In English, one of the ways to read $p \Rightarrow q$ is "p implies q". Here's where the trouble begins, for that last phrase is grossly misleading! Look at the last two rows of the truth table, in which p is F. The implication $p \Rightarrow q$ is listed as T in both rows. Students ask, "If p is false, how can it imply anything? Shouldn't $p \Rightarrow q$ be listed as *don't know* or something like that?" There seems to be something vaguely sinister about listing *anything* under $p \Rightarrow q$ when p is false. The problem, however, isn't in logic, but in English. The phrase "p implies q" suggests that p is an active agent, a causal force acting on q. We mentally picture p as standing over q with a gun and a whip, cruelly forcing q to be true against q's will. The crime is especially villainous if p is false, and therefore acting without lawful authority! But this is not the case. *Implication has nothing whatever to do with causality!* Propositions p and q are independent of each other. When we say that "it is true that p implies q", we simply say that, of the four possible cases, one (p true and q false) does not occur. If it occurs, then the proposition "p implies q" is false. In formula, this last definition becomes: $p \Rightarrow q$ is defined to be the same as $\sim(p \wedge \sim q)$. Of course, in specific instances relationships that *are* causal might be expressed as implications. Thunderstorms cause the grass to become wet, therefore thunderstorms $\Rightarrow$ wet grass. This implication is true even if it never rains.

This is another example, and in fact, the defining feature, of *degeneracy*. You should compare this with the degeneracy of the set-theoretic definitions you've already seen (Complements 5.2). Degeneracy is neither an arbitrary and capricious game, nor is it an illicit, unjustified shortcut. It's simply a technique in which definitions are considered to be satisfied in certain null or empty situations. If p is false, then $p \Rightarrow q$ is defined to be *degenerately* true. But remember: if p is false, it's the implication $p \Rightarrow q$, *not q itself,* that automatically becomes true.

6. Depending on the phrasing we were discussing, we gave four definitions of $\Rightarrow$, namely: $(\sim p) \vee (p \wedge q)$, $(\sim p) \vee q$, $\sim(p \wedge \sim q)$, and the declared truth table above. Complete the proof that these definitions are mutually equivalent.

7. Classical logic. Some rules of logic go back to Aristotle, and many more were added in the Middle Ages. Such rules were numerous and complicated, their names were even more abstruse, and their proofs took pages and pages of arguments, not all of which were convincing or even correct. Here are some such rules, given with their ancient names, but written in terms of modern formulae. In fairness, the first syllogism and the first modus are still in fashion, the others have faded away. (Observe the vowels of the names: they tell something about the structure of the formula. *Syn* (Greek) = together; *logos* = argument, discourse; *syllog-* = compound arguments.)

Syllogisms

BARBARA: $((q \Rightarrow r) \wedge (p \Rightarrow q)) \Rightarrow (p \Rightarrow r)$
CELARENT: $((q \Rightarrow \sim r) \wedge (p \Rightarrow q)) \Rightarrow (p \Rightarrow \sim r)$
DARII: $((q \Rightarrow r) \wedge \sim(p \Rightarrow \sim q)) \Rightarrow \sim(p \Rightarrow \sim r)$
FERIO: $((q \Rightarrow \sim r) \wedge \sim(p \Rightarrow \sim q)) \Rightarrow \sim(p \Rightarrow r)$
CESARE: $((r \Rightarrow \sim q) \wedge (p \Rightarrow q)) \Rightarrow (p \Rightarrow \sim r)$
CAMESTRES: $((r \Rightarrow q) \wedge (p \Rightarrow \sim q)) \Rightarrow (p \Rightarrow \sim r)$
FESTINO: $((r \Rightarrow \sim q) \wedge \sim(p \Rightarrow \sim q)) \Rightarrow \sim (p \Rightarrow r)$
BAROKO: $((r \Rightarrow q) \wedge \sim(p \Rightarrow q)) \Rightarrow \sim(p \Rightarrow r)$

Modi

MODUS PONENDO PONENS: $((p \Rightarrow q) \wedge p) \Rightarrow q$
MODUS TOLLENDO TOLLENS: $((p \Rightarrow q) \wedge \sim q) \Rightarrow \sim p$

Try to translate these few rules into ordinary language. Keep in mind that there are twice as many more. Think of proving all of them using ordinary language. Then you'll appreciate the simplicity of Boolean algebra. It may be a good exercise, however, to prove some of these rules, using the techniques of Boolean algebra, of course, or truth tables.

6.6 Two-Terminal Switching Networks

In Section 3.4.2, we considered an algebra of electrical switches. We can now elaborate on this idea and show that logic and switching algebra are intimately related. In fact, in most respects, they're one and the same thing—Boolean algebra.

A two-terminal switching network (2TSN) is shown in general in Figure 6.2. It consists of two terminals, marked *input* and *output*, connected by some network of switches, any of which can be *on* or *off*.

Figure 6.2. A two-terminal switching network (2TSN).

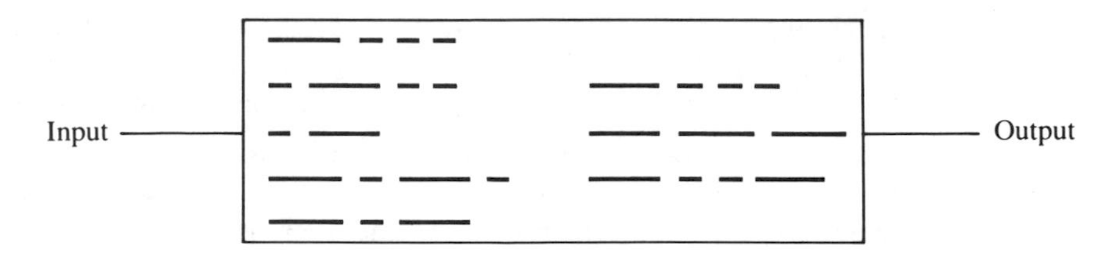

What we want to do is investigate the various types of switch networks, and the question we want to ask is the following: if an electrical current is introduced at the input terminal, will it flow through the network to the output terminal? Incidentally, if you're not familiar with electricity, just think of water flowing through a pipe network, and substitute valve for switch.

6.6.1 The Operators AND and OR

A single labelled switch, say p, is a 2TSN that passes current when "on" (or T or 1), and fails to pass current when "off" (or F or 0). Suppose now that x and y are 2TSNs, for instance, but not necessarily, two switches. We can combine two 2TSNs using two types of interconnection, series and parallel. In Figure 3.2 x and y are to be interpreted now as any 2TSNs, containing any number of switches each. It's nearly immediate that the series connection is the same as the logical operator $\wedge$ and that parallel connection is the same as the logical operator $\vee$. That is, series connection gives a network that is "on" if and only if both arguments x and y are "on," and parallel connection gives a network that is "on" if and only if either argument x or y is "on" or both are. This is already an indication that we may be in a situation similar to Boolean algebra. In order to complete the comparison we need two more things, self-defining networks and a NOT operation. The first thing we can do is to draw two **self-defining** 2TSNs:

A short circuit: Input ——•—— Output

An open circuit: Input —— —— Output

A **short circuit** *always* passes current, and an **open circuit** *never* passes current. An uninterrupted pipe and a clogged pipe are the hydraulic analogues, respectively. A short circuit is analogous to the self-defining logical proposition T; an open circuit is analogous to F.

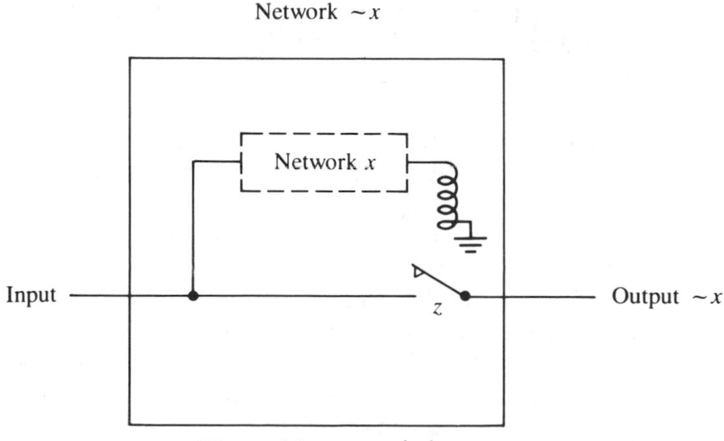

Figure 6.3. NOT-switch.

6.6.2 *The Operator* NOT

Negation in 2TSNs is a little more troublesome. Let's first consider the negation of an individual switch p. It's simple enough to imagine a switch with its "on" and "off" labels reversed, so that the switch, now labelled $\sim p$, passes current in its off position and not while on. A far more serious matter in 2TSNs is negating an entire network. For this, we need an entirely new device, called a **NOT-switch** (or **NOT-gate;** see Section 6.7) as in Figure 6.3. Such a device must be able to use the output of an arbitrary 2TSN, x, to operate a switch, z, according to the following rule: if there is current at the output of x, then z is thrown off; otherwise it's thrown on. Obviously, switch z needs a power source as its input, but the output of z is the same as the output of $\sim x$. There are many practical devices that perform this operation, relays, triodes, and transistors among them. All of them are rather complex devices, and in most electrical applications, such as house wiring, they are not used. Let's devote a few words to describe the role of a relay as illustrated in the figure. The output of network x sends current into a coil, creating a magnetic field, which attracts the bar of switch z and interrupts contact. Consequently, when x is on, switch z is off, and conversely. Early computers were relay computers, and relays are still used in many applications.

6.6.3 *Equivalent Networks*

Clearly, some 2TSNs are equivalent. For example, the 2TSNs shown as a1 and b1 in Figure 6.4 each reduce simply to the single 2TSN p. The same conclusion is valid for a2 and b2. Both facts are consequences of the formulae $(p \vee F) \Leftrightarrow (p \vee p) \Leftrightarrow (p)$, $(p \wedge T) \Leftrightarrow (p \wedge p) \Leftrightarrow (p)$. Also, 2TSN c1 in Figure 6.5 reduces to a short circuit, because $(p \vee \sim p) \Leftrightarrow T$, and d1 reduces to an open circuit, because $(p \wedge \sim p) \Leftrightarrow F$. The dashed lines indicate that these physical switches are "ganged," or operated together from one single lever. This is the equivalent, in ordinary algebra or in Boolean algebra, of a letter appearing several times in the same formula. In theory, there's no limit to how many times a single label may occur in a 2TSN, that is, how many switches are ganged. Of course, whether you could find such a device at your local hardware store is quite another matter! For our purposes, you needn't actually draw the dashed lines. Any group of switches with the same label, negated or not, will be assumed to be ganged. In Exercises 6.6 you'll be asked to find other such examples.

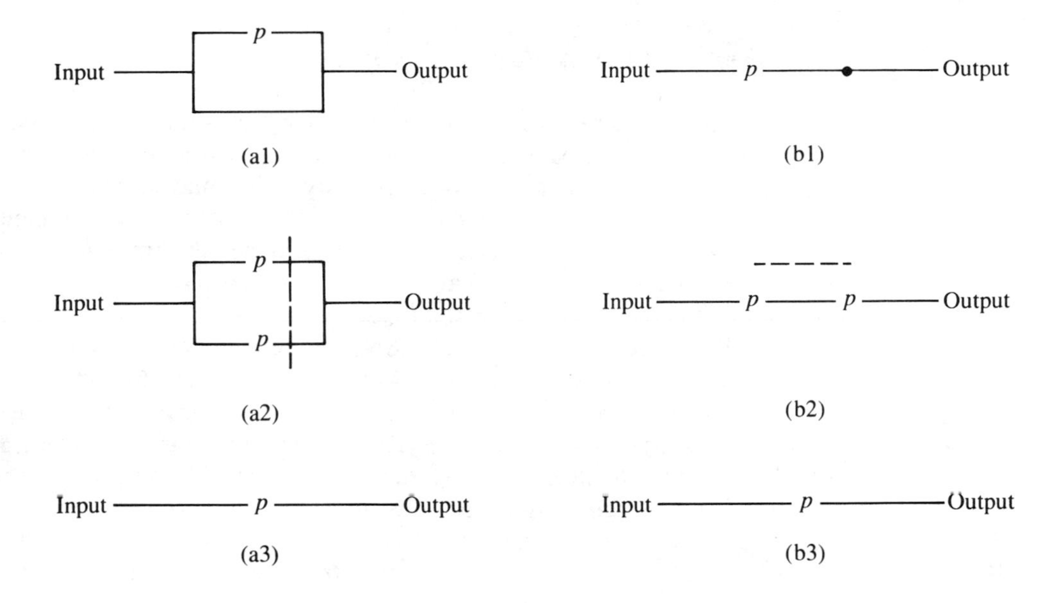

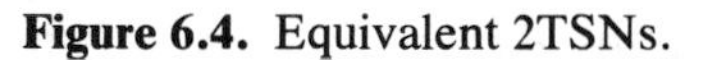
Figure 6.4. Equivalent 2TSNs.

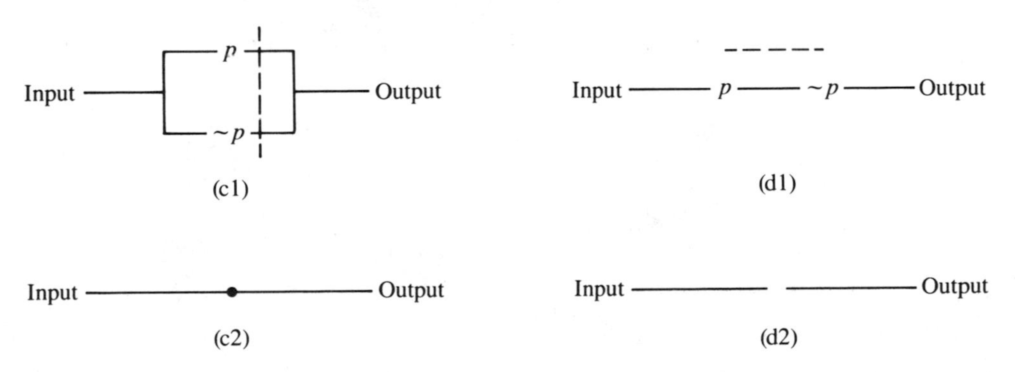

Figure 6.5. Equivalent 2TSNs. [Networks in the same column are equivalent to each other.]

6.6.4 Networks and Boolean Algebras

For now, we have already reached a very firm conclusion. For 2TSNs, the states of the network and of each of its components are two objects, off/on, or 0/1, or *F/T*, etc. We have two binary operations (parallel and series connection, the analogues of $\vee$ and $\wedge$) and a unary operation (the NOT-switch, the analogue of $\sim$ or $'$). The verification of the ten axioms of a Boolean algebra is nearly immediate in this concrete situation. We have already verified most of them. Then, if we consider 2TSNs as symbols for their states, we have a structure that *is* Boolean algebra. We can therefore apply to 2TSNs all the results of Boolean algebras.

Two-terminal switching networks are of interest primarily in two areas. First, in relatively simple electrical applications, like wiring light bulbs and toasters, all we need are simple, inexpensive switches. Negation is rarely needed. Second, in more complex applications like electronic computers, 2TSNs provide the abstract framework for designing the types of switching networks that actually are built. Note that 2TSNs use the flow of electricity (input to output of each device) to control a flow of information (effect of outputs onto the following stages of the network). It will be convenient, in many practical applications, to emphasize the flow of *information* through the network rather than the flow of *electricity*. This is achieved by presenting the theory of 2TSNs in a slightly different form, which goes under the name of *logic gates*. We consider logic gates in Section 6.7. Observe that the theory of logic gates will automatically be also a presentation of algebra of logic and of the elementary Boolean algebra.

Exercises 6.6

1. Consider each of the networks in the following figure, and write a logical proposition realized by that network.

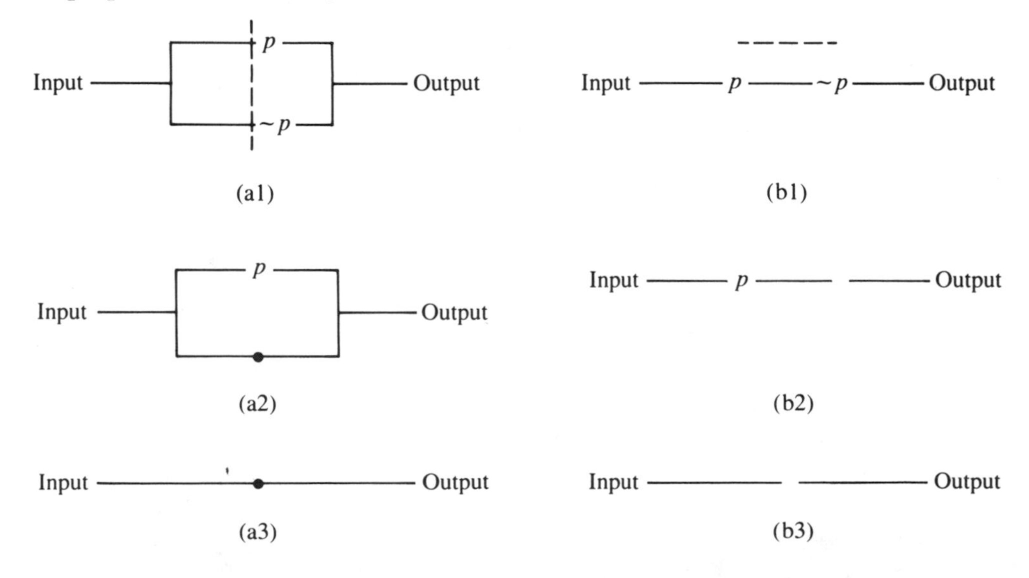

2. Draw a 2TSN equivalent to each of the following logical propositions: **(a)** $p \Rightarrow q$; **(b)** $p \mid q$ (that is, "p Sheffer stroke q"); **(c)** $(p \vee q) \wedge \sim(p \wedge q)$ (that is, p XOR q); **(d)** $(p \wedge q) \vee (p \wedge r) \vee (q \wedge r)$.

3. Suppose we wanted to construct a 2TSN equivalent to the formula $\sim((p \wedge q) \vee ((\sim r \vee s) \wedge t))$ without using a NOT-device other than a NOT-switch. We would apply DeMorgan's laws as many times as necessary to produce an expression in which only individual terms are negated, not entire expressions. Carefully verify for yourself the following logic algebra: $(\sim((p \wedge q) \vee ((\sim r \vee s) \wedge t))) \Leftrightarrow ((\sim(p \wedge q)) \wedge (\sim((\sim r \vee s) \wedge t))) \Leftrightarrow ((\sim p \vee \sim q) \wedge ((\sim(\sim r \vee s)) \vee \sim t)) \Leftrightarrow (((\sim p) \vee \sim q) \wedge ((r \wedge \sim s) \vee \sim t))$. Verify that an equivalent network is represented by

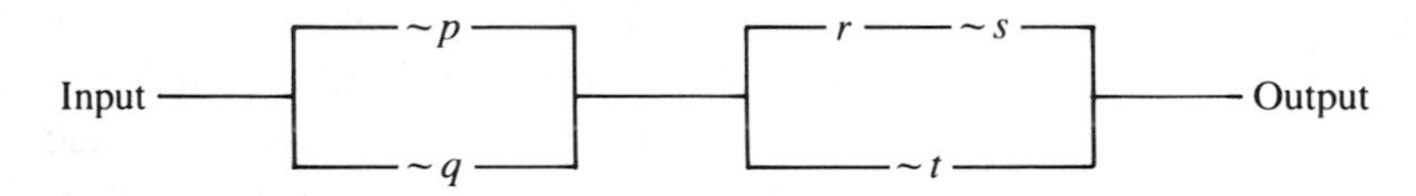

4. Draw ten two-terminal switching networks and their equivalents, representing the basic properties of 2TSNs. [Hint: The ten basic networks themselves come in five pairs.] State the formal names of these properties.

5. Draw five more pairs of 2TSNs and their equivalents, using Boolean algebra, set theory, and propositional calculus as precedents. State the formal names of these relationships. See Section 6.3.2.

Complements 6.6

1. Commonly in residential wiring the same light bulb is to be turned on and off independently from two different locations, say from the top and bottom of a flight of stairs, or from opposite ends of a hallway. Electricians usually solve this problem using a *three way switch* (although that name is inaccurate; it's also called variously a *T-switch* or *ABC-switch* or *double-throw switch*). Such a switch has three wires, labelled A, B, and C, and two positions. In the *up* position, contacts A and C are connected and B is disconnected; in the *down* position, B and C are connected and A is disconnected. In fact, C stands for *common*, the common connection. T-switches are *not* 2TSNs, since they have three wires. The circuit is then wired thus:

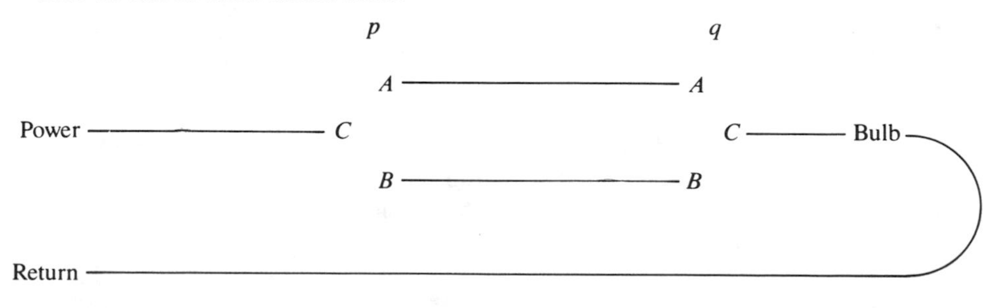

The truth table for this circuit is

p	q	Light
up	up	on
up	down	off
down	up	off
down	down	on

Verify that this corresponds to the proposition $p \Leftrightarrow q$. Write this proposition in terms of propositional calculus. (Take UP $:\Leftrightarrow T$ and DOWN $:\Leftrightarrow F$.) Draw a 2TSN for the light bulb circuit. The state of the final output is supposed to control the bulb; ignore the power return. What is the advantage of using *T*-switches over 2TSNs? That is, if we use T-switches, what 2TSN device can we *avoid* using?

2. We can also control a light from three locations by using a *four-way* or X-*switch,* having four wires, A, B, C, and D (so an X-switch is also *not* a 2TSN). In the *up* position, A and C are in contact, and B and D are in contact, but not with A and C; in the *down* position, A and D are in contact and B and C are in contact, but not with A and D. Then, for locations p, q, and r, we place T-switches at p and r and an X-switch at q. The circuit is then wired thus:

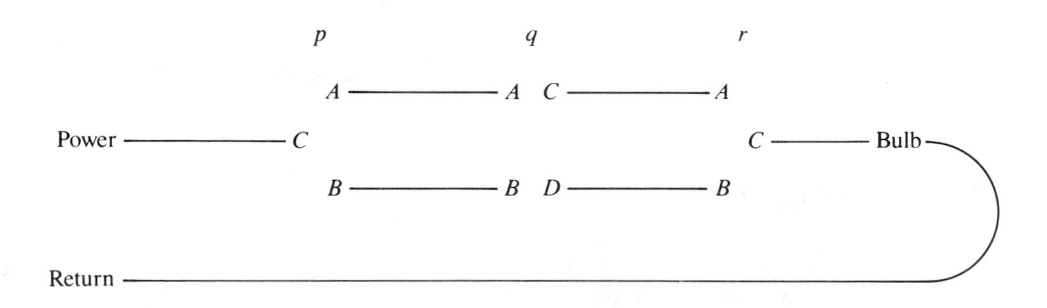

Construct a truth table for this circuit. Design an X-switch using only two ganged T-switches. Then diagram the three-location light circuit using T-switches only. Diagram the three-location light circuit using a 2TSN.

3. Show how a single light bulb could be controlled from four, five, or more locations using only T- and X-switches. Construct also equivalent 2TSNs.

6.7 Logic Gates

In Section 6.6 we saw that set theory, propositional calculus, and two-terminal switching networks are all simply syntactic variations of the same mathematical structure, Boolean algebra. As an abstract formalism, this equivalence is quite useful, and easy to see when the number of terms is half a dozen or so. But in very large switching circuits, like those in computers, the large number of interconnections makes 2TSNs cumbersome to use. Engineers still use propositional calculus to *design* these circuits, but they don't use physical switches to *implement* or *realize* their designs. Instead, they use an entirely new class of electronic devices called *gates*. Gates are simply a way of rapidly realizing a circuit expressed via propositional calculus. Gates have come to be almost synonymous with propositional calculus, so it's worthwhile to spend some time discussing them.

The term **logic gate** has the usual intuitive connotation: given one or more signals, a gate *controls*—or *gates*—them to produce one or more **output signals.** An output signal is the counterpart of the "output" of a switching network. The "input" of a switching network is disregarded here (consequently, gates are not 2TSNs). The given signals are the counterparts of the "arguments" or "labels" of a switching network and are called the **inputs** of the gate. Input and output signals may present different levels of voltage and current; this is another reason why gates are not 2TSNs. For our immediate purpose, we'll adopt two restrictions.

1. Each signal has one of two values, T or F (or 1/0, etc.)
2. Each gate has exactly one output.

We draw the devices using the accepted conventional electronic symbols, or *schematics,* without discussing how they work internally. The three basic gates are shown in Figure 6.6. The interpretation is the usual one: the AND-gate, for instance, produces a T signal if and only if both of its inputs are T, and so on. You must learn to recognize these gates by their schematic shapes alone. In electronic schematics, the symbols NOT, OR, AND are usually not shown, nor are the arrowheads (which in any event indicate the flow of *information,* not necessarily electricity). The algebraic expressions may or may not appear. For now, we'll drop the arrowheads and not show all intermediate expressions, but we will retain the individual gate symbols.

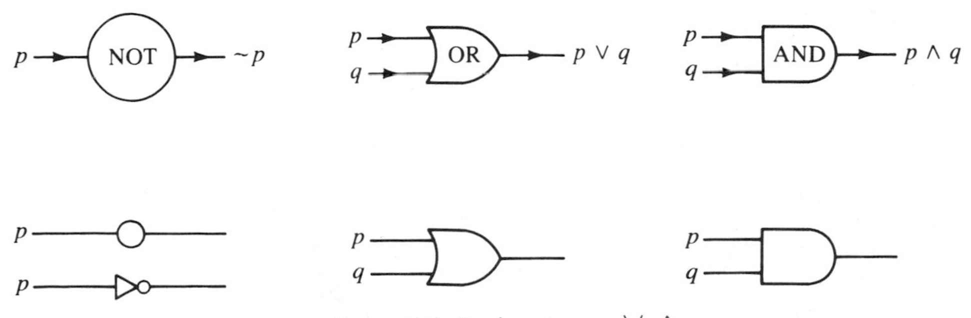

Figure 6.6. Basic gates: $\sim$, $\vee$, $\wedge$.

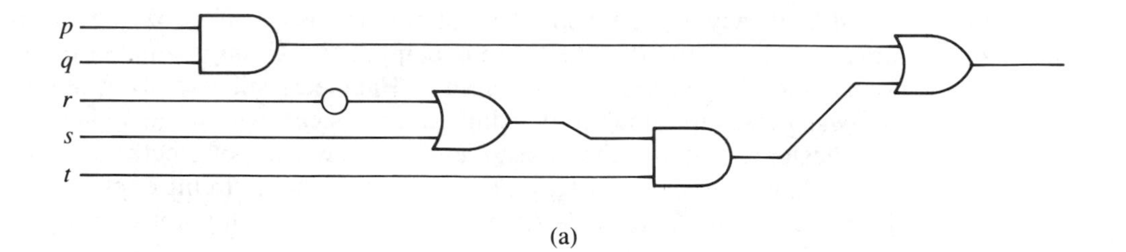

(a)

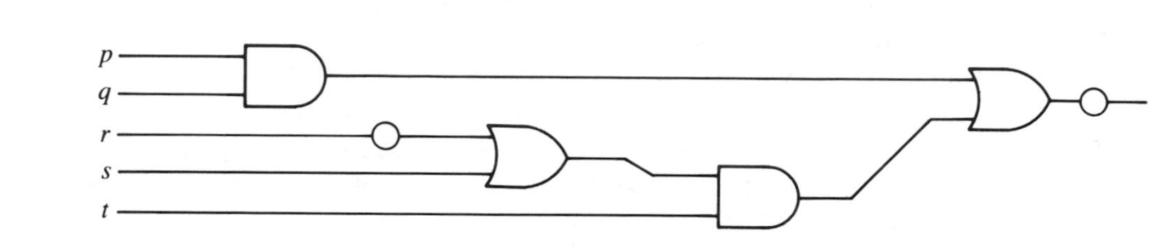

(b)

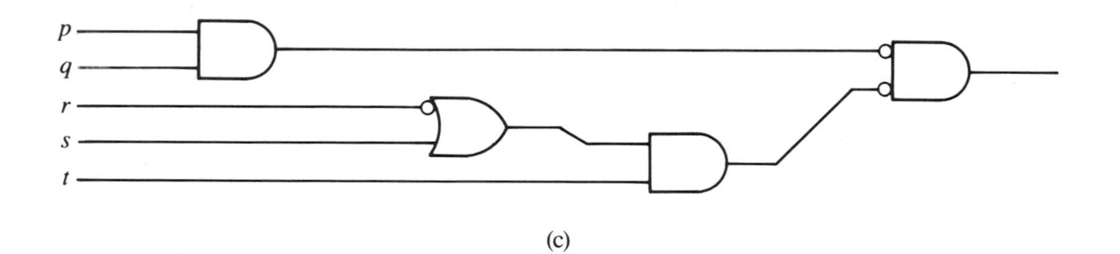

(c)

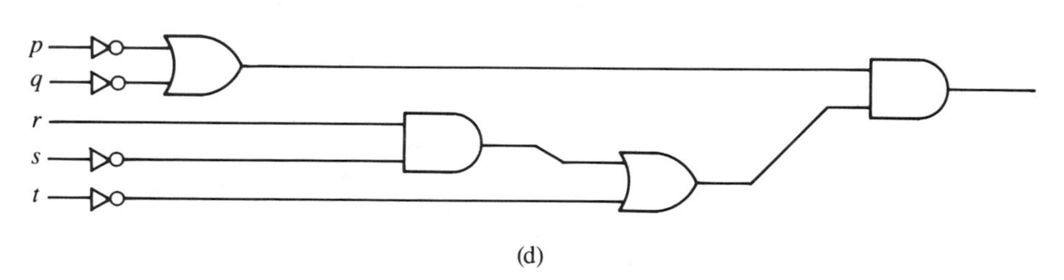

(d)

Figure 6.7. Examples of gate networks.

As expected, we can use Boolean algebra to handle gates. A few conventions are, however, necessary: (1) The objects of our Boolean algebra are the states of the input and output signals, that is F and T. (2) The three basic types of gates are the operations of the Boolean algebra, their inputs being the arguments of the operations. (3) A network of gates is, under the two restrictions given, an expression in Boolean algebra. (4) Although not necessary, two self-defining gates can be introduced through the appropriate formulae, for instance $p \wedge {\sim}p$, $p \vee {\sim}p$. These theoretical limitations, however, are more than offset by the ease of designing gate networks.

Let's illustrate the process with the logical expression from the third problem in Exercises 6.6, $(p \wedge q) \vee (((\sim r) \vee s) \wedge t)$. One possible *realization* of this expression would be the gate network in Figure 6.7a. If we now wish to negate this entire expression, to obtain ${\sim}((p \wedge q) \vee (((\sim r) \vee s) \wedge t))$, we need only add one additional NOT-gate to the output, yielding part b of the same figure. If we use some Boolean algebra, we can obtain equivalent realizations, for instance, Figures 6.7c and d.

There are many variations on and many compositions of the three basic gates. Some have multiple outputs. Some have NOT-gates built directly into some or all of their inputs or outputs. This is denoted by a small circle on the appropriate wire. For example, in Figure 6.8, NAND stands for "not AND" and NOR for "not OR". Note that, by DeMorgan's laws, the gates in the same row of the figure are equivalent to each other. Still other gates, called *memories*, retain their output signals even after the input signals have been removed. But those we've presented will serve to give you the flavor of how propositional calculus is brought to real-world fruition.

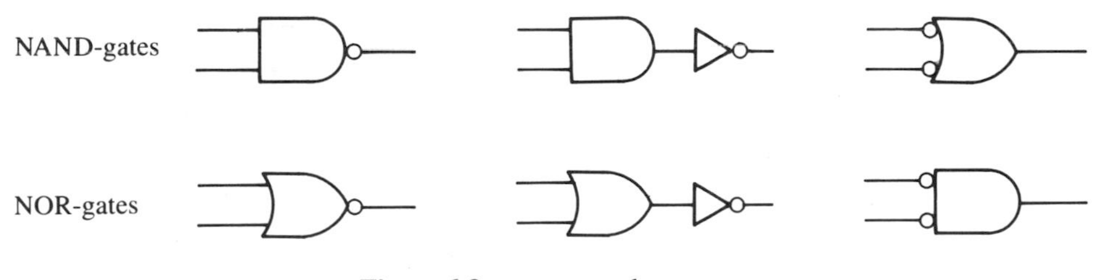

Figure 6.8. NAND- and NOR-gates.

Exercises 6.7

1. Give 2TSN and gate network realizations for each of the following formulae: **(a)** $p \wedge q \wedge r$; **(b)** $p \wedge (q \vee r)$; **(c)** $p \vee q \vee r$; **(d)** $\sim(p \vee q)$; **(e)** $(p \wedge q) \vee r$; **(f)** $(\sim p) \vee q$;**(g)** $p \wedge (\sim q)$. **(h)** Do the same for other formulae that you yourself propose.
2. Give a few equivalent realizations for each of the networks you constructed in problem 1.
3. For each of the following gate networks, write a formula that is realized by the given network.

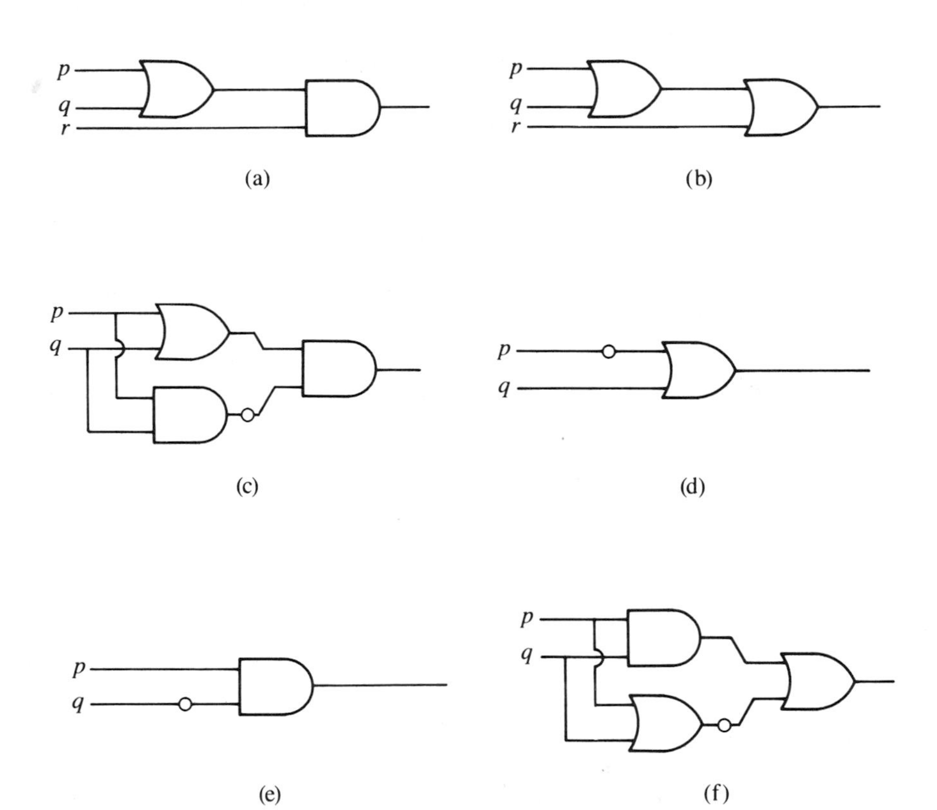

(a) (b) (c) (d) (e) (f)

(A branching "wire" as in parts c and f sends the same "signal" to two gates.)

4. Draw 2-input/1-output gate networks that realize each of the possible truth tables with two arguments. See problem 3 in Complements 6.5. Identify those for which we have used a specific name, such as XOR or Sheffer's stroke.

Complements 6.7

1. Referring to problems 1 and 2 in Complements 6.6, draw gate network realizations to control a light bulb from one, two, three, four, and five different independent signals.
2. In the figures, we have used a "horizontal" flow of information, left to right. This is not necessary, of course; gates and flow lines can be arranged in any positions, provided the flow of information is clear from the figure. But the horizontal scheme suggests a symbolism that has several advantages in many circumstances and which we'll see later in Section 12.1.3. Such a symbolism goes under the name of **postfix formula.** The basic idea is to place all symbols, signals and gates, in one single line of writing, with the convention that the symbol of each gate follows immediately the symbols for its input or inputs, whether these are single letters or compounds. The three basic examples are $p\sim$, $pq\vee$, and $pq\wedge$, in which p and q are each either a single letter or a postfix expression. They come from the three basic expressions $\sim p$, $p \vee q$, $p \wedge q$, respectively. Another example, for $(p \vee q) \wedge \sim r$, is $pq \vee r \sim \wedge$.
3. Find the postfix expressions for several formulae or gate networks given above.

6.8 Summary

The stress of this chapter is on two abstract structures, free semigroups (Section 6.1) and Boolean algebras (Section 6.3). We used the first of these in the *algebra of strings,* and the second in algebra of sets, algebra of propositions, two-terminal switching networks, and algebra of gates.

In *algebra of strings,* we started from an alphabet A, constructed the *free semigroup* of A (A^*, the set of all finite strings over A), and defined on A^* the basic operation of *concatenation* of strings (for example, xyz:zux := xyzzux; Section 6.1). The derived concepts and operations discussed are a *language* over A (any subset of A^*), *concatenation of languages* ($L{:}M := \{x{:}y \mid x \in L, y \in M\}$, *iteration* of a language L ($L^\star :=$ set of all concatenations of any finite number of elements of L, in any order, including the *null string* Λ), A^k (set of all k-strings over A) and, in particular, $A^0 := \{\Lambda\}$.

A *Boolean algebra* in general comprises a set A, two binary operations $\vee$ and $\wedge$ on A, a unary operation $\sim$ on A, and two special elements of A (0 and 1), such that these data satisfy the ten axioms of Section 6.3. Derived properties of Boolean algebras are given in the same section; two important ones are *DeMorgan's laws* and the *double complement* property.

Concrete cases of Boolean algebras are

- **Algebra of subsets** of a given universal set U: the operations are union, intersection, and complementation, with symbols $\cup$, $\cap$, and $'$; the identities are the empty set and the universal set, with symbols $\varnothing$ and U, respectively (Section 6.2).
- **Algebra of propositions** and **truth tables:** the operations are logical *and, or, not,* with symbols $\wedge$, $\vee$, $\sim$. The logical operator OR refers to the truth of either of two given propositions, the operator AND to the simultaneous truth of two given propositions, NOT is the negation of a given proposition. The identities are the *truth values false* and *true,* with symbols F and T, respectively (Sections 6.4, 6.5).
- **Two-terminal switching networks:** the operations are parallel, series, NOT-devices, with the same symbols of algebra of propositions; the identities are *off, on* (or 0, 1, etc.) and refer to the state of a switch or a device (Section 6.6).
- **Logic gates:** the operations are AND-, OR-, NOT-gates, with the same symbols used in algebra of propositions, or with the respective schematic symbols, which are common in diagrams; the functions of the basic gates are the same as the logical operators (Section 6.7).

7 Relations, Correspondences, and Functions

For students of mathematics and computer science, there are probably no two words that cause more confusion than *relation* and *function*. In everyday conversation, they appear to mean about the same thing. "The reliablity of the machine is related to the reliability of its parts" and "The reliability of the machine is a function of the reliability of its parts" both convey the same idea. On the other hand, students of high school algebra know that relations and functions are different. They know that $y = x^2 + 2x + 1$ is a function and that $y =$ (a number whose square is x) is a relation. But they're not taught exactly why, or exactly what the difference is. Indeed, some authors go so far as to say that relations and functions are totally different ideas, unconnected in any way. The reason for this supposed chasm is the applications to which relations and functions are put. In applied mathematics and computer science, relations are used to analyze one type of problems, functions another. We're going to take heed of this distinction; we devote Chapters 8, 12, and 14 to the study of discrete functions, and Chapter 10 to the study of discrete relations. Before we can do that, however, we have to know what relations and functions *are,* and how they're connected. As it happens, it's more convenient for us to begin by studying functions.

7.1 Examples of Functions and Correspondences

Let's start with some examples.

7.1.1 An Explicit Function

We said that $y = x^2 + 2x + 1$ is a function. What does that mean? For each real value of x we perform a specific operation to calculate a specific value of y, which we call $f(x)$, or fx, or just y. In this case, the operation is "calculate the sum of the square of x, twice x, and 1." For each x we obtain a single value as a result.

7.1.2 An Implicit Relation

Suppose we want to have for each possible integer x, a corresponding integer value y such that $y^2 = x$. In different form, this could be written as $y =$ (an integer whose square is x). For instance, if $x = 0$, then $y = 0$; if $x = 9$, then $y = 3$ or $y = -3$; if $x = -4$, then no y satisfes the condition. We still have an operation to be performed on x to obtain the corresponding y, if one exists; but now to some x there corresponds precisely one y, for some other x there may be two values of y, and for still others, no y at all.

7.1.3 Telephone Book

Consider the telephone book of a certain geographic area, white pages only, and ignoring the fact that some numbers are unlisted and that some listed numbers may be outside that area. Let A be the set of the names of the persons that reside in that area. Let B be the set of all seven-digit decimal numbers, that is $B := \{0, 1, \ldots, 9\}^7$. If we look up in the phone book any name in A, we may not find it (if that person has no phone), or we may find it only once (if that person has a single phone number), or we may find several entries. We have an operation to be performed on any element x of A, although this operation is not algebraic; it is a "table lookup." The "table" is the phone book, conceived as a list of pairs *(name, phone number)*. Such a list is a subset of $A \times B$.

7.1.4 The ZIP+4 Code

The ZIP+4 code is a nine-digit code that combines the original five-digit code used by the postal service to identify individual post offices with four additional digits that specify a very small delivery segment. Those last four digits, which are appended to the original code with a hyphen, may identify a particular business or apartment building, a single block, or some such unit. Most interesting for our purposes, the nine-digit code can be expressed as a bar code written only in long and short bars, and so, in effect, a binary code. The bar codes, which are readable by a mechanical sorting device, are most often used for pre-printed reply envelopes (Figure 7.1). Figure 7.2 gives the rules by which the ZIP+4 code is translated to bar code.

Figure 7.1. The ZIP+4 code in nine-digit and bar code forms. (From U.S. Postal Service "Bar Code and FIM," 1982)

When planning to print the envelopes, we must first decide on the nine digits of the ZIP+4 code and then translate the string of digits into a bar code. The preliminary step is to translate each of the decimal digits into a five-digit binary string. Here we have another example of correspondence. This particular correspondence can be solved through a "table lookup" (see the table within Figure 7.2). Note that, again, the table is a set of pairs (one-digit decimal number, 5-digit binary number). In other words, the pairs listed in column comprise a subset of $\{0, 1, 2, 3, 4, 5, 6, 7, 8, 9\} \times \{0, 1\}^5$.

7.1.5 Remarks

As usual, the first questions are: what do these explicit functions, implicit relations, the phone book, and the ZIP+4 bar code have in common? What is the essential goal we want to reach? The answer is that, for each x in a first set and for each y in a second set, we want to be able to decide whether or not y corresponds to x according to a specific type of assignment. Equivalently, for each ordered pair (x, y) of elements in appropriate sets, we want to decide whether the pair is acceptable or not. The phone book and the table of the ZIP+4 code answer the question directly, by listing the acceptable ordered pairs, although the sizes of the two lists are several orders of magnitude apart.

Any questions? "Yes!" we can hear you cry, "Surely you don't expect us to write sets of ordered pairs for functions like $y = x^2 + 2x + 1$ or correspondences like $y^2 = x$. Surely you're talking about a far different type of function than the ones we're used to! Listing these pairs could take forever!" Well, yes, it not only

Reading the ZIP + 4 Bar Code

The ZIP + 4 bar code issued to you represents your new ZIP + 4 code as it will appear in the address of your pre-printed reply envelopes. The bar code consists of the nine digits of your ZIP + 4 plus a correction character used by the bar code reader to identify reading errors.

The bar code consists of 52 bars as illustrated in Example 1 (not to scale). Each of the 10 digits contained between the frame bits consists of 2 long bars (read as 1's) and 3 short bars (read as 0's). Example 2 illustrates the same bar code, as it will appear on a mail piece [Figure 7.3].

Reading and understanding the bar code is simple. There are 10 combinations of 5 bars. each consisting of 2 long (1's) and 3 short (0's) bars. The digits 0 through 9 have been assigned to these combinations.

0	11000	4	01001	8	10010
1	00011	5	01010	9	10100
2	00101	6	01100		
3	00110	7	10001		

It is not necessary to memorize these ten combinations. Within the group of 5 bars, each position has a different value. From left to right 7, 4, 2, 1, and 0. Addition of the values in the two positions occupied by 1 bars gives the value of the combination, except in the case of 11000, which totals 11 and has been assigned as zero.

The sum of the 10 digits in the bar code must always be a multiple of 10. This determines the value of the correction character used. In Example 1, the sum of the nine digits of the ZIP + 4 is 45. Using a correction character of 5 makes the sum of all 10 characters 50, a multiple of 10. If the sum of the digits is not a multiple of 10, an error has been made and the bar code must not be used.

Figure 7.2. The ZIP + 4 bar code. (From U.S. Postal Service "Bar Code and FIM," 1982)

could, it would. But theoretically, every function, every correspondence could be expressed as a set of ordered pairs. Finding an element that corresponds to a given element of the first set is then simply a *table lookup* into that given set of pairs. In practice, in everyday use, there are two broad types of functions: those most easily represented in a list or table of ordered pairs, and those most easily represented by a formula to be calculated, like $f(x) = x^2 + 2x + 1$. The line between the two types is fuzzy and unimportant. In the ZIP + 4 example, the table lookup is clear, *and* a formula can be dug out from the description. We can view the telephone book as a set of ordered pairs (name, number), but no one would consider taking a name and passing it through a formula to calculate the phone number! If that were possible, think of how much money the phone company could save on

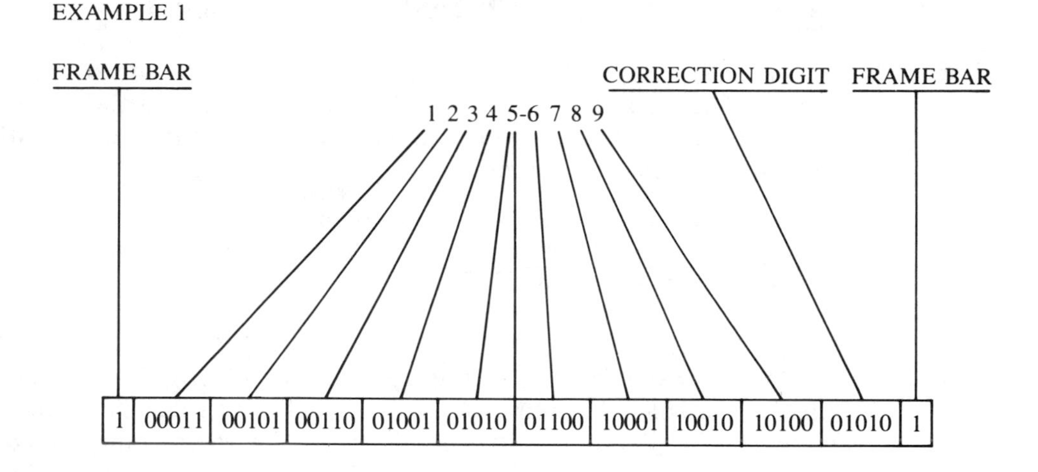

EXAMPLE 2

Figure 7.3. The ZIP + 4 code examples. (From U. S. Postal Service, "Bar Code and FIM," 1982)

books! The point is, you're used to equating the terms *function* and *correspondence* with *calculation* and *formula*. That's simply incomplete. A formula is just one possibility for expressing a function or correspondence. There are many correspondences in which a formula would be cumbersome or even impossible; enumeration is practically required. But the ideas we've presented work just as well in either case.

Exercises 7.1

1. Prove that each integer from 1 to 9 can be written, in one way and only one way, as the sum of precisely two of 0, 1, 2, 4, 7.
2. Find the cardinality (or an approximation of it) of each of the following sets: **(a)** $\{0, 1, 2, 3, 4, 5, 6, 7, 8, 9\}^7$; **(b)** $\{0, 1\}^5$; **(c)** $A :=$ set of the names in your telephone book; **(d)** $A \times \{0, 1\}^7$; **(e)** $\{0, 1, 2, 3, 4, 5, 6, 7, 8, 9\} \times \{0, 1\}^5$.

7.2 Correspondences, Relations, and Functions

Definition 1 Given sets A and B, a **correspondence g** from A to B is denoted $g: A \to B$ and is defined as a subset of $A \times B$. Set A is called the **domain** of the correspondence, and B is called the **codomain** of the correspondence. A **relation on** A is any correspondence from A to A.

The symbol $g: A \to B$ is used as a noun or as a full sentence. For instance: "Let $g: A \to B$ be the correspondence given in the ZIP+4 book," or "If $g: A \to B$, then $g(x)$ makes sense only if x is in A".

Examples Suppose $A := \{i, j, k\}$, $B := \{s, t, u, v, w\}$, and $r := \{(i, s), (i, t), (j, t), (j, u)\}$. Then r is a correspondence from A to B. Set A is thus the domain, B the codomain. Notice that element k ($k \in A$) doesn't appear anywhere in the pairs of r; nor do elements v, w ($v \in B$, $w \in B$). That is, k doesn't have any corresponding element, and neither of v, w corresponds to any elements. In tabular form, two representations of r are

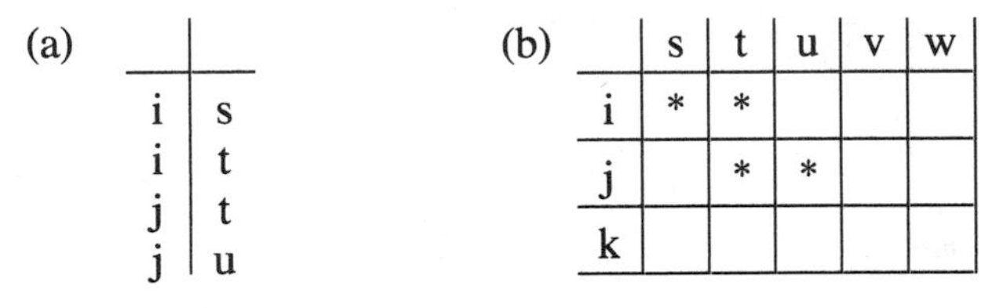

(a)

i	s
i	t
j	t
j	u

(b)

	s	t	u	v	w
i	*	*			
j		*	*		
k					

Table (a) is a list of ordered pairs, that is, the table we would look up. Table (b) emphasizes that this list is a subset of $A \times B$. Representation (a) can be considered as a very short phone book: i, j, k are "names" (even if k does not appear), s, t, u, v, w are "all possible numbers" (even if v and w do not appear).

Another pictorial but informal way to visualize the same correspondence is the following:

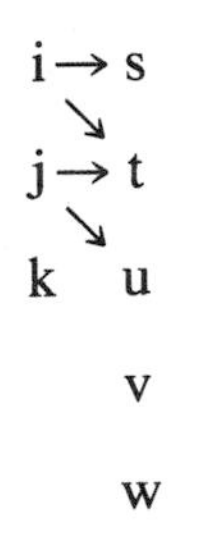

Each arrow represents one of the ordered pairs.

Consider another example: if $A := \{i, j, k\}$ and if $B := \{k, l\}$, then the set $\{(i, l), (k, l)\}$ is a correspondence: with i it associates l, with k it associates l, with j it does not associate any element.

Comparing the formal definition with these examples and the examples at the beginning of the chapter, note the efficiency of the *ordered pair notation*. It states exactly which elements correspond to which; no more, no less. Realize also that the ordered pair notation provides—if necessary—an operational approach. The operation that produces the elements of B (if any) that correspond to any given element x of A is: search the given subset g of $A \times B$, looking for those pairs for which the first element is x; the second elements of such pairs (if any) are the elements of B that correspond to the given x. Conversely, if you have a formula, then the ordered pair notation is $\{(x, y) \in A \times B \mid x, y \text{ satisfy the formula}\}$. In a correspondence, the roles of A and B *cannot* be reversed.

7.2.1 Definitions

The vocabulary connected with correspondences is formidable; we may as well give it to you all at once. Unfortunately in the language of correspondences many of the terms overlap partially in meaning, and others are exact synonyms. All of them, however, are in active use by the mathematics and computer science communities. Painful though it is, you'll have just to learn all of them. Furthermore, expect considerable variation among books. But don't let the vocabulary confuse you; aim at understanding the ideas, which are actually rather simple.

We'll illustrate the ideas on a concrete example: the set of *all* correspondences from set {0, 1} to set {u, v}. Figure 7.4 lists them in table form. For instance, correspondence h is the subset {(0, u), (0, v), (1, u)} of $A \times B$ = {0, 1} × {u, v}. Under h, 0 maps to u and v, 1 maps to u. As a check, in our case, $\#(A \times B) = 2 \cdot 2 = 4$, $\#\mathcal{P}(A \times B) = 2^4 = 16$; in fact we've listed 16 correspondences.

Two very particular correspondences are the **null correspondence,** which is the empty set as a subset of $A \times B$; and the **universal correspondence,** which is $A \times B$ as a subset of $A \times B$. In the null correspondence no element corresponds

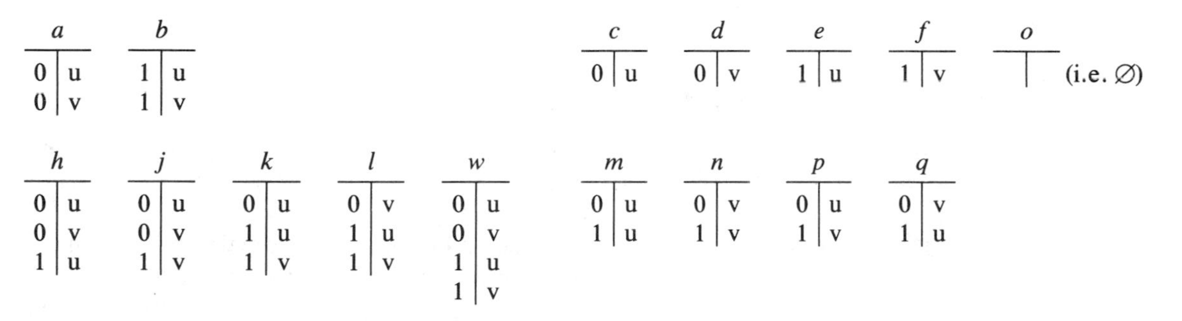

Figure 7.4. All correspondences from {0, 1} to {u, v}.

to any element; in the universal correspondence every element of B corresponds to each element of A. (In the example in Figure 7.4, o is the null correspondence and w is the universal correspondence.)

In the special case in which $A = B$ we have a correspondence from A to *itself,* that is, a relation *on* A. Usually, either A and B are disjoint, or $A = B$, but an example such as $A = \{\text{i, j, k}\}$, $B = \{\text{k, l}\}$, $g = \{(\text{i, l}), (\text{k, l})\}$, in which A and B are overlapping but unequal, is perfectly valid. When $A = B$, a special correspondence is $i := \{(x, x) \in A \times A \mid x \in A\}$; that is, for each x of A, the only corresponding element is x itself. Such a correspondence is called the **identity** on A. When there is danger of confusion, the symbol i_A is used instead of just i.

When $A \subseteq B$, there is another special correspondence, with the same definition as the identity: for each x in A, the only corresponding element is x itself, which is automatically an element of B. That is: $\{(x, x) \in A \times B \mid x \in A\}$. This correspondence is called *the* **injection** of A into B. Note the determinate article.

Suppose $g: A \rightarrow B$. The fact that $(x, y) \in g$ (that is, the fact that y corresponds to x) is often written $x \mapsto y$ (note the form of this arrow). In words, we use several phrases, among them the following: y corresponds to x; y is an (or the) image of x; g maps x to y; x maps to y under g; or x "goes to" y (quite colloquial). Often a symbol like $g: A \rightarrow B, x \mapsto \ldots$ is used to define a correspondence. The ellipsis here would actually be an explicit definition of the image (or an image) of x (understood to be in B) for an unspecified x in A. For example, $g: \mathbb{Z} \rightarrow \mathbb{Z}, x \mapsto$ an integer whose square is x; or $f: \mathbb{N} \rightarrow \mathbb{N}, t \mapsto t^3$.

The set of those elements of the codomain B which *participate* in g is the **image** or **range** of g; the set of those elements of the domain A which participate in g is called the **pre-image** of g. Equivalently, (image of g) $:= \{y \in B \mid$ for some x in A, $(x, y) \in g\}$, (pre-image of g) $:= \{x \in A \mid$ for some y in B, $(x, y) \in g\}$. Equivalently again, $y \in$ image if and only if $x \mapsto y$ for some x; $x \in$ pre-image if and only if $x \mapsto y$ for some y.

In the example in Figure 7.4, the pre-image of a is $\{0\}$, the image of a is $\{\text{u, v}\}$; the pre-image of m is $\{0, 1\}$, its image is $\{\text{u}\}$; the pre-image of k is $\{0, 1\}$, its image is $\{\text{u, v}\}$.

You should verify the following statements: For any correspondence $g: A \rightarrow B$, (1) $g \subseteq A \times B$; (2) (pre-image of g) $\subseteq$ (domain of g) $= A$; (3) (image of g) $\subseteq$ (codomain of g) $= B$; and (4) $g \subseteq$ (pre-image of g) $\times$ (image of g) $\subseteq A \times B$.

7.2.1.1 Total and Partial Correspondences

The major classification of correspondences into special cases is based on how many elements of B correspond to each element of A or on how many elements of A have the same corresponding element in B.

The correspondence $g: A \rightarrow B$ is said to be **total** if its domain and pre-image are the same set. That is, every x in A has an image. Equivalently, for any x in A, there exists a y in B such that $(x, y) \in g$. A correspondence that is not total is called **partial.** In the correspondences of Figure 7.4, m is total, because its pre-image is {0, 1}; every element of A has some image. Correspondence a is partial, because its pre-image is not {0, 1} (element 1 has no image). The top half of Figure 7.4 lists the partial correspondences, the bottom row the total ones.

7.2.1.2 Surjective and Nonsurjective Correspondences

The correspondence $g: A \to B$ is said to be **onto *B*** or **surjective** if its codomain and image are equal. That is, every y in B is the image of some element. Equivalently, for any y in B, there exists an x in A such that $(x, y) \in g$. In the example, a is surjective, because its image is {u, v}, each of u and v is the image of some element of A. Correspondence m is not surjective because its image is not {u, v} (no element of A has image v).

Thus, a total correspondence utilizes its full domain, but an onto correspondence (excuse the grammar, but that's what it's called; *surjective* is a better word, of course) utilizes its full codomain. By degeneracy, any correspondence with domain $\varnothing$ is total, and any correspondence with codomain $\varnothing$ is onto. Note that, if domain $= \varnothing$ or if codomain $= \varnothing$, then necessarily $g = \varnothing$. These notions—total or not, onto or not—are independent, so that four combinations are possible (list them, and give examples; Exercises 7.2).

7.2.1.3 Functions and One-to-Many Correspondences

If every x in A has no more than one corresponding element in B, then g is called a **function.** Alternative names are **map** and **mapping.** A correspondence that is not a function is said to be **one-to-many.** In the example of Figure 7.4, the right-hand part contains all the functions, that is, c, d, e, f, o, m, n, p, and q. The left-hand part contains the one-to-many correspondences. For instance, 0 maps to u *and* v under h or under j, 1 maps to u and v under k or w.

When $g: A \to B$ is a function, then the symbol $g(x)$ or gx denotes the element of B (if any) that corresponds to any one element x of A. The symbol x^g is sometimes used, as in x'. In the example we have, for instance, that $p(0) = \text{u}$, $p(1) = \text{v}$, $m(0) = \text{u}$, $m(1) = \text{u}$.

The set of all total functions $f: A \to B$ is denoted by B^A. In the example, the total functions appear in the bottom right-hand part of the figure. In other words, $\{\text{u}, \text{v}\}^{\{0, 1\}} = \{m, n, p, q\}$.

7.2.1.4 Many-to-One Correspondences

If there is a y in B that corresponds to more than one x in A, then g is said to be **many-to-one.** If g is not many-to-one, we simply say that; there's no specal term. A correspondence that is both one-to-many and many-to-one is said to be **many-to-many.** In the example, the many-to-one correspondences are h, j, k, l, w, m, and n. The many-to-many correspondences are h, j, k, l, and w.

7.2.1.5 Injections and Bijections

A function that is not many-to-one is said to be **one-to-one**, or simply 1-1. A total, one-to-one function is called an **injection**. That is, g is an injection if each z in A has precisely one image gx and if distinct elements have distinct images (that is, $x \neq y$ implies $gx \neq gy$ or, equivalently, $gx = gu$ implies $x = u$).

Finally, we turn our attention to a very special type of correspondence, the most restrictive we've encountered yet. A surjective injection is called a **bijection** or an **isomorphism**. If there exists an isomorphism from set A to set B, then the sets themselves are said to be **isomorphic.** In our example, the 1-1 functions are c, d, e, f, o, p, and q. Correspondences p and q are also injections. In fact, p and q are the only bijections.

Some of these terms may be more memorable if we look at their linguistic background. The root *ject* is from Latin meaning throw. The relevant prefixes give *inject* = throw into; *supra* (Latin) = on, over; so *surject* = throw onto; *bi-, bin-* (Latin) = twice, doubly; so *biject* = throw two ways. From Greek, *iso-* = same, *morph-* = form; so *isomorph-* = same form.

Note carefully in these definitions that it only takes *one* element of A to make a one-to-many correspondence (when that element has many corresponding elements) or a partial one (when it has none), and only one element of B to make a many-to-one correspondence or a nonsurjective one. A good example of a correspondence that is total and surjective, but is many-to-many, is the universal relation $u := A \times B$, when $\#A > 1$ and $\#B > 1$. A good example of a bijection is the identity on any set.

The two most important differences in terminology among different books are first that what we call *correspondence* is often called *relation;* and second that our *total function* is often called just *function*.

Did you notice how the definition of a function emerges from our discussion? A function is simply a correspondence that is not one-to-many; every element of a function's domain has *at most one* corresponding element in the codomain. Note the *at most!* A function can be partial: some element of A may have no correspondent in B.

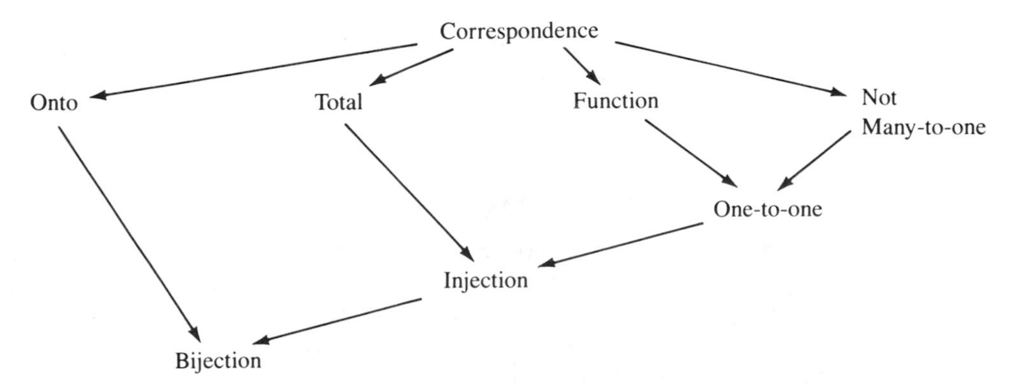

Figure 7.5. Some special types of correspondences.

Note 8 on Theorem Proving. In order to prove that a property is valid, we must prove that it holds under all allowable circumstances. A few examples are not sufficient (unless, of course, the set of examples exhausts all possible cases). When we have to prove that a property is *not* true, one single example in which the property does not hold is enough. Such an example is called a **counterexample.** In order to prove that the function f: $\mathbb{Z} \to \mathbb{Z}$, $x \mapsto x^2$ is not injective, we need only produce one instance of two distinct elements mapping onto the same element. A counterexample is, therefore, provided by 2 and -2, which are distinct but which both map onto 4.

7.2.1.6 Attributes of a Correspondence

The four primary attributes of a correspondence are thus whether or not it is *total, onto, not many-to-one,* and a *function*. These ideas are independent, and each is either satisfied or not satisfied by a given correspondence. A total of 2^4 $(= 16)$ combinations is possible. In the exercises, you'll be asked to provide examples of each. The diagram in Figure 7.5 shows what types of correspondences are special cases of more general types. The diagram is limited to a few of the possible choices of the attributes.

Example If $A := \{0, 1\}$ and if $B := \{u, v\}$, then we have already seen all correspondences g: $A \to B$ listed in Figure 7.4. Study them again.

7.2.2 Cardinalities

If A and B are finite sets, then we can count the numbers of correspondences that have one or the other attribute, or no attribute at all.

Theorem 2 The number of possible distinct correspondences form A to B is $2^{(\#A)(\#B)}$.

Proof: All we need is to apply a result from Section 5.8.2 and Theorem 27 of Chapter 5. Since a correspondence g is a subset of $A \times B$, the number of such subsets is $\#\mathcal{P}(A \times B)$, that is, $2^{\#(A \times B)}$. Since $\#(A \times B) = (\#A) \cdot (\#B)$, we have the claimed result. Q.E.D.

We could spend a lot of time asking how many possible distinct correspondences there are that satisfy each of the sixteen combinations of the four attributes in Section 7.2.1.6. It turns out that only three such results are of interest, all centering on functions.

Theorem 3 The number of possible distinct total functions f: $A \to B$ is $(\#B)^{\#A}$. If $A = B = \varnothing$, then the number is $0^0 := 1$. The number of possible distinct functions f: $A \to B$ is $((\#B) + 1)^{\#A}$. The notation B^A for the set of all total functions from A to B is very suggestive, because $\#(B^A) = (\#B)^{\#A}$.

Proof: Consider first the class of total functions. Each of the $\#A$ elements of A must have exactly one corresponding element among the $\#B$ elements of B. Thus we are selecting precisely $\#A$ elements from the set B with repetitions allowed. The process is thus selection (Section 4.2), and the number of such selections is $(\#B)^{\#A}$. If $A = B = \emptyset$, then $A \times B = \emptyset$, and $g = \emptyset$ is the only possibility.

Consider now the class of functions that are not necessarily total. For each element x of the $\#A$ elements of A there are $(\#B) + 1$ possibilities: x can map to an element of B (as before) or x can fail to map anywhere; that is, x might not participate in f. There are thus $((\#B) + 1)^{\#A}$ possible distinct functions f. Q.E.D.

Theorem 4 If A and B are isomorphic, then $\#A = \#B$ and there are $(\#A)!$ possible distinct isomorphisms between A and B (see Section 4.4).

Proof: Exercise.

7.2.3 Operations

We seem to have missed a large class of functions and correspondences, for instance functions of two variables, such as $z = f(u, v) := u^2 + v^2$ or $z = s(u, v) := u + v$. But we have not. Starting from u and v and obtaining z is not a new type of association; it still gives a *pair,* namely $((u, v), z)$. In general, a correspondence $h: A^k \to B$ is the analogue of a function of k variables. Such a correspondence has k **arguments,** which all are elements of A. Similar remarks apply to correspondences like $g: X \times Y \to B$. Such a correspondence has two arguments, one in X and one in Y.

There's also a deeper point to be made here. We began this book by talking about mathematics as a game, with objects and rules. The objects were generally sets or elements of sets; the rules were often operations on those sets, defined in terms of tables or formulae. But what we've called *operations* are really functions, and it's time now to equate these notions. For example, ordinary addition in $\mathbb{N}$ associates the result $x + y$ with a given pair of elements x and y. That is, the pair (x, y) maps to the value $x + y$. Thus we have a function $\mathbb{N} \times \mathbb{N} \to \mathbb{N}$.

Definition 5 Given a set A, an ***n*-ary operation on** A is a function $f: A^n \to A$. The natural number n is called the **arity** of the operation. (Some books call it *order*, but we prefer this horrible term, because there are too many other things that are also called *order*.) In particular, we have the following three cases: (1) A **binary** operation on A is a function $f: A \times A \to A$. In this case, the result $f(x, y)$ is usually represented in the form xfy as in $+: \mathbb{Z} \times \mathbb{Z} \to \mathbb{Z}$, $(x, y) \mapsto x + y$. (2) A **unary** operation on A is a function $f: A^1 \to A$ or $f: A \to A$. In this case, the distinction between A^1 and A is irrelevant. (3) A **zerary** operation on A is a function $f: \{\Lambda\} \to A$. Unless f is empty, this results in the assignment of a special element in A. That is, we have a **self-assigning** operation.

If $(x_1, \ldots, x_n) \mapsto z$, or if $(x_1, x_2) \mapsto z$, or if $(x_1) \mapsto z$, then the x_i are the **arguments** of the operation.

Again the terminology comes largely from Latin: the Latin *un-* = one, *bin-* = twice, *tern-* = three times, and *quatern-* = four times, give, respectively, *unary, binary, ternary,* and *quaternary*. *Zerary* comes, of course, from *zero,* which is not Latin but Arabic.

Examples The operations given by $f(x) = x^2 + 2x + 1$, addition, multiplication, and additive inverse; union, intersection, and complement; $\wedge$, $\vee$, and $\sim$ are all simply functions with different domains and codomains. The idea that an operation is a function is the thread that ties together all of the different arithmetics and algebras we've considered. Functions are the link between the objects (elements or sets of elements) and their rules (operations).

7.2.4 Applications of Functions

Many of the ideas we have seen before can be expressed much more compactly using the language of functions. Here are some instances.

7.2.4.1 Selections

A total function s: $\{1, 2, \ldots, k\} \to A$ means, intuitively: to 1 we assign a specific element $s(1)$ of A, to 2 we assign a specific element $s(2)$ of A, . . ., to k we assign a specific element $s(k)$ of A. Ultimately, we have a k-sequence of elements of A, namely $s(1), s(2), \ldots, s(k)$. This symbol is the most common in computer science; in more classical mathematics, the usual symbol is $s_1, s_2, \ldots, s_k$. Independently of the notation, we have a k-selection.

Thus a k-selection of elements of A is a total function s: $\{1, 2, \ldots, k\} \to A$. Then the set of all k-selections of elements of A is $A^{\{1, 2, \ldots, k\}}$. Instead of counting from 1 to k, we may count from 0 to $k - 1$. Then s: $\{0, 1, \ldots, k - 1\} \to A$ is the same as $(s_0, s_1, \ldots, s_{k-1})$. In this form, the set of all k-selections is $A^{\{0, 1, \ldots, k-1\}}$. Now recall that the set $\{0, 1, \ldots, k - 1\}$ is sometimes denoted just k; that is, k denotes a standard set with k elements. Then a symbol for the set of all k-selections from set A is set A^k, which we have seen in Sections 5.9.1 and 6.1.

7.2.4.2 Permutations

If the total function p: $\{1, 2, \ldots, k\} \to A$ is required to be one-to-one, then we obtain a k-selection without repetitions, that is, a k-permutation. In other words, a k-permutation of elements from set A is an injection p: $\{1, 2, \ldots, k\} \to A$ or an injection p: $\{0, 1, \ldots, k - 1\} \to A$. In particular, if A is finite and if $k = \#A$, then p is a bijection.

7.2.4.3 Combinations and Subsets

We have seen that a combination of elements of a given universal set U is a subset of U and a k-combination is a k-element subset. Assume U is a finite set and $\#U = n$. An intuitive way to choose a subset of U is to scan all the elements of U and, for each of them, decide whether or not to put it in the subset. But this is the same as having a total function $\chi: U \to \{0, 1\}$. Consequently, a subset S of U is determined by and determines a total function $\chi_S: U \to \{0, 1\}$ according to the following rules: (1) If S is given, then χ_S is defined by: $x \mapsto 0$ if $x \notin S$, $x \mapsto 1$ if $x \in S$. (2) If χ_S is given, then S is defined by $S := \{x \in U \mid \chi_S x = 1\}$. The function χ_S is called the **characteristic function** of S in U. The Greek letter χ *(chi)* stands for *ch*aracteristic. If you want to avoid Greek letters, another good symbol is CH_S.

Another consequence of this notation (because 2 is a symbol for $\{0, 1\}$) is that the set of all subsets of U, that is, the power set $\mathcal{P}U$, is essentially the same as the set of all total functions from U to $\{0, 1\}$. In other words, $\mathcal{P}U$ is essentially the same as $\{0, 1\}^U$, or 2^U. The phrase "essentially the same" is clarified by

Theorem 6 Define $f: \mathcal{P}U \to \{0, 1\}^U$ as follows: for each subset S of U, let fS be the function χ_S. Then f is a bijection. Consequently, $\#\mathcal{P}U = \#(\{0, 1\}^U) = 2^{\#U}$ (a result we saw before).

Proof: Exercise.

7.2.4.4 Combinations with Repetitions

The idea of characteristic function can be extended to cover a concept we haven't seen yet, *combinations with repetitions*. If we allow repetitions to appear in a string, but we maintain the fact that the order of the elements in the string is not essential, then we have a combination with repetitions. This process is equivalent to deciding, for each element of the set U from which we make our choices, how many times that element should be taken (including zero times, if we decide not to take it). If the resulting string is to comprise k elements (repetitions included), then the sum of those "numbers of times" must be k. This remark leads to a formal statement.

Definition 7 A **k-combination, with repetitions,** of elements of a finite set U is a total function $t: U \to \mathbb{N}$ such that $\Sigma_{x \in U}\, tx = k$. A combination with repetitions is called also a **multiset.**

Examples If $A := \{\text{a, b, c}\}$ and if $k = 2$, then a total function from A to $\mathbb{N}$ may be $\text{a} \mapsto 2$, $\text{b} \mapsto 0$, $\text{c} \mapsto 0$; the corresponding 2-combination with repetitions is aa. Another total function from A to $\mathbb{N}$ may be $\text{a} \mapsto 1$, $\text{b} \mapsto 0$, $\text{c} \mapsto 1$; the corresponding 2-combination with repetitions is ac or ca.

We wish now to count the k-multisets from a given set of n objects. Let's first see an example, the 4-multisets from set $\{\text{a, b, c}\}$ ($k = 4$, $n = 3$). Some are aaaa, aaab, aaac, aabb, aabc, and aacc. Although this is a very simple case, it's clear that we may easily miss a few. We need some systematic process. According to the definition, we want all sequences of three natural numbers that add up to 4, for instance (4, 0, 0), (3, 1, 0), (3, 0, 1), (2, 2, 0), (2, 1, 1), (2, 0, 2), ... But also this is messy. In those sequences, let's replace each term with as many ones as the value of that term. We obtain (1111, ,), (111, 1,), (111, , 1), (11, 11,), (11, 1, 1), (11, , 11), ... Now we have a pattern: within the parentheses, each sequence consists of four ones (in general, k ones) and two commas (in general, $n - 1$ separating marks). We are back to combinations: 4 out of 6 positions are ones, the others are marks. We know how to count and enumerate these: there are $C(6, 4)$ of them (that is, 15) and they are (1111, ,); (111, 1,); (111, , 1); (11, 11,); (11, 1, 1); (11, , 11); (1, 111,); (1, 11, 1); (1, 1, 11); (1, , 111); (, 1111,); (, 111, 1); (, 11, 11); (, 1, 111); (, , 1111). Translate each 1 into a or b or c and each comma into the instruction "change letter." Then the enumeration of all 4-multisets is aaaa, aaab, aaac, aabb, aabc, aacc, abbb, abbc, abcc, accc, bbbb, bbbc, bbcc, bccc, cccc. Generalizing this argument, we have the core of the proof of the following theorem.

Theorem 8 Let U be a finite set with n elements. The number of k-combinations with repetitions of elements of U is $C(n + k - 1, k)$.

Proof: Select any one n-permutation of U; that is, place the elements of U in a sequence $(u(1), u(2), \ldots, u(n))$. Each given function t as above gives then a sequence T of natural numbers $t(u(1)), t(u(2)), \ldots, t(u(n))$. Each element of this sequence denotes the number of times we want to take the corresponding element from U. For the purpose of this proof (and an algorithm that we'll see in a moment), let's replace sequence T with another sequence, say S, made up of $t(u(1))$ ones (possibly none), then a mark, then $t(u(2))$ ones (again, possibly none), then a mark, and so on to finally $t(u(n))$ ones (possibly none). The marks are needed to separate the substrings of ones. Then the number of the elements of string S is $\Sigma_{i=1}^{n} tu(i)$ [for the ones] + $(n - 1)$ [for the marks], that is, $k + n - 1$. Conversely, any string made up of k ones and $n - 1$ marks yields a function t, and different strings give different functions. Consequently, the number of k-combinations with repetitions is the same as the number of strings S. All we have to do to have a string S is to decide where to place the k ones in a sequence of $k + n - 1$ positions. The number of such choices is the same as the number of combinations of k elements (the positions for the ones) out of a set of $k + n - 1$ elements (the positions); that is, precisely the number given in the statement. Q.E.D.

Note that this is a typical example of "using another method" to solve a problem: a direct count of combinations with repetitions may be a nasty affair. We reduced it to a count of combinations *without* repetitions of another set, the positions of the ones among certain positions to be taken by marks and ones.

Note finally that this proof is *constructive,* because it gives a method (an algorithm) for enumerating the combinations with repetitions. We know how to enumerate combinations, so we know how to construct all strings S. Once such a string is given, the process is to start from the first element of U and to take each 1 in the string as the instruction "write down that element" and each mark as the instruction "take the next element in U." See Exercises 7.2.

7.2.4.5 Truth Tables

As an example, refer to the third table of Figure 6.1 in Section 6.5.2. With the pair (T, T) the table associates the result T, with the pair (T, F), the value F, and so on. Consequently, that table provides a total function $\{F, T\}^2 \to \{F, T\}$. If we replace F with 0 and T with 1, we can see that, in general, the truth table of a composite proposition with n arguments is a total function $\{0, 1\}^n \to \{0, 1\}$, that is, an n-ary operation, total on $\{0, 1\}$. A similar remark applies to two-terminal switching networks and to logic gates.

Exercises 7.2

1. **(a)** Given f: $\mathbb{Z}_5 \to \mathbb{Z}_5$, defined by $x \mapsto x^2 \pmod 5$, write the table of the ordered pairs of f. **(b)** Given $A := \{0, 1, 2\}$, let $\#: \mathcal{P}A \to \mathbb{N}$ be defined by $X \mapsto \#X$ = number of elements of subset X of A. Write the table of the ordered pairs of $\#$. **(c)** Given $U := \{0, 1, 2\}$, let $A: = \mathcal{P}U$ and let $\cup$: $A \times A \to A$ be defined by $(X, Y) \mapsto X \cup Y$. Write several ordered pairs from the table of $\cup$. **(d)** Given $U := \{0, 1, 2\}$, set $A := \mathcal{P}\ U$ and let $\cap$: $A \times A \to A$ be defined by $(X, Y) \mapsto X \cap Y$. Write several ordered pairs from the table of $\cap$.
2. For each set of data in the exercise above **(i)** verify that the given correspondence is a total function; **(ii)** state whether it is onto or 1-1; and **(iii)** justify your answers.
3. State whether each of the following functions is total or not, one-to-one or not, a surjection or not, a bijection or not. **(a)** f: $\mathbb{N} \to \mathbb{N}$, $x \mapsto 3x$; **(b)** g: $\mathbb{N} \to \mathbb{N}$, $x \mapsto x^2$; **(c)** g: $\mathbb{Z} \to \mathbb{Z}$, $x \mapsto x^2$; **(d)** h: $\mathcal{P}A \to \mathcal{P}A$, $x \mapsto A - x$ (A is any given set with two or more elements; **(e)** i: $A \to A$, $x \mapsto x$ (A is any given set); **(f)** CH: $\{0, 1, 2, 3\} \to \{0, 1\}$, CH is the characteristic function of $\{2, 3\}$ as a subset of $\{0, 1, 2, 3\}$; **(g)** t: $\mathbb{Z} \to \mathbb{Z}$, $x \mapsto$ remainder of division of x by 5; **(h)** u: $\mathbb{Z} \to \mathbb{Z}_5$, $x \mapsto$ remainder of division of x by 5; **(i)** v: $\mathbb{N} \to \mathbb{N}$, $x \mapsto$ a natural number whose square is x; **(j)** w: $\mathbb{N} \to \mathbb{Z}$, $x \mapsto$ an integer whose square is x; **(k)** z: $\mathbb{N} \to \{0, 1\}$, $x \mapsto$ truth value of the statement "$x < 5$".
4. Given sets $A := \{1, 2, 3\}$ and $B := \{4, 5\}$, give the tables of all total functions from A to B. Among them point out the surjective functions (if any).
5. For a correspondence, the two criteria *total* and *onto* give four combinations: both, total and not onto, onto and not total, neither. Find a concrete example of each. In your example, present A, B, and g by enumeration.
6. Let $A := \{0, 1, 2\}$ and $B := \{s, t\}$. [Hint: If you're careful, you won't do the same work more than once!] **(a)** How many possible distinct correspondences are there from A to B? List all of them. **(b)** How many possible distinct functions are there from A to B? List all of them. **(c)** How many possible distinct total functions are there from A to B? List all of them. **(d)** There are 16 combinations of the criteria of a correspondence being total, onto, not many-to-one, and a function. List all 16, and try to give an example of each using

sets A and B. (A few cases will be missing. Why?) The examples should be written as sets of ordered pairs of elements. **(e)** How many possible distinct isomorphisms are there on A? List them all.

This is a good exercise for the computer, even for larger sets. Watch the size of the output: calculate it in advance, by hand or by computer. When using the computer, near the beginning put a conditional *stop* to halt the execution if the output is too big.

7. Make a table for addition in $\mathbb{Z}_3$. Now express this addition as a function $+: \{0, 1, 2\}^2 \rightarrow \{0, 1, 2\}$. That is, enumerate the ordered pairs that define addition in $\mathbb{Z}_3$. Do the same for multiplication.

8. Let $A := \{a\}$. Enumerate, as sets of ordered pairs, all of the unary and binary operations that are total on $\mathcal{P}A$. [Hint: First calculate how many of each there are.]

9. Enumerate, as sets of ordered pairs, all of the unary and binary logical operations that are total on $\{T, F\}$.

10. Given $A := \{0, 1, 2\}$, list all the isomorphisms from $\{1, 2, 3\}$ to A.

11. For a nonempty set A, how many isomorphisms are there from the set of integers $\{1, 2, \ldots, \#A\}$ to A? Explain clearly what each isomorphism represents.

12. Prove Theorem 4 (on the number of isomorphisms between two finite sets). [Hint: What string formation process is being used here?]

13. Prove Theorem 6 (on the relationship between $\mathcal{P}U$ and 2^U).

14. Enumerate efficiently **(a)** all 3-combinations, with repetitions, from set {a, b, c}; **(b)** all 5-combinations, with repetitions, from set {a, b, c}; **(c)** all 5-combinations, with repetitions, from set {a, b}; **(d)** "all" 22-combinations, with repetitions, from set {a}.

Do the same problem on a computer, with larger sets and larger k; check sizes!

15. Review Section 6.5.2 on truth tables. Realize that the truth table of a proposition with n arguments is actually a function table for a total function from $\{0, 1\}^n$ to $\{0, 1\}$.

16. Suppose we consider a bi-directional relation between sets A and B. What would it be a subset of? What would be the universal bi-directional relation between A and B?

Complements 7.2

1. The method used in the text to count and enumerate the k-multisets of a given set with n elements also solves the problem of finding all n-term sequences of elements of $\mathbb{N}$ such that the sum of the terms of each sequence is k. **(a)** Find the number of 4-term sequences of natural numbers that add up to 234. **(b)** Enumerate all 4-term sequences of natural numbers that add up to 7.

2. Given that A and B are sets and that $f: A \to B$ is a given correspondence, define the correspondence $F: A \to \mathcal{P}B$ by $x \mapsto \{y \in B \mid (x, y) \in f\}$. Intuitively, for each x of A, Fx is the set of all elements of B that correspond to x under f. Show that F is a total function. In particular, when $A := \{0, 1, 2, 3, 4\}$ and $B := \{-2, -1, 0, 1, 2\}$ (both subsets of $\mathbb{Z}$), and fx = a number whose square is x, write by enumeration $F0$, $F1$, $F2$, $F3$, and $F4$. Furthermore, for each subset z of A, define $F^*(z)$ as $\bigcup_{x \in z} Fx$. Then F^* is a correspondence from which set to which set? Is it a function? Is it total? In the example above, write by enumeration $F^*\{0, 2\}$, $F^*\{0, 1\}$, and $F^*(A)$.

3. For some set A, suppose that f and g are binary operations, total on A. Operation f is said to be *commutative* if, for all choices of x and y in A, $f(x, y) = f(y, x)$. A similar definition holds for g. Satisfy yourself that this definition is equivalent to the usual definition of a commutative operator. Use function notation to state definitions for each of the following ideas (see Section 3.2). **(a)** Operation f is *associative* on A. **(b)** Operation f *distributes* over g. **(c)** Element u ($u \in A$) is an *identity* for f. **(d)** Element v ($v \in A$) is an *annihilator* for g. **(e)** Operation f is *idempotent* on A.

4. Let U be a given universal set. Look up the definition of the characteristic function, $\chi_S: U \to \{0, 1\}$ of a subset S of U. Recall the definitions of the Boolean operations $\vee$, $\wedge$, $\sim$ and of the set-theoretical operations of union, intersection, and complementation. Prove Theorem 9:

Theorem 9 Let S, T be subsets of U. For each element x of U: (1) $\chi_{S \cup T}x = (\chi_S x) \vee (\chi_T x)$; (2) $\chi_{S \cap T}x = (\chi_S x) \wedge (\chi_T x)$; (3) $(\chi_{\text{complement of } S}\, x) = \sim(\chi_S x)$.

This theorem allows us to "calculate" with subsets of the universal set. Experiment on a relatively small set, such as $U := \{a, b, c, d\}$. Give yourself a few subsets, calculate their characteristic functions, and shift to unions, intersections, complements and their characteristic functions. Note also that the three operations on characteristic functions are practically, in order, *bitwise* OR, *bitwise* AND, and *bitwise negation*. See also Complements 8.3.

7.3 Inverses

If a correspondence g takes us from a domain A to a codomain B, we're sometimes interested to know how to get from B back to A through g.

Examples Recall the example of the ZIP+4 code in Section 7.1.4. The computer of the post office must be able to read the ZIP+4 bar code, decide whether there are errors (more on this later), split the bar code into stretches of five bars each, interpret such 5-tuples of bars as five-digit binary numbers, and finally assign to each of them a decimal single-digit number. The last step concerns us now. It can be performed by looking up the same table (in Figure 7.2), but this time each line must be read right to left.

Given $g: \mathbb{Z}_5 \to \mathbb{Z}_5, x \mapsto x^2$, we have that $g = \{(0,0), (1,1), (2,4), (3,4), (4,1)\}$. [Function g just squares x and reduces mod 5.] The inverse is $g^{-1} = \{(0,0), (1,1), (4,2), (4,3), (1,4)\}$, which just inverts the ordered pairs, and answers the question "given y, what's an element (if any) whose square mod 5 is y?"

Definition 10 Let $g: A \mapsto B$. The **inverse** correspondence is denoted by $g^{-1}: B \to A$ and is defined as $g^{-1} := \{(y, x) \in B \times A \mid (x, y) \in g\}$. Clearly, $(g^{-1})^{-1} = g$. If $A = B$ and if $g = g^{-1}$, then g is said to be **involutory,** or **self-inverse,** or **symmetric.**

An example of self-inverse correspondence is the additive inverse in $\mathbb{Z}$, that is $h: \mathbb{Z} \to \mathbb{Z}, x \mapsto {}^-x$. The inverse would be (informally) ${}^-x \mapsto x$, but replacing ${}^-x$ with y gives $y \mapsto {}^-y$, which is the same as h.

The real purpose of this section is to ask the following question: If we have a correspondence $g: A \to B$, about which we know some of the four crucial attributes in Section 7.2.1.6, what can we say about the inverse correspondence $g^{-1}: B \to A$? The answer is summed up in the following theorem. Don't let the brevity of the theorem (or of this section) fool you. This theorem and its inherent machinations deserve loving and detailed attention!

Theorem 11 (1) g is total if g^{-1} is onto. (2) g is onto if g^{-1} is total. (3) g is not many-to-one if g^{-1} is a function. (4) g is a function if g^{-1} is not many-to-one. (5) g is an isomorphism if and only if g^{-1} is an isomorphism.

Proof: We only sketch the proof here; you'll finish it as an exercise. Consider part 1. You are to show that if g^{-1} is onto, then g is total. In this case, it turns out to be easier to prove the *contrapositive* statement, that is, if g is not total, then g^{-1} is not onto. In parts 2, 3, and 4, a similar remark is valid. Part 5 is then nearly immediate.

Like some of the proofs we've seen before, this one is important not only for the result in its own right, but also for its technique.

Exercises 7.3

1. Give explicit definitions of the inverses of the correspondences given in Section 7.1. First rephrase the data in correct terms. Then define the inverse independently, not just as "the inverse of" the original.
2. Do the same for several of the correspondences we have seen. Beware: some very simple examples produce sometimes tricky inverses; for instance, $+: \mathbb{N}^2 \to \mathbb{N}, (x, y) \mapsto x + y$.
3. Prove Theorem 11 on the inverse of a correspondence.

Complements 7.3

1. In the ZIP+4 code, the nine digits of the actual code are followed by a *check digit,* calculated so that the sum of the ten digits is a multiple of 10. Compare this rule with the ISBN number in Section 3.3. Note the similarities and the differences. In particular, an erroneous interchange of two digits in the ZIP code is *not* detected. Try for the ZIP+4 code some exercises similar to those in Exercises 3.3.

7.4 Composition

Suppose we have correspondences $f: A \to B$ and $g: B \to C$. It's possible, then, to map from A to C by first using f to map from A to B, and then using g to map from B to C. The final result is a correspondence from A to C which no longer even mentions B! This process is called **composition,** and the resultant correspondence is called a **composite** correspondence.

Suppose $A := \{1, 2, 3\}$, $B := \{a, b, c, d\}$, $C := \{x, y, z\}$. Let $f: A \to B$ be $f := \{(1, a), (2, c), (3, b), (3, d)\}$ (so f is *not* a function). Let $g: B \to C$ be $g := \{(a, y), (b, x), (b, z), (d, y)\}$ (so g is neither total nor a function). Then the composition $g \circ f: A \to C$ would be $g \circ f := \{(1, y), (3, x), (3, y), (3, z)\}$. A picture will probably help here. In this table we're using the informal notation with arrows. Follow the arrows: $1 \mapsto a \mapsto y$ under f and g, respectively, so $1 \mapsto y$ under $g \circ f$; $2 \mapsto c$ under f, c has no image under g, so 2 has no image under $g \circ f$; and so on. The second table represents the correspondence $g \circ f$.

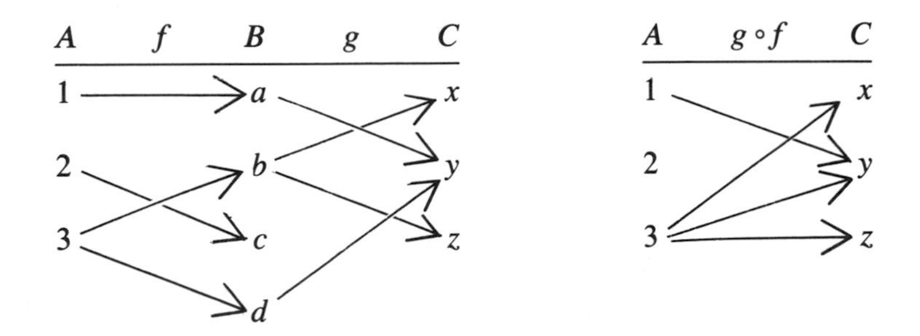

Definition 12 Let $f: A \to B$ and $g: B \to C$. The **composition** of f and g is denoted by $g \circ f: A \to C$ and is defined as $g \circ f := \{(x, y) \in A \times C \mid \text{there exists } v \text{ in } B \text{ such that } (x, v) \in f \text{ and } (v, y) \in g\}$.

The **first component** of the composition is f, the **second component** of the composition is g.

Note carefully that, in the composition notation $g \circ f$, we apply f first to map from A to B, then apply g to map from B to C. When "operating with operators," we always work from right to left. The easy way to remember this is to think of function notation, even though f and g might not be functions: if $x \in A$, then $g \circ f(x) = g(f(x))$.

Examples

- In ordinary algebra, given that $f(x) = x^2 + 1$ and $g(x) = 2x - 3$, we have that

$$\begin{aligned} g \circ f(x) &= g(f(x)) \\ &= g(x^2 + 1) \\ &= 2(x^2 + 1) - 3 \\ &= 2x^2 - 1 \end{aligned} \qquad \text{and} \qquad \begin{aligned} f \circ g(x) &= f(g(x)) \\ &= f(2x - 3) \\ &= (2x - 3)^2 + 1 \\ &= 4x^2 - 12x + 10. \end{aligned}$$

- "Tunnel ahead. Clearance 12 ft. Overheight vehicles take exit 9." There are two functions involved in this highway sign: (1) height: {vehicles} $\rightarrow$ {positive numbers}, (2) decision: {positive numbers} $\rightarrow$ {"exit", "noexit"}. The composite function decision $\circ$ height (*vehicle*) results in "*vehicle* must or must not exit."
- Of course, it's possible that $A = B$ or $A = C$ or $B = C$ or even $A = B = C$. In this example, we'll use $A = B = C := \{1, 2, 3, 4\}$, so that both compositions $g \circ f$ and $f \circ g$ are defined. You should carefully work through the tabulations of $g \circ f$ and $f \circ g$ to see how they were obtained. Now we're using a formal table representation, but in an abbreviated way. When dealing at the same time with several functions with the same domain, it is customary to list the left-hand side of every table just once, as a first column. The right-hand sides of the tables are listed in the next columns, one for each function.

A	f	g	$g \circ f$	$f \circ g$
1	2	3	1	3
2	4	1	2	2
3	3	4	4	1
4	1	2	3	4

- In this example, suppose $A := \{a_1, a_2\}$, $B := \{b_1, b_2, b_3, b_4\}$, $C := \{c_1, c_2\}$. Also suppose $f: A \rightarrow B$ and $g: B \rightarrow C$ are given by $f := \{(a_1, b_1), (a_1, b_4), (a_2, b_2), (a_2, b_4)\}$, $g := \{(b_1, c_1), (b_2, c_2), (b_3, c_1), (b_3, c_2)\}$. In pictures,

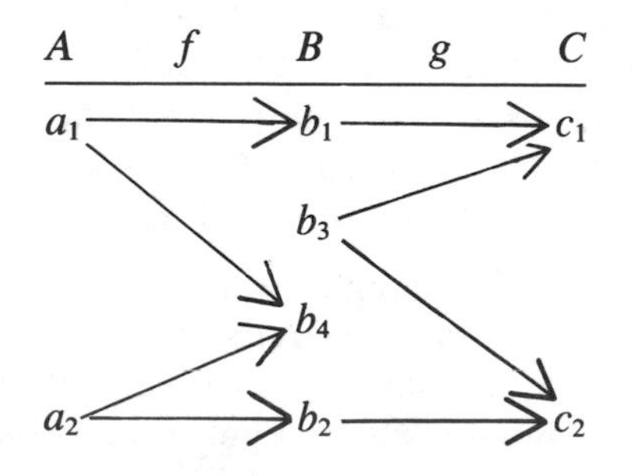

So f is total and many-to-one, but neither onto B nor a function; g is many-to-one and onto C, but neither total nor a function. But the composition $g \circ f$ is $\{(a_1, c_1), (a_2, c_2)\}$, which is an isomorphism between A and C! The "problem" with this example is the set B. It is the codomain of f and the domain of g. But that isn't precise enough to tell us much. What we really have to compare are the two subsets of B, the *image* of f and the *pre-image* of g. The image of f is $\{b_1, b_2, b_4\}$, the pre-image of g is $\{b_1, b_2, b_3\}$. The intersection of these is $\{b_1, b_2\}$, which is the only portion of B we actually use to compose f and g. The exercises ask you to experiment with these ideas.

This last example is designed to illustrate a rather subtle point. Suppose we know the attributes of f and g individually—whether each is total, onto, not many-to-one, a function. Then, for each instance, we could easily work out the attributes of $g \circ f$. But in general, it's difficult to state exactly what the attributes of $g \circ f$ will be. There are, however, a few cases where modest conclusions can be simply stated.

Theorem 13 Let $g: A \to B$ be a correspondence. If $g^{-1} \circ g$ is the identity on A, then g is total and not many-to-one. If $g \circ g^{-1}$ is the identity on B, then g is a function onto B. If both conditions are satisfied, then g is a bijection.

Let $f: A \to B$, $g: B \to A$ be correspondences. If $g \circ f$ is the identity on A, then f is total. If $f \circ g$ is the identity on B, then f is onto B. If both conditions are satisfied, then f and g are both bijections and each is the inverse of the other.

Proof: We prove the first statement and leave the others as exercises. Suppose x is an arbitrary element of A. Since $g^{-1} \circ g = i_A$, we know that $g^{-1} \circ gx = x$. Consequently, there is a u in B such that g maps x to u and g^{-1} maps u to x. This implies that x has an image and that g is total. Suppose that g maps x *and* y to the same u. Because of the definition of g^{-1}, g^{-1} must map u to each of x *and* y (and possibly to other elements). Then $g^{-1} \circ g$ maps x to x *and* to y (at least). Since $g^{-1} \circ g = i_A$, we know that $g^{-1} \circ g$ maps x to x only. Therefore $y = x$ and g is not many-to-one. Q.E.D.

Two more cases in which we can reach some conclusions about the attributes of a composite correspondence are handled better in terms of multicorrespondences, for which we give a definition first.

Definition 14 Let $A_0, A_1, \ldots, A_n$ be given sets $(n > 1)$. For each i $(1 \leq i \leq n)$, let g_i: $A_{i-1} \to A_i$. Then the composition $g_n \circ g_{n-1} \circ \cdots \circ g_2 \circ g_1: A_0 \to A_n$ is defined as $g_n \circ g_{n-1} \circ \cdots \circ g_2 \circ g_1 := \{(x, y) \in A_0 \times A_n \mid$ there exist elements z_1 (in A_1), z_2 (in A_2), $\ldots$, z_{n-1} (in A_{n-1}) such that $(x, z_1) \in g_1$, $(z_1, z_2) \in g_2, \ldots$, $(z_{n-2}, z_{n-1}) \in g_{n-1}$, $(z_{n-1}, y) \in g_n\}$.

Obviously, the composition can't be total unless g_1 is total: you can't hit a home run if you can't get to first base, regardless of how easily you could get to second or third. Similarly, you can't hit a home run if you can't get to home plate, regardless of how easily you got to third base the composition can't be onto A_n unless g_n is onto A_n.

Theorem 15 (1) A composition is total only if its first component is total. (2) A composition is onto only if its last component is onto.

Proof: Exercise.

Note carefully that these are merely *necessary (only if)* conditions. A composition can fail to be total even if its first component is total and can fail to be onto even if its last component is onto.

Definition 16 With reference to Definition 14, we can have, in particular, that all the A_i are equal to a same set A $(0 \le i \le n)$ and all the g_i are equal to a same correspondence g $(0 < i \le n)$; then the composition is denoted g^n and is called the n-th **iteration** of g. In other words, $g^2 := g \circ g$, $g^3 := g \circ g \circ g$, and so on. Furthermore, $g^1 := g$, $g^0 := i_A =$ identity on A, $g^{-m} := (g^{-1})^m$ $(m > 0)$.

There is a slight conflict of notations with a common usage in ordinary algebra. Suppose $g(x) := x + 3$. According to our definition, $g^2(x) = g \circ g(x) = g(g(x)) = g(x + 3) = (x + 3) + 3 = x + 6$. In ordinary algebra, another interpretation is $g^2(x) = [g(x)]^2 = (x + 3)^2 = x^2 + 6x + 9$. When we mean the latter interpretation, we'll write $[g(x)]^n$.

Exercises 7.4

1. Given $f: \mathbb{Z} \to \mathbb{Z}, x \mapsto x^3$ and $g: \mathbb{Z} \to \mathbb{Z}, x \mapsto x + 1$, find a formula for $g \circ f$, $f \circ g$, $f \circ f$, and $g \circ g$.
2. Given $h: \mathbb{Z}_5 \to \mathbb{Z}_5, x \mapsto x^3 \pmod 5$, find a formula for $h \circ h$ and $h \circ h \circ h$.
3. Given $g: \mathbb{Z} \to \mathbb{Z}, x \mapsto x^2 + 1$ and $h: \mathbb{Z} \to \mathbb{Z}, x \mapsto x + 5$, give a formula for each of the following: **(a)** $g \circ h$; **(b)** $h \circ g$; **(c)** $g \circ g$; **(d)** $h \circ h$; **(e)** g^3; **(f)** h^3; **(g)** $g \circ h \circ g$; **(h)** $h \circ g \circ h$; **(i)** $h \circ h \circ g$.
4. Given $f: \mathbb{Z} \to \mathbb{Z}, x \mapsto 2x$, calculate each of the following: **(a)** $f(1)$; **(b)** $f \circ f(1)$; **(c)** $f \circ f \circ f(1)$; **(d)** $f^4(1)$; **(e)** $f^{-1}(4)$; **(f)** $f^{-3}(32)$.
5. Given $f: \mathbb{Z} \to \mathbb{Z}, x \mapsto x + 1$, find a formula for $f^2, f^3, f^0 f^{-1}, f^{-2}$, and f^n $(n \in \mathbb{Z})$.
6. Prove Theorems 13 and 15.

Complements 7.4

1. Prove the following theorem:

Theorem 17 (1) If $f: A \to B$ and $g: B \to C$ are bijections, then $g \circ f: A \to C$ and $f^{-1}: B \to A$ are bijections. (2) The identity on A (that is, $i_A: A \to A, x \mapsto x$) is a bijection. (3) Suppose $g: A \to A$ is a correspondence. Then (a) if $g^{-1} \subseteq g$, then $g^{-1} = g$; and (b) if $g^2 \subseteq g$ and if $i_A \subseteq g$, then $g^n \subseteq g$ for every natural n.

7.5 Summary

A *correspondence* g from a set A (the *domain*) to a set B (the *codomain*) is a specific rule according to which each element of A has none, one, or more corresponding elements in B. We say that x (in A) *maps* to y (in B), that y *corresponds* to x, or simply $x \mapsto y$. The *pre-image* of g is the set of those elements of A (if any) that have some corresponding element in B. The *image* of g is made up of those elements of B (if any) that correspond to some element of A (Section 7.2.1).

Correspondences are classified according to several criteria, all based on how many elements of each set map to or correspond to each element of the other set (Section 7.2.1.6). The more important cases in this classification are the following: in a *bijection*, each element of A maps to precisely one element of B and each element of B corresponds to precisely one element of A; in an *injection*, each element of A maps to precisely one element of B and distinct elements of A map to distinct elements of B; in a *function*, each element of A maps to no more than one element of B; in a *total function*, each element of A maps to precisely one element of B; in a *total correspondence*, each element of A maps to at least one element of B; in a *surjective correspondence*, each element of B corresponds to at least one element of A. The set of all total functions from A to B is denoted B^A. In the finite case, it has cardinality $(\#B)^{\#A}$ and the set of all functions from A to B has cardinality $((\#B) + 1)^{\#A}$. Special correspondences are the *operations;* see Section 7.2.3. A *binary* operation on A is a function from A^2 to A, a *ternary* operation on A is a function from A^3 to A, etc.

Suppose correspondence $g: A \to B$ maps x to y. Then the *inverse* correspondence operates from B to A and maps y to x. A symbol for the inverse of g is g^{-1} (Section 7.3).

If $f: A \to B$ maps x to y and $g: B \to C$ maps y to z, then the *composite* correspondence $g \circ f$ operates from A to C and maps x to z. If, in particular, $A = B = C$, then $f \circ f$ is denoted f^2, and similarly for multiple *iterations* of f (Section 7.4).

Some applications of the idea of function (Section 7.2.4) give new approaches to the ideas of selection, permutation, combination, and truth table. In particular, we obtain a definition of *multiset* (or *combination with repetitions*).

Algebraic Structures

In the introduction to this book, we observed that a mathematical structure could most readily be compared to a game, with game objects and rules specifying what to do with the objects. In general, a structure consists of one or more sets, one or more operations or correspondences or relations on or among those sets, and possibly some distinguished elements in those sets. An example is the structure of the integers. We have the set of the integers, the operations of addition and multiplication on it, and the special elements 0 and 1, which are the respective identities for those operations. Two more examples among the many we have seen are the free semigroup of an alphabet (Section 6.1) and Boolean algebras (Section 6.3).

Our study of such a general topic as algebraic structures will be a little more systematic if we ask the following questions about a structure and consider the answers in order from simplest to more complicated:

1. How many operations are there? If there are more than one, do they interact in any peculiar ways? We'll start by considering only single operations, and then move up to multiple, interrelated operations.
2. Of what *arity* is any one operation? Is it *unary? binary?* As it happens, unary and binary operations are the only ones of interest to us now.
3. Does a given operator have any interesting features, such as associativity, distributivity, and so on?
4. What are the most significant derived properties?
5. Is the structure a special case of a more general structure? Can it be upgraded to a richer structure?
6. In an application, does the structure have the "algebra" we need for that application?

As we go through some of these possibilities, keep in mind the remark we made when speaking of Boolean algebras: The general definitions of structures are given at an abstract level. The nature of elements and operations is totally unspecified, and the derived properties and definitions are connected with the *syntax* of the structure at hand. The concrete *facts* will surface when we consider specific *realizations* of the structure. But there is much advantage in considering an abstract setting first. We avoid tedious repetitions, and we're able to apply the same general abstract results to many different cases. An example of this process is the application of the general properties of Boolean algebra to algebra of propositions, two-terminal switching networks, and logic gates.

Practically, if you know in any one concrete situation that you are working within a semigroup, or a Boolean algebra, or the like, then you know what "algebra" you can use. You have a set of formal rules that you can apply automatically.

In this chapter we devote our attention to a class of structures that go under the name *algebraic structures*. There are several other broad types of structures, such as *order structures* and *incidence structures,* which we'll see later. Each of these denominations is a matter of taste, not of definition. The nature of each specific structure *is* instead a matter of very rigorous definition and for once there is general agreement on these definitions among different books and even different sciences. The two main sections of this chapter are devoted, respectively, to structures with essentially one operation and to structures with essentially two operations; the names are *groupoids* and *semirings,* respectively. Other sections are devoted to important special cases.

8.1 Structures with One Operation

In this section we consider structures with essentially one binary operation. There may be a second operation in some of them, but this second operation will not really have any meaning of its own apart from the binary operation.

8.1.1 Groupoid

A *groupoid* (an awkward name for a very simple idea) is just a set with a binary operation. The most typical example is $(\mathbb{N}, +)$, that is, the structure that comprises the set of the natural numbers and the operation of ordinary addition. Note that at the present time we are ignoring additional features of the structure of the natural numbers, such as the existence of a multiplication.

Definition 1 A **groupoid** is a pair $(A, *)$, where A is a set, and $*$ is a total binary operation on A. As usual for binary operations, the element $*(x, y)$ is denoted $x * y$. A groupoid is said to be **abelian** (uh-BEE-lee-un) or **commutative** if the binary operation is commutative, that is, if $y * x = x * y$ for every choice of x and y in A.

The term *abelian* comes from Niels Henrik Abel (UH-b'l), the Norwegian mathematician (1802–1829) who first studied these structures.

Example Let A be the set of all finite strings over a given nonempty alphabet. Let $*$ denote "concatenation with cancellation." That is, let x and y be any two strings, then $x * y$ is the usual concatenation $x{:}y$ if the last element of x is not equal to the first element of y (or if either is not there); if such elements are equal, then they are both cancelled and the remaining strings are concatenated according to the usual process. Examples: abc $*$ de := abcde, abcc $*$ ccd := abccd. Note that this operation is *not* associative: (ab $*$ c) $*$ cb = abc $*$ cb = abb, but ab $*$ (c $*$ cb) = ab $*$ b = a. Clearly, the operation isn't commutative either.

8.1.2 Semigroup

Essentially, a semigroup is a set with an associative binary operation. The structure $(\mathbb{N}, +)$, which we have seen to be a groupoid, is also a semigroup because addition is associative.

Definition 2 A **semigroup** is a pair $(A, *)$, where

- A is a set,
- $*$ is a total binary operation on A, and
- $*$ is *associative*. That is, $(x * y) * z = x * (y * z)$ for every choice of x, y, and z in A.

Equivalently, $(A, *)$ is a groupoid, and operation $*$ is associative.

Example Let A be the set of the finite strings over an alphabet, as before. Let $*$ be "concatenation with a mark." That is, $x * y := x{:}(\text{-}){:}y$, where : is the usual concatenation and - is an assigned symbol of the alphabet. Example: -bc $*$ de = -bc-de. Clearly, there is no identity element for $*$.

8.1.3 Monoid

A monoid is a set on which we have a binary operation that is associative and has an identity. The structure $(\mathbb{N}, +)$, which we saw before as a groupoid and as a semigroup, is also a monoid, because 0 is the identity for addition. Formally,

Definition 3 A **monoid** is a triple $(A, *, u)$, where

- A is a set,
- $*$ is a total binary operation on A,
- $*$ is associative,
- $u \in A$, and
- u is an **identity** for operation $*$, that is $u * x = x$ and $x * u = x$ for every choice of x in A.

Equivalently, $(A, *)$ is a semigroup, $u \in A$, and u is an identity for operation $*$.

Example Let A be the set of the finite strings over an alphabet, as before. Let : be the usual concatenation (that is, without cancellation or insertion of a mark). Let Λ be the null string. Then $(A, :, \Lambda)$ is a monoid.

In Section 6.1.2.2 we called this structure a *free semigroup*. Every monoid is a semigroup, of course, and this justifies the name *semigroup* we used then. The adjective *free* refers to the fact that when we concatenate strings we can never "simplify" after the concatenation. This is rather uncommon. In ordinary algebra, for instance, multiplying x and $1/x$ gives 1, and neither x nor $1/x$ is there any more; there has been some simplification. In the free semigroup on an alphabet, whenever you concatenate any two strings, the original strings are still there after the concatenation.

There are many properties that are valid in all monoids. Here are some.

Theorem 4 Let $(A, *, u)$ be a monoid. The only identity for $*$ is u.

Proof: Let v be an identity, that is, suppose $v * x = x$, $x * v = x$ for every choice of x in A. Replacing x with u in the last equation gives $u * v = u$. Replacing x with v in the first formula in the definition gives $u * v = v$. Consequently, $u = v$. Q.E.D.

Definition 5 In a monoid, for each element x of A and for each natural number k, the k-th **power** of x is defined by

$$x^0 := u,$$

$$x^1 := x,$$

$$x^k := x * \cdots * x \, (k \text{ “factors”}) \qquad \text{if} \qquad k > 1.$$

Theorem 6 ***Rules of Exponents.*** Let $(A, *, u)$ be a monoid. If $x \in A$ and if h and k are any natural numbers, then

$$x^h * x^k = x^{h+k} \qquad \text{and} \qquad (x^h)^k = x^{h \cdot k}.$$

Proof: Exercise.

8.1.4 Group

A group is a monoid in which the basic operation has an inverse operation. More specifically, let u be the identity and let x be any element. There may be an element y such that $x * y = u$. If this occurs for each element x, then we have a unary operation that associates the element y with the respective element x. If we denote y as x', then (according to our notations for correspondences) we have a correspondence $' : A \to A, x \mapsto x'$. For example, the structure $(\mathbb{N}, +)$ which we have seen several times is *not* a group, because addition does not have an inverse in $\mathbb{N}$.

Definition 7 A **group** is a quadruple $(A, *, ', u)$, where

- A is a set,
- $*$ is a total binary operation on A,
- $*$ is associative,
- $u \in A$,
- u is an identity for $*$,
- $'$ is a total unary operation on A, and
- $x * (x') = u$ for every x in A.

Equivalently, $(A, *, u)$ is a monoid and $'$ is a total unary operation on A such that $x * (x') = u$ for every x in A. The binary operation of a group is often called group **composition.** The term applies to groupoids in general, but less frequently. For each x in A, the element x' is called the **inverse** of x with respect to operation $*$.

Example Let S be any set. Let A be the set of all bijections from S to S. Let $\circ$ denote composition of functions. Let i_S be the identity on S. Let f^{-1} denote the inverse of f. Then $(A, \circ, ^{-1}, i_S)$ is a group. This special group A is called the **symmetric** group on S, or the group of **permutations** on S, or the **total** group on S. Be sure to understand the meaning of the symbols. See also Theorem 7.17 and Complements 8.1. In this example, the binary operation is actually **composition;** this explains the reason for the term in general.

Theorem 8 Let $(A, *, ', u)$ be a group. (1) For each x in A, $(x')' = x$; that is, $' \circ ' = i_A$. (2) $': A \to A$ is a bijection. (3) For each x in A, $x' * x = u$. (4) For each x in A, if $w * x = u$ or if $x * w = u$, then $w = x'$.

Proof: (1) Define $y := x'$, $z := y'$. Then $(x')' = z$. By the definition of inverse, $x * y = u$ and $y * z = u$. Consequently, the identity property implies $(x * y) * z = u * z = z$. Invoking associativity and identity property again, $(x * y) * z = x * (y * z) = x * u = x$. Consequently, $z = x$, that is, $(x')' = x$. Part 1 is proved. (2) Consider the "two" functions $'$ and $'$, and apply Theorem 5.11. Then $'$ is a bijection. Part 2 is proved. (3) Consider again the formulae in the first part of this proof. Since $y * z = u$ and $z = x$, we have $x' * x = u$. Part 3 is proved. (4) For part 4, multiply both sides of $w * x = u$ by x' on the right, or $x * w = u$ by x' on the left and obtain $w = x'$. Q.E.D.

Definition 9 In addition to Definition 5, we set, for every x in A, (1) $x^{-1} := x'$, and (2) $x^{-k} := (x')^k$ for every positive integer k. Consequently, x^n is defined for every integer n, and x^{-1} is another symbol for x'.

Theorem 10 Let $(A, *, ', u)$ be a group. (1) The **rules of exponents** stated in Theorem 6 are valid when h and k are any integers. (2) If x and y are any elements of A, then $(x * y)^{-1} = (y^{-1}) * (x^{-1})$. (3) **Cancellation rule:** if x, y, and z are any elements of A, then $x * z = y * z$ or $z * x = z * y$ implies $x = y$.

Proof: Exercise.

8.1.5 Groups of Symmetry

Interesting examples of groups of permutations are those associated with symmetry of geometric figures.

Group of the Rectangle. Let 1234 be a nonsquare rectangle as in Figure 8.1.

We can perform various rigid motions that move the rectangle but put it back to its original position. We are not interested in the trajectory the rectangle may have taken, but only in the final locations of the vertices. For example, we can "flip" vertex 1 to vertex 4 (and necessarily also 4 to 1, 2 to 3, and 3 to 2). The ultimate result is the permutation

$$\begin{pmatrix} 1 & 2 & 3 & 4 \\ 4 & 3 & 2 & 1 \end{pmatrix}.$$

This permutation, which is interpreted, top to bottom, 1 to 4, 2 to 3, 3 to 2, 4 to 1, reflects the top-bottom symmetry of the rectangle. We can also flip the rectangle right-to-left and obtain

$$\begin{pmatrix} 1 & 2 & 3 & 4 \\ 2 & 1 & 4 & 3 \end{pmatrix}.$$

This reflects its left-right symmetry. Or, we can do both flips, one after the other and (recalling that composition is done "second component first") obtain

$$\begin{pmatrix} 1 & 2 & 3 & 4 \\ 2 & 1 & 4 & 3 \end{pmatrix}\begin{pmatrix} 1 & 2 & 3 & 4 \\ 4 & 3 & 2 & 1 \end{pmatrix} = \begin{pmatrix} 1 & 2 & 3 & 4 \\ 3 & 4 & 1 & 2 \end{pmatrix}.$$

Or, we can do nothing. That is, consider the identity permutation

$$\begin{pmatrix} 1 & 2 & 3 & 4 \\ 1 & 2 & 3 & 4 \end{pmatrix}.$$

If we set, for short,

$$u := \begin{pmatrix} 1 & 2 & 3 & 4 \\ 1 & 2 & 3 & 4 \end{pmatrix}, \qquad a := \begin{pmatrix} 1 & 2 & 3 & 4 \\ 4 & 3 & 2 & 1 \end{pmatrix}, \qquad b := \begin{pmatrix} 1 & 2 & 3 & 4 \\ 2 & 1 & 4 & 3 \end{pmatrix},$$

then we have the four permutations u, a, b, and ab. If we compose them, two at a time, in all possible ways, could we obtain more? The answer is negative, as we

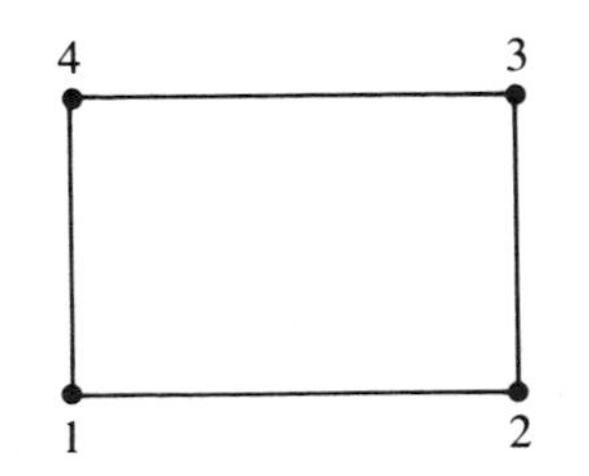

Figure 8.1 The rectangle.

can see in either of two ways. Geometrically, once we know the destination of vertex 1, then everything is uniquely determined; so we obtain four permutations (the possible destinations of vertex 1) and no more; and four we have. Algebraically, just compose u, a, b, ab, two at a time, and you'll obtain the composition table

	u	a	b	ab
u	u	a	b	ab
a	a	u	ab	b
b	b	ab	u	a
ab	ab	b	a	u

which shows that combining permutations gives nothing new. We have, therefore, a special group of permutations called the **group of the rectangle.**

Group of the Tetrahedron. Let 1234 be a regular tetrahedron as in Figure 8.2. The four faces are equilateral triangles of the same size.

Again, we can consider the rigid motions that bring the tetrahedron back to its original position and consider the resulting permutation on the vertices (ignoring the intermediate trajectory of the solid). Now we have a different situation: when we know the final position of vertex 1, we can still "rotate" the solid about the "height" through the new position of vertex 1 and obtain three different permutations. As an example, if 1 goes to 1, then we have the three permutations

$$u := \begin{pmatrix} 1 & 2 & 3 & 4 \\ 1 & 2 & 3 & 4 \end{pmatrix}, \qquad a := \begin{pmatrix} 1 & 2 & 3 & 4 \\ 1 & 3 & 4 & 2 \end{pmatrix}, \qquad a' := \begin{pmatrix} 1 & 2 & 4 & 3 \\ 1 & 4 & 3 & 2 \end{pmatrix}.$$

Observe that $aa' = u$, that is, a' is the inverse of a. Also, $a' = a^2$, that is, a' is also the "square" of a.

Geometric reasons imply then that we have twelve permutations (vertex 1 can go to four different places, and we have three possibilities for each of the four locations of vertex 1).

Suppose vertex 1 goes to 2; then edges 12, 13, 14 must go—all together—to edges 21, 23, 24 and the orientation must be maintained (we exclude mirror images

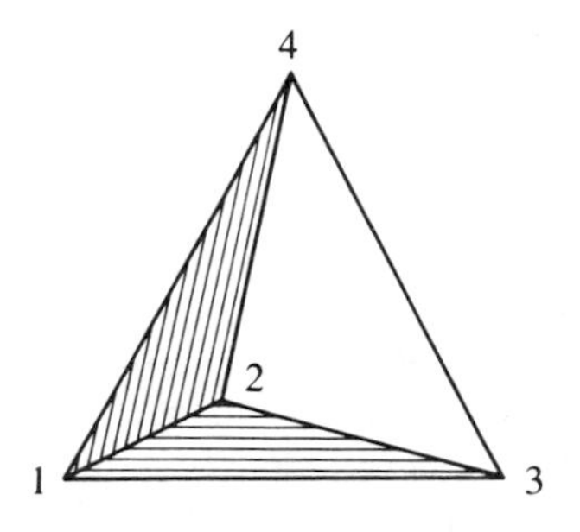

Figure 8.2 The regular tetrahedron.

from consideration). One of the permutations we obtain this way is, for instance,

$$b := \begin{pmatrix} 1 & 2 & 3 & 4 \\ 2 & 1 & 4 & 3 \end{pmatrix}.$$

Let's try now to conclude algebraically. By just performing compositions of permutations the hard way, we can easily verify that $b^2 = u$, that u, a, a^2, b, ab, a^2b, ba, ba^2, and aba are all distinct from each other, and that $ba^2b = aba$. After that, we can do much just algebraically, without actually calculating the permutations. An example of such calculations is: $a^2ba^2 = a(aba)a = a(ba^2b)a = (aba)(aba) = (ba^2b)(ba^2b) = ba^2b^2a^2b = ba^4b = bab$. Ultimately, we obtain the composition table

u	a	a^2	b	ba	ba^2	ab	aba	aba^2	a^2b	a^2ba	a^2ba^2
a	a^2	u	ab	aba	aba^2	a^2b	a^2ba	a^2ba^2	b	ba	ba^2
a^2	u	a	a^2b	a^2ba	a^2ba^2	b	ba	ba^2	ab	aba	aba^2
b	ba	ba^2	u	a	a^2	a^2ba^2	a^2b	a^2ba	aba	aba^2	ab
ba	ba^2	b	a^2ba^2	a^2b	a^2ba	aba	aba^2	ab	u	a	a^2
ba^2	b	ba	aba	aba^2	ab	u	a	a^2	a^2ba^2	a^2b	a^2ba
ab	aba	aba^2	a	a^2	u	ba^2	b	ba	a^2ba	a^2ba^2	a^2b
aba	aba^2	ab	ba^2	b	ba	a^2ba	a^2ba^2	a^2b	a	a^2	u
aba^2	ab	aba	a^2ba	a^2ba^2	a^2b	a	a^2	u	ba^2	b	ba
a^2b	a^2ba	a^2ba^2	a^2	u	a	aba^2	ab	aba	ba	ba^2	b
a^2ba	a^2ba^2	a^2b	aba^2	ab	aba	ba	ba^2	b	a^2	u	a
a^2ba^2	a^2b	a^2ba	ba	ba^2	b	a^2	u	a	aba^2	ab	aba

We have obtained twelve permutations. Geometric reasons told us before that we couldn't have obtained more with this method. We have therefore the **group of the tetrahedron.**

8.1.6 Comments and Examples

Note that mathematical convention requires that, in the tuple that describes a structure, we list first the set, then the binary operations, then the unary operations, then special elements such as identities. It's just done that way.

By now, you may be thoroughly confused by all the new terms. Have patience and go through the following examples slowly and carefully. You need not memorize everything at once. When you need to work with any one of these structures, you can look it up, and in few minutes you'll be able to use its "algebra." Our aim is to show that groupoid, semigroup, monoid, and group, as well as the structures we'll see later, are each distinct and useful ideas. Also, in most cases, the symbols for the operations will be special symbols. The abstract symbol $*$ is seldom used in a concrete setting (except, in many computer languages, for ordinary multiplication). For instance, $*$ is replaced by $+$, $\oplus$, or $\vee$ when the operation has an *additive* connotation, or by $\cdot$, $\circ$, or $\wedge$ when the operation has a *multiplicative* connotation. The symbol $'$ is replaced by $^-$, $^{-1}$, $\sim$, or one of various other symbols.

We have already seen many structures with two binary operations. In studying any of those structures, nothing prevents us from considering only one of the two operations and disregarding the other, at least for a while. This we'll do in many of the following examples.

Examples

1. Consider $(\mathbb{N}, -)$ or $(\mathbb{Z}, -)$, where $-$ denotes subtraction, a binary operation. Subtraction is not associative $(9 - (5 - 1) \neq (9 - 5) - 1)$, so neither $(\mathbb{N}, -)$ nor $(\mathbb{Z}, -)$ is a semigroup. In fact, $(\mathbb{Z}, -)$ is a groupoid, but $(\mathbb{N}, -)$ isn't even a groupoid, since substraction isn't total.
2. Structures $(\mathbb{N}, +, 0)$, $(\mathbb{Z}, +, 0)$, $(\mathbb{Q}, +, 0)$, and $(\mathbb{R}, +, 0)$ are all abelian monoids.
3. Let $^-$ denote the *unary minus* (change sign, additive inverse, +/− in many calculators, etc.). Then $(\mathbb{Z}, +, ^-, 0)$, $(\mathbb{Q}, +, ^-, 0)$, $(\mathbb{R}, +, ^-, 0)$ are abelian groups, but $(\mathbb{N}\ +, ^-, 0)$ is not a group, because $^-$ is defined for 0 only.
4. If the centered dot $\cdot$ denotes multiplication, then $(\mathbb{N}, \cdot, 1)$, $(\mathbb{Z}, \cdot, 1)$, $(\mathbb{Q}, \cdot, 1)$, and $(\mathbb{R}, \cdot, 1)$ are all abelian monoids. None, however, is a group. For $\mathbb{N}$ and $\mathbb{Z}$, the reason is obvious: most integers have no *integer* multiplicative inverse. But what of $\mathbb{Q}$ and $\mathbb{R}$? What's the multiplicative inverse of 0? Remember, to have a group, the inverse operation must be total. We can, however, say that $(\mathbb{Q} - \{0\}, \cdot, ^{-1}, 1)$ and $(\mathbb{R} - \{0\}, \cdot, ^{-1}, 1)$ are abelian groups, where $^{-1}$ is the usual inversion $x \mapsto 1/x$.
5. In modular arithmetic, $(\mathbb{Z}_m, +, ^-, 0)$ is an abelian group if ^-x is defined as $^-x \pmod{m}$. The multiplicative structure $(\mathbb{Z}_m, \cdot, 1)$ is a monoid when $m > 1$, but (recalling Chapter 3) multiplicative inverses can be defined only if m is prime. If that's the case, then $(\mathbb{Z}_m - \{0\}, \cdot, ^{-1}, 1)$ is an abelian group.
6. Let U be a given, nonempty set and recall the algebra of the subsets of U. The structures $(\mathcal{P}U, \cup, \emptyset)$ and $(\mathcal{P}U, \cap, U)$ are abelian monoids. Neither $\cup$ nor $\cap$ has an inverse, so there's no chance of upgrading either of these structures to a group.
7. In a Boolean algebra $(A, \vee, \wedge, \sim, 0, 1\}$, both $(A, \vee, 0)$ and $(A, \wedge, 1)$ are abelian monoids. Since neither $\vee$ nor $\wedge$ has an inverse, we can't strengthen these to be groups.

Exercises 8.1

1. For each of the four examples in Sections 8.1.1–8.1.4, verify that the given example actually satisfies the conditions of the definition above it but that it cannot be "upgraded."

2. Each of the following semigroups is given by means of set S and a binary operation on S. See if the structure is actually a monoid (if so, what's the

identity?), a group (if so, what's the inverse?), or an abelian semigroup. Justify your answers. **(a)** $S := \mathbb{Z}_5$, operation + mod 5. **(b)** $S := \mathbb{Z}_5 - \{0\}$, operation $\cdot$ mod 5. **(c)** $A := \{0, 1\}$, $S := A^A$, operation $\circ$. **(d)** $S := \{0, 1\}$ (Boolean), operation $\vee$. **(e)** $S := \{0, 1\}$ (Boolean), operation $\wedge$. **(f)** $A := \{0, 1, 2\}$, $S := \mathcal{P}A$, operation $\cup$. **(g)** $A := \{0, 1, 2\}$, $S := \mathcal{P}A$, operation $\cap$. **(h)** $S := \{00, 01, 10, 11\}$, operation BITWISE OR. **(i)** $S := \{00, 01, 10, 11\}$, operation BITWISE AND.

3. Let A be a suitable set of two-terminal switching networks. Discuss the possible structures that could be constructed using 2TSN and the operations of series and parallel connection.
4. Prove Theorem 6 (the rules of exponents in a monoid).
5. Prove Theorem 10 (the rules of exponents in a group).

Complements 8.1

1. Let A be a nonempty set. Let $C(A)$ be the set whose elements are all the partitions of A (see Complements 5.7). If $S \in C(A)$ and if $T \in C(A)$, then let $S * T$ be the cross-partition obtained from S and T. Prove that $(C(A), *, u)$ is an abelian monoid. Note that u must be a specific partition; which one? Is there an annihilator, that is, a partition v such that $S * v = v$ for every partition S?

2. For a given set U, consider an operation represented by +, the **symmetric difference;** that is, if $A \in \mathcal{P}U$ and if $B \in \mathcal{P}U$, then $A + B := ((A \cap B') \cup (A' \cap B))$. Then $(U, +)$ is a groupoid. Show that this structure is an abelian group. Don't forget to prove associativity, which is not obvious in this case. What is the identity? What is the inverse of an element?

3. For a given set S, define $A := \mathcal{P}(S \times S)$ = the set of all correspondences $g{:}S \to S$ from S to itself. Let $\circ$ be the composition of correspondences. Consider the groupoid $(A, \circ)$. Is this a semigroup? A monoid (if so, what is the identity)? A group (if so, what is the inverse)? Is it abelian? Address the same questions for the groupoid $(S^S, \circ)$, in which the elements are all the total functions from S to itself and the operation is composition of functions. For each structure, review the definition of g^k and show that the rules of exponents are valid (for which values of the exponents?).

4. Suppose now A is defined as the set of all bijections from set S to S, that is, the set of all permutations of S. Is $(A, \circ)$ now a semigroup? A monoid? A group? Is it abelian? See the example in Section 8.1.4.

5. Suppose S in complement 4 is "small," for instance, $S := \{1, 2, 3\}$. Then it is customary to represent an element of A in the form

$$\begin{pmatrix} 1 & 2 & 3 \\ a & b & c \end{pmatrix},$$

that is, as a table written *across* rather than *down*. Equivalently,

$$\begin{pmatrix} 1 & 2 & 3 \\ a & b & c \end{pmatrix}$$

is the bijection defined by $1 \mapsto a, 2 \mapsto b, 3 \mapsto c$. In this case,

$$A = \left\{ \begin{pmatrix} 1 & 2 & 3 \\ 1 & 2 & 3 \end{pmatrix}, \begin{pmatrix} 1 & 2 & 3 \\ 1 & 3 & 2 \end{pmatrix}, \begin{pmatrix} 1 & 2 & 3 \\ 2 & 1 & 3 \end{pmatrix}, \begin{pmatrix} 1 & 2 & 3 \\ 2 & 3 & 1 \end{pmatrix}, \begin{pmatrix} 1 & 2 & 3 \\ 3 & 1 & 2 \end{pmatrix}, \begin{pmatrix} 1 & 2 & 3 \\ 3 & 2 & 1 \end{pmatrix} \right\}.$$

Write the "composition table" for A, that is, the table for the operation $\circ$. [Hints: For an analogy, see Figures 2.1, 2.4, 2.5, and 3.3. For instance, one entry of the table will represent the composition

$$\begin{pmatrix} 1 & 2 & 3 \\ 1 & 3 & 2 \end{pmatrix} \circ \begin{pmatrix} 1 & 2 & 3 \\ 2 & 1 & 3 \end{pmatrix},$$

which, working from right to left, is

$$\begin{pmatrix} 1 & 2 & 3 \\ 3 & 1 & 2 \end{pmatrix}.]$$

Find the inverse of each element of A.

Answer the same questions when $S := \{1, 2\}$ and when $S := \{1\}$.

6. Write the elements of the symmetric group on $\{0, 1, 2, 3\}$. Write the inverses of several elements of the group. For several pairs of elements of the group, calculate their composition in each of the two possible orders.

7. Using the computer, for each of the preceding complements from problem 1 to 4, assign concretely the underlying set (such as A, or U, or S). Produce the composition table of the structure, and find the inverse of each element, when one exists. Watch out for combinatorial explosion. This exercise is seldom possible by hand because the power set or the symmetric group of a set gets big quickly in terms of the cardinality of the set.

8. Complete the details of Section 8.1.5 (group of the rectangle and group of the tetrahedron).

9. In analogy to Section 8.1.5, define and construct the composition table of the *group of the triangle* (for an equilateral triangle), the *group of the square,* the *group of the cube.*

10. Suppose $(A, *)$ is a semigroup. Suppose that for any elements a and b in A there exist x and y in A such that $a * x = b$ and $y * a = b$. **(a)** Show that $(A, *)$ must possess an identity and thus must be a monoid. **(b)** Show that every element of A must possess an inverse, and thus $(A, *)$ must be a group. **(c)** Use these results to state an alternative definition of a group.

11. Given two groupoids (A, f) and (B, g), a **homomorphism** from (A, f) to (B, g) is a total function $h: A \to B$ such that

(1) for all x, y in A, $h(xfy) = (h(x))g(h(y))$;
(2) if f has identity u and if g has identity v, then $h(u) = v$.
Note that it is possible that (image of A) $= B$, but it is also possible that (image of A) $\subset B$. (Greek *homo-* = similar, *morph-* = form, hence *homomorph-* = similar form.)

(a) Show that, if the two groupoids are groups, then condition (2) above is automatically satisfied. **(b)** Show that the function $x \mapsto (x \pmod m)$ is a homomorphism from $(\mathbb{N}, +, 0)$ to $(\mathbb{Z}_m, + \bmod m, 0 \bmod m)$. What is the mathematical term for this homomorphism? **(c)** Show that the correspondence $h: \mathbb{R} \to \mathbb{R}$, $x \mapsto 2^x$ is a homomorphism from the monoid $(\mathbb{R}, +, 0)$ to the monoid $(\mathbb{R}, \cdot, 1)$.

12. Given two groupoids (A, f) and (B, g), an **isomorphism** from (A, f) to (B, g) is defined as a bijective homomorphism. If such an isomorphism exists, then we say that (A, f) and (B, g) are **isomorphic. (a)** Show that, under these conditions, $zgy = h((h^{-1}z)f(h^{-1}y))$. **(b)** Show that if $(A, f, {}', u)$ and $(B, g, {}^{\bullet}, v)$ are groups and if h is an isomorphism from the first one to the second one, then $h(x') = (h(x))^{\bullet}$ for every x in A. Equivalently: ${}^{\bullet} = h \circ {}' \circ h^{-1}$.

13. Show that the correspondence $h: \mathbb{R} \to \{x \in \mathbb{R} \mid x > 0\}$, $x \mapsto 2^x$ is an isomorphism from the group $(\mathbb{R}, +, {}^{-}, 0)$ to the group $(\{x \in \mathbb{R} \mid x > 0\}, \cdot, {}^{-1}, 1)$.

14. Consider the monoids, $(A, \vee, 0)$, $(A, \wedge, 1)$ from a Boolean algebra $(A, \vee, \wedge, \sim, 0, 1)$. Find an isomorphism from $(A, \vee, 0)$ to $(A, \wedge, 1)$. What would you call such an isomorphism?

8.2 Structures with Two Operations

To discuss mathematical structures with two operations, we normally begin with two single-operation structures (groupoids, semigroups, monoids, or groups) defined on the same set. For instance, addition and multiplication are both defined for integers. The two structures need not have the same degree of richness; in fact, they usually don't. For this discussion, we'll use a set A, and two operations, which we'll denote in the abstract by $\oplus$ and $*$. Remember, though, that these are abstractions. You should not worry (except in concrete examples) what $\oplus$ and $*$ really are. They aren't really anything—yet. In particular, we're going to treat $\oplus$ as though it were an *additive* or *addition-like* operation, and we'll use 0 to denote its identity (if it has one). Similarly, we'll treat $*$ as a *multiplicative* or *multiplication-like* operation, with identity 1 (if it has one). But $\oplus$ and $*$ are *not* necessarily addition and multiplication, although they are often read as if they were. Also, 0 and 1 are *not* necessarily "zero" and "one," although they are usually read so. The notation is merely suggestive, not definitive.

There are two distinct classes of two-operation structures we want to discuss, rings in one direction and Boolean algebras (which we've already seen) in the other direction. They differ in a fundamental way, as we'll see. But they have a common point of departure, which we present now as the most basic two-operation system, a semiring.

8.2.1 Semiring

Briefly, a *semiring* is a set A with an "addition" and a "multiplication"; the addition is commutative and associative and has an additive identity; the multiplication is associative (but not necessarily commutative) and distributes over addition. For example, the structure $(\mathbb{Z}, +, \cdot, 0)$ (that is, the set of the integers with ordinary addition and multiplication) is a semiring in which the additive identity is 0. This semiring has also a multiplicative identity, which is 1.

Definition 11 A **semiring** is a quadruple $(A, \oplus, *, 0)$, where

- A is a set,
- $\oplus$ is a total binary operation on A,
- $\oplus$ is associative,
- $\oplus$ is commutative,
- $0 \in A$,
- 0 is an identity for operation $\oplus$,
- $*$ is a total binary operation on A,
- $*$ is associative, and
- $*$ distributes over $\oplus$; that is, for every choice of x, y, and z in A,

$$z * (x \oplus y) = (z * x) \oplus (z * y), \quad \text{and}$$
$$(x \oplus y) * z = (x * z) \oplus (y * z).$$

Equivalently, $(A, \oplus, 0)$ is an abelian monoid, $(A, *)$ is a semigroup, and $*$ distributes over $\oplus$. A semiring is called **commutative** if operation $*$ is commutative. A semiring is said to have an **identity** (understood, for operation $*$) if operation $*$ has an identity.

Example Let A be the set of all 2×2 matrices with entries that are natural numbers, that is, the set of all the arrays $\begin{pmatrix} a & b \\ c & d \end{pmatrix}$ in which a, b, c, d are natural numbers. Let $\oplus$ denote matrix addition, that is,

$$\begin{pmatrix} a & b \\ c & d \end{pmatrix} \oplus \begin{pmatrix} e & f \\ g & h \end{pmatrix} := \begin{pmatrix} a+e & b+f \\ c+g & d+h \end{pmatrix}$$

where the usual symbol $+$ has the usual meaning of integer addition. Let 0 stand for the matrix $\begin{pmatrix} 0 & 0 \\ 0 & 0 \end{pmatrix}$. Let $*$ denote matrix multiplication. That is,

$$\begin{pmatrix} a & b \\ c & d \end{pmatrix} * \begin{pmatrix} e & f \\ g & h \end{pmatrix} := \begin{pmatrix} ae + bg & af + bh \\ ce + dg & cf + dh \end{pmatrix}$$

where a symbol like $ae + bg$ has the usual arithmetic meaning. You may have to study this for a while. Do several concrete numerical examples to get used to matrix multiplication. An example is

$$\begin{pmatrix} 1 & 2 \\ 0 & 1 \end{pmatrix} * \begin{pmatrix} 1 & 0 \\ 2 & 4 \end{pmatrix} = \begin{pmatrix} 5 & 8 \\ 2 & 4 \end{pmatrix}$$

Then we have a semiring with identity $\begin{pmatrix} 1 & 0 \\ 0 & 1 \end{pmatrix}$. Note that $*$ is not commutative.

Note carefully the lack of richness in $(A, *)$. It need not be abelian and need not possess an identity.

8.2.2 Ring

A semiring whose "addition" has an inverse is called a *ring*. Example. The structure $(\mathbb{Z}, +, \cdot, 0)$ that we saw before as a semiring is actually a ring because each element has an additive inverse. In fact, this is a commutative ring because multiplication is commutative. Formally,

Definition 12 A **ring** is a semiring for which the operation $\oplus$ has an inverse. If we denote by ^-x the inverse of x, then a ring is a quintuple $(A, \oplus, *, ^-, 0)$, where $(A, \oplus, *, 0)$ is a semiring and $(A, \oplus, ^-, 0)$ is an abelian group.

Example We can alter the example of the semiring of 2×2 matrices (Section 8.2.1) slightly by allowing a, b, c, d to be integers (not just natural numbers) and by defining the additive inverse of a matrix as

$$^-\begin{pmatrix} a & b \\ c & d \end{pmatrix} := \begin{pmatrix} ^-a & ^-b \\ ^-c & ^-d \end{pmatrix}$$

where ^-a denotes the usual "negative a," $-a$. Then we have a ring with identity $\begin{pmatrix} 1 & 0 \\ 0 & 1 \end{pmatrix}$, in which $*$ is still noncommutative.

Theorem 13 Let $(A, \oplus, *, ^-, 0)$ be a ring. Then,

(1) **Annihilator property.** If $x \in A$, then $0 * x = 0$ and $x * 0 = 0$.

(2) **Rules of signs.** If $x \in A$ and if $y \in A$, then (2a) $(^-x) * y = {}^-(x * y)$, (2b) $x * {}^-y = {}^-(x * y)$, (2c) $(^-x) * (^-y) = x * y$.

(3) **Distinctness of the identities.** If $A \neq \{0\}$ and if the ring has identity 1, then $0 \neq 1$.

Proof: (1) The distributive and the identity property imply $x * 0 + x * x = x * (0 + x) = x * x = 0 + x * x$. Part 3 of Theorem 10 implies $x * 0 = 0$. A similar argument is valid for the other part of the statement. The annihilator property is so proved.

(2) The distributive and the identity property, together with the annihilator property, imply $x * y + ((^{-}x) * y) = (x + {}^{-}x) * y = 0 * y = 0$ and $x * y + (x * {}^{-}y) = x * (y + {}^{-}y) = x * 0 = 0$. Then part 4 of Theorem 8 implies the first two rules of signs, that is, (2a) and (2b). Replacing y with ^{-}y in (2a) and using (2b) proves (2c).

(3) If $1 = 0$ and if x is an element of A, then $x = x * 1 = x * 0 = 0$, so $x = 0$ and the only element in A is 0. We have so proved the contrapositive of statement 3 and therefore also statement 3 itself. Q.E.D.

8.2.3. Boolean Algebras

The two things you must remember about rings both revolve around inverses for $\oplus$ and $*$. First, our constant effort is to define inverses for these operations. Second, and most important, *we always fail in one crucial case*. The $\oplus$ identity 0 *never* has an inverse under $*$. Thus, we must always distinguish $\oplus$ and $*$ not only in their meaning, but also in their structure. They differ in syntax (form) as well as semantics (structure).

Now consider the second type of semiring we want to study, typified by set theory, propositional calculus, and two-terminal switching networks. Here we're interested, not in the definition of inverses, but rather of *complements* (set complements, statement negations, and NOT-networks). Rings have no notion of complement. Similarly, Boolean structures lack inverses. It's really a case of having one or the other. The clearest example of this type of structure, propositional calculus, was also the first to be studied, by the British mathematician George Boole (1815–1864); the abstraction thus bears his name. The amazing thing is that rings and Boolean algebras, two very different types of semirings, have anything in common at all. The semiring is thus of far more fundamental importance than simply being a stepping-stone to either arithmetics or Boolean algebras individually.

We've already studied Boolean algebra and its typical applications in Chapter 6. For the abstract definition and some of the derived properties see Section 6.3. Note that a Boolean algebra is a semiring and that the monoids $(A, \vee, 0)$ and $(A, \wedge, 1)$ have the same "richness," being "dual" of each other. Actually, a Boolean algebra is a semiring in two ways: $(A, \vee, \wedge, 0, 1)$ is a commutative semiring with identity 1, and $(A, \wedge, \vee, 1, 0)$ is a commutative semiring with identity 0.

Exercises 8.2

1. Can semirings or rings be constructed using the sets $\mathbb{N}$, $\mathbb{Z}$, $\mathbb{Z}_m$, $\mathbb{Q}$, and $\mathbb{R}$, and their usual forms of addition and multiplication? Discuss all possibilities.
2. For the two examples in Sections 8.2.1 and 8.2.2, verify that each example actually satisfies the conditions of the definition above it. Verify that the first structure cannot be "upgraded" to a ring.

3. Each of the following structures with two operations is given by means of set S and two binary operations on S. See if the structure is actually a semiring (If so, what's the additive identity?), a ring, a ring with (multiplicative) identity, a commutative ring. Justify your answers. **(a)** $S := \mathbb{Z}_5$, operation $+$ mod 5, operation $\cdot$ mod 5. **(b)** $S := \{0, 1\}$ (Boolean), operation $\vee$, operation $\wedge$. **(c)** $A := \{0, 1, 2\}$, $S := \mathcal{P}A$, operation $\cup$, operation $\cap$. **(d)** $S := \{00, 01, 10, 11\}$, operation BITWISE OR, operation BITWISE AND.

4. In a ring, define $x - y := x \oplus (^-y)$. The **binary minus** is so defined in terms of the **unary minus.** Prove that, for any elements x, y, and z of a ring **(1)** $x - x = 0$; **(2)** $x - (^-y) = x \oplus y$; **(3)** $x - (y \oplus z) = (x - y) - z$; **(4)** $x * (y - z) = (x * y) - (x * z)$; **(5)** $(y - z) * x = (y * x) - (z * x)$.

5. Is the statement $x^2 - y^2 = (x - y) * (x + y)$ true in every ring? Discuss the two possibilities.

6. The goal of this exercise is to become familiar with different kinds of "algebras." Expand each of the following expressions; reduce like terms. Assume that you are operating in a commutative ring. Then do the same problem all over again by assuming that you are in a noncommutative ring. There will be discrepancies! Note that the "multiplication" symbol $*$ is omitted here, as customary when there is no ambiguity. **(a)** $x + y - x - y$; **(b)** $xy - yx$; **(c)** $(x + y)^2$; **(d)** $(x - y)(x + y)$; **(e)** $(x + y)^3$; **(f)** $x^2(x + y^2)xy^2$; **(g)** $(x + y)^2 - (x - y)^2$. In some calculations, you may have written something like $2xy$. What does it really mean? See Definition 5 and adapt it to the additive case; see also Complements 3.4

7. Let K be a given alphabet. Define $H := K^*$ (that is, the free semigroup of K = {finite strings over K}). Define $A := \mathcal{P}H$ (= {languages over K}); see Section 6.1. Note that two special elements of A are the empty language $\varnothing$ and the language $\{\Lambda\}$ comprising only the null string Λ. On A, we've defined two binary operations between languages: *union* (symbol $\cup$) and *concatenation* (of languages; symbol :). Verify that $(A, \cup, :, \varnothing)$ is a semiring with identity $\{\Lambda\}$.

8. Let $\{0, 1, a\}$ be a set, on which we have the appropriate operations to obtain a commutative ring. The element a is distinct from 0 and 1. How must the operations be defined? [Hint: the definitions of the operations must satisfy the properties of a commutative ring. There isn't much choice!]

9. Answer the same question as in problem 8 for the set $\{0, 1\}$.

10. Review Section 6.3.

11. Let $\{\{0, 1, a, b\}, \vee, \wedge, ', 0, 1\}$ be a Boolean algebra, where a and b are two elements distinct from each other and from 0 and 1. The fact that we have a Boolean algebra is given, but the operations $\vee$, $\wedge$, and $'$ are not given explicitly. How must such operations be defined? [Hint: as a consequence of the axioms of a Boolean algebra, there are *very* few possibilities.]

Complements 8.2

1. *Homomorphisms* and *isomorphisms* between semirings are defined in analogy to the case of groupoids (see Complements 8.1), but the condition expressed in complement 11 of that section must be valid for both operations. Give an explicit definition of each morphism in the case of semirings or more special structures. When $m > 1$, find a homomorphism from commutative ring $(\mathbb{Z}, +, \cdot, ^-, 0)$ with identity 1 to commutative ring $(\mathbb{Z}_m, +, \cdot, ^-, 0)$ with identity 1 (mod m). For Boolean algebras, find an isomorphism from semiring $(\{0, 1\}, \vee, \wedge, 0)$ to semiring $(\{0, 1\}, \wedge, \vee, 1)$. Note the order of the symbols!
2. Extend the matrix example of Section 8.2.1 to the following cases: **(a)** The entries of the matrices are elements of the Boolean algebra $(\{0, 1\}, \vee, \wedge, \sim, 0, 1)$. [Hint: replace addition with $\vee$, multiplication with $\wedge$.] **(b)** The entries of the matrices are elements of any Boolean algebra. **(c)** The entries of the matrices are subsets of a given nonempty set U. [Hint: replace addition with $\cup$, multiplication with $\cap$.] In each case, investigate operations, identities, inverses, etc. Do we obtain semirings or rings? Are they commutative? Do they have an identity? In cases b and c, select a concrete Boolean algebra or a concrete set U and in all three cases practice concrete operations.

8.3 Matrices

Informally, a matrix is a rectangular or square table like any of these four:

$$\begin{bmatrix} a_{11} & a_{12} & a_{13} \\ a_{21} & a_{22} & a_{23} \end{bmatrix} \qquad \begin{bmatrix} a_{00} & a_{01} & a_{02} \\ a_{10} & a_{11} & a_{12} \end{bmatrix}$$

$$\begin{bmatrix} a_{11} & a_{12} & \cdots & a_{1k} \\ a_{21} & a_{22} & \cdots & a_{2k} \\ \cdots & \cdots & \cdots & \cdots \\ a_{h1} & a_{h2} & \cdots & a_{hk} \end{bmatrix} \qquad \begin{bmatrix} a_{00} & a_{01} & \cdots & a_{0,k-1} \\ a_{10} & a_{11} & \cdots & a_{1,k-1} \\ \cdots & \cdots & \cdots & \cdots \\ a_{h-1,0} & a_{h-1,1} & \cdots & a_{h-1,k-1} \end{bmatrix}$$

The entries a_{ij} are usually numbers, but not always. Such a table can be conceived as a *sequence of sequences*—either as a sequence of h rows, each of which is a sequence of k elements, or as a sequence of k columns, each of which is a sequence of h elements. A usual name for this setting is *double array*. We can formalize the idea in a way similar to that for sequences in Section 7.2.4.1.

Definition 14 Given a set S and two positive natural numbers h and k, let h represent also the set $\{x \in \mathbb{N} \mid x < h\}$; let k represent also the set $\{x \in \mathbb{N} \mid x < k\}$. Then an **$h \times k$ matrix over S** is a total function $a: h \times k \to S$. When $i \in h$ and $j \in k$: $a(i, j)$ is called the **entry in position (i, j)** and is often written also $a_{i,j}$ or a_{ij}. The sequence $(a(x, j))_{x \in h}$ is **column j**; the sequence $(a(i, y))_{y \in k}$ is **row i**. The matrix a itself is denoted $(a(i, j))_{i \in h,\, j \in k}$ or $(i, j) \mapsto a(i, j)$ or similarly. If $h = k = n$, then matrix a is said to be a square matrix of **order n**, or of **size n**.

The most interesting cases occur when the underlying set S has some algebraic structure, for instance when $S = \mathbb{N}$ or $S = \mathbb{Z}$. Then we can define algebraic operations on matrices.

Example Given

$$a = \begin{bmatrix} 5 & 3 & 0 \\ 4 & 1 & 2 \\ 0 & 0 & 2 \end{bmatrix}$$

we have $h = 3, k = 3$ and the entries of matrix a are integers, for instance, $a(1, 1) = 5$, $a(1, 2) = 3$, $a(2, 1) = 4$, $a(2, 2) = 1$; row 1 is $[5 \quad 3 \quad 0]$, column 3 is

$$\begin{bmatrix} 0 \\ 2 \\ 2 \end{bmatrix},$$

and so on. In this case, we can define addition and multiplication of matrices. An example of addition is

$$\begin{bmatrix} 5 & 3 & 0 \\ 4 & 1 & 2 \\ 0 & 0 & 2 \end{bmatrix} + \begin{bmatrix} 1 & 2 & 1 \\ 0 & 1 & 3 \\ 1 & 1 & 2 \end{bmatrix} = \begin{bmatrix} 6 & 5 & 1 \\ 4 & 2 & 5 \\ 1 & 1 & 4 \end{bmatrix}.$$

We have simply added together entries that occupy corresponding positions in the two given matrices, and we have written the result in the same position in the resulting matrix. Clearly, we can do this operation only if the two matrices have the same sizes.

Multiplication is a little less evident. An example is

$$\begin{bmatrix} 3 & 5 & 7 \\ 1 & 0 & 3 \\ 5 & 1 & 2 \end{bmatrix} * \begin{bmatrix} 1 & 5 & 3 \\ 3 & 1 & 0 \\ 1 & 2 & 5 \end{bmatrix}$$

$$= \begin{bmatrix} 3\cdot 1 + 5\cdot 3 + 7\cdot 1 & 3\cdot 5 + 5\cdot 1 + 7\cdot 2 & 3\cdot 3 + 5\cdot 0 + 7\cdot 5 \\ 1\cdot 1 + 0\cdot 3 + 3\cdot 1 & 1\cdot 5 + 0\cdot 1 + 3\cdot 2 & 1\cdot 3 + 0\cdot 0 + 3\cdot 5 \\ 5\cdot 1 + 1\cdot 3 + 2\cdot 1 & 5\cdot 5 + 1\cdot 1 + 2\cdot 2 & 5\cdot 3 + 1\cdot 0 + 2\cdot 5 \end{bmatrix}$$

$$= \begin{bmatrix} 25 & 34 & 44 \\ 4 & 11 & 18 \\ 10 & 30 & 25 \end{bmatrix}.$$

This requires some detailed explanation. How did we obtain, for instance, entry (1, 3) of the result? We considered row 1 of the first matrix and column 3 of the second matrix:

$$\begin{matrix} 3 & 5 & 7 \end{matrix} \qquad \begin{matrix} 3 \\ 0 \\ 5 \end{matrix}$$

Then we multiplied together the first entries $(3 \cdot 3)$, the second entries $(5 \cdot 0)$, and the third entries $(7 \cdot 5)$ of the row and column in the diagram above. Finally, we added together the three results $(3 \cdot 3 + 5 \cdot 0 + 7 \cdot 5 = 44)$.

A case in which we have the same process in everyday activities is the following. Suppose there are three items that can be ordered by mail: item a costs \$3, item b costs \$5, and item c costs \$7. You want to order 3 items a, no item b, and 5 items c. What's the total cost? Clearly, $3 \cdot 3 + 5 \cdot 0 + 7 \cdot 5$; it's precisely the same operation.

See also the examples in Sections 8.2.1 and 8.2.2.

Clearly, multiplication of matrices can be done only if the number of columns of the first matrix equals the number of rows of the second matrix.

Such operations can be extended to much more general cases. What we need is "addition" and "multiplication" of entries. Since the entries are taken from set S, we need an algebra on S. A useful structure for S is a semiring, or a richer structure. We'll devote our attention to this case and limit ourselves for the time being to square matrices only. Furthermore, we'll use for the operations in S the usual symbols $+$ and $\cdot$ although the operations will not always be addition and multiplication. For the operations on matrices we'll use the symbols $\oplus$ and $*$, although (in most concrete cases) the usual symbols are $+$ for addition and none for multiplication.

Definition 15 Given a semiring $T := (S, +, \cdot, 0_S)$ and a positive natural number n, let n denote also the set $\{x \in \mathbb{N} \mid x < n\}$, i. e., $\{0, 1, \ldots, n - 1\}$. Then $S^{n \times n}$ is the set of all **$n \times n$ matrices** over S. If a and b are any matrices in $S^{n \times n}$, then the **sum** of a and b is denoted $a \oplus b$ and is defined as the matrix $(i, j) \mapsto a(i, j) + b(i, j)$. That is, entry (i, j) of the matrix sum is the sum of entries (i, j) of the addend matrices. The **product** of a and b is denoted $a * b$ and is defined as the matrix $(i, j) \mapsto \sum_{t=0}^{n-1} a(i, t) \cdot b(t, j)$. The *null* matrix, or *zero* matrix, is denoted 0_n and is defined as the matrix $(i, j) \mapsto 0_S$. If, in particular, semiring T has an additive inverse$^-$, then the **additive inverse** of matrix a is denoted ^-a and is defined as the matrix $(i, j) \mapsto {}^-a(i, j)$. If semiring T has multiplicative identity u_S, then the **identity matrix** is denoted I_n and is defined as the matrix $(i, j) \mapsto u_S$ if $i = j$ and $(i, j) \mapsto 0_S$ if $i \neq j$.

The definition of product of matrices may offer some difficulty at first. Try to visualize the operation: to construct entry (i, j) of the result, we use only row i of the first factor and column j of the second factor

$$\begin{matrix} & b(0, j) \\ & b(1, j) \\ a(i, 0) \quad a(i, 1) \quad \ldots \quad a(i, n - 1) & \vdots \\ & b(n - 1, j) \end{matrix}$$

and we perform the *row-by-column* operation $a(i, 0) \cdot b(0, j) + a(i, 1) \cdot b(1, j) + \cdots + a(i, n-1) \cdot b(n-1, j)$, just as we did in the example of a product of matrices over the integers.

Theorem 16 If $T := (S, +, \cdot, 0_S)$ is a semiring, then $(S^{n\times n}, \oplus, *, 0_n)$ as defined above is a semiring. If, in addition, operation $+$ has an inverse $^-$, then $(S^{n\times n}, \oplus, *, ^-, 0_n)$ is a ring. If, in addition, operation $\cdot$ has an identity u_S, then ring $(S^{n\times n}, \oplus, *, ^-, 0_n)$ has identity I_n.

Proof: Exercise. All you have to do is look up the definitions and patiently check the required properties.

The rest of Section 8.3 is devoted to examples. In these examples, we'll often use the symbol $+$ for matrix addition (instead of $\oplus$) and no symbol for matrix multiplication.

8.3.1 Integer and Real Matrices

The "usual" cases are those in which T is the commutative ring of the integers or of the reals. Then the structure of the $n \times n$ matrices is a ring, in which the null matrix 0_n contains all zeroes and the identity matrix I_n all zeroes except for ones in positions (i, i). See the example after Definition 14.

8.3.2 Boolean Matrices

A rather simple—but perhaps confusing—case occurs when T is a Boolean algebra, for instance the elementary Boolean algebra $(\{0, 1\}, \vee, \wedge, \sim, 0, 1)$. Then the entries of the matrices are Boolean zeroes and ones. An example of "addition" follows.

$$\begin{bmatrix} 1 & 0 & 1 \\ 1 & 1 & 1 \\ 0 & 1 & 0 \end{bmatrix} + \begin{bmatrix} 1 & 1 & 0 \\ 1 & 0 & 1 \\ 0 & 0 & 0 \end{bmatrix} = \begin{bmatrix} 1 \vee 1 & 0 \vee 1 & 1 \vee 0 \\ 1 \vee 1 & 1 \vee 0 & 1 \vee 1 \\ 0 \vee 0 & 1 \vee 0 & 0 \vee 0 \end{bmatrix} = \begin{bmatrix} 1 & 1 & 1 \\ 1 & 1 & 1 \\ 0 & 1 & 0 \end{bmatrix}$$

An example of "multiplication" is the following:

$$\begin{bmatrix} 1 & 0 & 1 \\ 1 & 1 & 1 \\ 0 & 1 & 0 \end{bmatrix} \begin{bmatrix} 1 & 1 & 0 \\ 1 & 0 & 1 \\ 0 & 0 & 0 \end{bmatrix}$$

$$= \begin{bmatrix} (1\wedge 1)\vee(0\wedge 1)\vee(1\wedge 0) & (1\wedge 1)\vee(0\wedge 0)\vee(1\wedge 0) & (1\wedge 0)\vee(0\wedge 1)\vee(1\wedge 0) \\ (1\wedge 1)\vee(1\wedge 1)\vee(1\wedge 0) & (1\wedge 1)\vee(1\wedge 0)\vee(1\wedge 0) & (1\wedge 0)\vee(1\wedge 1)\vee(1\wedge 0) \\ (0\wedge 1)\vee(1\wedge 1)\vee(0\wedge 0) & (0\wedge 1)\vee(1\wedge 0)\vee(0\wedge 0) & (0\wedge 0)\vee(1\wedge 1)\vee(0\wedge 0) \end{bmatrix}$$

$$= \begin{bmatrix} 1 & 1 & 0 \\ 1 & 1 & 1 \\ 1 & 0 & 1 \end{bmatrix}$$

8.3.3 Matrices mod m

Another case (to be contrasted with the previous one, in spite of the similar appearance) occurs when T is the commutative ring of the integers mod 2. The entries of the matrices are still zeroes and ones and the operations are nearly the same *except* that $1 + 1 = 0$ now. In the previous example, $1 + 1$ (that is, $1 \vee 1$) was 1. Examples of addition and multiplication of matrices mod 2 follow.

$$\begin{bmatrix} 1 & 0 & 1 \\ 1 & 1 & 1 \\ 0 & 1 & 0 \end{bmatrix} + \begin{bmatrix} 1 & 1 & 0 \\ 1 & 0 & 1 \\ 0 & 0 & 0 \end{bmatrix} = \begin{bmatrix} 0 & 1 & 1 \\ 0 & 1 & 0 \\ 0 & 1 & 0 \end{bmatrix}$$

$$\begin{bmatrix} 1 & 0 & 1 \\ 1 & 1 & 1 \\ 0 & 1 & 0 \end{bmatrix} \begin{bmatrix} 1 & 1 & 0 \\ 1 & 0 & 1 \\ 0 & 0 & 0 \end{bmatrix} = \begin{bmatrix} 1 & 1 & 0 \\ 0 & 1 & 1 \\ 1 & 0 & 1 \end{bmatrix}$$

In general, when $m > 1$, we can consider the ring of the matrices with entries in $\mathbb{Z}_m$.

8.3.4 Path Matrices

Path matrices are interesting for two reasons: theoretically, they present a case in which algebraic, set-theoretical, and string structures are layered one on the top of each other; and practically, they will be a tool to solve *path* problems in structures like *graphs* (see Section 11.2). Let K be a given, nonempty alphabet. Let $A := K^*$ = set of all finite strings of elements of K, including the null string Λ. Let $S := \mathcal{P}A$ = collection of all subsets of A = set of all languages over K, including the empty set $\varnothing$. Let L and M be any two elements of S, that is, any two subsets of A or, equivalently, any two languages over K. We are going to consider two operations we have already seen: (1) union, $L \cup M := \{x \in A \mid x \in L \vee x \in M\}$ and (2) concatenation, $L{:}M := \{x{:}y \in A \mid x \in L \wedge y \in M\}$.

$$\text{Then } (S, \cup, :, \varnothing) \text{ is a semiring with identity } \{\Lambda\}. \tag{1}$$

The verification is straightforward; do it as an exercise. Consequently, we can consider $n \times n$ matrices with entries in S. As strange as it may sound, their entries will be sets of strings, that is, languages. For instance, assuming $K :=$ {a, b, c, . . ., z},

$$\begin{bmatrix} \{\text{aa}, \text{abc}\} & \{\text{a}, \text{bx}\} & \{\Lambda\} \\ \varnothing & \{\text{abx}, \text{xaa}\} & \varnothing \\ \varnothing & \{\Lambda\} & \varnothing \end{bmatrix}$$

In particular, 0_n will have all entries equal to $\varnothing$ and I_n will have entries $\{\Lambda\}$ when $i = j$ and $\varnothing$ otherwise. Then $(S^{n \times n}, \oplus, *, 0_n)$ is a semiring with identity I_n. The proof of the last statment is just a consequence of statement (1) and Theorem 16. (When you already possess several basic properties, many proofs of advanced statements become this quick!)

Exercises 8.3

1. Practice numerical operations on 3×3 and 4×4 matrices over the integers, over $\mathbb{R}$, over $\mathbb{Z}_m$.
2. Let a, b, c be unspecified square matrices of the same size over the same unspecified semiring. Practice the following algebraic manipulations on them. Recall that you are operating in a semiring, so that addition is commutative and associative and has a zero; multiplication is associative but not commutative; multiplication distributes over addition; and expressions like a^2, a^3 (but not a^{-1} or a^{-2}) are "legal." Verify that **(a)** $a^2(a + b) = a^3 + a^2b$; **(b)** $(a + b)^2 = a^2 + ab + ba + b^2$ (Watch out!); **(c)** $a^2a^3 = a^5$; **(d)** $(a^2)^3 = a^6$. Expand **(e)** $(a + b + c)^2$; **(f)** $(a + b)^3$. Factor **(g)** $a^3b^2a^3 + ab^2a^2$, **(h)** $a^3b^2 + a^2b$.
3. Prove Theorem 16 (on a semiring of matrices).
4. Practice several calculations with concretely given matrices in each of the cases of Sections 8.3.1 to 8.3.4. For instance, calculate

(a) (over $\mathbb{Z}$) $\begin{bmatrix} 2 & 3 \\ 1 & 5 \end{bmatrix} + \begin{bmatrix} 1 & 0 \\ 1 & 7 \end{bmatrix}$; $\begin{bmatrix} 2 & 3 \\ 1 & 5 \end{bmatrix} * \begin{bmatrix} 1 & 0 \\ 1 & 7 \end{bmatrix}$.

(b) (over $\mathbb{Z}$) $\begin{bmatrix} 1 & 0 \\ 1 & 3 \end{bmatrix} + \begin{bmatrix} 1 & 1 \\ 1 & 2 \end{bmatrix}$; $\begin{bmatrix} 1 & 1 \\ 1 & 5 \end{bmatrix} * \begin{bmatrix} 0 & 1 \\ 1 & 7 \end{bmatrix}$.

(c) (over $\mathbb{Z}$) $\begin{bmatrix} 0 & 1 & 5 \\ 1 & 0 & 2 \\ 0 & 0 & 2 \end{bmatrix} * \begin{bmatrix} 2 & 3 & 4 \\ 1 & 1 & 0 \\ 0 & 1 & 1 \end{bmatrix}$.

(d) (over $\mathbb{Z}_5$) $\begin{bmatrix} 1 & 0 \\ 1 & 3 \end{bmatrix} + \begin{bmatrix} 1 & 1 \\ 1 & 4 \end{bmatrix}$; $\begin{bmatrix} 1 & 1 \\ 1 & 3 \end{bmatrix} * \begin{bmatrix} 0 & 1 \\ 1 & 2 \end{bmatrix}$.

(e) (over $\mathbb{Z}_5$) $\begin{bmatrix} 2 & 1 \\ 1 & 3 \end{bmatrix} + \begin{bmatrix} 1 & 3 \\ 1 & 4 \end{bmatrix}$; $\begin{bmatrix} 2 & 1 \\ 3 & 3 \end{bmatrix} * \begin{bmatrix} 0 & 1 \\ 1 & 2 \end{bmatrix}$.

(f) (over $\mathbb{Z}_5$) $\begin{bmatrix} 0 & 2 & 2 \\ 1 & 2 & 0 \\ 0 & 1 & 0 \end{bmatrix} * \begin{bmatrix} 2 & 0 & 3 \\ 1 & 1 & 0 \\ 0 & 0 & 1 \end{bmatrix}$.

(g) (Boolean) $\begin{bmatrix} 0 & 1 \\ 1 & 1 \end{bmatrix} \vee \begin{bmatrix} 1 & 1 \\ 1 & 0 \end{bmatrix}$; $\begin{bmatrix} 0 & 1 \\ 1 & 1 \end{bmatrix} * \begin{bmatrix} 0 & 1 \\ 1 & 0 \end{bmatrix}$.

(h) (Boolean) $\begin{bmatrix} 1 & 1 \\ 1 & 1 \end{bmatrix} \vee \begin{bmatrix} 1 & 0 \\ 1 & 0 \end{bmatrix}$; $\begin{bmatrix} 0 & 0 \\ 1 & 1 \end{bmatrix} * \begin{bmatrix} 1 & 1 \\ 1 & 1 \end{bmatrix}$.

(i) (over alphabet $K := \{x, y, z\}$)

$$\begin{bmatrix} \varnothing & \{x\} \\ \{x\} & \varnothing \end{bmatrix} \cup \begin{bmatrix} \{z\} & \varnothing \\ \varnothing & \varnothing \end{bmatrix}; \begin{bmatrix} \{z\} & \{x\} \\ \{x\} & \varnothing \end{bmatrix} * \begin{bmatrix} \{z\} & \{x\} \\ \{x\} & \varnothing \end{bmatrix}.$$

(j) (over alphabet $K := \{x, y, z\}$)

$$\begin{bmatrix} \varnothing & \{x,y\} \\ \{x,y\} & \varnothing \end{bmatrix} \cup \begin{bmatrix} \{z\} & \varnothing \\ \varnothing & \varnothing \end{bmatrix}; \begin{bmatrix} \{z\} & \{x,y\} \\ \{x,y\} & \varnothing \end{bmatrix} * \begin{bmatrix} \{z\} & \{x,y\} \\ \{x,y\} & \varnothing \end{bmatrix}.$$

(k) (over alphabet $K := \{x, y, z\}$)

$$\begin{bmatrix} \varnothing & \{z\} & \{y\} \\ \{z\} & \varnothing & \{x\} \\ \{y\} & \{x\} & \varnothing \end{bmatrix} * \begin{bmatrix} \varnothing & \{z\} & \{y\} \\ \{z\} & \varnothing & \{x\} \\ \{y\} & \{x\} & \varnothing \end{bmatrix}.$$

5. Prove statement (1). See also problem 7 in Exercises 8.2.

Complements 8.3

1. Consider the following path matrix over the English alphabet:

$$p := \begin{bmatrix} \varnothing & \{x\} & \varnothing \\ \{x\} & \varnothing & \{y, z\} \\ \varnothing & \{y, z\} & \varnothing \end{bmatrix}$$

(a) Calculate p^2 and p^3. **(b)** For each path matrix a define the natural number matrix $N(a)$ as the matrix $(i, j) \mapsto \#a(i, j)$. Recall that $a(i, j)$ is a set and that the symbol $\#a$ means "the number of elements of a". Calculate $N(p)$, $[N(p)]^2$, $[N(p)]^3$. Verify that, in this case, $[N(p)]^2 = N(p^2)$, $[N(p)]^3 = N(p^3)$. **(c)** For each natural number matrix m, define the Boolean matrix $B(m)$ as the matrix $(i, j) \mapsto (m(i, j) > 0)$, that is, $(i, j) \mapsto 0$ if $m(i, j) = 0$, $(i, j) \mapsto 1$ if $m(i, j) > 0$. **(d)** Calculate $B(N(p))$, $[B(N(p))]^2$, $[B(N(p))]^3$. Verify that $[B(N(p))]^2 = B(N(p^2))$, $[B(N(p))]^3 = B(N(p^3))$. **(e)** Verify that $B \circ N$: $\{n \times n$ path matrices$\} \to \{n \times n$ Boolean matrices$\}$ is a homomorphism, with respect to the matrix semirings defined above. **(f)** Verify that B: $\{n \times n$ integer matrices$\} \to \{n \times n$ Boolean matrices$\}$ is a homomorphism with respect to the matrix semirings defined above. **(g)** Check that N is not, in general, a homomorphism.

2. Bitwise operations. (See also Exercises 6.3.) Given the positive integer n, define $S_n := \{0, 1\}^n$ = set of all n-sequences of Boolean zeroes and ones. An element p of S_n is therefore a function $i \mapsto p(i)$, $(0 \le i < n, p(i) \in \{0, 1\})$. Let the operations of the elementary Boolean algebra on set $\{0, 1\}$ be denoted now OR, AND, NOT (we'll need the usual symbols for another purpose). Given that p and q are such sequences, define **(1)** $p \vee q$ as the sequence $i \mapsto p(i)$ OR $q(i)$; **(2)** $p \wedge q$ as the sequence $i \mapsto p(i)$ AND $q(i)$; and **(3)** $\sim p$ as the sequence $i \mapsto$ NOT $p(i)$. The usual names for these operations are, respectively, BITWISE OR, BITWISE AND, BITWISE NEGATION.

For example, $01011 \vee 11001 = 110011$, $01011 \wedge 11001 = 01001$, $\sim 01011 = 10100$.

Prove that (S_n, $\vee$, $\wedge$, $\sim$, n-sequence of all zeroes, n-sequence of all ones) is a Boolean algebra with 2^n elements. This Boolean algebra is called the **direct product** of n elementary Boolean algebras.

3. Let data be as in the previous complement. Let p XOR q be defined as $((p \vee q) \wedge \sim(p \wedge q))$. This operation is the *bitwise exclusive* OR. For example, 01011 XOR $11001 = 10010$. Let I be the identity correspondence; that is, $I(x) =$

x for every x. Prove that (S_n, XOR, $\wedge$, I, n-sequence of all zeroes) is a commutative ring with identity n-sequence of all ones.

4. Let B_n be the set of all n-sequences of zeroes and ones, interpreted as elements of $\mathbb{Z}_2$. Given that p and q are such sequences, define **(1)** p PLUS q as the sequence $i \mapsto p(i) + q(i) \bmod 2$; **(2)** p TIMES q as the sequence $i \mapsto p(i)q(i) \bmod 2$; and **(3)** NEG p as the sequence p. These operations are, respectively, *bitwise addition mod 2, bitwise multiplication mod 2,* and identity. Note that bitwise addition can be interpreted as binary addition without carrying. Examples: 01011 PLUS 11001 = 10010, 01011 TIMES 11001 = 01001, NEG 01011 = 01011.

Prove that (B_n, PLUS, TIMES, NEG, n-sequence of all zeroes) is a commutative ring with the identity n-sequence of all ones.

Let: g: $B_n \to S_n$ be defined as the "natural injection," that is, the function that maps a sequence of zeroes and ones onto the same sequence, except for the interpretation of the symbols (integers mod 2 or Boolean). Prove that g is an isomorphism from the ring (B_n, PLUS, TIMES, NEG, n-sequence of all zeroes) to the ring (S_n, XOR, $\wedge$, I, n-sequence of all zeroes). Practice all of these types of operations on several concrete 4- or 5-bit strings.

8.4 Summary

An *algebraic structure* comprises a *set* and some *operations,* for instance *additive* and *multiplicative operations*. Such operations satisfy some rules and produce a type of algebra on the given set. There are several types of algebraic structures, and each type has its own algebra. We considered some basic structures with one operation and some with two operations. Some structures with essentially one operation are groupoids (Section 8.1.1), with a binary operation $*$, semigroups, (Section 8.1.2), in which operation $*$ is associative, monoids (Section 8.1.3), in which operation $*$ also has an identity, and groups (Section 8.1.4), in which operation $*$ also has an inverse.

A basic structure with essentially two operations is a *semiring* (Section 8.2.1). We considered one of the two operations as an *additive operation* (addition, for short), the other as a *multiplicative operation* (multiplication, for short). The crucial point is that the additive structure is an abelian monoid, the multiplicative structure is a semigroup, and multiplication distributes over addition. If we introduce additive inverses in a semiring, then we obtain a *ring* (Section 8.2.2). If we introduce complements in a semiring and if each operation has an identity, then we obtain Boolean algebras (Sections 6.3 and 8.2.3).

For each structure of these types we considered some of the derived properties, examples, and applications.

Finally, we considered *matrices* whose entries are elements of a given semiring (Section 8.3) and operations on matrices, such as addition and multiplication.

Recursion and Induction

The terms *induction* and *recursion* occur repeatedly throughout mathematics and computer science. As usually applied, they describe two distinct techniques, and in fact, students generally learn about induction, if at all, in a mathematics course, and about recursion, if at all, in a computer science course. But the two are closely related in two important ways. First, as we'll see momentarily, they are basically the same process, which goes under the name of induction when used for proving theorems and recursion when used in constructions or definitions. Second, they have the unfortunate image among students of being vaguely illicit, nonrigorous short-cuts, rather than the solid, rigorous mathematics that they are. In light of their history—especially that of induction—this attitude is perhaps not so surprising.

Therein lies a tale to be told. In philosophical discourse and rhetoric, a sharp distinction has traditionally been made between *deductive* and *inductive* logic. Deductive logic moves "from the general to the specific." That is, specific conclusions are drawn from application of general principles. Within the realm of deductive logic lies the classical syllogism:

All men are mortal.
Socrates is a man.
Therefore, Socrates is mortal.

Deductive logic is solid, rigorous, and unambiguous. It forms the basis for all of the mathematics you've learned thus far. The algebra of propositions we saw is a major part of deductive logic.

Inductive logic, on the other hand, moves "from the specific to the general." Conclusions are drawn, based not on the application of global principles, but on the preponderance of evidence. Inductive logic cannot possibly be as rigorous and conclusive as deductive logic, since it lacks those global principles. On the other hand, if we required the rigor and clarity of mathematics in every aspect of life, we would find ourselves able to conclude practically nothing. In real life, information is neither perfect nor free. Most criminal trials use inductive logic; mathematicians and juries do indeed use the word "proof" in different senses. The great watershed in the history of logic was the medieval debate over whether the sun revolves around the Earth, or vice versa. When Copernicus, Kepler, and Galileo advocated the latter view, they cited as their proof no grand philosophical absolutes, but rather reams of precise, detailed measurements. Against them were arrayed not merely those with political and religious biases, but also the orthodox (and sincere) scientific community, which was as yet unprepared to accept inductive arguments as valid. The vindication of inductive logic was slow, painful, and bitter.

In the midst of this controversy, there arose a technique known today as *mathematical induction*. The terminology is most unfortunate, since mathematical induction has nothing whatever to do with inductive logic; mathematical induction is a deductive technique, not an inductive one. It is as rigorous as the syllogism. It *is* syllogism. The confusion arises innocently enough; induction *seems* to draw general conclusions from specific examples. In fact, though, what mathematical induction does is to rigorously verify conclusions already drawn, perhaps inductively.

In later centuries, mathematics came to depend upon strict and clear definitions of all terms used in its discourse. In cases where explicit definitions were cumbersome or even impossible, mathematicians familiar with formal induction began to devise *recursive* definitions, using similar techniques. Sure enough, recursive definitions became tarred with the same brush: critics decried them as nonrigorous. Once again, acceptance was slow. But recursive definitions are as rigorous as any others and are in some cases a good deal more convenient.

Our discussion of mathematical induction and recursion begins by considering some basic recursive definitions. The aim here is to establish the vocabulary of recursive definitions and to demonstrate that these definitions are as precise and unambiguous as their explicit counterparts. Then, using the same ideas, we'll move on to the more complicated notion of mathematical induction.

9.1 Recursion

In mathematics recursion usually refers to a method of definition. To underscore this, we'll give examples of recursive definitions. You've seen all of these definitions before, but sometimes in explicit rather than recursive form. As you read them, try not to recall or apply what you've learned from previous chapters; that would defeat the purpose here. Rather, ask yourself, "Does this definition tell me *everything* that I need to know all by itself?"

Definition 1 For a positive real number x and a natural number n, the n-th power of x is denoted x^n and is defined by

(1) $x^0 := 1$.
(2) $x^{h+1} := x^h \cdot x$ when $h \in \mathbb{N}$.

Note carefully that this recursive definition occurs in two parts: a **recurrence condition** such as (2) above and an **initial condition** such as (1) above.

The recurrence condition, of course, is the source of the words *recursive* and *recursion* (Latin, *re-* = back, again; *curr-* = run; recur = run back again). Once a power is defined, the next one can be defined in terms of those already done. At first blush, we seem to be defining all of these powers in terms of each other, a clear violation of good logic. But the initial condition makes this valid because the first object, x^0, is defined explicitly as being 1, and all others are defined recursively. For instance, $x^1 := 1 \cdot x$, $x^2 := x \cdot x$, $x^3 := (x \cdot x) \cdot x$. No ambiguity can arise. The initial condition is thus the link between the recurrence condition and the real world.

Definition 2 For a natural number n, the **factorial** of n is denoted $n!$ and is defined by

(1) $0! := 1$.
(2) $(h + 1)! := h!(h + 1)$ when $h \in \mathbb{N}$.

Here again, we have a recursive definition consisting of a recurrence condition (1) and an initial condition (2).

Our next example contains two initial conditions.

Definition 3 Let n and k be natural numbers such that $0 \le k \le n$. The value of the **combinatoric function** for the given values of n, k is denoted $C(n, k)$ and is defined by

(1) $C(n, 0) := 1$ when $n \in \mathbb{N}$.
(2) $C(n, n) := 1$ when $n \in \mathbb{N} - \{0\}$.
(3) $C(h + 1, k) := C(h, k - 1) + C(h, k)$ when $h \in \mathbb{N}$, $k \in \mathbb{N}$, and $0 < k \le h$.

The initial conditions (1) and (2) define explicitly not only $C(0, 0)$, but also $C(1, 0)$, $C(2, 0)$, . . ., $C(1, 1)$, $C(2, 2)$, ... The recurrence condition is (3); it is recurrent on h, for every allowable k. Then, too, notice that we've said nothing about the definition of $C(n, k)$ when $n < 0$ or $k < 0$ or $k > n$. They're simply left undefined. We could modify the definition to make these values zero, but that's unnecessary. Finally, of course, the recurrence condition is the same one we've seen before in Pascal's triangle (Sections 3.6 and 4.6).

All three of these examples involve the definition of a function based on its arguments. In actual practice, we could also say that we *evaluate* these functions recursively. Some computer programming languages, such as ALGOL and Pascal, offer a *recursive subprogram* feature, allowing a subprogram to invoke its own execution repeatedly. (Do not assume this feature in all programming languages!) Here again, it's crucial to provide correct initial conditions, or else the program will loop indefinitely. In programming a computer, the fact that a function can be defined recursively doesn't mean that it must be programmed recursively. For instance, calculation of powers is seldom programmed recursively. The reason is that subprogram recursion is an expensive technique, best reserved for those few cases where explicit evaluation is truly difficult.

Not all recursions are so complicated, however. Here is a simple example. Let function $f(n)$ be defined recursively by the initial condition $f(0) := 1$ and the recursive condition $f(h + 1) := 2f(h) + 1$ ($h \in \mathbb{N}$). Then the first few steps of the recursion are: $f(0) = 1$ (given); $f(1) = 2f(0) + 1 = 2 \cdot 1 + 1 = 3$; $f(2) = 2f(1) + 1 = 2 \cdot 3 + 1 = 7$; $f(3) = 2f(2) + 1 = 2 \cdot 7 + 1 = 15$; and so on.

9.1.1 The String Recognition Problem

There is an area of programming, however, in which recursion is used extensively: the programming languages themselves. Recursion is used both to design the various syntactic constructs which that language will deem "valid" and also to implement those designs in a compiler that will recognize and compile them. We're going to examine a narrow variation of this process, the **string recognition problem.** As in Chapter 4 and in Section 6.1, we begin with an alphabet A and the free semigroup A^* over A. A string is defined as any element of A^*. The semigroup possesses the associative concatenation operator, with the null string Λ as its identity. When we defined a subset of A^* as a language (that is, a set of valid strings), we stated explicitly which strings over the alphabet we would accept as valid. We then went about counting and enumerating them. Now we're going to use recursion to approach the definition of a language L in a different way, by addressing two related questions:

- how to define valid strings, that is, elements of L, in terms of their general syntactic descriptions; and
- how to determine whether a given string satisfies that syntactic description.

This second notion is quite new for us. In our previous combinatoric treatment of strings, we were concerned only with generating and counting the elements of L. Given an arbitrary string, there was no problem ascertaining whether or not the string was valid. We examined its contents combinatorically, position-by-position. Now, using syntactic rather than combinatoric descriptions, two things change. On the one hand, we're really no longer concerned with counting and enumerating the valid strings. In most cases, the string length will be unrestricted, so there will be infinitely many anyway. On the other hand, given an arbitrary string, it's not now obvious whether or not that string is valid. **String recognition** is the problem of determining whether a string is valid.

We're dealing with the same problem as before—strings over an alphabet—but with a decidedly different emphasis. The terminology should reflect this. To bring our vocabulary into line with the usual mathematical usage, we're going to use the term **well-formed formula** instead of *valid string*. This is usually abbreviated WFF, and pronounced "wif" or "woof" (as in "proof"). Consequently, we define a **language** L as the set of the WFFs. Note that the definition of a WFF and of the respective language is a very specific one in each given context. Our first task is now to define a language L *syntactically* rather than combinatorically. Let's consider an example that has been simplified almost (but not quite) to the point of being meaningless. You're familiar with the use of parentheses in arithmetic expressions. For example, the expression ((a + b) · c) uses parentheses correctly, while the string)(a + b)(· c) is gibberish. To address only the question of parentheses, let's eliminate everything else, and just consider the parentheses alphabet $A := \{\,),\,(\,\}$ and the strings in the free semigroup A^*. The string (( )) is deemed a WFF, but the string)( )( ) is not, and so on in the usual intuitive way. How do we state clearly and unambiguously just which strings are WFFs? How do we determine whether a given string of parentheses is a WFF? In the first case, at least, recursion comes to the rescue.

Definition 4 Consider alphabet $A := \{\,)\,,\,(\,\}$ and the free semigroup A^*, which contains the null string Λ. The set of WFFs in A^* is defined as follows:

(1) The null string Λ is a WFF.
(2) If x is a WFF, then (x) is a WFF. This operation is called **enclosure.**
(3) If x and y are WFFs, then $x{:}y$ is a WFF, where : denotes **concatenation.**
(4) There are no WFFs in A^* except as defined above.

For example, consider the string (( )( ))( ). Since Λ is a WFF so is ( ), which is the enclosure of Λ. Then, by concatenation, ( )( ) is a WFF, and by enclosure, so is (( )( )). Finally, by concatenation again, (( )( ))( ) is a WFF. In other words, complex WFFs are built up from simpler WFFs. Enclosure and concatenation are the recurrence conditions. The sole initial condition is that Λ is a WFF. The final restriction (4) is unrelated to recursion per se, but is needed nonetheless. It's the only way we can say that)( )( ) is not a WFF.

How can we determine whether a given string in A^* is a WFF? The simple answer is to try to trace backwards the steps in its construction. If we can show how a string x could have been constructed starting from Λ, then x must be a WFF. Generally speaking, that's more easily said than done. For the parenthesis language, though, there's a method that is both simple and educational. It proceeds as follows:

1. Consider a natural number n, the *parenthesis counter*. Initially, set n to 0.
2. Scan the parenthesis string from left to right. For each left parenthesis encountered, add 1 to n. For each right parenthesis, subtract 1 from n.
3. The string is a WFF if and only if
- n is always nonnegative, and
- at the end of the string, n is once again exactly 0.

Thus, for the string (( )( ))( ), we have the sequence of values of n as 0, 1, 2, 1, 2, 1, 0, 1, 0, so that (( )( ))( ) is a WFF. String)( )( ) has sequence 0, -1, . . ., so we know immediately that)( )( ) is not a WFF. For any string that is a WFF, it's possible to use the sequence of ns to reconstruct the steps by which the string was constructed (see Exercises 9.1). But do not expect such a simple rule to work in general.

You say you want something a little more useful? All right, how about arithmetic expressions? It's best for now if we limit ourselves to a small set of expressions such as might be found in the elementary arithmetic of a programming language, expressions such as $((a + b) * c)/d$. We could proceed like this.

Definition 5 The following sets are given: **Variables** := $\{a, b, c, \ldots, z\}$, **Operators** := $\{+, -, *, /\}$, **Parentheses** := {) , (}, Alphabet A := **Variables** ∪ **Operators** ∪ **Parentheses,** A^* := free semigroup of A.

Then the language L ($L \subseteq A^*$) is defined as follows:

(1) Every variable is in L.
(2) If $x \in L$, then $(x) \in L$. This operation is called **enclosure.**
(3) If $x \in L$, then $-x$ and $+x$ are in L. These are the **unary operations.**
(4) If $x \in L$ and if $y \in L$, then $x + y$, $x - y$, $x * y$, and x/y are in L. These are the **binary operations.**
(5) There are no WFFs in L except as defined above.

As before, enclosure and the unary and binary operations constitute the recurrence condition, and the fact that all variables are WFFs is the initial condition. You should note here that Λ is not a WFF, nor, in general, is the concatenation of two WFFs. In fact, although concatenation and its identity Λ are the defining features of A^*, they appear only indirectly in the definition of WFFs within A^*. Looking again at our example string $((a + b) * c)/d$, we see that a, b, c, and d are WFFs by themselves. Then, we can build up $a + b$ and $(a + b)$ by, respectively, binary addition and enclosure. Then comes $(a + b) * c$, $((a + b) * c)$, and finally $((a + b) * c)/d$ by multiplication, enclosure, and division. Thus, $((a + b) * c)/d$ is a WFF. The definition of WFFs yields a few surprises. First, by historical accident, the symbols + and − each denote both unary and binary operators. We could have used here $^+$, +, $^-$, −, but this is not normally done. Thus $a + -b$ and $a - +b$ are both WFFs, read as "a plus negative b" and "a minus positive b" respectively. Second, some important things are missing. The strings $a + 3$ and cos(a) are invalid; that is, they are not in L since we said nothing about constants or functions. Third, multicharacter variables such as $a1$ or $b2$ have been forgotten. Fourth, final, and most important, there is no way to decide whether a string like $a + b * c$ was obtained by constructing $a + b$ by addition and then $a + b * c$ by multiplication or $b * c$ by multiplication and then $a + b * c$ by addition. The present definition does not *require* parentheses in either case. Both constructions are syntactically correct and give the same syntactically correct result. The semantics would, in this case, be the arithmetical meaning, which is different in the two constructions; but this is beyond the goal of the present definition. We'll consider this problem in Chapter 12. In the exercises below, you'll be asked to elaborate on some of these points.

9.1.2 Syntax-Directed Compilation

The technique of syntactically recognizing a string as belonging to a predefined language L within A^* is known to computer scientists as **syntax-directed compilation.** The process of examining a given string to reconstruct its components is called **parsing.** Obviously, parsing an expression string is much more complicated than parsing the parenthesis strings as we did above. We must keep track not only of a simple counter, but of the entire syntax of the string. We will not delve into parsing algorithms here, because the efficient ones are quite complicated. Many involve recursive programming techniques. Still, the heart of syntax-directed compilation is the recursive definition of valid expressions (WFFS) from a programming language's alphabet.

9.1.3 Patterns of Recursive Definitions

The Basic Pattern. To define objects O_n ($n \in \mathbb{N}$)

1. Initial condition: define O_0.
2. Recurrence condition: when $h \in \mathbb{N}$ and under the hypothesis that O_h has been defined, define O_{h+1}.

A More General Pattern. Let $a, b, \ldots, c$ be finitely many consecutive concretely given integers. To define objects O_n ($n \in \mathbb{Z}, n \geq a$)

1. Initial conditions: define $O_a, O_b, \ldots, O_c$.
2. Recurrence condition: for every integer h ($h \in \mathbb{Z}$, $h \geq c$), and under the hypothesis that O_j has been defined ($a \leq j \leq h$), define O_{h+1}.

Example (1) $x_1 := 1$, $x_2 := 1$; (2) when $h \in \mathbb{Z}$ and $h \geq 2$, $x_{h+1} := x_{h-1} + x_h$.
This sequence is called the **Fibonacci sequence.** Its first few terms are 1, 1, 2, 3, 5, 8, 13, ...

Other Patterns. There are many variations of the patterns above. We could have *finite* recursion: the recurrence condition poses such restrictions that the process can be performed only finitely many times. We could have a *descending* recursion: O_{h-1} is defined from O_h. The recursion may proceed by a step other than 1, even by a step of variable length.

Once the general pattern is understood, there are usually no essential difficulties in adapting it to special circumstances.

Exercises 9.1

1. Write the first eight terms of each of the following recursively defined functions x_n. In each case, identify the initial condition(s) and the recurrence condition. **(a)** $x_0 := 0$, $x_{h+1} := x_h + 1$ ($h \in \mathbb{N}$). **(b)** $x_0 := 7$, $x_{h+1} := 2x_h - 4$ ($h \in \mathbb{N}$). **(c)** $x_1 := 5$, $x_{h+1} := 3 \cdot x_h + 4$ ($h \in \mathbb{N} - \{0\}$). **(d)** $x_{-3} := 10$, $x_{h+1} := x_h + h$ ($h \in \mathbb{Z}, h \geq -3$). **(e)** $x_0 := 1$, $x_1 := 1$; $x_{h+1} := x_{h-1} + x_h + 1$ ($h \in \mathbb{Z}, h > 0$).
2. For every natural number n and for every natural number k such that $k \leq n$, define the symbol $F(n, k)$ by $F(n, k) := 2$ if $k = 0$ or if $k = n$, $F(n + 1, k) :=$

$F(n, k) + F(n, k - 1)$ if $0 < k \leq n$. Calculate the values of $F(n, k)$ when $n \leq 4$ and for all possible values of k.

3. On the Boolean algebra over $A := \{00, 01, 10, 11\}$, with the bitwise operations, define the function $f:\mathbb{N} \to A$ by $f0 := 1$, $f(h + 1) := \sim((fh) \vee 10)$ $(h \in \mathbb{N})$. Calculate explicitly all fn $(n \in \mathbb{N})$.

4. The given alphabet is $A := \{\text{r, t, u, v}\}$. Define strings s_n over A by $s_0 := (\text{tu})$, $s_{h+1} := (s_h\text{:ru})$ if s_h contains the symbol r fewer than two times, $s_{h+1} := (s_h\text{:}\Lambda)$ otherwise $(h \in \mathbb{N})$. Calculate explicitly all s_n $(n \in \mathbb{N})$.

5. State recursive definitions for the following: **(a)** The n-fold cartesian product of a set A. **(b)** The n-fold concatenation of a string x. **(c)** $\bigvee_{i=1}^{n} x_i$, $\bigwedge_{i=1}^{n} x_i$ in a Boolean algebra. **(d)** R^n in $A \times A$, where R is a relation on A.

6. In each of the following cases, use the definition in the text to construct a few of the objects. Stop when you understand how things work. **(a)** 2^n $(n \in \mathbb{N})$. **(b)** $n!$ $(n \in \mathbb{N})$. **(c)** $C(n, k)$ $(n \in \mathbb{N}, 0 \leq k \leq n)$. **(d)** Parenthesis strings. **(e)** Arithmetic expressions, when **Variables** $:= \{a, b, c, \ldots, z\}$, **Operators** $:= \{+, -, *, /\}$, **Parentheses** $:= \{\,),\,(\,\}$. (This problem is suitable for computer solution. Do not construct too many objects.)

7. For the parenthesis language of Definition 4, state an algorithm that will reconstruct an entire parenthesis string from the sequence of values of the parenthesis counter. For instance, given the sequence 0, 1, 2, 1, 2, 1, 0, 1, 0, the algorithm should reconstruct the string (( )( ))( ). Test your algorithm on some examples. (This problem is suitable for computer solution.)

8. Extend the capability of Definition 5 to correctly allow for **(a)** arithmetic constants; **(b)** the functions sqrt(x) and cos(x); **(c)** the combinatoric function $C(n, k)$.

9. Using Definition 5 as a model, state a recursive definition for syntactically correct expressions **(a)** in set theory; **(b)** in any given Boolean algebra; **(c)** in propositional calculus; **(d)** in A^A, for any given nonempty set A, with the binary operator "composition of functions."

10. State a recursive definition of two-terminal switching networks (2TSNs).

Complements 9.1

1. The theoretical principles that underlie all of the constructions by recurrence derive ultimately from the following theorem:

Recursion Theorem. Let C be a nonempty set. Let $g:C \to C$ be a total function. Let b be a given element of C. Then there exists precisely one total function $f:\mathbb{N} \to C$ such that (1) $f0 = b$; (2) if $h \in \mathbb{N}$, then $f(h + 1) = gfh$.

For example, let C be the set of the positive reals. Let b be 1. Let x be any one given positive real. For each y in C, define gy as $y \cdot x$. Then fn turns out to be x^n.

We'll not attempt the proof of the theorem. We should mention, however, that it is based on the definition of $\mathbb{N}$ (or ω) as indicated in Complements 5.6, and on the *principle of mathematical induction*. See Complements 9.2.

9.2 Induction

Mathematical induction is a technique for proving theorems. Thus, it differs from recursion, in that recursion is primarily a definition technique. Note that there is no such thing as an inductive theorem, just theorems which lend themselves to inductive proof. Every theorem that can be proved by induction could also, in theory, be proved some other way. What, then, is the advantage of inductive proofs? The key to induction lies in realizing that we deal here, not with individual theorems, but with entire *families* of theorems. Using standard deductive techniques, we would have to prove each one singly. Using induction, we prove all at once. This is a heady idea, so let's look first at an example.

9.2.1 The Arithmetic Series

Theorem 6 For any natural number n, $\sum_{i=0}^{n} i = n(n+1)/2$.

You've seen this theorem before and proved it using arithmetic series (Section 3.5). But for purposes of illustration, we'll take an inductive approach. What we have here is a family of theorems, namely:

theorem T_0 (when $n = 0$): $\displaystyle\sum_{i=0}^{0} i = 0(0+1)/2;$

theorem T_1 (when $n = 1$): $\displaystyle\sum_{i=0}^{1} i = 1(1+1)/2;$

theorem T_2 (when $n = 2$): $\displaystyle\sum_{i=0}^{2} i = 2(2+1)/2;$

.

.

.

theorem T_h (when $n = h$): $\displaystyle\sum_{i=0}^{h} i = h(h+1)/2;$

theorem T_{h+1} (when $n = h + 1$): $\displaystyle\sum_{i=0}^{h+1} i = (h+1)(h+1+1)/2;$

.

.

.

How can we prove them all at once? An inductive proof proceeds in two general stages. For Theorem 6, our goal is to prove that

- T_0 is true, and
- for every natural number h, theorem T_h implies theorem T_{h+1}.

Is this circular logic? No! The first theorem, T_0, must be proved explicitly, that is, deductively. This serves as the initial step. Having achieved that, we must prove that T_h implies T_{h+1}, again deductively. Note that we need *not* prove at this stage either T_h or T_{h+1}; we prove only the implication $T_h \Rightarrow T_{h+1}$. The final consequence is recursive: we have proved T_0; we have proved that T_0 implies T_1; so T_1 is true (from algebra of propositions, recall that, if p and if $(p \Rightarrow q)$, then q); we have proved T_1; we have proved that T_1 implies T_2; so T_2 is true; and so on. Notice the similarity between this and a recursive definition. Let's return to our specific example.

Proving the initial step is quite easy. Statement T_0 is simply the statement $\sum_{i=0}^{0} i = 0(0 + 1)/2$, which is trivially true, because both sides are 0. Now consider exactly what we want to prove in the induction step: T_h implies T_{h+1}. More specifically, we want to prove if T_h, then T_{h+1}. The part "if T_h" is called the **inductive hypothesis.** First of all, what is T_h? what is T_{h+1}? From the statement of the theorem, T_h is the statement $\sum_{i=0}^{h} i = h(h + 1)/2$; T_{h+1} is the statement $\sum_{i=0}^{h+1} i = (h + 1)(h + 1 + 1)/2$. In formula, a proof of the induction step is:

From the inductive hypothesis,

$$\sum_{i=0}^{h} i = h(h + 1)/2.$$

Adding $h + 1$ to both sides gives

$$\left(\sum_{i=0}^{h} i\right) + (h + 1) = h(h + 1)/2 + (h + 1). \tag{1}$$

The left-hand side of equation (1) is the sum of terms i when i ranges from 0 to h and then of $h + 1$. This gives the sum of term i when i ranges from 0 to $h + 1$, that is the left-hand member of the equation below. The right-hand side of equation (1) is equivalent to $h(h + 1)/2 + (2h + 2)/2$, that is, $(h^2 + 3h + 2)/2$, which can be rewritten as $(h + 1)(h + 2)/2$, which is the right-hand side of the equation below. We can so conclude that $\sum_{i=0}^{h+1} i = (h + 1)(h + 2)/2$, which is precisely T_{h+1}.

Since we have now established T_{h+1}, we're finished! The entire family of theorems T_n has been proved for all natural numbers n.

It is crucial in inductive proofs to be neat and clear. You should develop the habit of doing inductive proofs in four steps, the middle two being for your own benefit:

1. A clear statement and explicit deductive proof of the initial step T_0.
2. A clear statement of the induction hypothesis T_h.
3. A clear statement of the goal T_{h+1}.
4. An explicit deductive proof that $T_h \Rightarrow T_{h+1}$.

Putting this all together, let's see how our proof would look under normal circumstances—that is, when we want to present the proof without all the explanations we gave before.

Proof: Proof is by induction on n.

The initial step T_0 is the statement $\sum_{i=0}^{0} i = 0(0 + 1)/2$. The induction hypothesis T_h is the statement $\sum_{i=0}^{h} i = h(h + 1)/2$. The conclusion of the induction step, T_{h+1}, is the statement $\sum_{i=0}^{h+1} i = (h + 1)(h + 1 + 1)/2$. The induction step is $T_h \Rightarrow T_{h+1}$.

The statement of the initial step is trivially true.

The proof of the induction step goes thus: From the inductive hypothesis, $\sum_{i=0}^{h} i = h(h + 1)/2$. Adding $h + 1$ to both sides, we obtain $\sum_{i=0}^{h} i + (h + 1) = h(h + 1)/2 + (h + 1)$. Algebraic manipulations give $\sum_{i=0}^{h+1} i = (h + 1)(h + 2)/2$, which is precisely T_{h+1}. Since we have now established T_{h+1}, the theorem is proved. Q.E.D.

If it seems like a lot of work, keep in mid that we're proving an infinite number of theorems!

9.2.2 The Combinatoric Function

As a second example, we consider again the recursive Definition 3 of the combinatoric function $C(n, k)$. Does this definition agree with the factorial formula you've already learned?

Theorem 7 Let $C(n, k)$ be as in Definition 3. Then, for all natural numbers n and k such that $0 \leq k \leq n$, $C(n, k) = n!/(k!(n - k)!)$.

Proof: Proof is by induction on n.

Initial steps: since the initial condition of Definition 3 is valid when $k = 0$ or $k = n$, these two cases will both be proved explicitly.

Induction hypothesis: if $h \in \mathbb{N}$ and if $0 \leq k \leq h$, then $C(h, k) = h!/(k!(h - k)!)$.

Conclusion of the induction step: if $h \in \mathbb{N}$ and if $0 < k \leq h$, then $C(h + 1, k) = (h + 1)!/(k!((h + 1 - k)!)$. Note that we do not have to prove the conclusion when $k = 0$ or $k = h + 1$, because these cases will have been proved in the initial step.

Induction step: show that the inductive hypothesis above implies the conclusion above.

Proof of the initial steps. By definition, $0! := 1$. Clearly, $n!/n! = 1$ for every natural number n. Consequently, $n!/(0!(n - 0)!) = 1$ and $n!/(n!(n - n)!) = 1$ for every n. Since $C(n, 0) := 1$ and $C(n, n) := 1$ by definition, the initial steps are proved.

Proof of the induction step. The definition of C implies, under the stated conditions,

$$C(h + 1, k) := C(h, k) + C(h, k - 1).$$

Using the inductive hypothesis on each of the addends on the right-hand member, we obtain

$$C(h+1, k) = h!/(k!(h-k)!) + h!/((k-1)!(h-(k-1))!).$$

The right-hand side is

$$\frac{h!}{k!(h-k)!} + \frac{h!}{(k-1)!(h-(k-1))!}.$$

These two fractions are reduced to common denominator by multiplying numerator and denominator by $h - k + 1$ in the first fraction and by k in the second fraction (why?). The sum of the numerators is then $h!(h-k+1) + h!k = h!(h-k+1+k) = h!(h+1) = (h+1)!$ The sum of the two fractions reduces to

$$\frac{(h+1)!}{k!(h-k+1)!}.$$

The conclusion is so proved, and so is the theorem. Q.E.D..

Inductive proofs are not necessarily difficult. But they do require, even more than deductive proofs, two things: understanding what you're doing, and writing down what you're trying to show.

9.2.3 Patterns of Proofs by Induction

Basic Pattern. To prove statement T_n for every natural number n

1. Initial step: prove T_0.
2. Induction step: if $h \in \mathbb{N}$, prove that T_h implies T_{h+1}.

In order to clarify step 2, it is convenient to state clearly each of T_h and T_{h+1}.

A More General Pattern. Let $a, b, \ldots, c$ be finitely many consecutive concretely given integers. To prove statement T_n ($n \in \mathbb{Z}, n \geq a$)

1. Initial steps: prove $T_a, T_b, \ldots, T_c$.
2. Induction step: for every integer h ($h \geq c$), and under the hypotheses that T_j have been proved ($a \leq j \leq h$), prove T_{h+1}. In other words, prove $(\bigwedge_{j=a}^{h} T_j) \Rightarrow T_{h+1}$.

Other Patterns. As in the case of recurrence, there are many variations of the patterns above. We could have *finite* induction: the induction step poses such restrictions that the process can be performed only finitely many times. We could have a *descending* induction: T_{h-1} is proved on the ground of T_h. The induction may proceed by a step other than 1, even by a step of variable length.

Note 9 on Theorem Proving. Clearly, proof by induction is an important and powerful technique for proving theorems, provided that the nature of the theorem allows for this type of proof.

Exercises 9.2

1. Prove each of the following theorems by induction. You should verify each one with a few examples first. **(a)** If $n \in \mathbb{N}$, then $\Sigma_{i=0}^{n} i^2 = n(n+1)(2n+1)/6$. **(b)** If $n \in \mathbb{N}$, then $\Sigma_{i=0}^{n} i^3 = (n(n+1))^2/4$. **(c)** The sum of the first n odd natural numbers is equal to n^2. [Hint: What is the kth odd natural number? Clearly state the theorem in formula before you attempt to prove it.] **(d)** The sum of the first n even positive natural numbers is equal to $n(n+1)$. **(e)** $\Sigma_{k=0}^{n} C(n, k) = 2^n$. **(f)** If $n > 0$, then $\Sigma_{k=0}^{n} (-1)^k C(n, k) = 0$.

2. Principle of inclusion and exclusion. This theorem is a generalization of the last part of Theorem 5.21, which may be needed in this proof.

Let U be a given finite universal set. Let B be a collection of subsets of U. Set $n := \#B$. For each k ($k \in \mathbb{N}, 0 < k \leq n$) let $C(k)$ be the set of all k-combinations of elements of B. Prove

$$\#\bigcup_{A \in B} A = \sum_{k=1}^{n} (-1)^{k+1} \sum_{S \in C(k)} \#\bigcap_{X \in S} X.$$

For example, when $n = 2$ and $B := \{A_1, A_2\}$, the statement is $\#(A_1 \cup A_2) = \#A_1 + \#A_2 - \#(A_1 \cap A_2)$ (see Theorem 5.21). When $n = 3$ and $B := \{A_1, A_2, A_3\}$, the statement is $\#(A_1 \cup A_2 \cup A_3) = \#A_1 + \#A_2 + \#A_3 - \#(A_1 \cap A_2) - \#(A_2 \cap A_3) - \#(A_3 \cap A_1) + \#(A_1 \cap A_2 \cap A_3)$.

3. Derangements. Given $A := \{1, 2, \ldots, n\}$ ($n > 1$). Let $x := (x_1, x_2, \ldots, x_n)$ be a permutation of A. Permutation x is called a **derangement** of A if $x_j \neq j$ for every j in A. For example, when $n = 4$, $\{2, 1, 4, 3\}$ or $\{4, 3, 2, 1\}$ is a derangement of A. There aren't any when $n = 1$.

Prove that the number of derangements of n objects is $n!\, \Sigma_{i=0}^{n} (-1)^i\, 1/i!$.

4. Euclidean algorithm. The following algorithm produces the greatest common divisor (GCD, for short) of two given, positive integers a and b. This is an example of finite recursion. Recall the definition of integer division with remainder. Given a and b, the Euclidean algorithm is

(1) *Initial conditions:* $a_1 := a$, $a_2 := b$ (when $a \neq b$, it is practically convenient to take a_1 as the larger of a and b, and a_2 as the smaller).

(2) *Recursion step:* if $h \in \mathbb{N}$, if $h > 1$, and if $a_h \neq 0$, then a_{h+1} := remainder of division of a_{h-1} by a_h.

(3) If $a_h = 0$ for a certain h, then a_{h-1} is the GCD and the process stops.

It is essential to note that since $a_2 > a_3 > \cdots \geq 0$, the process must end after finitely many steps. The algorithm needs a proof of correctness, of course. Produce one.

5. Use the Euclidean algorithm to find the GCD of each of the following pairs: **(a)** 10, 55; **(b)** 972, 153; **(c)** 87, 175. (This problem is suitable for computer solution.)

6. Prove that the positive integers a and b have GCD equal to 1 if and only if there exist integers x, y such that $xa + yb = 1$.
7. In each of the following cases find integers x and y that safisty the given equation. **(a)** $87x + 175y = 1$; **(b)** $163x + 109y = 1$; **(c)** $3x + 4y = 1$.

Complements 9.2

1. Mathematical induction finds its logical justification in the axioms of set theory, especially the axiom of infinity, and is based on the definition of $\mathbb{N}$ (or ω) we mentioned in Complements 5.6. The crucial property is the following principle:

   Principle of Mathematical Induction. Let S be a subset of $\mathbb{N}$. Let S satisfy the conditions **(1)** $0 \in S$, **(2)** if $h \in S$, then $h + 1 \in S$. Then $S = \mathbb{N}$.

   We shall not attempt the proof here. For a few comments, see Complements 5.6.

9.3 Difference Equations

Every recursively defined function can be viewed as having the set $\mathbb{N}$ of natural numbers as its domain. That is, every such function gives rise to a sequence of numbers. For instance, the recursively defined function

$$x_0 := 5 \tag{2}$$

$$x_h := 3x_{h-1} \tag{3}$$

produces the sequence $x_0 = 5, x_1 = 15, x_2 = 45, x_3 = 135, x_4 = 405, \ldots$, which is a total function $\mathbb{N} \to \mathbb{R}$, $n \mapsto x_n$. In this context, the element x_n is called the n-th **component** of the sequence.

What is perhaps a bit less obvious is that this function also has an explicit representation, that is, an expression in which the only argument is the natural number n from the domain. The function

$$x_n := 5 \cdot 3^n \tag{4}$$

produces exactly the same sequence as the recursive definition. Of the three forms of this function—recursive definition, sequence (at least the first several terms), and explicit definition—the explicit definition is clearly preferable, at least from some computational viewpoint. If the only thing we really want is the thousandth component of the sequence, or every five-hundredth component, it's much better to be able to calculate that number explicitly rather than having to work through each and every sequence component.

But if we really want an explicit definition of a function, why didn't we say that in the first place? Why did we bother to write a recursive definition at all? Put simply, we might not have *known* the explicit form. Mathematical formulae, especially those for physical phenomena, don't arise from thin air. We must first observe the phenomenon itself and characterize its behavior. Such observation might easily take the form of a recursive function definition. Then we're faced with the problem of turning that recursive form into an explicit one. This isn't always possible; some functions have only a recursive form and no explicit form at all. Others may have explicit forms, but they are, as yet, unknown. Still others have explicit forms for some initial conditions but not for others.

Trying to find an explicit form for a recursively defined function puts us into the *theory of difference equations*. Equation 3 is known as a **difference equation** because, if we rewrite it slightly as $x_h - 3 \cdot x_{h-1} = 0$, then what we're defining is a sort of a "difference" between x_h and x_{h-1}. Since x_h is expressed in terms of only the next most recent component x_{h-1}, equation 3 is called a **first-order** difference equation. We'll consider higher-order equations in a moment. The explicit form, expression 4, is called the **solution** of the difference equation. (Reflect on that terminology for a moment: The difference equation is now the problem to be solved, the recursively defined function x_n is the unknown, and the explicit expression of the function that satisfies that equation is the problem's solution. The actual function values, the sequence, are really now just a side issue. Don't get confused as to what we mean by problem and solution.) Note that the initial condition is *not* part of the difference equation; that's still considered a separate entity.

We have to delineate two types of solutions. If we momentarily ignore the initial condition (equation 2), then *any* function of the form

$$x_n := C \cdot 3^n \tag{5}$$

will satisfy the difference equation 3, with C being any constant at all. Thus, expression 5 is called the **general solution** of equation 3. (This is a crucial observation, and you should take a moment to verify it. After all, all that equation 3 says is that each new component is three times the last.) But if the initial condition (equation 2) must also be satisfied, then formula 4, with $C = 5$, is the only solution, and is referred to as the **particular solution**. Thus, comparing difference equations to recursive functions, a much sharper distinction is made between the difference equation and the initial condition. A general solution, involving one or more arbitrary constants, satisfies only the difference equation itself. A particular solution, with all constants assigned particular values, satisfies the difference equation *and* the initial condition(s).

9.3.1 First-Order Difference Equations

Theorem 8 Every first-order difference equation of the form $x_h - A \cdot x_{h-1} = 0$, where A is a constant, has a general solution $x_n = C \cdot A^n$, for some arbitrary constant C. If, in addition, Y is a given constant and the initial condition $x_0 = Y$ must be satisfied, then, in the particular solution, $C = Y$, so that $x_n = Y \cdot A^n$.

Proof: The technique is to successively express each component of the sequence in terms of its predecessor.

$$
\begin{aligned}
x_n &= A \cdot x_{n-1} \\
&= A(A \cdot x_{n-2}) = A^2 x_{n-2} \\
&= A^2(A \cdot x_{n-3}) = A^3 x_{n-3} \\
&= A^3(A \cdot x_{n-4}) = A^4 x_{n-4} \\
&\vdots \\
&= A^j x_{n-j} \\
&\vdots \\
&= A^n x_0 \\
&= C \cdot A^n
\end{aligned}
$$

The last step yields an arbitrary constant C because, for the general solution, the initial term x_0 is arbitrary. To satisfy the initial condition, we simply evaluate the general solution when $n = 0$ to obtain $C = Y$. (As an exercise, replace this proof with a proof by induction.) Q.E.D.

So our example simply has $A = 3$ and, for the particular solution, $C = 5$. For another example, the difference equation $x_h + 9 \cdot x_{h-1} = 0$ has general solution $x_n = C \cdot (-9)^n$. For initial condition $x_0 = -50$, the particular solution becomes $x_n = -50 \cdot (-9)^n$.

9.3.2 Second-Order Difference Equations

If each component of the sequence is expressed in terms of the previous two components, then we have a **second-order difference equation,** of the form

$$x_h - A \cdot x_{h-1} - B \cdot x_{h-2} = 0, \tag{6}$$

where A and B are given constants. (We also now expect that the first *two* components must be explicitly given as initial conditions. That's correct, but let's first try to obtain a general solution.) We could try some variation of the proof of Theorem 8, but the algebra would be terrible and the result of questionable value. Instead, let's use our imagination and try to guess what the *form* of the solution might be. For first-order equations, we had a general solution of the form $x_n = C \cdot A^n$, where A depended upon the difference equation, and C upon the initial conditions. Now, for second-order equations, let's suppose that the general solution is going to look something like $x_n := C \cdot Z^n$, where Z somehow depends on the equation, and C on the initial conditions. We also have to assume, for the moment, that both C and Z are nonzero, since otherwise we would have only the trivial solution $x_n = 0$ for all n. For given conditions, we might wind up with $C = 0$, but that's only in the particular solution. In general, C is nonzero. We don't know that this is going to work, but let's substitute it into expression 6 just to see what we get.

$$C \cdot Z^h - A(C \cdot Z^{h-1}) - B(C \cdot Z^{h-2}) = 0,$$
$$C \cdot Z^{h-2}(Z^2 - AZ - B) = 0.$$

But, since both C and Z are nonzero, $C \cdot Z^{h-2} \neq 0$, so that

$$Z^2 - AZ - B = 0. \tag{7}$$

Equation 7 is called the **characteristic equation** for equation 6. Its left-hand member is called the **characteristic polynomial** of the equation. By the quadratic formula, the value of Z must be $Z = (A \pm \sqrt{A^2 + 4B})/2$. Now we have a problem. Depending upon the values of A and B, the characteristic equation will have either two real roots, a single real root, or two complex roots. (If you don't know what complex numbers are, relax and read on anyway. In this case, it won't make much difference.) Let's organize our strategy into three mutually exclusive and exhaustive cases.

Case I. $A^2 + 4B > 0$. If we set $Z_1 := (A + \sqrt{A^2 + 4B})/2$ and $Z_2 := (A - \sqrt{A^2 + 4B})/2$, then the general solution is

$$x_n = C_1Z_1^n + C_2Z_2^n. \tag{8}$$

We also now have two arbitrary constants C_1 and C_2 to resolve for any particular solution. We'll see in a moment how this is done.

Case II. $A^2 + 4B < 0$. This case is identical to Case I, except that both Z_1 and Z_2 will be complex. It's almost certain, therefore, that C_1 and C_2 will also be complex. Allowing for that, the general solution is simply equation 8.

Case III. $A^2 + 4B = 0$. Now we have a real puzzlement. There is only one possible value of Z, namely $Z = A/2$. But we must have two arbitrary constants in order to satisfy the two initial conditions in the particular solution (and to maintain compatibility with Cases I and II). It turns out that we can use the index n as a *multiplier* in this case. Try the function $C \cdot n \cdot x^n$ as a solution for the recursive condition. It works: substitution into the equation gives $ChZ^h - AC(h-1)Z^{h-1} - BC(h-2)Z^{h-2} = CZ^{h-2}[hZ^2 - A(h-1)Z - B(h-2)]$. Since $Z = A/2$ and $4B = -A^2$, the part in brackets is $hA^2/4 - (h-1)A^2/2 + A^2(h-2)/4$, which is 0. So we have a second solution nCZ^n and obtain as a general solution

$$x_n = C_1Z^n + n \cdot C_2Z^n = (C_1 + nC_2)(A/2)^n \tag{9}$$

We have so proved

Theorem 9 Every second-order difference equation of the form of equation 6, where A and B are constants, has a general solution of the type in equation 8 if $A^2 + 4B \neq 0$, where Z_1 and Z_2 are the distinct roots of the characteristic equation 7, or of the type in equation 9 if $A^2 + 4B = 0$, where Z ($= A/2$) is the only root of the characterist[c] equation 7. In each case, C_1 and C_2 are arbitrary constants. If, in addition, Y_0 and Y_1 are given constants, then there exists one particular solution that satisfies the initial conditions $x_0 = Y_0$ and $x_1 = Y_1$.

As an example, let's consider a Fibonacci sequence (Section 9.1.3), that is, a sequence that solves the difference equation $x_h := x_{h-1} + x_{h-2}$. The original Fibonacci sequence satisfies also the initial conditions $x_0 = 1$, $x_1 = 1$. Each component (after two initial ones) is the sum of the previous two. The sequence begins 1, 1, 2, 3, 5, 8, 13, 21, 34, 55, 89, 144, 233, ... But what is, say, the 250th term in the sequence? As a difference equation, the Fibonacci equation looks like

$$x_h - x_{h-1} - x_{h-2} = 0$$

and has characteristic equation $Z^2 - Z - 1 = 0$, with $A = B = 1$, so that $A^2 + 4B = 5$, and the characteristic roots are $Z_1 := (1 + \sqrt{5})/2$, $Z_2 := (1 - \sqrt{5})/2$. The general solution is thus $x_n = C_1[(1 + \sqrt{5})/2]^n + C_2[(1 - \sqrt{5})/2]^n$. For the initial conditions $x_0 = 1$, $x_1 = 1$, we evaluate the general solution for $n = 0$ and $n = 1$, and find

$$\begin{aligned} &\text{for } n = 0, && x_0 = 1 = C_1 + C_2 \\ &\text{for } n = 1, && x_1 = 1 = C_1(1 + \sqrt{5})/2 + C_2(1 - \sqrt{5})/2, \end{aligned}$$

which gives us a linear system in two unknowns, C_1 and C_2.

$$\begin{aligned} C_1 + \qquad\qquad\qquad\qquad C_2 &= 1 \\ [(1 + \sqrt{5})/2]C_1 + [(1 - \sqrt{5})/2]C_2 &= 1, \end{aligned}$$

which yields $C_1 = (1 + \sqrt{5})/(2\sqrt{5})$, $C_2 = -(1 - \sqrt{5})/(2\sqrt{5})$. Substituting these back into the general solution gives us the particular solution

$$x_n = (1/\sqrt{5})([(1 + \sqrt{5})/2]^{n+1} - [(1 - \sqrt{5})/2]^{n+1}). \tag{10}$$

We can now use a hand calculator to answer our original question: $x_{250} \simeq 1.28 \times 10^{52}$. You should take a moment to verify that this particular solution actually does produce all of the sequence components listed above.

Another example of a second-order difference equation is $x_h - 4x_{h-1} + 4x_{h-2} = 0$, which produces characteristic equation $Z^2 - 4Z + 4 = 0$. Since now $A = 4$ and $B = -4$, we have $A^2 + 4B = 0$, and $Z = 2$ is the only possibility. The general solution is thus $x_n = (C_1 + nC_2)2^n$. If we now impose initial conditions $x_0 = 0$, $x_1 = 6$, the first two evaluations yield the linear system

$$\begin{aligned} &\text{for } n = 0, && x_0 = 0 = (C_1 + 0 \cdot C_2) \cdot 1 \\ &\text{for } n = 1, && x_1 = 6 = (C_1 + 1 \cdot C_2) \cdot 2, \end{aligned}$$

which yields $C_1 = 0$ and $C_2 = 3$. The particular solution is thus $x_n = 3n2^n$. As a check, the next two terms in the sequence are $x_2 = 24$ and $x_3 = 72$. You should verify that both the difference equation with initial condition and the particular solution yield these values. But only the particular solution allows us to quickly calculate that $x_{100} \simeq 3.80 \times 10^{32}$.

Difference equations are usually not taught as such any more in most undergraduate programs in mathematics and computer science. The reason is that most of the analyses and techniques used for difference equations are formally the same as for their continuous cousins, *differential equations*. Since differential equations are of interest primarily to physicists and engineers, discrete mathematicians and computer scientists often avoid them as irrelevant to their own studies. This is unfortunate, because discrete mathematicians and computer scientists need difference equation theory just as much as physicists and engineers need differential equation theory.

9.3.3 Classification of Difference Equations

Obviously there are many other types of difference equations, and we've hardly begun to scratch the surface of their many forms. Some of these equations can be solved (i.e., general and particular solutions obtained) using methods only modestly more involved than the ones we've used. Others require considerably more elaborate techniques. For still other equations, solutions may exist, but nobody's been able to find them as yet. Finally, some difference equations do not have general or particular solutions at all. You can see, then, that in terms of complexity and difficulty, the theory of difference equations runs quite a gamut. You should at least be aware of some broad classes of difference equations, if only to delineate those you know how to solve from those you don't. We present now three classifications of difference equations that you should learn to recognize. We will not attempt solution of any of them. Along with the *order* we studied above, these three give a fairly exact characterization of equations.

First, we note that the coefficients A and B in equation 6 are constants that do not depend on the index h. So our examples above are said to have *constant coefficients*. On the other hand, the factorial equation, which we can write as $x_h - h \cdot x_{h-1} = 0, x_0 = 1$ (see Section 9.1), is a first-order difference equation with a *nonconstant coefficient*. The difficulty of solving such equations depends upon the complexity of these coefficient functions.

Second, all the difference equations we've studied involve only the sequence components themselves and multiplicative coefficients. Because of this, they're said to be **homogeneous.** Difference equations may, however, contain additive terms that are either constant or dependent on h but do not involve sequence components. These terms are collected on the right side of the equality, and the equation is said to be *inhomogeneous*. For instance,

$$x_h - 6x_{h-1} + 8x_{h-2} = 0$$

is a second-order homogeneous equation with constant coefficients. The equation

$$x_h - 6 \cdot x_{h-1} + 8 \cdot x_{h-2} = 3$$

is again second order with constant coefficients, but it is inhomogeneous. The same applies to

$$x_h - 6x_{h-1} + 8x_{h-2} = 3h.$$

Solving inhomogeneous difference equations involves a few more steps but is usually not especially difficult.

Third and last, we've never had sequence components multiplying each other, or raised to powers. All of our equations were **linear.** In comparison, such equations as $x_h = (x_{h-1})^2$ and $x_h = x_{h-1} + x_{h-2}x_{h-3}$ are examples of *nonlinear* equations. Nonlinear difference equations are by far the most difficult to work with.

In formal terminology, then, what we studied in this section were *first- and second-order linear homogeneous* difference equations with *constant coefficients*.

Exercises 9.3

1. For each of the following first-order difference equations and initial conditions, write general and particular solutions. For each equation, state the first ten sequence components and component 100. **(a)** $x_h - 5x_{h-1} = 0$; $x_0 = 1$. **(b)** $x_h - 4x_{h-1} = 0$; $x_0 = -4$. **(c)** $x_h - 9x_{h-1} = 0$; $x_0 = 0$.
2. Give an inductive proof of Theorem 8.
3. For each of the following second-order difference equations and initial conditions, write general and particular solutions. For each equation, state the first ten sequence components and component 100. **(a)** $x_h = -x_{h-1} + 6x_{h-2}$; $x_0 = 2, x_1 = -1$. **(b)** $x_h = x_{h-1} + 6x_{h-2}$; $x_0 = 7, x_1 = -4$. **(c)** $x_h = 6x_{h-1} - 9x_{h-2}$; $x_0 = 2, x_1 = 12$. **(d)** $x_h = 2x_{h-1} - 26x_{h-2}$; $x_0 = 0, x_1 = 0$.
4. Use the particular solution of the Fibonacci sequence (equation 10) to calculate directly approximations of sequence components 50, 100, 150, and 200.
5. The following are initial conditions for Fibonacci sequences. For each pair, write the particular solution and calculate approximations of the first ten sequence components and components 50, 100, 150, 200, and 250. **(a)** $x_0 = 0, x_1 = 1$. **(b)** $x_0 = 1, x_1 = 0$. **(c)** $x_0 = 2, x_1 = 0$. **(d)** $x_0 = 2, x_1 = 2$. **(e)** $x_0 = 0, x_1 = 0$.
6. Prove that for any first-order linear homogeneous difference equation with constant coefficients, the initial condition $x_0 = 0$ yields only the trivial particular solution $x_n = 0$ for all n.
7. Prove that for any second-order linear homogeneous difference equation with constant coefficients, the initial conditions $x_0 = 0, x_1 = 0$ yield only the trivial particular solution $x_n = 0$ for all n.
8. When a general solution cannot be found and the initial conditions are given, a recursive calculation is the only practical possibility. The computer is then the ideal tool, because of its iteration capabilities. **(a)** In several of the exercises above, calculate by computer a few (ten or twenty) of the first components using iteration. Then compare with the results given by the explicit formula.

   In a few of the cases for which we did not give a method to find the general solution, use the initial conditions and iteration to calculate the first few components (four or five by hand, ten or so by computer). For instance, **(b)** $x_h - 6x_{h-1} + 8x_{h-2} = 3h$, $x_0 = 1, x_1 = 0$; **(c)** $x_h = (x_{h-1})^2$, $x_0 = .5$; **(d)** $x_h = x_{h-1} + x_{h-2}x_{h-3}$, $x_0 = 2, x_1 = 1, x_2 = 1$.

9. The following linear homogeneous constant coefficient difference equations are *third-order* and were not discussed explicitly in the text. Try to extend the methods in the text and write general solutions for them. As hints, the characteristic equations have been factored for you.

(a) $x_h = 2x_{h-1} + 5x_{h-2} - 6x_{h-3}$ $\quad (Z^3 - 2Z^2 - 5Z + 6 = (Z-1)(Z+2)(Z-3))$.

(b) $x_h = 3x_{h-1} - 4x_{h-3}$ $\quad (Z^3 - 3Z^2 + 4 = (Z-2)^2(Z+1))$.

(c) $x_h = 3x_{h-1} - 3x_{h-2} + x_{h-3}$ $\quad (Z^3 - 3Z^2 + 3Z - 1 = (Z-1)^3)$.

Complements 9.3

1. Loan amortization. One excellent example of the application of difference equations occurs in the everyday life of most Americans—paying off a loan. Most large consumer loans, such as those for a house or car, are of a class known as *declining balance loans*. The customer borrows a lump sum of money and then pays it back via a set of periodic (usually monthly) payments. The payments are all the same amount, and when the last payment has been made, the entire loan, interest as well as principal, has been repayed. Moreover, with each payment, the balance of remaining principal becomes smaller and interest for the month is charged *only* on this remaining balance. A repayment process is called *amortization*, and our goal is to express this special process mathematically.

We all remember the formula for *simple interest*, $I = Prt$, where $I :=$ interest, $P :=$ principal, $r :=$ rate of interest per unit of time, and $t :=$ time. But this formula works only if the principal P remains constant over time.

Definition 10 The **period** of a loan is defined as the largest unit of time over which the principal P cannot change, so that, within each period, simple interest is charged. We denote by n the **number of periods** for the entire loan, and denote by r the **simple interest rate per period.**

For most consumer loans, the period is one month. For a two-and-a-half year car loan, then, $n = 30$ months. Note carefully that r is the simple rate per month. Interest rates are usually advertised as *annual percentage rates* or A.P.R.'s, since most people prefer that. But for an A.P.R. of 18 percent, the figure we're really after is $r = 0.015$, that is, 1.5 percent per month. (Note that r is a decimal fraction, not a percentage.)

We're ready now to characterize this process mathematically. We'll retain P as the principal, n as the number of periods, and r as the simple interest rate per period. We denote the monthly payment by Y, as yet unknown. Each payment Y consists of some amount of interest and some amount by which the principal is to be reduced. Thus we define $I_j :=$ interest charged for month j, and $P_j :=$ principal reduction for month j, for $j = 1, 2, \ldots, n$. (Note that P_j is *not* the balance, but the amount by which the balance is reduced!) Thus,

$$Y = I_j + P_j \qquad (j = 1, 2, \ldots, n) \tag{11}$$

and

$$\sum_{j=1}^{n} P_j = P. \tag{12}$$

Finally, since interest for each period is charged only on the unpaid balance,

$$I_j = \left(P - \sum_{k=1}^{j-1} P_k\right) \cdot r \tag{13}$$

and for the first period, $I_1 = Pr$. These three equations (11, 12, and 13) are called the **amortization equations.** The first step in analyzing amortization is to rewrite equation 13 for month $j - 1$ as

$$I_{j-1} = \left(P - \sum_{k=1}^{j-2} P_k\right) \cdot r = \left(P - \sum_{k=1}^{j-1} P_k\right) \cdot r + P_{j-1} \cdot r = I_j + P_{j-1} \cdot r,$$

so that $I_j = I_{j-1} - P_{j-1} \cdot r$. We next substitute this into equation 11: $P_j = Y - I_j = Y - (I_{j-1} - P_{j-1} \cdot r) = (Y - I_{j-1}) + P_{j-1} \cdot r = P_{j-1} + P_{j-1} \cdot r = (1 + r) \cdot P_{j-1}$. This last equality gives now

$$P_j = (1 + r) \cdot P_{j-1}, \tag{14}$$

which is a first-order linear homogeneous difference equation with a constant coefficient $(1 + r)$. The general solution is

$$P_j = P_1 \cdot (1 + r)^{j-1} \tag{15}$$

(Note that the exponent is now $j - 1$ instead of j, since the initial evaluation is for $j = 1$ rather than $j = 0$.) The only remaining problem is to find P_1 or to find Y. The two problems are equivalent, because $Y = I_1 + P_1$.

We can substitute the general solution into equation 12: $P = \sum_{j=1}^{n} P_j = P_1 \sum_{j=1}^{n} (1 + r)^{j-1}$. The summation is simply a geometric series. Provided $r > 0$, we can immediately conclude (Section 3.5) that $\sum_{j=1}^{n} (1 + r)^{j-1} = [(1 + r)^n - 1]/r$. Solving for P_1,

$$P_1 = Pr/[(1 + r)^n - 1] \tag{16}$$

and the particular solution becomes $P_j = Pr(1 + r)^{j-1}/[(1 + r)^n - 1]$. Finally, the payment Y is given by

$$Y = I_1 + P_1 = P(r + r/[(1 + r)^n - 1]).$$

Let's look at an example of how this might work. We'll assume a \$1000 loan for four months at 1 percent per month. By equation 16, $P_1 = \$246.28$ (rounding to the penny). The monthly payment is $P_1 + I_1 = P_1 + Pr = 246.28 + 10 = 256.28$. We can further list month by month the breakdown of the payments into interest and principal components. Such a list is called an **amortization schedule.**

j	I_j	Principal	Y
1	10.00	246.28	256.28
2	7.54	248.74	256.28
3	5.05	251.23	256.28
4	2.53	253.75	256.28
Total	25.12	1,000.00	1,025.12

Recall now how the procedure worked. Our analysis gave rise to a difference equation (equation 14), for which we obtained a general solution (equation 15). This solution is then used to continue the analysis, in this case giving rise to a geometric series. We probably wouldn't have been able to continue the analysis with the P_j expressed recursively. Our analysis would have halted before an algebraic stone wall! This illustrates the point that, although recursively defined functions are less desirable than their explicit equivalents, they arise naturally in analysis. When they do, we try to solve them for explicit solutions. In the case of loan amortization, it might have been possible to calculate the payment amount Y by completely different and more direct means, but the difference equation method is fast, efficient, and easy to apply. In other situations, it may well be the only tool available.

2. Solve several problems like the example above. That is: **(1)** Assign the amount of a loan (e.g., $25,000). **(2)** Assign a given repayment time (e.g., 10 years; reduce it to months). **(3)** Assign a given annual interest rate (e.g., 8.7%; reduce it to a monthly rate as a decimal fraction). **(4)** Assuming monthly payments, calculate the monthly payment and an amortization schedule.

9.4 Summary

In summary, *recursive definition* is a technique by which an entire set of objects can be defined via a *recurrence condition,* new objects being defined in terms of old ones. The process is started with one or more *initial conditions,* defining some small core of objects explicitly. The general patterns are presented in Section 9.1.3. Two common applications of recursion are in the definition of functions and the syntactical specification of valid strings in a language.

Mathematical induction is a technique by which an entire set of theorems can be proved via a proof by *induction,* new theorems being proved in terms of old ones. The process begins with one or more *initial steps,* proving some small core of statement explicitly. The general patterns are presented in Section 9.2.3.

These techniques find an application in the solution of *difference equations* (Section 9.3). Starting with the recurrence condition, we find (when possible) its *general solution,* that is, a formula involving the recursion index and one or more *arbitrary constants* that satisfies the recurrence condition for every choice of values for the arbitrary constants. The initial conditions then determine the values of the arbitrary constants and so provide a *particular solution*. We concentrated on *first-order homogeneous differential equations with constant coefficients*.

We've now reached a crucial point in our study. We've just completed the presentation of the mathematical tools we'll need to study applied structures. From now on, things will be somewhat easier, in the sense that we'll *use* the tools we've built and see how they produce concrete, applicable results.

Order Structures and Equivalence Relations

In Chapter 7 we defined a relation on a set A as any subset of $A \times A$. Most of that discussion was directed toward correspondences, relations that imply some sort of transformation from one set, the domain, to another set, the codomain. There was a clear distinction between these roles, and the fact that domain and codomain might sometimes be the same set did not change this. In this chapter, we eliminate that distinction. We consider relations on a set A in which there is no connotation of domain versus codomain. Certain elements of A are related to certain others, and we ask what sorts of structures are possible, and even useful, in this much more relaxed setting. Several structures arise frequently in applied mathematics and computer science. Here we whall explore two of these in detail, *order structures* and *equivalence relations*. As you read, it's important that you not be seduced by the everyday English usage of the word "relation". For instance, in everyday usage, the statement "x is related to y" implies (informally) that "y is therefore related to x". In mathematics, this isn't necessarily the case. Conversely, if we're speaking about people's blood relatives, we might say "John is related to Mary, hence Mary is related to John", but we usually wouldn't say "John is related to himself". The idea of self-relation seems awkward and useless. Nonetheless, it arises frequently in mathematics. As you read, remember that a relation is simply an explicit set of ordered pairs, no more, no less.

10.1 Examples

In the numeric structures $\mathbb{N}$, $\mathbb{Z}$, $\mathbb{Q}$, and $\mathbb{R}$, we've seen that there is a natural ordering of the numbers. In $\mathbb{N}$, for example, we have $0 < 1$, $1 < 2$, $2 < 3$, $3 < \ldots$. This natural ordering is, of course, a relation on $\mathbb{N}$, usually denoted $x < y$. As a set of ordered pairs, this relation is, formally, the set $\{(x, y) \in \mathbb{N} \times \mathbb{N} \mid x < y\}$ and is clearly a subset of $\mathbb{N} \times \mathbb{N}$. Such relations in numeric structures have many important properties. For instance, if $x < y$ and if $y < z$, then

$x < z$. This property goes under the name of *transitive* property. The list of such properties is long; you've seen and used many of them before. We'll explore them in detail here. We've also discussed instances of order relations on nonnumeric structures. For instance, suppose A, B, and C are all subsets of a given universal set U. Again, if $A \subseteq B$ and if $B \subseteq C$, then $A \subseteq C$. Although the context is quite different, the structure still has the transitive property. But there are differences. In $\mathbb{N}$, any two numbers must be related by one of $x < y$, $x = y$, $y < x$. That is, x and y are always *comparable;* they can be compared. In set theory, if A and B are both subsets of U, it may be that $A \subset B$, $A = B$, and $B \subset A$ are all false. It may be that A and B cannot be compared; that they are *noncomparable*. (Note that *non.* Publicity hounds say *in*comparable; mathematicians say *non*comparable.) Although an ordering relation exists in set theory, it is *nontotal;* see Section 10.6.3.

Even in numeric structures, there are examples of nontotal order relations. In $\mathbb{N} - \{0\}$, consider the relation given by "x is a divisor of y". It is often denoted $x\backslash y$. Thus, $3\backslash 6$, $2\backslash 4$, $5\backslash 10$, and so on. For any n in $\mathbb{N} - \{0\}$, $1\backslash n$ and $n\backslash n$ are always true. But not every pair of natural numbers are related. For instance, 5 and 6 are unrelated, as are 3 and 5, 2 and 3, and so forth. In this case, the fact that x is a divisor of y does *not* imply that y is a divisor of x; on the contrary, this is certainly *not* true, unless $x = y$. We can even make a diagram of the divisor relation by using an arrow $x \to y$ to denote that x is a proper divisor of y. We've done that in Figure 10.1 for the set $\{1, 2, 3, 4, 5, 6, 7, 8, 9, 10, 11, 12\}$. Note that we have omitted arrows which are a *consequence* of other arrows: for instance, 3 divides 6 and 6 divides 12, therefore 3 divides 12 (which we have not marked in the diagram; the transitive property takes care of that).

If $A :=$ {living human beings} and the relation is "x and y have a common ancestor, not necessarily living", then we have the usual sense of blood relation, but it includes now the possibility that x is related to x (after all, I and I do have a common ancestor). In other words, the *reflexive* property is valid: x is related to x for every x in A. Also the *symmetric* property is valid here: if x is related to y, then also y is related to x.

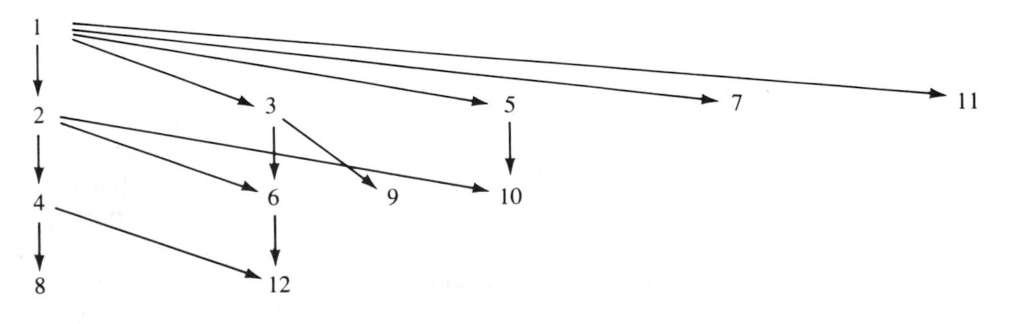

Figure 10.1 Divisibility of integers.

Let $A := \mathbb{Z}$ and let the relation be $x = y \bmod 5$. In this relation, reflexive, symmetric, and transitive property are all valid.

You've no doubt noticed the tendency to write relations as sentences, either in mathematical form, like $x \subseteq y$, or in nearly ordinary language, like "x and y have a common ancestor". On the other hand, a relation R on A is a subset of $A \times A$. The two notations are not contradictory: we'll use systematically the "sentence" xRy as equivalent to the sentence $(x, y) \in R$. It may sound strange, at first, to use the symbol of a set, R, as a verb. Even stranger are two implications. First, the negation of xRy is $x \not R y$. Second, observe that, in general, $R = \{(x, y) \in A \times A \mid xRy\}$; out of consistency, we must then write, for instance, $< = \{(x, y) \in \mathbb{N} \times \mathbb{N} \mid x < y\}$, the set of all pairs for which the first element is less than the second element. Now the symbol $<$ of a verb has been used as a set. The key observation here is that any notation of the form xRy is a *logical statement,* an assertion that, under the appropriate conditions, is either true or false. In the statement xRy, the symbol R (or a similar symbol) is maintained only in general contexts. In most specific cases it's replaced by more usual symbols, as in $x < y$, $x \leq y$, $x = y$, $x \neq y$, $x \subseteq y$, $x \Rightarrow y$. Some other symbols are used in general. For instance, $\sim$ or $\equiv$ when there is an idea of "somewhat equal", $\preccurlyeq$ or $\prec$ when there is an idea of "smaller" or "before".

We must therefore do two things: First, we must consider in detail some implications of properties like the reflexive, symmetric, and transitive properties. Second, we must define special types of relations, such as order and equivalence.

10.2 Universal Relation, Identity and Null Relation

There are three very special relations on any set A. We've already seen them, but let's point out their roles in the present context.

The **universal relation on *A*** is $u_A := A \times A$. Every element is related to every element.

The **identity on *A*** is $i_A := \{(x, x) \in A \times A \mid x \in A\}$. It is also called the **diagonal of *A*.** Every element is self-related; distinct elements are not related. This is the relation of *equality*. We must point out that this is true, unequivocal, perfect equality: x is related to y if and only if x is the same object as y. In other contexts, we have seen and we will see "equality with respect to something", such as numbers having the same remainder under division by 5 or sets having the same number of elements. Such relations are important, but they are not the identity.

The **empty** or **null relation on *A*** is $\varnothing$. No element is related to any element.

None of these three relations is really important in practice, because they are all too simple. But it has to be kept in mind that their use simplifies many statements and formulae. They represent cases of degeneracy, similar to many we've seen before.

10.3 Reflexive and Irreflexive Relations

Definition 1 A relation R on a set A is said to be **reflexive** if xRx for every x in A. A relation R on a set A is said to be **irreflexive** if $x \not R x$ for every x in A.

In other words, a relation is reflexive if every element is related to itself, it is irreflexive if no element is related to itself.

Example On any set A, the identity is reflexive and the null relation is irreflexive. In particular, if $A = \varnothing$, then $i_A = \varnothing$ and the relation is both reflexive and irreflexive. On $\mathbb{N}$, $<$ is irreflexive, $\leq$ is reflexive. On $\mathcal{P}U$ (for any given universal set U), $\subseteq$ is reflexive, $\subset$ is irreflexive. On $\mathbb{N} - \{0\}$, "x is a divisor of y" is reflexive and "x is a proper divisor of y" is irreflexive. On the set of the living human beings, "x is a blood relative of y" (in the usual sense of the phrase) is irreflexive; "x and y have a common ancestor, not necessarily living" is reflexive.

In set-theoretical terms, clearly R is reflexive if and only if $i_A \subseteq R$, and R is irreflexive if and only if $i_A \cap R = \varnothing$. Recall also the definition of R^k for any integer k (see Section 7.4). In particular, recall that $R^0 = i_A$ and $R^1 = R$.

The examples suggest that there is a close connection between any relation (such as $\leq$, $\subseteq$, or "divisor of") and the respective *strict* relation (such as $<$, $\subset$, or "proper divisor of"). For instance, consider again the relation $<$ in $\mathbb{N}$. We can easily define $\leq$ as "$<$ or $=$". More precisely, $x \leq y$ is defined to be the same as "$x < y$ or $x = y$". Conversely, if we are given the relation $\leq$, then we can define the relation $<$ as "$\leq$ but $\neq$". More precisely, $x < y$ is defined as "$x \leq y$ and $x \neq y$". We want to extend this process to the general case.

Definition 2 Let R be a relation on a set A. The **reflexive closure** of R is defined as the relation $R^{[r]} := R \cup R^0$ or, equivalently, as $R \cup i_A$. The **strict** relation of R is defined as $R^{[\text{strict}]} := R - R^0$ or, equivalently, as $R - i_A$.

Examples The examples of reflexive and irreflexive relations above were given in pairs: a strict relation and its reflexive closure. In practice, the shift from a reflexive relation to the strict relation or from an irreflexive relation to the reflexive closure is quite natural and rather irrelevant, except for formal purposes. Each of the two relations uniquely determines the other. Selecting one or the other is a matter of convenience in the presentation.

Exercises 10.3

1. In each of the following cases, find the strict relation and the reflexive closure of the given relation. **(a)** Boolean algebra on set A, relation $\Rightarrow$; that is, "$((\sim x) \vee y) = 1$". **(b)** Finite strings over an alphabet Y, relation "length of

$u <$ length of v''. **(c)** Finite strings over an alphabet Y, relation ``$u = v = \Lambda$ or (first element of u) = (first element of v)''.

10.4 Symmetric and Antisymmetric Relations

Definition 3 A relation R on a set A is said to be **symmetric** if, for every choice of x and y in A, xRy implies yRx. A relation R on a set A is said to be **antisymmetric** if, for every choice of *distinct* elements x and y in A, xRy implies $y \not R x$.

In other words, a relation is symmetric if elements are related or not independently of their order. A typical example is ``blood relative''. A relation is antisymmetric when two distinct elements, if comparable, are related in only one of the two possible orders. A typical antisymmetric relation is ``ancestor of''.

Examples (1) On any set A, the universal relation, the identity, and the null relation are symmetric. The identity and the null relation are also antisymmetric, although degenerately so. (2) On $\mathbb{N}$, $<$ is antisymmetric, and so is $\leq$. (3) On $\mathcal{P}U$ (for any given nonempty universal set U), $\subseteq$ and $\subset$ are antisymmetric. (4) On $\mathbb{N} - \{0\}$, ``$x$ is a divisor of $y$'' and ``x is a proper divisor of y'' are antisymmetric. (5) On the set of the living human beings, ``$x$ is a blood relative of $y$'' (in the usual sense of the phrase) is symmetric, and so is ``x and y have a common ancestor, not necessarily living''.

Theorem 4 Let R be a relation on set A. In set-theoretical terms, (1) R is symmetric if and only if $R^{-1} \subseteq R$; (2) R is symmetric if and only if $R^{-1} = R$; (3) R is antisymmetric if and only if $R \cap R^{-1} \subseteq i_A$; (4) R is antisymmetric if and only if $(R^{[\text{strict}]})^2 \cap i_A = \varnothing$.

Proof: We'll prove the first part and leave the other parts as exercises. According to the definition of set inclusion, we want to prove that, if $(x, y) \in R^{-1}$, then $(x, y) \in R$. Suppose $(x, y) \in R^{-1}$. By the definition of R^{-1}, we know that our pair has been obtained by inverting a pair of R, and such a pair is necessarily (y, x). Therefore $(y, x) \in R$. By the definition of symmetry, the last formula implies $(x, y) \in R$. Q.E.D.

As in the case of reflexivity, any relation can be ``made'' symmetric. Formally,

Definition 5 Let R be a relation on a set A. The **symmetric closure** of R is defined as the relation $R^{[s]} := R \cup R^{-1}$. The **reflexive-symmetric closure** of R is defined as the relation $R^{[rs]} := R \cup R^0 \cup R^{-1}$ or, equivalently, as $(R^{[r]})^{[s]}$.

Examples Among numbers, the symmetric closure of $<$ is $\neq$, the symmetric closure of $\leq$ is the universal relation; the reflexive-symmetric closure of $<$ is the universal relation. Among sets, the reflexive or the reflexive-symmetric closure of either $\subseteq$ or

$\subset$ is *not* in general the universal relation. (Exceptions occur when the universal set is "small"; how small?) The same remark is valid for divisibility among positive integers.

There is no such thing as the "antisymmetric closure" of a relation, because a relation cannot be made antisymmetric by extending it. We should "take away" something from the relation, but this is not a clearly defined process, except in special cases.

Exercises 10.4

1. State whether each of the relations in exercise 1 of Exercises 10.3 is symmetric, or antisymmetric, or neither. Find the reflexive, symmetric, and reflexive-symmetric closure.

2. Let R be a relation on a set A. Prove **(a)** $R^{[\text{strict}]}$ is symmetric if and only if $R^{[r]}$ is symmetric; **(b)** $R^{[\text{strict}]}$ is antisymmetric if and only if $R^{[r]}$ is antisymmetric.

3. Complete the proof of Theorem 4, the set-theoretical form of symmetry and antisymmetry.

10.5 Transitive Structures

Definition 6 A relation R on a set A is said to be **transitive** if, for every choice of x, y and z in A, the fact that xRy and yRz implies that xRz. A pair (A, R) in which R is a transitive relation on set A is called a **transitive structure.**

Examples The relations $<$, $\leq$ (between numbers) or $\subset$, $\subseteq$ (between sets) are all transitive. Blood relation is not: a cousin of a cousin is not necessarily a blood relative. Inequality ($x \neq y$) is not transitive: if $x \neq y$ and if $y \neq z$, we cannot conclude that $x \neq z$.

If you recall from Section 7.4 the definition of R^2, you'll see immediately that R is transitive if and only if $R^2 \subseteq R$. By Theorem 7.17, this occurs if and only if, for every positive k, $R^k \subseteq R$. As in the case of reflexivity and symmetry, any relation R can be made transitive.

Consider again the example given in Figure 10.1 (divisibility of positive natural numbers; recall that $x \backslash y$ is a symbol for "x divides y"). The diagram states pictorially that $1\backslash 2$, $1\backslash 3$ . . ., $2\backslash 4$, $2\backslash 6$, $4\backslash 8$, $4\backslash 12$, $6\backslash 12$ . . .. The fact that $2\backslash 4$ and $4\backslash 8$ implies that $2\backslash 8$ (by transitivity); the fact that $2\backslash 6$ and $6\backslash 12$ implies that $2\backslash 12$, and so on. The last two conclusions ($2\backslash 8$ and $2\backslash 12$) are not explicitly given in the diagram; they are tacitly implied via transitivity. In other words, if we are given a relation and want to make it transitive, we must "adjoin" the pairs that are obtained by applying transitivity to pairs that are known to be related. This is what the next definition does formally and in general.

Definition 7 Let R be a relation on a set A. The **transitive closure** of R is defined as the relation

$$R^{[t]} := \bigcup_{k \in \mathbb{N} - \{0\}} R^k.$$

The **reflexive-transitive closure** of R is defined as the relation

$$R^{[rt]} := \bigcup_{k \in \mathbb{N}} R^k$$

or, equivalently, as $(R^{[r]})^{[t]}$. The **symmetric-transitive closure** of R is defined as the relation

$$R^{[st]} := \bigcup_{k \in \mathbb{Z} - \{0\}} R^k$$

or, equivalently, as $(R^{[s]})^{[t]}$. The **reflexive-symmetric-transitive closure** of R is defined as the relation

$$R^{[rst]} := \bigcup_{k \in \mathbb{Z}} R^k$$

or, equivalently, as $((R^{[r]})^{[s]})^{[t]}$.

Examples (1) The relations $<$, $\leq$, $\subset$, $\subseteq$, and "divisor of" are already transitive, and their transitive closures coincide with the given relations. (2) The transitive closure of "blood relation" is a complicated one: starting from any one person, we should go "up" to an ancestor, then "down" to a descendant of such ancestor, then "up" to an ancestor, and so on, for a finite number of times. No wonder that inheritance laws nearly always avoid such a situation. (3) Among integers, the relation "is the immediate predecessor of" is not a transitive one: 3 is the immediate predecessor of 4, and 4 is the immediate predecessor of 5, but 3 is not the immediate predecessor of 5. The transitive closure is $<$ (think a moment: why?). (4) In $\mathbb{Z}_3$, the relation $y = x + 1 \pmod 3$ leads to a strange conclusion. By definition, $0R1$, $1R2$, $2R0$; a two-step process gives $0R2$, $1R0$, $2R1$; a three-step process gives $0R0$, $1R1$, $2R2$. There are nine ordered pairs, with repetitions; we've obtained nine ordered pairs that are related. Consequently, the transitive closure of the given relation is the universal relation. Perhaps this is more than what we want.

This last example warns us that, if we want some kind of order, or "before/after", we do want transitivity, but we want to stay away from a "circular" situation. This remark leads to the concept of *order*, which we shall consider next.

Exercises 10.5

1. For each of the examples of problem 1 in Exercises 10.3, determine whether the given relation is transitive. If not, find its transitive closure.
2. Each of the following pairs of data comprises a set A and a relation S on A. For each pair: **(i)** if S is antisymmetric, find the reflexive-transitive closure R of S; **(ii)** in either case, find the reflexive-symmetric-transitive closure T of S. A diagram or a table may help. A table is a suitable form for the answer. **(a)** $A := \{v, w, x, y, z\}$, $S := \{(v, x), (w, y), (y, z)\}$. **(b)** $A := \{v, w, x, y, z\}$, $S := \{(v, x), (y, w), (z, y)\}$. **(c)** $A := \{u, v, w, x, y, z\}$, $S := \{(v, v), (w, w), (z, z)\}$. **(d)** $A := \{u, v, w, x, y, z\}$, $S := \{(v, x), (y, w), (z, y), (w, y)\}$.
3. Let Y be a given alphabet and Y^* the free semigroup of Y (Section 6.1), that is, the set of all finite strings of elements of Y, including the null string Λ, with the operation of concatenation of strings. On Y^*, define a relation "initial substring" or **prefix,** that is, u **prefix** v if and only if there exists a string z in Y^* such that $u{:}z = v$. For instance, aba **prefix** abaca, because (aba):(ca) = (abaca). Prove that **prefix** is a transitive relation.

Complements 10.5

1. In Section 5.7 and Complements 8.1 we've considered partitions of a nonempty set A and some of their properties. We're going to see here an additional concept, *refinement* of a partition.

Definition 8 Let S and T be partitions of the same nonempty set A. Then S is called a **refinement** of T if, for each block u of S, there is a block v of T such that $u \subseteq v$. (Recall that u and v are each subsets of A.) If, for at least one u, it happens that $u \subset v$, then S is called a **strict refinement** of T. The symbols we're going to use for the statements "S is a refinement of T" or "S is a strict refinement of T" are, respectively, $S \preccurlyeq T$ and $S < T$.

Let's illustrate this with an example. Let $A := \{a, b, c, d\}$, and let $S := \{\{a\}, \{b, c\}, \{d\}\}$ and $T := \{\{a\}, \{b, c, d\}\}$ be partitions of A. The idea is that block $\{b, c, d\}$ of T has been broken down, or refined, into blocks $\{b, c\}$, and $\{d\}$ of S. The definition states that every block of S should be a subset of some block of T. You can easily verify this in our example. Moreover, every block of S is a subset of *one and only one* block of T. The definition doesn't say this explicitly because it doesn't have to. It's always true anyway. Prove this statement in general. If we now let $U := \{\{a, b\}, \{c, d\}\}$ be another partition of A, you can easily verify that neither S nor T is a refinement of U, nor is U a refinement of S or T.

Now let's look at the degenerate aspects of partition refinement. First, every partition is a refinement of itself. Second, recall that there is a one-block partition of A composed of the single block A: $I := \{A\}$. Every partition of A is a refinement of I above. Third, recall that there is one partition of A into blocks that are all singletons: $J := \{\{x\} \mid x \in A\}$. Partition J is a refinement of every partition of A. In our example $J = \{\{a\}, \{b\}, \{c\}, \{d\}\}$. Expressed as a formula, for every partition S of A, $J \preccurlyeq S \preccurlyeq I$. These observations all hold degenerately when $A = \varnothing$ because there are no partitions.

There is an alternative, and much "cleaner", definition of partition refinement. Not only is it shorter than Definition 8, but it provides a practical and efficient way of testing whether one partition is a refinement of another. As with any alternative definition, we must first state it as a theorem and prove that it is equivalent to the original. Recall the definition of cross-partition $V * W$ of any two partitions V and W of the same set A. Prove the following:

Theorem 9 Let S and T be partitions of set A. Let $\preccurlyeq$ denote partition refinement. Let $*$ denote cross-partition. Then $S \preccurlyeq T$ if and only if $S = S * T$.

For a set A, we have denoted $C(A)$ the set whose elements are all the partitions of A. Interestingly, the set $C(A)$ exhibits a structure within itself. The basis of this structure is the partition refinement. Prove that $(C(A), \preccurlyeq)$ is a transitive structure and that $\prec$ is its strict relation.

2. Find several refinements of each of the following partitions: **(a)** Integers into {even integers} and {odd integers}. **(b)** Five-digit binary numbers into {those having an even number of ones} and {those having an odd number of ones}. **(c)** Octal strings into {those that have no 0 digit} and {those that have at least one 0 digit}. **(d)** The symbols in Figure 10.3 into {those involving a closed region (like P or Q)} and {those involving no closed region (like *X* or *Y*)}. **(e)** The symbols in Figure 10.3 into {those that have some loose end (like A or X)} and {those with no loose end (like O or, in the specific physical symbols in the figure, B or D)}.

3. A very general method that produces transitive structures is the following. We've just seen some examples of it.

Definition 10 Let $(G, *)$ be a semigroup (Section 8.1.2), that is, a set G with an associative total binary operation $*$. Let H be a subset of G. Set H is said to be **closed** under operation $*$ if, whenever $x \in H$ and $y \in H$, also $x * y \in H$.

Let H be closed under operation $*$. For any two elements u and v of G, define $u \preccurlyeq v$ if and only if there exists a w in H such that $u * w = v$. In other words, define the relation $\preccurlyeq$ on G by $\preccurlyeq := \{(u, u * w) \in G \times G \mid u \in G \text{ and } w \in H\}$. Similarly, define relation $\succcurlyeq := \{(u * w, u) \in G \times G \mid u \in G \text{ and } w \in H\}$. Prove that, under these conditions, $\preccurlyeq$ and $\succcurlyeq$ are both transitive relations.

4. In each of the following cases, show that the given relation can be generated by the construction in Definition 10 above. In each case, identify G, H, and the operation $*$. **(a)** Relation "divisor of" on $\mathbb{N} - \{0\}$. **(b)** Relation "initial substring of" on the free semigroup of an alphabet. **(c)** Relation "refinement of" on the set of all partitions of a given nonempty set. **(d)** Relation $\leq$ on $\mathbb{N}$. **(e)** Relation $\leq$ on $\mathbb{Z}$. **(f)** Relation $\subseteq$ on $\mathcal{P}U$, where U is a given, nonempty, universal set.

5. In each of the following cases, verify that the given set H is closed under the given operation. Then describe in usual terms the transitive relation generated according to Definition 10 above. **(a)** Given a nonempty universal set U, define $G := \mathcal{P}U$, $H := G$, and let the operation be $\cap$. **(b)** $G :=$ Boolean set $\{0, 1\}$, $H := G$, operation $\wedge$. **(c)** $G := \mathbb{Z}$, $H := \mathbb{N}$, operation $+$.

6. Generalize the construction in Definition 10 above. Consider the relations $\{(u, w * u) \mid u \in G \text{ and } w \in H\}$, $\{(u, w * u * t) \mid u \in G \text{ and } w \in H \text{ and } t \in H\}$, $\{(w * u, u) \mid u \in G \text{ and } w \in H\}$ $\{(w * u * t, u) \mid u \in G \text{ and } w \in H \text{ and } t \in H\}$. Examples are "final substring" and "substring," respectively. In each of the cases presented in complements 4 and 5 above, consider the new relations. Are they the same as the one considered previously? Are they different? How can they be described directly?

10.6 Order Structures

Order structures are those for which there is a concept of larger/smaller, before/after, greater/less or the like. Several types of order structures are extremely useful in theoretical and practical applications.

10.6.1 Partial Order

A very basic type of order relation is the *partial order*. Intuitively, a partial order is a specifically given relation involving a notion like "before", or "part of", or "less than", and so on.

Definition 11 Let A be a set and R a relation on A. The relation R is called a **partial order on** set A if it is reflexive, antisymmetric, and transitive. A pair (A, R) in which R is a partial order on A is called an **order structure,** or a **partially ordered set,** or a **poset.**

Examples (1) Let A be the collection of the subsets of set $\{a, b, c\}$, that is, $A = \{\emptyset, \{a\}, \{b\}, \{c\}, \{b, c\}, \{c, a\}, \{a, b\}, \{a, b, c\}\}$. Let the relation be $\subseteq$, that is, set inclusion. Then $(A, \subseteq)$ is a partial order. (2) The most typical example of partial order is $(\mathbb{N}, \leq)$. This order is also *total;* any two elements are comparable. (3) For any set A, the null relation and the identity are always partial orders, possibly in a degenerate way (for instance, when $A = \emptyset$). If A has more than one element, the universal relation is never a partial order, because *xRy and yRx* for every pair of elements.

It is important to remember that under partial order nothing guarantees that all pairs of elements are related. A partial order might well be just that—partial.

The adjective "partial" does not exclude, however, that the order be also "total", that is, that every pair of elements be comparable.

For order relations, there is a practical advantage in focusing on the *core* relation. As an example, consider the relation $\leq$ on $\mathbb{N}$. The essential part of the relation is $0 \leq 1$, $1 \leq 2$, $2 \leq 3$, ... When we know that, we are able to conclude (by transitivity) that $0 \leq 2$, $1 \leq 3$, . . ., $0 \leq 3$, ... We also know (by reflexivity) that $0 \leq 0$, $1 \leq 1$, ... In general,

Definition 12 Let R be a partial order on a finite set A. The **core relation** of R is defined as $R^{[\text{core}]} := R^{[\text{strict}]} - \bigcup_{k \in \mathbb{N} - \{0, 1\}} (R^{[\text{strict}]})^k$.

Theorem 13 Let R be a partial order on a finite set A with at least two elements. Then

(1) $R^{[\text{core}]} = R^{[\text{strict}]} - \bigcup_{k=2}^{\#A-1} (R^{[\text{strict}]})^k$.

(2) $R^{[\text{core}]} = (R^{[\text{strict}]})^{[\text{core}]}$.

(3) $R^{[\text{strict}]} = (R^{[\text{core}]})^{[\text{t}]}$.

(4) $R = (R^{[\text{core}]})^{[\text{rt}]}$.

(5) Consequently, R, $R^{[\text{strict}]}$, and $R^{[\text{core}]}$ each uniquely determines the other two.

(6) A relation on A is the strict relation of a partial order if and only if it is irreflexive, antisymmetric, and transitive.

Proof: Exercise.

Note that the idea of core relation cannot be extended to every type of relation and, even when it can, the definition above may not be the appropriate one. The present definition is valid for *order structures*. As a counterexample, try to apply it to the universal relation on a set with more than two elements; it would yield the null relation, which would contradict part 4 of Theorem 13.

10.6.2 Order-Preserving Functions

Let's consider the order structure analogues of homomorphism and isomorphism.

Definition 14 Let (A, R), and (A^*, R^*) be partially ordered sets, and let $f{:}A \to A^*$ be a total function. The function f is called **order-preserving** if $xRy \Rightarrow (fx)R^*(fy)$ for every choice of x and y in A. It is called **order-inverting** if $xRy \Rightarrow (fy)R^*(fx)$ for every choice of x and y in A. If f is an order-preserving bijection, then f is called an **order-isomorphism** and the two order structures are said to be **order-isomorphic.** If f is an order-inverting bijection, then f is called an **order-antiisomorphism** and the two order structures are said to be **order-anti-isomorphic.**

Example (1) Consider the function $x \mapsto 3x$ on the poset $(\mathbb{N}, \leq)$. It's order-preserving, because any pair (x, y) such that $x \leq y$ maps to the pair $(3x, 3y)$, and $3x \leq 3y$. (2) Consider the posets $(\mathbb{N}, \leq)$ and $(\mathbb{N}, \geq)$. Let $f{:}\mathbb{N} \mapsto \mathbb{N}$ be defined by $x \mapsto x$. Then f is an order-inverting function. It is actually an order-antiisomorphism.

Saying that two posets are order-isomorphic or order-antiisomorphic is a very powerful and restrictive assertion. It says, in effect, that the two posets have identical structures and differ only in the "labelling" of their elements and perhaps in the "direction" of the relation. In particular, if two posets are order-isomorphic, then their diagrams are identical except for this labelling. In the exercises, you'll be asked to produce examples of this.

10.6.3 Total Order

Definition 15 Let (A, R) be a partial order. The relation R is called a **linear order** or **simple order** or **total order** if, for every choice of x and y in A, at least one of xRy, yRx is true. If that is the case, then the structure (A, R) is called a **linearly ordered set,** or **loset.** An order relation that is partial but not linear is called a **nontotal order.**

Typical examples are relations "less than or equal to" between numbers or alphabetical order among strings on an ordinary alphabet.

The term *linear* is a bit misleading, since it's often confused with linear functions (like the straight line $y = mx + b$) in continuous structures. *Total* order is a more accurate term, and more consistent with our usage. Nonetheless, *linear order* is by far the most common term, so be careful. Another common area of confusion arises as to the relations $<$ and $>$ (strict inequalities) on $\mathbb{N}$, $\mathbb{Z}$, $\mathbb{Q}$, and $\mathbb{R}$, and the *trichotomy axiom* on these structures. Resolving the confusion isn't difficult, but it does require careful language.

Definition 16 A relation R on a set A is said to be **trichotomic** if, for all x and y in A, precisely one of xRy, $x = y$, or yRx is true.

With this definition, the correct relationship between trichotomy axioms and linear orders can be stated as a theorem.

Theorem 17 Let R be a relation on set A. Then $R^{[\text{strict}]}$ is irreflexive, trichotomic, and transitive if and only if $R^{[r]}$ is a linear order.

Proof: Exercise.

Examples On $\mathbb{N}$, $\mathbb{Z}$, $\mathbb{Q}$, and $\mathbb{R}$, the relations $<$ and $>$ are each irreflexive, trichotomic, and transitive. Their reflexive closures are, respectively, $\leq$ and $\geq$, each of which is a linear order.

10.6.4 Lexicographic Order

Lexicographic order is just a generalization of the idea of alphabetizing words. In order to be able to do that systematically and in general, we need some formalism.

Definition 18 Let Y be a finite nonempty alphabet. Let $\leq$ be a given total order on Y. In simple cases, Y will be a usual alphabet and $\leq$ the alphabetical order of its letters. Let Y^* be the set of the finite strings of elements of Y, as in Section 6.1, with concatenation. The **lexicographic order** on Y^* is the relation $\preccurlyeq$ given by the following definition, which is recursive on the length of a string. Given that u and v are any two elements of Y^*,

1. If $u = \Lambda$, then define $u \preccurlyeq v$;
2. If $u \neq \Lambda$ and if $v \neq \Lambda$, let x be the first element of u and let y be the first element of v. In the case in which $x \neq y$, define $u \preccurlyeq v$ if $x \leq y$ in Y, and define $v \preccurlyeq u$ if $y \leq x$ in Y;
3. With the same notations as in part 2, and in the case in which $x = y$, let u^* and v^* be the strings such that $u = (x){:}u^*$ and $v = (y){:}v^*$. Then define $u \preccurlyeq v$ if $u^* \preccurlyeq v^*$ and define $v \preccurlyeq u$ if $v^* \preccurlyeq u^*$.

We have to prove, of course, that what we obtain is a total order on Y^*. Do that as an exercise.

Exercises 10.6

1. Draw a graph similar to the one in Figure 10.1 for the core of the order structure $(\mathcal{P}U, \subseteq)$ in each of the following cases: $U := \{a\}$, $U := \{a, b\}$, $U := \{a, b, c\}$, $U := \{a, b, c, d\}$. Such graphs are called **Hasse diagrams.**
2. For several of the posets we saw before, draw the Hasse diagram of the core relation.
3. For each of the relations R given in exercise 2 of Exercises 10.5: if relation R is a partial order, check whether it is a total order, and draw a Hasse diagram.
4. Given alphabet $X := \{\text{A, B, C}\}$, $Y := X^1 \cup X^2 \cup X^3$, and an order $\preccurlyeq$ on Y defined by $x \preccurlyeq y$ if and only if x is a substring of y, draw a Hasse diagram of the given order.
5. There is one case in which the universal relation on a nonempty set will be a partial order. What is that case?
6. A given alphabet {3, *, 1} is ordered totally according to the core relation $3 < * < 1$. Given the strings **13*, **11*, **1*1, 13*3*, 13**1, 13*31, place them in lexicographic order.
7. The given alphabet {%, *, 1} is ordered totally according to the core relation $\% < 1 < *$. List the strings 1%**1, **1%*, **11*, 1%*%*, 1%*%1, **1*1, in lexicographic order.
8. Consider the straight line in the xy-plane, given by $y(x) = mx + b$, where $m > 0$. Show that this function $y{:}\mathbb{R} \to \mathbb{R}$ preserves the order $<$ on $\mathbb{R}$. Describe what happens when $m = 0$ or when $m < 0$.
9. Prove that the integers and the odd integers are order-isomorphic.

10. Prove that the posets $(\{1, 2, 3, 5, 6, 10, 15, 30\}, \text{DIVIDES})$ and $(\mathcal{P}\{a, b, c\}, \subseteq)$ are order-isomorphic. Draw a diagram for each and compare them.
11. Prove that the posets $(\{0, 1\}, \Rightarrow)$ (Boolean) and $(\mathcal{P}\{a\}, \subseteq)$ (set-theoretical) are order-isomorphic. Draw a diagram for each and compare them.
12. Prove Theorem 13 on the core of a partial order.
13. Prove Theorem 17 on trichotomy.
14. Prove that the lexicographic order is a total order (see statement after Definition 18).
15. Consider all possible partial orders on set $A := \{1, 2, 3, 4\}$. Place together those that are order-isomorphic to each other. Use separate lists for those that are not. Identify those pairs that are order-antiisomorphic. (This problem has many practical applications but becomes extremely difficult when the set grows big. If the order of magnitude of $\#A$ is in the hundreds, the problem is nearly intractable even with present-time high speed computers. For small sets, you can try a computer solution.)
16. Find the number of all total orders on a given, finite set. [Hint: First show that a total order is practically a permutation of the set.]
17. On $\mathbb{N}$, find at least two total orders that are not order-isomorphic nor order-antiisomorphic. [Hint: something like $0 \preccurlyeq 2 \preccurlyeq 4 \preccurlyeq \cdots \preccurlyeq 1 \preccurlyeq 3 \preccurlyeq 5 \preccurlyeq \cdots$.]
18. Prove that, for any finite poset (A, R), the identity on A is an order-antiisomorphism from (A, R) to (A, R^{-1}).

Complements 10.6

1. Show that each of the following structures is a partial order. **(a)** $(\mathbb{N}, \leq)$; **(b)** $(\mathbb{Z}, \leq)$; **(c)** $(\mathbb{Q}, \leq)$; **(d)** $(\mathbb{R}, \leq)$; **(e)** $(\mathbb{N} - \{0\}, \text{"divisor of"})$; **(f)** $(\mathcal{P}U, \subseteq)$, where U is a given nonempty universal set; **(g)** $(\{0, 1\}, \Rightarrow)$, where $\Rightarrow$ is the Boolean operation of implication; **(h)** $(C(A), \preccurlyeq)$, where $C(A)$ is the set of partitions of a set A and $\preccurlyeq$ is the relation of partition refinement.

10.7 An Application: PERT

We begin here to consider a practical application of the idea of partial order and, later, of graph. The application we have in mind is the coordination of the several phases of a complex project. Let's start with a fairly simple project, building a rudimentary toy wagon with a wood body and metal wheels. The metal hardware (wheels, axles, etc.) is to be purchased ready made, and its precise dimensions will be known only after the purchase. The body is to be made out of wood, which is to be cut after the dimensions of the hardware are known, and then glued together and varnished. All the necessary tools are already available.

The problem is to identify and coordinate the separate actions that contribute to the project. According to current jargon, we'll call **event** a stage at which some **activities** (one or more) have been completed and a next activity can be considered. The key aspect is that each event (except an initial one) can take place only after certain other event or events. We then have a relation, "event x must take place before event y". For instance, we must wait until the varnish is dry and the wheel assembly has been put together before we can mount the wheels on the body; we must wait until the glue is dry before we can start varnishing; and so on. Is such a relation a partial order? The relation is clearly irreflexive, but that is no problem: we can always consider a strict partial order. That's what we'll try to do. Very likely the relation, as given, is not transitive. For instance, we didn't say that the glue must be dry before we mount the wheels, but that is understood by transitivity. In other words, whichever relation is given explicitly, we'll consider its transitive closure in order to check whether we have a partial order. The important point is whether the transitive closure is antisymmetric. If it is not, then we find ourselves in a situation in which a certain event must come "before" itself. Such a situation may be highly undesirable in most cases, such as our example of the wagon. In other cases, it may be highly desirable: an event x may trigger a sequence of other events which ultimately influence x (or a recurrence of x). As an example, a robot that is supposed to tighten a nut grabs the nut, "senses" how good the grab is, then (if necessary) grabs it better, possibly several times, and ultimately tightens the nut. This situation is called *feedback.* We'll limit our considerations to the cases in which there is no feedback, that is, to the cases in which we have a partial order. We'll see in Section 11.6 how to establish whether the transitive closure of the given relation is antisymmetric and how to handle the problem of coordinating several activities. At that time, we'll consider a specific technique called PERT.

For the time being, let's look in detail at the events we can identify in our project.

0. Ready to start.
1. Approximate design is ready.
2. Wheels, axles, etc. have been purchased.
3. Lumber and glue have been purchased.
4. Ready to measure, cut, and glue wood parts.
5. Varnish has been purchased.
6. Glue is dry.
7. Ready to varnish.
8. Wheels, axles, etc. have been put together to form the wheel assembly.
9. Varnish is dry.
10. Ready to mount wheel assembly onto body.
11. Wheel assembly is mounted. Wagon is ready.

The core of our relation is described by the activities of the project and is, very likely, the following (see Figure 10.2 for a Hasse diagram).

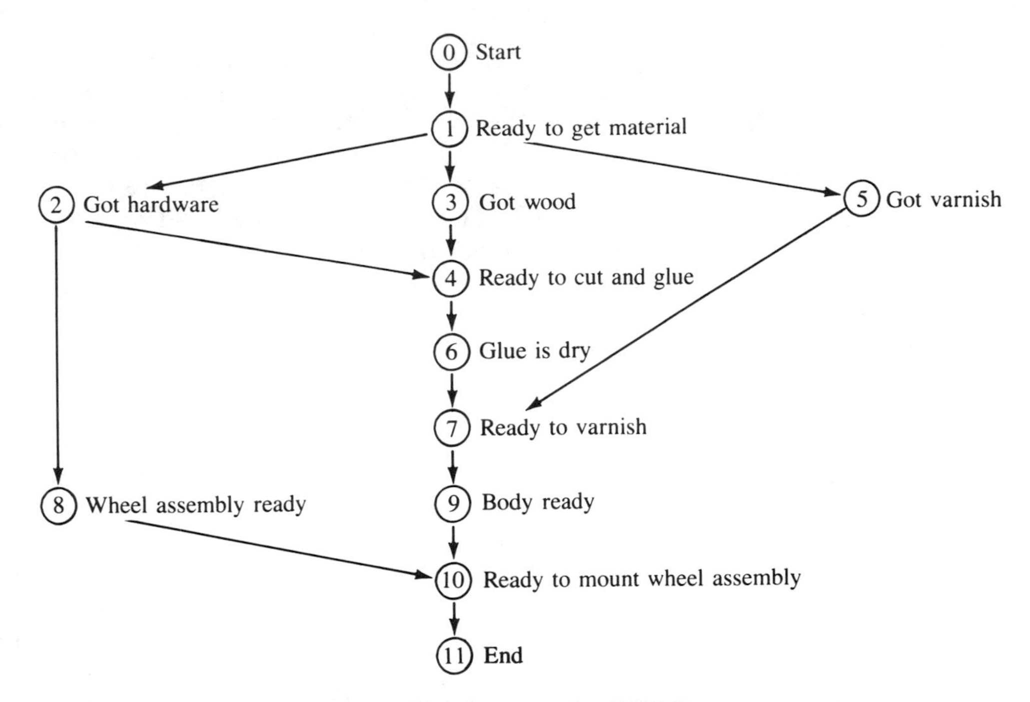

Figure 10.2 An example of PERT.

$0 \to 1$ Do approximate design.
$1 \to 2$, $1 \to 3$, $1 \to 5$ Get the material.
$2 \to 4$, $3 \to 4$ No activity.
$2 \to 8$ Put together the wheel assembly.
$4 \to 6$ Measure, cut, and glue wood parts; wait for glue to dry.
$6 \to 7$, $5 \to 7$ No activity.
$7 \to 9$ Varnish body; wait until dry.
$8 \to 10$, $9 \to 10$ No activity.
$10 \to 11$ Mount wheel assembly onto body.

The **zero time activities** (denoted above as "no activity") are, occasionally, convenient devices to point out precisely what must be done before what. In this simple case, it's rather easily checked that the transitive closure of our core relation is antisymmetric (the Hasse diagram may help). In other words, we have the core relation of a partial order. In more complicated cases, we may need theory and methods. That's why we may want to use PERT. We'll do that in Section 11.6.

Exercises 10.7

1. On set $A := \{0, 1, 2, 3, 4, 5, 6, 7, 8, 9, 10, 11\}$, where the numbers refer to the events in the text for the toy wagon), consider the relation S defined by the pairs explicitly listed as $x \to y$. Find all event sequences $(0, u, v, \ldots, 11)$ such that $0 \to u \to v \to \cdots \to 11$.

2. Construct the transitive closure R of the given relation S. Note that we do not need all S^k for every k. A few will do. Verify that R is actually the strict relation of a partial order.

10.8 Lattices

In this section, we examine in more detail some properties of order structures. We'll pay special attention to an aspect that is somewhat obvious in the "usual" orders but that presents some unexpected twists in general. Consider the special case of the partial order $(\mathbb{N}, \leq)$. The set $\mathbb{N}$ has a "smallest" element, and so does every nonempty subset of $\mathbb{N}$. It is only natural to call such a "smallest" element a *minimum* or a *least* element. However, $\mathbb{N}$ does not have a "largest" element. Some subsets of $\mathbb{N}$ have a largest element (such as the set of the numbers that can be represented with up to five digits in any one given base), some don't (for instance, the set of the even numbers). It's clear that infinite sets present some difficulties, but even for finite sets we'll have problems when the order is not total. What we intend to do next is study those elements generically called *extreme*, which will be defined specifically in each case.

10.8.1 Extreme Elements

We'll use a partially ordered set $(A, \preccurlyeq)$ with the understanding that A is simply a set, and $\preccurlyeq$ is a partial order, not necessarily *less than or equal to*. This notation will make the discussion a bit clearer. When needed, we'll use the symbol $<$ for the strict relation.

Intuitively, a *maximal* element of a set is an element that is not "below" any other element of the set, and a *minimal* element is an element that is not "above" any other element of the set. This idea produces the following general definition.

Definition 19 Let $(A, \preccurlyeq)$ be a partially ordered set. Element v $(v \in A)$ is said to be a **minimum** or a **minimal element** if, for all x in A, $x \preccurlyeq v$ implies $x = v$. That is, the only element x in A satisfying $x \preccurlyeq v$ is v itself; there is nothing in A "less than" v. Similarly, w $(w \in A)$ is a **maximum** or a **maximal element** if, for all x in A, $w \preccurlyeq x$ implies $w = x$; there is nothing in A "greater than" w.

For example, consider the relation *divides* on the set $A := \{2, 3, 4, 6, 8\}$ in $\mathbb{N}$. The diagram of the *core* relation for this is shown below.

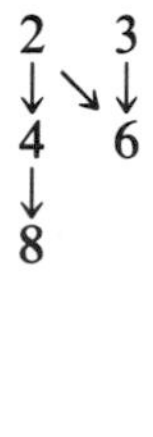

$$\begin{array}{ccc} 2 & & 3 \\ \downarrow & \searrow & \downarrow \\ 4 & & 6 \\ \downarrow & & \\ 8 & & \end{array}$$

Elements 2 and 3 are both minimal, but they are noncomparable to each other. Elements 6 and 8 are both maximal, but they are noncomparable to each other. An important idea is illustrated here. *A partially ordered set can have multiple minima and multiple maxima*. In fact, for the identity relation, since each element is related only to itself, each element is *both minimal and maximal*.

It's important to distinguish minimal/maximal elements from *least/greatest* elements. Intuitively, a least element is one that is "below" all other elements of the set, a greatest element is one that is "above" all other elements of the set. Note the difference with maximum and minimum: a least element is below all others, a minimal element is not above any other.

Definition 20 Let $(A, \preccurlyeq)$ be a partially ordered set. Element v $(v \in A)$ is said to be a **least element** in A if $v \preccurlyeq x$ for all x in A. Element w $(w \in A)$ is said to be a **greatest element** in A if $x \preccurlyeq w$ for all x in A.

If A has a least element v, then v must be a minimum. The difference is that, for v to be least, *v must be related to every other element in A*. This is why 2 and 3 in the example above are minimal, but *neither is least*. There is no least element. Similarly, for w to be greatest, it must be a maximum, *and be related to every other element in A*. So, above, 6 and 8 are maxima, but neither is greatest. Thus, not every poset has least/greatest elements. In the exercises, you'll be asked to show that least and greatest elements, if they exist, are unique. In the toy wagon example (Section 10.7) "ready to start" is a least element and also the only minimum, and "wagon is ready" is a greatest element and also the only maximum. Suppose we eliminate the somewhat artificial events 0 and 1. Then our project would start with the three "purchases", that is, with events 2, 3, and 5. These events would all be minimal events, but none of them would be a least event.

But does every poset have minimal/maximal elements? Not necessarily. First, $(\varnothing, \varnothing)$ is degenerately a poset but has neither minimal nor maximal elements. Second, in the poset $(\mathbb{Z}, \leq)$, where $\leq$ here *does* refer to "less than or equal to", there exists neither a largest nor a smallest integer. Thus $(\mathbb{Z}, \leq)$ has neither minima nor maxima. It's even possible for a poset to have a *unique* minimum (maximum) element, yet fail to have a least (greatest) element. Consider the set $A :=$ {odd natural numbers} $\cup$ {2}, and the DIVIDES relation. Since 2 doesn't divide any odd natural number, 2 is maximal; in fact, it's the only maximum. But A clearly has no greatest element. The moral here is that infinite sets, like degenerate cases, must always be treated carefully. Excluding these two cases, however, we may draw a conclusion:

Theorem 21 Let $(A, \preccurlyeq)$ be a partially ordered set. (1) If A is nonempty and finite, then A must contain at least one minimum and at least one maximum. (2) If, in addition, the order is total, then there exists a unique minimum and a unique maximum and they coincide, respectively, with the only least element and the only greatest element.

Proof: Exercise. [Hints: For part 1, proceed by contradiction and suppose that a maximum does not exist. Then start from an element x_0 of A, which must exist, because A is nonempty. Proceed by recursion: for each x_h ($h \in \mathbb{N}$), we have assumed that it is not a maximum and there is therefore an x_{h+1} such that $x_h < x_{h+1}$. By transitivity, all the previously constructed x_i are distinct from x_{h+1}. Consequently, we have an infinite sequence of distinct elements of A. For part 2, suppose x and y are maxima. Then $x \preccurlyeq y$ or $y \preccurlyeq x$ (because the relation is total) and the maximality condition implies what?]

Except in special cases, however, there is no way to guarantee the existence of least/greatest elements. We have already mentioned that such an element, if existent, is unique.

10.8.2 Infima and Suprema

All partial orderings of A—including linear orderings—are described as sets of pairs of elements of A. If we made a list of every possible ordered pair (x, y) in $A \times A$, including with each pair the truth value of the statement $x \preccurlyeq y$, we would completely determine the order relation. As thorough as such a list would be, it would also be cumbersome to deal with, and, moreover, wouldn't tell us *easily* what we wanted to know. But, in many instances, what we really want to know is *how x and y relate to the rest of A*. We could derive this information laboriously from the list, but for a certain class of posets, there's a better way. Intuitively, we give the "closest" element that lies "before" both x and y and the "closest" element that lies "after" x and y. Posets for which this is possible are called *lattices*. Not every poset can be defined as a lattice, and for those that can, the definition requires a fair amount of foundation. But the increased power of a lattice over a poset, and the fact that most of the important examples of posets are also lattices, make the added work well worth the trouble. We begin with a poset $(A, \preccurlyeq)$, where set A is nonempty, finite or infinite. The definitions that follow occur in pairs. We've already seen two examples of this pairing on posets: minimal (maximal) elements, and least (greatest) elements. We're simply going to continue this convention.

To see intuitively the idea of *bound*, consider the fact that some element of A may be "below" (or equal to) every element of a given subset B of A. For instance, consider the relation $\leq$ among natural numbers; zero is below every positive natural number; then we say that zero is a *lower bound* of the set of the positive natural numbers. But 1 is also a lower bound for the same set because it is less than or equal to any positive natural number. Actually, 1 is the greatest number that satisfies this condition; that is, 1 is the *greatest lower bound* for the set of the positive natural numbers.

Definition 22 Let $(A, \preccurlyeq)$ be a poset. Let B be a nonempty subset of A. Then an element v of A is called a **lower bound (upper bound)** for B if $v \preccurlyeq x$ for every x in B ($x \preccurlyeq v$ for every x in B). If the set of lower bounds (upper bounds) for B has a greatest (least) element a, then a is called the **greatest lower bound** of B or the **infimum** of B (**least upper bound** of B or **supremum** of B). These are usually abbreviated as **glb** B or **inf** B (**lub** B or **sup** B).

Example Let $A := \mathbb{N}$. Let the order be $\leq$. Let B be the subset $\{3, 4, 5, 6\}$. Then 0, 1, 2, and 3 are each a lower bound for B. In fact, these elements are all the lower bounds of B. Also, there is a greatest one among them, that is, 3. Consequently, $3 = \textbf{inf}\, B$.

Before applying Definition 22, we should remark on some of the subtleties involved, some things that the definition allows and others it does not allow. We'll state them in terms of lower bounds and infima, although similar remarks apply to upper bounds and suprema:

1. For a to be a lower bound for B, it must be that $a \preccurlyeq x$ for *all* x in B. If $a \preccurlyeq x$ for some, but not all, x in B then a is not a lower bound for B.
2. A lower bound for B might or might not be an element of B itself. In fact, the infimum of B, if it exists, might or might not be an element of B. But B can contain at most one of its own lower bounds. To see this, suppose that b and b^* are both lower bounds for B and are both elements of B. By definition, $b \preccurlyeq b^*$ and $b^* \preccurlyeq b$. Thus, by antisymmetry, $b = b^*$. If B does contain b as a lower bound, then b must be the infimum of B.
3. Set B might have many lower bounds in A, but for B to have an infimum, those lower bounds must have one unique greatest element. If the set of lower bounds for B has no single greatest element, then B has no infimum. Put another way, B may have many lower bounds, but can have at most one infimum.
4. Thus, there are two ways in which B can fail to have an infimum. On the one hand, B might have no lower bound(s) at all. For instance, the set of the even integers in the set of all the integers, with respect to the order $\leq$. On the other hand, the set of lower bound(s) might not have a unique greatest element. For instance, consider the poset defined by a diagram below.

   2 3
   ↓↙↘↓
   12 18

   That is, $A := \{2, 3, 12, 18\}$ and the order is DIVIDES. If we take $B := \{12, 18\}$, then 2 and 3 are lower bounds for B, because $2 \preccurlyeq 12$, $3 \preccurlyeq 12$, $2 \preccurlyeq 18$, $3 \preccurlyeq 18$. No other elements are lower bounds for B, because 12 and 18 are noncomparable. But the set $\{2, 3\}$ of the lower bounds has no *greatest* element, because 2 and 3 are noncomparable. Thus, there is no infimum of B.
5. As a special case, let a be any one given element of A and let $B := \{a\}$. Then B has element a as both infimum and supremum. This follows from reflexivity: $a \preccurlyeq a$. Thus, B can fail to have infimum or supremum only if $\#B > 1$.
6. If $\#B$ is 1 or 2, say $B := \{a, b\}$, then $a \preccurlyeq b$ is equivalent to each of $a = \textbf{inf}\, B$ or $b = \textbf{sup}\, B$. It may be the case, however, that a and b are noncomparable. But if they are comparable or, in particular, if the order is total, then either a or b is necessarily the infimum of B and the other is the supremum.

7. The definition can be degenerately applied to the case in which $B = A$. Thus, if A has a least element, that element is also the infimum of A.

Let's look at two more cases in which the ideas of infimum and supremum fail. First, consider the poset $(A := \mathbb{Q}, \leq)$ and $B = \mathbb{Z}$. (In Section 10.11 we'll raise some objections to the statement $\mathbb{Z} \subseteq \mathbb{Q}$, but for now we'll allow the statement in its naive meaning.) Here, the set B has no lower or upper bounds at all under $\leq$, hence there is neither infimum nor supremum. Second, within the set $A := \mathbb{Q}$ of all rationals, still with the order $\leq$, the set B of those positive rationals whose squares are greater than 2 has infinitely many lower bounds, but there is no greatest lower bound. This may be hard to prove, but is not important for our immediate purposes.

10.8.3 Lattice

Definition 23 A partially ordered set $(A, \preccurlyeq)$ is called a **lattice** if the set $B := \{x, y\}$ has both an infimum and a supremum for each choice of x and y in A. The poset $(\varnothing, \varnothing)$ is degenerately a lattice. The element $\mathbf{inf}\{x, y\}$ is denoted, in this context, also $x \downarrow y$, and $\mathbf{sup}\{x, y\}$ is denoted also $x \uparrow y$.

An example of a lattice is the poset ({1, 2, 3, 5, 6, 10, 15, 30}, DIVIDES), which we've studied before. The diagram for this is

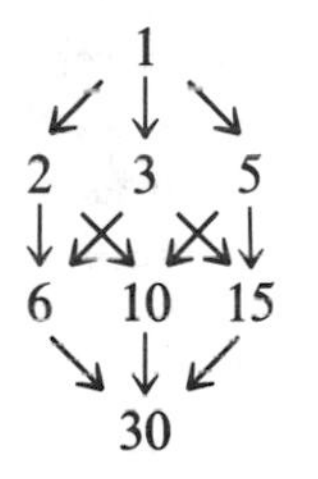

Then, for instance, $2 \downarrow 6 = 2$, $2 \uparrow 6 = 6$, $2 \downarrow 3 = 1$, $2 \uparrow 3 = 6$, $3 \downarrow 10 = 1$, $3 \uparrow 10 = 30$, and so on.

"Well," you say, "Okay, I suppose, but so what? What have we gained?" Two things. First, we're going to show that a lattice has a structure much richer than its definition might suggest. In particular, the strong body of mathematics of binary operations can now be brought to bear on lattices. The second point is even more important and is somewhat unexpected, a serendipity. We required that a lattice have infima/suprema for any set B consisting of a pair of elements. As we'll see, they will exist for *any* finite nonempty subset B of A. But let's simplify our notation first. Because both infimum and supremum are defined for every pair of elements, the natural next step is to consider them as binary operations $\downarrow : A \times A \to A$, $\uparrow : A \times A \to A$. They are clearly total operations. The exercises will ask you to tabulate the functions $\downarrow$ and $\uparrow$ for the lattice ({1, 2, 3, 5, 6, 10, 15, 30}, DIVIDES). It's probably a good idea for you to do that now, so that you'll observe in practice what the remainder of the text talks about in theory. In order to use $\downarrow$ and $\uparrow$ as binary operations, we need more convenient and explicit definitions for them. As usual, alternative definitions must first be stated and proved as theorems.

The idea of the next theorem is this: the infimum of the pair $\{x, y\}$ is an element z that is "below" each of x and y, but is also the "highest" element that satisfies that condition. In other words, if there is an element that is "below" both x and y, then such an element is "below" z (or equal to it).

Theorem 24 Let $(A, \preccurlyeq)$ be a lattice. Let x and y be any elements of A. Then (1) $\{z \in A \mid (z \preccurlyeq x)$ and $(z \preccurlyeq y)$ and (if $u \in A$, if $u \preccurlyeq x$, and if $u \preccurlyeq y$, then $u \preccurlyeq z)\} = \{x \downarrow y\}$. (2) $\{z \in A \mid (y \preccurlyeq z)$ and $(y \preccurlyeq z)$ and (if $u \in A$, if $x \preccurlyeq u$, and if $y \preccurlyeq u$, then $z \preccurlyeq u)\} = \{x \uparrow y\}$.

Proof: Exercise.

At this point we introduce an example that is very important on its own and that also will clarify many of the ideas we shall see next. We refer here to *any* Boolean algebra (Section 6.3), not just to the elementary Boolean algebra with two elements $\{0, 1\}$. A Boolean algebra does not have an order by its definition, but (as we have already remarked) its structure automatically defines an order.

Theorem 25 Let $(A, \vee, \wedge, \sim, 0, 1)$ be a Boolean algebra as defined in Section 6.3. For every choice of x and y in A, define $x \preccurlyeq y$ if and only if $x \wedge y = x$. Then, (1) $(A, \preccurlyeq)$ is a lattice; (2) $x \preccurlyeq y$ if and only if $x \vee y = y$; (3) operations $\downarrow$ and $\uparrow$ coincide with $\wedge$ and $\vee$, respectively; (4) $\mathbf{inf}\, A = 0$ and $\mathbf{sup}\, A = 1$.

Proof: Exercise. Simply recall the definitions and general properties of Boolean algebras. As a sample, let's see the proof of part 1. We must show that the definition of $\preccurlyeq$ in this Boolean algebra satisfies the conditions of a lattice (that is, we have a partial order and the conditions of Definition 23 are satisfied). For order, we must show reflexivity, antisymmetry, and transitivity (that is, $x \preccurlyeq x$; if $x \preccurlyeq y$ and if $y \preccurlyeq x$, then $x = y$; if $x \preccurlyeq y$ and if $y \preccurlyeq z$, then $y \preccurlyeq z$). In this case, $x \preccurlyeq x$ means (by definition) $x \wedge x = x$, which is true in a Boolean algebra; $x \preccurlyeq y$ and $y \preccurlyeq x$ mean that $x \wedge y = x$ and $x \wedge y = y$, which imply $x = y$; also, $x \preccurlyeq y$ and $y \preccurlyeq z$ mean that $x \wedge y = x$ and $y \wedge z = y$, which imply $x \wedge (y \wedge z) = x$, $(x \wedge y) \wedge z = x$, $x \wedge z = x$, that is, $x \preccurlyeq z$. So, we have a partial order. To prove the lattice properties, let x and y be any two elements; we want to show that $z := x \wedge y$ is actually $\mathbf{inf}\{x, y\}$. Since $z \wedge x = (x \wedge y) \wedge x = x \wedge y = z$, we have that $z \preccurlyeq x$ (and similarly for the comparison of z and y). If $u \preccurlyeq x$ and if $u \preccurlyeq y$ we want to show that $u \preccurlyeq z$. This is true, because $u \wedge x = u$, $u \wedge y = u$, $u \wedge z = u \wedge (x \wedge y) = (u \wedge x) \wedge (u \wedge y) = u \wedge u = u$, which implies $u \preccurlyeq z$.

Example Consider the Boolean algebra on set $\{00, 01, 10, 11\}$, for which the operations are BITWISE OR, BITWISE AND, and BITWISE NEGATION. See Exercises 6.3. The core of the resulting order is then $00 \preccurlyeq 01$, $00 \preccurlyeq 10$, $01 \preccurlyeq 11$, $10 \preccurlyeq 11$. A diagram is

$$\begin{array}{ccc} 00 & \rightarrow & 01 \\ \downarrow & & \downarrow \\ 10 & \rightarrow & 11 \end{array}$$

Although every Boolean algebra is a lattice, the converse is not true: some lattices are not Boolean algebras. The properties that may be missing are the existence of a 0 and a 1, the existence of complements, and distributivity.

Definition 26 A lattice $(A, \preccurlyeq)$ is said to be **bounded** or **complete** if **inf** A and **sup** A exist in A.

But, you're probably thinking, *bounded* is the same as *finite*, right? Wrong, or at least half wrong. We'll see soon that any finite lattice is bounded, but some infinite lattices are also bounded. Some are not. For instance, the lattice of any Boolean algebra is bounded with infimum 0 and supremum 1, even if the Boolean algebra is infinite, such as the Boolean algebra of the subsets of an infinite universal set. Another example is provided by the order $\leq$ on the set of the reals between 0 and 1, 0 and 1 included. Here too 0 is the infimum and 1 the supremum. On the other hand, if we consider only the reals between 0 and 1 *excluded*, then the lattice is *not* bounded, because the extrema 0 and 1 exist in $\mathbb{R}$, but not in A. That distinction is important. Of course, $(\mathbb{R}, \leq)$ has no bounds at all.

What we need, then, is some characteristic criterion, a test to determine whether or not a lattice is bounded. Simply testing finite versus infinite cardinality of A will not do. Well, out of chaos comes order—this time, anyway. Remember the confusion in partial orders regarding minimum versus lower bound versus infimum, and maximum versus upper bound versus supremum. For lattices, the distinction vanishes! *Any* minimum will be unique and will automatically be a lower bound and the infimum of A, and *any* maximum will automatically be an upper bound and the supremum of A.

The following theorem is important not only for its content, but also for its technique. Study if thoroughly.

Theorem 27 Let $(A, \preccurlyeq)$ be a lattice, that has minimal and maximal elements. Then: (1) the minimal and maximal elements are unique; (2) the minimal element is **inf** A, and the maximal element is **sup** A; and (3) $(A, \preccurlyeq)$ is bounded.

Proof: We give the proof only for the minimal element; the maximal element proof is similar. To show that the minimal element is unique, we use this time a *nonconstructive uniqueness argument*. That is, we don't actually demonstrate (construct) a minimal element; we merely show that there can't be more than one. This is a standard technique, with which you must become familiar.

Suppose both a and a^* are minimal elements in A. Set $c := a \downarrow a^*$. Since $(A, \preccurlyeq)$ is a lattice, c exists. Also, $c \preccurlyeq a$ and $c \preccurlyeq a^*$ by Theorem 24. But a and a^* are minimal, so by Definition 19, $c = a$ and $c = a^*$, and therefore $a = a^*$. So there is really only one minimum. To show that this minimum is actually **inf** A, we must show that it's a lower bound, i.e., that it's related to every element of A. Let b be an arbitrary element of A, and define $c := a \downarrow b$. Once again, c must exist, and $c \preccurlyeq a$; thus $c = a$ since a is minimal. But now we have $a = a \downarrow b$, which implies that $a \preccurlyeq b$. That $(A, \preccurlyeq)$ is bounded now follows immediately from Definition 26. Q.E.D.

Now we can handle finite lattices as a special case.

Theorem 28 If $(A, \preccurlyeq)$ is a lattice and if A is finite, then $(A, \preccurlyeq)$ is bounded.

Proof: We need only show that A has a minimal and a maximal element. Pick any element a from A. If a is a minimal, we're finished. If not, let B be the subset of A containing elements b such that $b \preccurlyeq a$ but $b \neq a$. Since $a \notin B$, $\#B < \#A$. Repeat the process: pick any element of B, and so on. Since A is finite, this can't go on forever. You can supply the details, as well as the proof that a maximal element exists also. Note how this is a proof by induction on $\#B$. Q.E.D.

We're going to "shift gear" now, and to look more closely at $\downarrow$ and $\uparrow$ as binary operations. You may want to look up Section 6.3.

Theorem 29 For a lattice $(A, \preccurlyeq)$ the functions $\downarrow$ and $\uparrow$ are associative and commutative. If $(A, \preccurlyeq)$ is bounded, then **inf** A is the unique identity for $\downarrow$ and **sup** A is the unique identity for $\uparrow$.

Proof: Exercise. For associativity, one way is to first show that each of $(a \downarrow b) \downarrow c$ and $a \downarrow (b \downarrow c)$ can be characterized as the only element z in A such that $z \preccurlyeq a$, $z \preccurlyeq b$, $z \preccurlyeq c$, and $u \preccurlyeq z$ for every u in A such that $u \preccurlyeq a$, $u \preccurlyeq b$, $u \preccurlyeq c$. Then compare the two expressions to each other. The other assertions are readily proved.

Theorem 29 has some immediate consequences. First, since the order relation $\preccurlyeq$ has really been replaced by the functions $\downarrow$ and $\uparrow$ (which carry the same information), some authors denote a lattice by the triple $(A, \downarrow, \uparrow)$ rather than simply $(A, \preccurlyeq)$. Furthermore, in analogy to the case of a Boolean algebra, operation $\downarrow$ is often denoted $\wedge$ and $\uparrow$ is denoted $\vee$, so that the lattice is denoted $(A, \wedge, \vee)$. Second, we now know that $(A, \downarrow)$ and $(A, \uparrow)$ are each abelian semigroups (that is, sets with an associative binary operation that is also commutative), and, if $(A, \preccurlyeq)$ is bounded, then each has an identity and is thus a monoid. However, a little thought (and an exercise) will convince you that inverses for $\downarrow$ and $\uparrow$ cannot exist in general. (What does that tell us?) Third, with the associativity of the operations $\downarrow$ and $\uparrow$ comes some notational convenience as well. Since both operations are total as binary operations and are associative, they are also total when composed any finite number of times. Thus, it's perfectly valid to use expressions like $\downarrow_{i=1}^{n} x_i$ and $\uparrow_{i=1}^{n} x_i$, since this is nothing more than successive pairwise evaluations. A little more problematical are expressions like $\downarrow_{x \in B} x$ when B is not a finite set. A better notation in this case is just **inf** B. The problem here is the same one we discussed above: if B is infinite, the infimum (supremum) might not exist. If it does, then the notation is perfectly valid.

Let's consider now the possibility of the operations $\downarrow$ and $\uparrow$ distributing over each other.

Definition 30 A lattice $(A, \preccurlyeq)$ is said to be **distributive** if $a \downarrow (b \uparrow c) = (a \downarrow b) \uparrow (a \downarrow c)$ and $a \uparrow (b \downarrow c) = (a \uparrow b) \downarrow (a \uparrow c)$ for every choice of a, b, and c in A.

It would be nice, of course, if it turned out that every lattice were distributive, but nature is not quite so kind. Consider the lattice in this diagram:

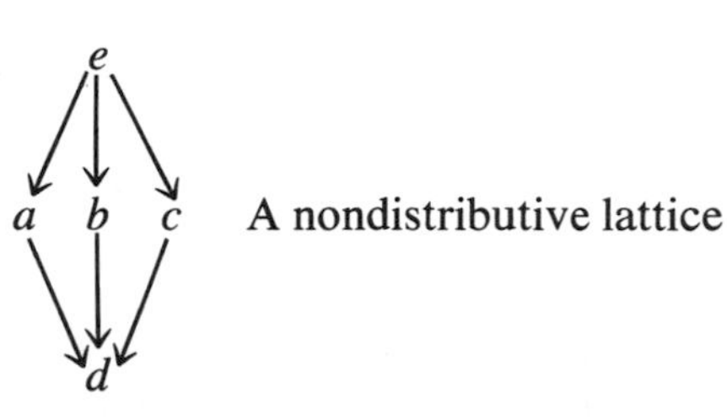

A nondistributive lattice

This lattice is not distributive, since $a \downarrow (b \uparrow c) \neq (a \downarrow b) \uparrow (a \downarrow c)$. In fact, $a \downarrow (b \uparrow c) = a \downarrow d = a$ and $(a \downarrow b) \uparrow (a \downarrow c) = e \uparrow e = e$. Still, many important examples of lattices *are* distributive, for instance, all Boolean algebras.

With distributivity, we now have enough machinery to strengthen certain lattices to semirings. Recall that a semiring is a set with two binary operations, both associative; one of them ("addition") is commutative and has an identity; the other ("multiplication") distributes over "addition." In our special case, each operation will be commutative, and each will distribute over the other.

Theorem 31 Every bounded distributive lattice $(A, \downarrow, \uparrow)$ is a semiring.

Proof: See Exercises 10.8. Simply recall the definitions.

And now we face a decision. From a semiring, we could attempt to strengthen the structure to be a ring, but this requires that at least one, and possibly both, of $\downarrow$ and $\uparrow$ have inverses. In the exercises, you'll show that this is not possible. On the other hand, we know that we could try to strengthen the lattice to be a Boolean semiring by introducing complement elements. This is much more promising; the symmetry of $\downarrow$ and $\uparrow$ resembles the Boolean regime much more than the arithmetic one.

Definition 32 A lattice $(A, \preccurlyeq)$ is said to be **complemented** if it is bounded and if, for each a in A, there exists an element a^* in A such that $a \downarrow a^* =$ **inf** A and $a \uparrow a^* =$ **sup** A.

No unbounded lattice can be complemented, since we must first have **inf** A and **sup** A. However, a bounded but infinite lattice could be complemented. Remember, the key is boundedness, not finiteness. For instance, every Boolean algebra, finite or not, is bounded and complemented. As with distributivity, nature wasn't kind enough to give every lattice a complement, nor to make every complemented lattice distributive. You can easily verify that the nondistributive lattice in the figure above is complemented; in fact, a, b, and c each have both of the other *two* as a complement. On the other hand, you can easily verify that the lattice in this diagram

$$a \rightarrow b \rightarrow c$$

is distributive, and $a =$ **inf** A, $c =$ **sup** A. However, b has no complement b^*, since $b \downarrow x = a$ is satisfied only by $x = a$ and $b \uparrow x = c$ is satisfied only by $x = c$. For what it's worth, a and c are each other's complements. But, lacking a workable b^*, the lattice is not complemented. In other words, distributivity and complementation are independent attributes of a lattice. Any given lattice might satisfy neither, one or the other, or both.

Suppose now that we have established $(A, \preccurlyeq)$ as a distributive and complemented lattice. What exactly have we got? We have a set A, with two binary operations $\downarrow$ and $\uparrow$, each of which is associative and commutative and each of which has an identity. Further, $\downarrow$ and $\uparrow$ distribute over each other, and each a in A has a complement with respect to both $\downarrow$ and $\uparrow$. These five pairs of properties, associativity, commutativity, distributivity, existence of identities, and existence of complements, are by now depressingly familiar. We have, of course, thus proved the following theorem:

Theorem 33 Every distributive complemented lattice is a Boolean semiring, that is, a Boolean algebra.

The distinction between a Boolean semiring and a Boolean algebra is purely semantic; mathematically they are identical. Although *Boolean algebra* is rather the more commonly used term, especially among computer scientists, *Boolean semiring* offers two psychological and linguistic advantages. First, it emphasizes the fact that this structure fits into a family of structures, starting with groupoids, splitting off at semirings, and so on. It doesn't simply occur out of the blue, in isolation. This is important because the steps needed to identify a Boolean semiring are precisely the same as those needed to define one: define binary operations and their identities, show associativity, commutativity, and distributivity, etc. Second, and more important, the term *Boolean algebra* is often used, especially by computer programmers, as if it were synonymous with *propositional calculus*. A Boolean algebra is an abstract structure. The propositional calculus is an example of a Boolean algebra, as are set theory, 2TSNs, and many others. Using the term *Boolean semiring* tends to reinforce this. There are, of course, many other conclusions we could draw about complemented distributive lattices—precisely those we can derive in a Boolean algebra (see Section 6.3). Nonetheless, just for practice, the exercises ask you to prove them explicitly. If you stop and consider for a minute, you'll realize that the proofs are quite easy.

Exercises 10.8

1. Prove Theorem 21 on the existence of maxima and minima.
2. Prove that a partially ordered set $(A, \preccurlyeq)$ can have at most one least element and at most one greatest element.
3. For the poset ($A := \{1, 2, 3, 5, 6, 10, 15, 30\}$, DIVIDES), tabulate $x \downarrow y$ and $x \uparrow y$ for all x and y in A. Identify $\mathbf{inf}\, A$ and $\mathbf{sup}\, A$.
4. For the poset ($A := \mathcal{P}\{1, 2, 3\}, \subseteq$), tabulate $x \downarrow y$ and $x \uparrow y$ for all x and y in A. Identify $\mathbf{inf}\, A$ and $\mathbf{sup}\, A$. What set-theoretical operations are performed by $\downarrow$ and $\uparrow$?

5. Prove Theorem 24 on a characterization of $x \downarrow y$, $x \uparrow y$.

6. Prove Theorem 25, that Boolean algebras are lattices.

7. Prove Theorem 28, that every finite lattice is bounded.

8. Prove Theorem 29 on commutativity and associativity of $\downarrow$ and $\uparrow$ in a lattice.

9. Let $(A, \preccurlyeq)$ be a lattice, and let x be any element of A. **(a)** State the criterion for an element x^{-1} of A to be the inverse of x under $\downarrow$. For what x in A is it possible for x^{-1} to exist? **(b)** State the criterion for an element x^{-1} of A to be the inverse of x under $\uparrow$. For what x in A is it possible for x^{-1} to exist?

10. Using Venn diagrams, show that for any set U, the poset $(\mathcal{P}U, \subseteq)$ is distributive.

11. Prove Theorem 31 on lattices and semirings.

12. Define $[0, 1] := \{x \in \mathbb{R} \mid 0 \leq x \leq 1\}$. Consider the linear order $([0, 1], \leq)$. Show that this is a bounded, infinite, distributive lattice, but is not complemented. [Hint: Carefully define the operations $\downarrow$ and $\uparrow$.]

13. Let $(A, \preccurlyeq)$ be a finite, totally ordered set. Prove **(a)** $(A, \preccurlyeq)$ is bounded; **(b)** $(A, \preccurlyeq)$ is distributive; **(c)** $(A, \preccurlyeq)$ is complemented if and only if $\#A \leq 2$, and degenerately so if $\#A < 2$.

14. Let $(A, \preccurlyeq)$ be a complemented distributive lattice, and let x and y be elements of A. Prove each of the following theorems explicitly, and state and explicitly prove the theorem's dual, if there is one. Identify the name under which the theorem/dual pair is commonly known. **(a)** $x \downarrow x = x$. **(b)** If $x \downarrow z = x$ for all z in A, then $x = \textbf{inf}\, A$. **(c)** $x \downarrow (\textbf{inf}\, A) = \textbf{inf}\, A$. **(d)** If $x \uparrow y = \textbf{sup}\, A$ and $x \downarrow y = \textbf{inf}\, A$, then $y = x^*$. **(e)** $(x^*)^* = x$. **(f)** $(\textbf{sup}\, A)^* = \textbf{inf}\, A$. **(g)** $(x \uparrow y)^* = x^* \downarrow y^*$.

15. Each of the following diagrams represents the core of a partial order. The heavy dots stand for the distinct elements of set A. For each of the diagrams, verify that the relation is actually a partial order and decide whether the structure is **(1)** a lattice, **(2)** a distributive lattice, **(3)** a bounded lattice, **(4)** a complemented lattice, **(5)** a Boolean algebra.

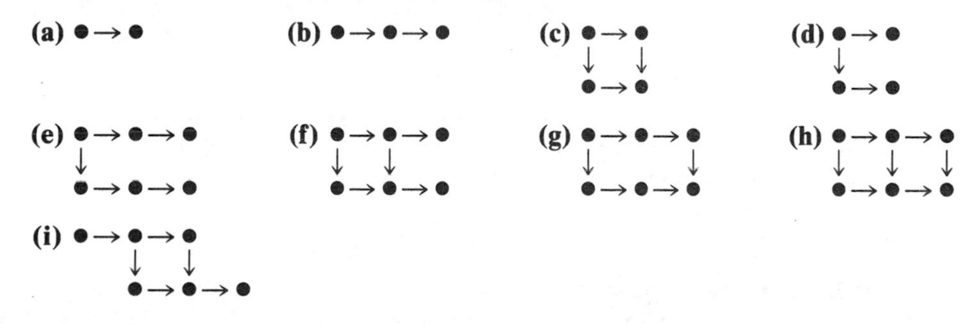

16. Given: alphabet $X := \{\text{A, B, C, D}\}$, $Y := X^1 \cup X^2 \cup X^3$, an order $\preccurlyeq$ on Y defined by $x \preccurlyeq y$ if and only if x is a substring of y. For each element v of Y such that the length of v is 2 and $(\text{B}) \prec v$, find (if possible) an upper bound and a lower bound of set $\{(\text{A}), v\}$.

17. Let A be one of the sets given below. The letters b, c, d, and z denote distinct elements. In each case, find all possible partial order structures (if any) that make A a bounded lattice such that $b = \textbf{inf } A$ and $z = \textbf{sup } A$. For each such bounded lattice, decide whether it is distributive, complemented, or both. **(a)** $A = \{b, z\}$; **(b)** $A = \{b, c, z\}$; **(c)** $A = \{b, c, d, z\}$.

10.9 Equivalence Relations

The single most important type of relation that can be defined on a set is the equivalence relation.

Definition 34 A relation R on a set A is called an **equivalence relation** if it is reflexive, symmetric, and transitive.

The first few exercises for this section ask you to determine which of the relations given as examples are reflexive, symmetric, or transitive, and which are equivalence relations. You might find it helpful to do this now.

Definition 35 Let A be a nonempty set and R an equivalence relation on A. For each element x of A, the **equivalence class of x with respect to R** is denoted x/R and is defined by $x/R := \{y \in A \mid xRy\}$.

In other words, the equivalence class of x is the set of those elements that are equivalent to x. Some alternative notations you're liable to run across are $[x]$, x^*, x_R, and $x \bmod R$. But x/R is by far the most convenient.

Let's consider the example of the integers and their equality mod 5. In other words, define $A := \mathbb{Z}$, $R := \{(x, y) \in \mathbb{Z} \times \mathbb{Z} \mid x = y \bmod 5\}$. Then the class of 0 is $\{\ldots, -10, -5, 0, 5, 10, \ldots\}$, that is, the set of those integers that are equivalent to 5; in this special case, they are the multiples of 5. The class of 1 is $\{\ldots, -9, -4, 1, 6, \ldots\}$ that is, the set of those integers that are equivalent to 1, in this case, we have those integers whose difference from 1 is a multiple of 5 (in jargon, "the multiples of 5 plus 1"). The class of 2 is $\{\ldots, -8, -3, 2, 7, \ldots\}$; the class of 3 is $\{\ldots, -7, -2, 3, 8, \ldots\}$; the class of 4 is $\{\ldots, -6, -1, 4, 9, \ldots\}$, and there are no more classes because all the integers are already accounted for.

Some properties of equivalence classes are immediately obvious, some are a little less evident. The following theorem shows how equivalence classes interrelate. Are they disjoint? Do they overlap? Do they agree? Remember that there is an equivalence class for each x in A.

Theorem 36 Let R be an equivalence relation on a nonempty set A. For each x in A, let x/R be the equivalence class of x with respect to R. (1) For each x in A, $x \in x/R$. Consequently, $x/R \neq \varnothing$ and $\bigcup_{x \in A} x/R = A$. (2) Let x, y be any two elements of A. The following three statements are logically equivalent: (a) xRy; (b) $x/R \cap y/R \neq \varnothing$; (c) $x/R = y/R$. (3) Any two equivalence classes are either coincident or disjoint.

Proof: For part 1, observe that xRx by the reflexive property of R. By the definition of equivalence class, x is one of the elements of x/R. This proves the first statement of part 1. The rest of part 1 is then immediate.

For part 2, we must show that each of the three statements implies the other two. We'll do that by proving that (b) implies (a), (a) implies (c), and (c) implies (b).

(b) $\Rightarrow$ (a): Let z be an element of $x/R \cap y/R$. By the definition of equivalence class, xRz and yRz. By symmetry and transitivity, xRy, which proves (a).

(a) $\Rightarrow$ (c): Assuming (a), we have to prove that (c1) $x/R \subseteq y/R$ and (c2) $y/R \subseteq x/R$. Let z be any element of x/R. By the definition of equivalence class, xRz. By hypothesis, commutativity, and transitivity, yRz. By the definition of equivalence class again, $z \in y/R$. We have proved that $z \in x/R$ implies $z \in y/R$. Part (c1) is so proved. Since (a) and symmetry imply that yRx, interchanging the roles of x and y in the arguments above proves (c2). After that, (c1) and (c2) together prove (c).

(c) $\Rightarrow$ (b): The hypothesis implies $x/R \cap y/R = x/R \cap x/R = x/R$. Part 1 implies $x/R \neq \varnothing$. These two facts imply (b). Part 2 is so proved.

Part 3 is an immediate consequence of part 2. Q.E.D.

The important idea to glean from this theorem is that if two equivalence classes overlap at all, then they overlap entirely, they're the same set. Thus each element of A participates in precisely one equivalence class.

Definition 37 Let R be an equivalence relation on a nonempty set A. The **quotient set** of A with respect to R is denoted A/R and is defined as the set whose elements are the equivalence classes of the elements of A; that is,

$$A/R := \{x/R \in \mathcal{P}A \mid x \in A\}.$$

The corrrespondence $s\colon A \to A/R$ defined by $x \mapsto x/R$ is a total function by Theorem 36 and surjective by definition. It is called the **natural surjection** of R.

Note carefully that the equivalence classes are *subsets* (not elements) of A; they are *elements* (not subsets) of A/R. Also, the description of A/R in Definition 37 lists each equivalence class once for each of its elements. This may imply—and normally does—that in that description each equivalence class appears several times. But recall that repetitions are irrelevant in a set. Consequently, A/R ultimately contains each equivalence class just once. As an example, recall the relations that we used in this section as an example after Definition 35. We defined $A := \mathbb{Z}$, $R := \{(x, y) \in \mathbb{Z} \times \mathbb{Z} \mid x = y \bmod 5\}$. Then the equivalence classes were $0/R = \{\ldots, -10, -5, 0, 5, 10, \ldots\}$, $1/R = \{\ldots, -9, -4, 1, 6, \ldots\}$, $2/R = \{\ldots, -8, -3, 2, 7, \ldots\}$, $3/R = \{\ldots, -7, -2, 3, 8, \ldots\}$, and $4/R = \{\ldots, -6, -1, 4, 9, \ldots\}$. Thus $A/R = \{0/R, 1/R, 2/R, 3/R, 4/R\}$. Another example is again one we've seen before, although in different terms. Let U be some finite set, and let $A := \mathcal{P}U$. If $x \in A$ and if $y \in A$, then x and y are said to be **equipotent** if there exists a bijection $f\colon x \to y$. The nature of bijections ensures that equipotency is an equivalence relation. (You may want to go back to Chapter 7 and review why this is true; see Theorem 7.17.) In fact, the equivalence classes for equipotency are those collections of subsets that have the same number of elements. This is the basic idea behind cardinality of a finite set. For instance, if $U := \{a, b, c\}$, then $A = \mathcal{P}U = \{\varnothing, \{a\}, \{b\}, \{c\}, \{b, c\}, \{c, a\}, \{a, b\}, \{a, b, c\}\}$; $A/R = \{\{\varnothing\}, \{\{a\}, \{b\}, \{c\}\}, \{\{b, c\}, \{c, a\}, \{a, b\}\}, \{\{a, b, c\}\}\}$, that is, the four equivalence classes are $\{\varnothing\}$, $\{\{a\}, \{b\}, \{c\}\}$, $\{\{b, c\}, \{c, a\}, \{a, b\}\}$, $\{\{a, b, c\}\}$. The first equivalence class comprises those subsets with no elements, the second class those with one element, and so on.

Note that the universal relation and the identity on any nonempty set A are equivalence relations. For the universal relation, there is only one equivalence class, which is made up of all the elements of A. For the identity, each equivalence class consists of one element only, and there are as many equivalence classes as there are elements in A. Obviously, not every relation will be reflexive or symmetric or transitive. But every relation R can be made reflexive or symmetric or transitive through the process of *closure* we've seen before. In this case, we're referring to reflexive-symmetric-transitive closure (or one or two of them only, if the original correspondence already has some of the required properties).

Often in applications, what we want to do is to specify the *core* of a relation, that is, some clear but incomplete statement of what the relation is intended to look like. Then, more or less mechanically, we extend it to include some combination of reflexivity, symmetry, and transitivity.

Let's consider how this might work in practice. Let $A := \mathbb{Z}$, and take as our "core" relation the relation "x is the immediate predecessor of y," that is, $(xRy) :\Leftrightarrow (y = x + 1)$. Then, for instance, we have $1R2$, $2R3$, $7R8$, and $8R9$. In general $xR(x + 1)$ $(x \in \mathbb{Z})$. By forming the transitive closure $R^{[t]}$, we obtain such statements as $1R3$ and $7R9$, as well as $1R7$, $2R7$, etc. In other words, $R^{[t]}$ is now the relation $x < y$. If we also include the identity $i_{\mathbb{Z}}$, we get $R^{[rt]} = \leq$. If we now included also the symmetric closure, we would have $R^{[rst]} = \mathbb{Z} \times \mathbb{Z}$, the universal relation on $\mathbb{Z}$. This is probably more than we had in mind, given the core relation $y = x + 1$, so we wouldn't do it.

In short, given a "core" relation, the closure operations allow us to conveniently extend it to a more powerful structure, but to do so in a systematic manner. The problem of constructing a core relation for a given relation is a different one. We've already solved it for partial orders, and we've indicated that it is rather unmanageable in general. For an equivalence relation on a finite set, we can, however, construct a core. A new feature is that a core relation will not be uniquely determined. See Complements 10.9.

Exercises 10.9

1. Let $A := \{a, b, c, d, e\}$ and represent a relation R by the table

	a	b	c	d	e
a	*	*		*	
b	*	*		*	
c			*		*
d	*	*		*	
e			*		*

with the understanding that xRy if and only if entry (x, y) in the table above is marked with an asterisk. Show that R is an equivalence relation. List the equivalence classes.

2. Section 10.3 gives as examples five pairs of relations. For each of these ten relations, prove (or disprove via a counterexample) that the relation is **(1)** reflexive, **(2)** symmetric, **(3)** transitive, **(4)** an equivalence relation. For those relations that are not reflexive, alter the definitions slightly to make them reflexive. Are they now equivalence relations? For all equivalence relations, identify the equivalence classes and quotient set.

3. Let $A :=$ {straight lines in a plane which all pass through a given point P}. Let $R := \{(x, y) \in A \times A \mid x \perp y\}$. Modify the definition of R to make it reflexive. Prove that, with this modification, R is an equivalence relation.

4. Let $A := \mathbb{Z} \times \mathbb{Z}$. Prove that each of the following is an equivalence relation on A. For each, try to give an informal description of the equivalence classes. **(a)** $(x, y)\, R\, (u, v) :\Leftrightarrow x + y = u + v$; **(b)** $(x, y)\, R\, (u, v) :\Leftrightarrow x + v = y + u$; **(c)** $(x, y)\, R\, (u, v) :\Leftrightarrow |x + y| = |u + v|$.

5. Prove that, for any set A, the identity relation and the universal relation are both equivalence relations. Identify the equivalence classes and quotient set in each case. Under what circumstances will the null relation be an equivalence relation? Identify its equivalence classes and quotient set.

6. Prove that if A is finite and R is an equivalence relation on A, then $\#A = \sum_{X \in A/R} \#X$.

7. Note that the representation of a relation R as a table like the one in exercise 1 is really a total function CH: $A \times A \to \{0, 1\}$, for which CH(x, y) is 1 or 0 according as xRy is true or not.

Among the relations we saw, consider those on "small" sets. Represent them in table form. Consider, in particular, the diagonal relation and the universal relation on a set with four or five elements. In general, find some "visual" rules to recognize from the table whether the relation is reflexive, symmetric, or transitive. For reflexivity, you'll see why the diagonal relation has this name.

8. For each of the relations T found in problem 2 of Exercises 10.5, recall that T is necessarily an equivalence relation. Give its equivalence classes by enumeration.

9. Let $A := \mathbb{Z}$ and $R := \{(x, y) \in \mathbb{Z} \times \mathbb{Z} \mid x - y \text{ is even}\}$. Construct all the closures of R, and identify the equivalence classes and quotient set of $R^{[rst]}$.

10. For any nonempty set A, there is precisely one relation R on A such that R is both an equivalence relation and a partial order. What is it?

11. For a relatively small set $A := \{1, \ldots, n\}$ and for a given relation R ($R \subseteq A \times A$), suitable computer exercises are **(a)** recognize whether R is an equivalence relation; **(b)** if not, replace R with $R^{[rst]}$. **(c)** Find the equivalence classes. **(d)** Draw a Hasse diagram for a core relation.

Complements 10.9

1. When A is finite, the function CH in exercise 7 above can be interpreted as a $\#A \times \#A$ matrix B with Boolean entries. Prove that B^k is the matrix of R^k for every k in $\mathbb{N}$. If B^{-1} denotes the transpose of B (this is *not* customary!), prove that the statement above is valid for every k in $\mathbb{Z}$. [The **transpose** of matrix B is defined as the matrix for which $(i, j) \mapsto B(j, i)$.] Find the representation of the various closures of R in terms of matrix B.
2. The basic idea of a core relation of a partial order is to artificially establish a linear order within each equivalence class. In order to do that, prove

Theorem 38 Let R be an equivalence relation on a finite set A. Let B be an equivalence class with at least two elements, if such a class exists. Let $m := \#B$. Let $(b_1, \ldots, b_m)$ be any one permutation of B. Let $r(B) := \{(b_{l-1}, b_l) \mid 1 < l \leq m\}$. In other words, order each equivalence class according to any one linear order and consider pairs of consecutive elements in that order. Then $\bigcup_{(B \in A/R) \wedge (\#B > 1)} r(B)$ is a relation on A whose reflexive-symmetric-transitive closure is R.

Example (1) Let R be the universal relation on set $A := \{x, y, z, t\}$. Then the only class is $\{x, y, z, t\}$. Consider the permutation (x, y, z, t). Then the only $r(B)$ is $\{(x, y), (y, z), (z, t)\}$. Let $R^* := \{(x, y), (y, z), (z, t)\}$, which is a core of R. Let's check that. The transitive closure of R^* is $\{(x, y), (x, z), (x, t), (y, z), (y, t), (z, t)\}$. The symmetric closure of this is $\{(x, y), (x, z), (x, t), (y, z), (y, t), (z, t), (y, x), (z, x), (t, x), (z, y), (t, y), (t, z)\}$. The reflexive closure of this is $\{(x, y), (x, z), (x, t), (y, z), (y, t), (z, t), (y, x), (z, x), (t, x), (z, y), (t, y), (t, z), (x, x), (y, y), (z, z), (t, t)\}$, which is the original R. (2) In the case of any equivalence relation on a finite set, we would perform the construction above on each of the equivalence classes, and then take the union of the results. (3) In the case of the identity, each equivalence class comprises one element, and therefore the core relation, as defined above, is empty, that is $R^* = \varnothing$. The symmetric-transitive closure of R^* is still empty. The reflexive-symmetric-transitive closure of R^* is then $(R^*)^{[rst]} := \varnothing \cup i_A = i_A$, which is precisely what we wanted.

10.10 Three Aspects of Equivalence Relations

By now it's probably occurred to you that there is a close interaction between the ideas of equivalence relations on a set and partitions of the set. To formalize this let's consider, not just one equivalence relation, but the entire set of equivalence relations on a set.

Definition 39 Let A be a nonempty finite set and let $n := \#A$. Denote by $E_k(A)$ the set of equivalence relations R on A which have precisely k equivalence classes.

Clearly, $1 \leq k \leq n$ and $k = \#(A/R)$. Denote by $E(A)$ the set of all equivalence relations on A, regardless of the cardinality of A/R; that is, $E(A) := \bigcup_{k=1}^{n} E_k(A)$. By degeneracy, if A were empty, then we would have $E(A) = \varnothing$.

Example Let's consider all equivalence relations on set $\{1, 2, 3\}$. There is one with one class only, namely the universal relation: all pairs are related, both ways. This is the only element of $E_1(A)$. There are three relations that have precisely two classes. Necessarily, one class consists of just one element; the other of two elements. In other words, one element (1, or 2, or 3) is related only to itself, and the other two elements are related to each other (both ways) and to themselves. The equivalence classes are, respectively, $\{1\}\ \{2, 3\}$, $\{2\}\ \{1, 3\}$, $\{3\}\ \{1, 2\}$. These three relations are the elements of $E_2(A)$. Finally, there is one with three classes; this is the identity: each element is related to itself only. The classes are $\{1\}\ \{2\}\ \{3\}$. This relation is the only element of $E_3(A)$. Consequently, $E(A)$ comprises five elements, the five equivalence relations on set A.

Recall the definitions of set $C(A)$ (the set of all partitions of A) and of sets $C_k(A)$ (the sets of partitions of A having precisely k blocks). See Section 5.7 and Complements 10.5. Then we can state the following theorem:

Theorem 40 Let A be any finite nonempty set. Define $n := \#A$. Then: (1) the sets $C(A)$ and $E(A)$ are isomorphic, that is, there exists a bijection between them; (2) for any k $(1 \leq k \leq n)$, the sets $C_k(A)$ and $E_k(A)$ are isomorphic; (3) the number of equivalence relations on A with exactly k classes is $S[n, k]$, the Stirling number defined in Complements 4.6; (4) the number of equivalence relations on A is $\sum_{k=1}^{n} S[n, k]$.

In the example, the five partitions are $\{\{1, 2, 3\}\}$, $\{\{1\}, \{2, 3\}\}$ $\{\{2\}, \{1, 3\}\}$, $\{\{3\}, \{1, 2\}\}$, $\{\{1\}, \{2\}, \{3\}\}$.

Proof: What we have to show is that every equivalence relation completely determines a partition, that every partition completely determines an equivalence relation, and that the cardinalities of partition and quotient set are always equal. That the set of the classes of each equivalence relation is a partition we've already seen, as an immediate consequence of Theorem 36. To show that every partition determines an equivalence relation, suppose S is a partition of A. On A define the relation

$$R := \{(x, y) \in A \times A \mid \text{for some block } b \text{ of } S,\ x \in b \text{ and } y \in b\}. \tag{1}$$

There is little difficulty in showing that R is an equivalence relation. We have so actually constructed a total function $f\colon E(A) \to C(A)$ and a total function $g\colon C(A) \to E(A)$. Specifically, for each equivalence relation R, we define fR as the set A/R; for each partition S, we define gS as the relation R given by formula 1. From the construction of these functions it is clear that $g \circ f$ is the identity on $E(A)$ and $f \circ g$ is the identity on $C(A)$. By Theorem 7.13, f and g are both bijections and are inverses of each other. This proves part 1. From the construction again, it is clear that each of f, g maps $E_k(A)$ bijectively to $C_k(A)$ and conversely. This proves part 2. Parts 3 and 4 are direct consequence of parts 1 and 2 and Complements 5.7. Q.E.D.

We've just seen that an equivalence relation and the respective partition are two aspects of the same idea. The third aspect is the following. We have already observed that there is the natural surjection $s: A \to A/R$. The next theorem reverses the situation, that is, shows how a total surjective function determines a partition or, equivalently, an equivalence relation.

Theorem 41 Let A be a nonempty set and $g: A \to W$ a total surjective function. Then (1) the set $S^* := \{\{x \in A \mid gx = z\} \in \mathcal{P}A \mid z \in W\}$ is a partition of A, (2) the relation $R^* := \{(x, y) \in A \times A \mid gx = gy\}$ is an equivalence relation on A; (3) S^* and R^* are obtained from each other according to the process given in the proof of Theorem 40.

Proof: Exercise.

In ordinary words, the partition S^* is obtained by taking as blocks the sets of those elements of A that map on the same element of W and the equivalence relation R^* is obtained by calling equivalent two elements of A that map on the same element of W.

A B C D E F G H I J K L
M N O P Q R S T U V W X Y Z

Figure 10.3 A specific font of the English alphabet.

Note that a practical aspect of this idea is a classification of the type "__ and __ have the same __". For instance, call two 5-strings of binary digits equivalent if they contain the same number of ones, or call two letters from Figure 10.3 equivalent if they have the same number of loose ends, or the same number of enclosed regions, or both. In most real-life applications, the function g is onto a set of numerical intervals. For instance, an egg is classified "jumbo" if its weight is more than a given value, "extra large" if its weight is less than or equal to that value, but larger than another given value, and so on.

Exercises 10.10

1. Consider the set of five-digit binary strings, and the partition S based on the number of ones occurring in each string. Write down the equivalence relation corresponding to this partition. Just a core of the relation will do; don't write the reflexive-symmetric-transitive closure. A Hasse diagram is a convenient form for the result. (This exercise is suitable for computer solution.)

2. Write down a core or a Hasse diagram of the equivalence relation obtained by classifying each of the following sets according to the given criterion. **(a)** Numbers $\{0, 1, 2, \ldots, 55\}$ according to the value of the sum of their decimal digits. **(b)** The first twenty words of your dictionary according to the number of a's in them. **(c)** The 26 symbols in Figure 10.3 according to the number of straight-line segments that appear in them. **(d)** The binary numerals with one to five digits according to the value of the numeral.

3. Prove Theorem 41, that a surjection determines a partition.

Complements 10.10

1. Let $A := \mathbb{N}^{\mathbb{N}}$ and let f and g be elements of A. Define $f \sim g$ if $fx = gx$ for all x in $\mathbb{N}$, except possibly for finitely many x's. Show that $\sim$ is an equivalence relation on A. Define fRg if $fx = gx$ for all x in $\mathbb{N}$, except *one* x. Prove that $R^{[rst]} = \sim$.

2. Let A be a nonempty set. Let $E(A)$ denote the set of all equivalence relations on A. Prove that, for any relation R on A,

$$R^{[rst]} = \bigcap_{X \in E(A) \text{ and } R \subseteq X} X.$$

3. Prove that, for any finite set A, the sets $C_k(A)$, $(k = 1, \ldots, \#A)$, constitute a partition of $C(A)$, and that the sets $E_k(A)$ likewise constitute a partition of $E(A)$. Refer to the notations in Definition 39 and immediately following it. Also, see Complements 10.5.

4. Classification of a collection of objects is quite simple in theory, but in practice it is a difficult and unresolved problem, which probably cannot be solved at all. Classification amounts to finding a partition of a set of objects so that some specified practical criteria are satisfied. There are two problems with classification: First, for any given object, determining whether the criteria are met may be subtle and vague. Second, the criteria themselves may not give rise to a clear and distinct partition of the objects.

Let's look first at an example of the second problem. Consider the classification of library books by subject matter. At first glance, this seems easy: history, economics, mathematics, physics, biology, and so forth. On second thought, though, where does history end and economics begin? Where does economics end and mathematics begin? So we end up with classes such as economic history, mathematical physics, and biochemistry. This problem is usually dealt with by cross-partitioning, although the librarians who actually do it probably don't call it by that name.

Rather more difficult is the problem of determining whether the critera are met. Whenever a concrete problem is treated as a mathematical problem, some margin for approximation must be made. Sometimes this is easy, other times quite hard. Consider, for instance, the classification of computers according to the number of bits they use to represent an integer. We hear terminology like

"32-bit machine" and "36-bit machine." Here the classification is quite obvious and simple, unless we go into many details of complex machines. We have a surjective mapping from the set of computers onto a subset $\mathbb{N}$ (actually a small subset of $\mathbb{N}$), and thus a partition of the set of computers. This partition determines an equivalence relation. Any 32-bit computer uses exactly the same number of bits for an integer as any other 32-bit computer. Although they may differ widely in other aspects, in this they are equivalent.

Our last example of difficult classification is without doubt one of the hardest: the classification of goods and services in a free-market economy according to "economic value." Suppose we wish to compare the economic value of widgits versus doohickies. Widgits and doohickies are assumed to have the same *economic value* (or *market value*) if there are two people who are willing to trade one widgit for one doohicky. (Of course, this definition bypasses some psychological and sociological details, but that's part of the approximation.) There's nothing inherent about a widgit that makes it worth one doohicky; it's sufficient that two persons are mutually willing to trade. Other people might not be; fine. If we consider widgits and doohickies to be related by this definition, then the relation is rather clearly reflexive and clearly symmetric. The question is whether it's transitive. If there are two people willing to trade widgits and doohickies, and two others willing to trade doohickies and thingamajigs, does it follow that there are two people willing to trade widgits and thingamajigs? Ideally, the answer is yes, assuming that the society is fluid, and that communications and transportation are perfectly reliable. In case of major calamities, such a condition is clearly *not* satisfied: one may get food for nothing (from the Red Cross, for instance) or one may have to pay for it in gold (if one has it); both painful episodes to witness.

Thus, this definition of economic value is an equivalence relation, and every good and service has a place in one of the blocks, according to its "intertradability". However, the only way to know where a particular good or service fits in is to place it in the market. There is no absolute concept of economic value. In modern market systems, the surjection from goods and services to economic value is quantified using money. That is, **price:** {goods and services} $\rightarrow$ {prices}. This is much more convenient than using shells, or cattle, or barter. In the terminology of economics, this use of money to characterize economic value is called a *standard of value*. (Money has three other functions: as a *medium of exchange* (spending), as a *store of value* (saving), and as a *commodity* (a good)). Note carefully that price does not determine value. Value is determined in the market and is conveniently characterized as a price.

10.11 Number Systems

An important application of equivalence relations arises in the theoretical problem of defining the integers. "Defining the integers?," you say. "Aren't the integers axiomatic? How do we *define* integers?" The natural numbers $\mathbb{N}$ are axiomatic, but the notion of negative integers is not. In fact the whole idea of negative numbers developed, in mathematical terms, quite recently, about 500 years ago. As German mathematician Leopold Kronecker (1823–91) put it, "God made the natural numbers; all else is the work of man". We'll obtain the integers by extending the structure of the natural numbers. Then we'll extend the integers to the rationals, the rationals to the reals, and the reals to the complex numbers. The entire process falls outside the main goals of this book and will not be needed in any other part of the book. On the other hand, this process provides good examples of a combined use of equivalence relations and correspondences. We'll give only a compact sketch; other courses will go into the details.

10.11.1 Integers

For all of our reluctance to take them on faith, the derivation of the integers is quite straightforward. It rests on the observation that any integer, positive, negative, or zero, can be expressed as the difference of two natural numbers. For instance, $+3 = 8 - 5 = 12 - 9 = 12345 - 12342$, and so on. $0 = 5 - 5 = 12 - 12$, etc. For negative numbers $-3 = 5 - 8 = 9 - 12 = 12342 - 12345$. But this observation is circular logic! This is the way we *want* subtraction to work, but the last set of subtractions *doesn't* work in $\mathbb{N}$. What we have to do is go back into $\mathbb{N}$ and *define* a set $\mathbb{Z}$ and a subtraction (and addition and multiplication) that do work as we want.

10.11.1.1 Construction of Structure $\mathbb{Z}$

Definition 42 In $\mathbb{N} \times \mathbb{N}$, define the equivalence relation $\sim$ as follows. Two ordered pairs (x, y) and (u, v) in $\mathbb{N} \times \mathbb{N}$ are equivalent, denoted as $(x, y) \sim (u, v)$, if and only if $x + v = y + u$, where $x, y, u,$ and v are in $\mathbb{N}$. Then the set $\mathbb{Z}$ of integers is defined as the quotient set $\mathbb{Z} := \mathbb{N} \times \mathbb{N}/\sim$.

(Obviously, the criterion we're really after is that $x - y = u - v$. But this assumes that this subtraction works in $\mathbb{N}$. It doesn't. The fact that this is indeed an equivalence relation is left as an exercise.)

What we need next is to define operations of addition and multiplication on $\mathbb{Z}$ that preserve as much as possible the properties of the operations in $\mathbb{N}$. But note carefully that $\mathbb{Z}$ is composed of the equivalence classses of $\sim$. By our definition, an integer is a class $(x, y)/\sim$. In $\mathbb{N} \times \mathbb{N}$, (x, y) is simply an ordered pair. Thus, our operations in $\mathbb{Z}$ will look a bit bizarre at first and must be clearly distinguished from those in $\mathbb{N}$. The procedure is embodied in the following definitions and theorems. The details of the proofs are left as exercises.

Definition 43 Let $p: \mathbb{N} \times \mathbb{N} \to \mathbb{Z}$ be the natural surjection, that is, the total function defined by $(x, y) \mapsto (x, y)/\sim$. The function p uniquely determines a function $q: (\mathbb{N} \times \mathbb{N}) \times (\mathbb{N} \times \mathbb{N}) \to \mathbb{Z} \times \mathbb{Z}$ by $((x, y), (u, v)) \mapsto (p(x, y), p(u, v))$. Let

$a:((\mathbb{N} \times \mathbb{N}) \times (\mathbb{N} \times \mathbb{N})) \to \mathbb{N} \times \mathbb{N}$ be the total function defined by $((x, y), (u, v)) \mapsto (x + y, u + v)$, where the symbol $+$ denotes ordinary addition in $\mathbb{N}$. Addition in $\mathbb{Z}$ is denoted by $+_{\mathbb{Z}}$ and is defined as the correspondence $+_{\mathbb{Z}}$: $\mathbb{Z} \times \mathbb{Z} \to \mathbb{Z}$ given by $+_{\mathbb{Z}} := p \circ a \circ q^{-1}$.

Once again, the intuitive idea is that $(x - y) + (u - v) = (x + u) - (y + v)$.

Theorem 44 The correspondence $+_{\mathbb{Z}}$ has the following properties: (1) $+_{\mathbb{Z}}$ is a total function. That is, it is a total binary operation and $\mathbb{Z}$ is closed under $+_{\mathbb{Z}}$. (2) $+_{\mathbb{Z}}$ is associative. (3) $+_{\mathbb{Z}}$ is commutative. (4) $+_{\mathbb{Z}}$ has identity $(0, 0)/\sim$ in $\mathbb{Z}$. (5) Every $(x, y)/\sim$ in $\mathbb{Z}$ has, in $\mathbb{Z}$, an additive inverse, which is $(y, x)/\sim$. (6) The structure $(\mathbb{Z}, +_{\mathbb{Z}}$, additive inverse, $(0, 0)/\sim)$ is an abelian group.

Proof: Exercise. The only difficult step is in part 1. If we start from pairs $(x, y)/\sim$ and $(u, v)/\sim$ and apply q^{-1}, we obtain infinitely many pairs, such as $(x + h, y + h)$ (and the analogue for $(u, v)/\sim$). Consequently, we should expect several "results" for the correspondence $p \circ a \circ q^{-1}$. Verifying that in fact we obtain just one result is proof of the fact that this correspondence is a total function. Proof of all the other parts is a matter of patient and long verification.

The fact that addition in $\mathbb{Z}$ has an additive inverse allows us to immediately define both negation and subtraction in $\mathbb{Z}$. Following the usual pattern, we use the symbol $-_{\mathbb{Z}}$ for both. Consequently, $-_{\mathbb{Z}}(x, y)/\sim\ = (y, x)/\sim$, $(x, y)/\sim\ -_{\mathbb{Z}} (u, v)/\sim\ := (x + v, y + u)/\sim$.

The definition of multiplication in $\mathbb{Z}$ follows the same pattern, except for the usual absence of an inverse. The intuitive idea is that $(x - y) \cdot (u - v) = (x \cdot u + y \cdot v) - (x \cdot v + y \cdot u)$.

Definition 45 As before, let $p:\mathbb{N} \times \mathbb{N} \to \mathbb{Z}$ be the natural surjection and q the function q we saw before. Let $\mathbf{m}:((\mathbb{N} \times \mathbb{N}) \times (\mathbb{N} \times \mathbb{N})) \to \mathbb{N} \times \mathbb{N}$ be the total function defined by $((x, y), (u, v) \mapsto (x \cdot u + y \cdot v, x \cdot v + y \cdot u)$, where the symbols $+$ and $\cdot$ signify ordinary addition and multiplication in $\mathbb{N}$. Multiplication in $\mathbb{Z}$ is denoted by $\cdot_{\mathbb{Z}}$ and is defined as the correspondence $\cdot_{\mathbb{Z}}:\mathbb{Z} \times \mathbb{Z} \to \mathbb{Z}$ given by $\cdot_{\mathbb{Z}} := p \circ \mathbf{m} \circ q^{-1}$.

Theorem 46 The correspondence $\cdot_{\mathbb{Z}}$ has the following properties: (1) $\cdot_{\mathbb{Z}}$ is a total function. That is, it is a total binary operation and $\mathbb{Z}$ is closed under $\cdot_{\mathbb{Z}}$. (2) $\cdot_{\mathbb{Z}}$ is associative. (3) $\cdot_{\mathbb{Z}}$ is commutative. (4) $\cdot_{\mathbb{Z}}$ has identity $(1, 0)/\sim$ in $\mathbb{Z}$. (5) The structure $(\mathbb{Z}, \cdot_{\mathbb{Z}}, (1, 0)/\sim)$ is an abelian monoid.

Proof: Exercise.

Theorem 47 In $\mathbb{Z}$: (1) multiplicaton $\cdot_{\mathbb{Z}}$ distributes over addition $+_{\mathbb{Z}}$; (2) there are no divisors of zero; that is, if the product of two elements of $\mathbb{Z}$ is zero, then at least one of the two elements is zero; (3) the structure $(\mathbb{Z}, +_{\mathbb{Z}}, \cdot_{\mathbb{Z}}$, additive inverse, $(0, 0)/\sim)$ is thus a commutative ring with identity $(1, 0)/\sim)$ and has no divisors of zero.

Proof: Exercise.

Definition 48 In general, a commutative ring with multiplicative identity and with no divisors of zero is called an **integral domain.**

Thus, the structure ($\mathbb{Z}$, $+_{\mathbb{Z}}$, $\cdot_{\mathbb{Z}}$, additive inverse, $(0, 0)/\sim$)with identity $(1, 0)/\sim$ is an integral domain, the integral domain $\mathbb{Z}$ of the **integers.**

10.11.1.2 Injection of Structure $\mathbb{N}$ into Structure $\mathbb{Z}$

We've now reached a point where our intuition is going to need a little rigorous help. We're used to thinking of the natural numbers $\mathbb{N}$ as being a subset of the integers $\mathbb{Z}$. By our derivation of $\mathbb{Z}$, this can't be true, since the elements of $\mathbb{Z}$ are equivalence classes of ordered pairs of natural numbers. We can, however, do the next best thing. We can demonstrate that there is an injective mapping from $\mathbb{N}$ into $\mathbb{Z}$ that not only is consistent with the definition of $\mathbb{Z}$, but that also preserves operations from $\mathbb{N}$, that is, it yields consistent results. This process is called *injecting $\mathbb{N}$ into $\mathbb{Z}$ isomorphically*.

Theorem 49 Let the mapping $\mathbf{j}:\mathbb{N} \to \mathbb{Z}$ be defined by $n \mapsto (n, 0)/\sim$ for all n in $\mathbb{N}$. Then (1) the mapping $\mathbf{j}$ is injective, but not surjective. (2) The image of $\mathbb{N}$ under $\mathbf{j}$ is a subset of $\mathbb{Z}$, which we'll denote temporarily $\mathbb{N}^*$. Set $\mathbb{N}^*$ is called the **non-negative part** of $\mathbb{Z}$. (3) For all m and n in $\mathbb{N}$, $\mathbf{j}(m + n) = \mathbf{j}(m) +_{\mathbb{Z}} \mathbf{j}(n)$ and $\mathbf{j}(m \cdot n) = \mathbf{j}(m) \cdot_{\mathbb{Z}} \mathbf{j}(n)$. (4) $\mathbf{j}(0) = (0, 0)/\sim$ and $\mathbf{j}(1) = (1, 0)/\sim$. (5) Consequently, $\mathbf{j}$ is an isomorphism from the structure ($\mathbb{N}$, $+$, $\cdot$, 0) with multiplicative identity 1 to the structure ($\mathbb{N}^*$, $+_{\mathbb{Z}}$, $\cdot_{\mathbb{Z}}$, $(0, 0)/\sim$) with multiplicative identity $(1, 0)/\sim$.

Proof: Exercise.

The mapping $\mathbf{j}$ thus preserves addition and multiplication, complete with identities, from $\mathbb{N}$ into $\mathbb{Z}$. The key idea here is that, although $\mathbb{N}$ is not a subset of $\mathbb{Z}$, *its isomorphic image under* $\mathbf{j}$ *is*. This idea is so natural to us that most people, including mathematicians, simply treat $\mathbb{N}$ as if it were indeed a subset of $\mathbb{Z}$. Here again, you should know better but insist on it only when necessary.

Another shorthand form used in dealing with $\mathbb{Z}$ concerns the labelling of its elements. The integer $+3$ is in reality the equivalence class $(3, 0)/\sim$, and the integer 0 is really $(0, 0)/\sim$. So far, this is simply the application of the injection $\mathbf{j}$. But every element of $\mathbb{Z}$ has an additive inverse, so that for $+3 := (3, 0)/\sim$ there is an inverse $-3 := (0, 3)/\sim$. In fact, every element of $\mathbb{Z}$ can be characterized as precisely one of $(n, 0)/\sim$, $(0, 0)/\sim$, or $(0, n)/\sim$, where n is a positive natural number. In the exercises, you'll be asked to prove this and to show how it is used to define an order relation on $\mathbb{Z}$.

10.11.2 Rational Numbers

Having defined the integers $\mathbb{Z}$ by extending the natural numbers $\mathbb{N}$ and by injecting $\mathbb{N}$ into $\mathbb{Z}$, we can now define the rational numbers $\mathbb{Q}$. Logically, the process is almost identical: we'll extend $\mathbb{Z}$ to $\mathbb{Q}$ and inject $\mathbb{Z}$ into $\mathbb{Q}$. The problem sounds familiar: a rational number would be intuitively defined as the quotient of two integers (requiring, as usual, a nonzero denominator), if we had a division operation in $\mathbb{Z}$. But we don't. If we did, we would note that $(x/y) = (u/v)$ if and only if $xv = yu$. That, and the outline that follows, are the only hints you're going to get; the exercises ask you to complete the process. Furthermore, $\mathbb{Q}$ will possess a multiplicative inverse for every element except the additive identity, so that $\mathbb{Q}$ will not be merely an integral domain, but a *field* (see the definition of *field* below).

Outline of the Extension from the Integers to the Rationals. Define $S := \mathbb{Z} \times (\mathbb{Z} - \{0\})$, and $S^\# := (\mathbb{Z} - \{0\}) \times (\mathbb{Z} - \{0\})$. Clearly, $S^\# \subset S \subset \mathbb{Z} \times \mathbb{Z}$. On S, define the equivalence relation $\sim$ as follows: Two ordered pairs (x, y) and (u, v) in S are equivalent, written $(x, y) \sim (u, v)$, if and only if $x \cdot v = y \cdot u$, where x, y, u, and v are in $\mathbb{Z}$. Prove that this is indeed an equivalence relation. Again, the intuitive idea is that $x/y = u/v$. Then the set $\mathbb{Q}$ of rationals is defined as the quotient set $\mathbb{Q} := S/\!\sim$. Let $p{:}S \to \mathbb{Q}$ be the natural surjection, that is, the total function defined by $(x, y) \mapsto (x, y)/\!\sim$. Let q be defined in analogy to the last number extension.

Let $\mathbf{a}{:}S \times S \to S$ be the correspondence defined by $((x, y), (u, v)) \mapsto (xv + yu, yv)$, where the symbol $+$ denotes addition in $\mathbb{Z}$ and multiplication in $\mathbb{Z}$ has no symbol. Prove that $\mathbf{a}$ is a total function. Addition in $\mathbb{Q}$ is denoted by $+_\mathbb{Q}$ and is defined as the correspondence $+_\mathbb{Q}{:}\mathbb{Q} \times \mathbb{Q} \to \mathbb{Q}$ given by $+_\mathbb{Q} := p \circ \mathbf{a} \circ q^{-1}$. Prove that $+_\mathbb{Q}$ has the following properties: (1) $+_\mathbb{Q}$ is a total function; (2) $+_\mathbb{Q}$ is associative; (3) $+_\mathbb{Q}$ is commutative; (4) $+_\mathbb{Q}$ has identity $(0, 1)/\!\sim$ in $\mathbb{Q}$; (5) every $(x, y)/\!\sim$ in $\mathbb{Q}$ has, in $\mathbb{Q}$, an additive inverse, which is $(-x, y)/\!\sim$; (6) the structure $(\mathbb{Q}, +_\mathbb{Q}$, additive inverse, $(0, 1)/\!\sim)$ is an abelian group.

Define negation and subtraction in $\mathbb{Q}$, using the symbol $-_\mathbb{Q}$ for both, as $-_\mathbb{Q}(x, y)/\!\sim\; := (-x, y)/\!\sim$, and $(x, y)/\!\sim\; -_\mathbb{Q} (u, v)/\!\sim\; := (xv - yu, yv)/\!\sim$.

Let $\mathbf{m}{:}S \times S \to S$ be the correspondence defined by $((x, y), (u, v)) \mapsto (xu, yv)$. Prove that it is a total function. The operation of multiplication in $\mathbb{Q}$ is denoted by $\cdot_\mathbb{Q}$ and is defined as the correspondence $\cdot_\mathbb{Q}{:}\mathbb{Q} \times \mathbb{Q} \to \mathbb{Q}$ given by $\cdot_\mathbb{Q} := p \circ \mathbf{m} \circ q^{-1}$. Define the correspondence $\mathbf{r}{:}S^\# \to S^\#$ as $(x, y) \mapsto (y, x)$, where $x \in \mathbb{Z} - \{0\}$ and $y \in \mathbb{Z} - \{0\}$. Prove that it is a bijection. Define the correspondence ${}^{-1}{:}\ \mathbb{Q} - \{(0, 1)/\!\sim\} \to \mathbb{Q} - \{(0, 1)/\!\sim\}$ as ${}^{-1} := p \circ \mathbf{r} \circ p^{-1}$. Prove that it is a bijection. Prove that $\cdot_\mathbb{Q}$ has the following properties (1) $\cdot_\mathbb{Q}$ is a total function; (2) $\cdot_\mathbb{Q}$ is associative; (3) $\cdot_\mathbb{Q}$ is commutative; (4) $\cdot_\mathbb{Q}$ has identity $(1, 1)/\!\sim$ in $\mathbb{Q}$. (5) The structure $(\mathbb{Q}, \cdot_\mathbb{Q}, (1, 1)/\!\sim)$ is an abelian monoid. (5) The structure $(\mathbb{Q} - \{(0, 1)/\!\sim\}, \cdot_\mathbb{Q}, {}^{-1}, (1, 1)/\!\sim)$ is an abelian group.

Prove that, in $\mathbb{Q}$, multiplication $\cdot_\mathbb{Q}$ distributes over addition $+_\mathbb{Q}$. The structure $(\mathbb{Q}, +_\mathbb{Q}, \cdot_\mathbb{Q}$, additive inverse, $(0, 1)/\!\sim)$ is a commutative ring with identity $(1, 1)/\!\sim$ and every nonzero element has a multiplicative inverse.

Definition 50 In general, a commutative ring with multiplicative identity and a multiplicative inverse for each nonzero element is called a **field.** It is easily checked that every field in an integral domain.

Thus the structure we just constructed on $\mathbb{Q}$ is a field, the field of the **rational numbers.** According to common use, denote the class $(x, y)/\!\sim$ as x/y (a fraction).

Outline of the Isomorphic Injection of the Integers into the Rationals. To *inject* $\mathbb{Z}$ into $\mathbb{Q}$ *isomorphically* use the following outline: Let the mapping $\mathbf{j}:\mathbb{Z} \to \mathbb{Q}$ be defined by $n \mapsto n/1$ for all n in $\mathbb{Z}$. Prove that (1) the mapping $\mathbf{j}$ is injective but not surjective; (2) the image of $\mathbb{Z}$ under $\mathbf{j}$ is a subset of $\mathbb{Q}$, which we'll denote temporarily $\mathbb{Z}^*$; (3) for all m and n in $\mathbb{Z}$, $\mathbf{j}(m + n) = \mathbf{j}(m) +_{\mathbb{Q}} \mathbf{j}(n)$ and $\mathbf{j}(mn) = \mathbf{j}(m) \cdot_{\mathbb{Q}} \mathbf{j}(n)$; (4) $\mathbf{j}(0) = 0/1$ and $\mathbf{j}(1) = 1/1$; (5) Consequently, $\mathbf{j}$ is an isomorphism from the integral domain ($\mathbb{Z}$ +, $\cdot$, additive inverse, 0) with identity 1 to the structure ($\mathbb{Z}^*$, $+_{\mathbb{Q}}$, $\cdot_{\mathbb{Q}}$, additive inverse, 0/1) with multiplicative identity 1/1.

Define an order relation in $\mathbb{Q}$ by using the set $\mathbb{Q}^+ := \{x/y \mid x \geq 0 \text{ and } y > 0\}$ and the method of Definition 10. Prove that the mapping $\mathbf{j}$ is order-isomorphic between the given order structures on $\mathbb{Z}$ and $\mathbb{Z}^*$.

10.11.3 Real Numbers

Next, we consider the question of extending the rationals $\mathbb{Q}$ to and injecting $\mathbb{Q}$ into the real numbers $\mathbb{R}$. Structure $\mathbb{R}$, of course, contains irrational numbers, such as π, Euler's e, and $\sqrt{2}$. The question is whether every irrational number (or any of them, for that matter) can be constructed from only the rationals $\mathbb{Q}$, and algebraic and set-theoretic operations in $\mathbb{Q}$. This problem has a long history. The ancient Greeks at first thought that $\mathbb{Q}$ was "all there was", that is, that all numbers were rational. They believed that all numbers, including π and $\sqrt{2}$ (they knew nothing of e), were complicated combinations of integers through addition, multiplication, subtraction, and division. When Pythagoras proved that the square root of two was indeed irrational, it so astounded him that he thought it was a religious revelation, and forbade his followers from revealing the fact! Indeed, this is the source of our terminology: it was "not rational" to believe that the square root of two was not some combination of integers. But similar proofs for π and e followed in the modern era. The question was finally settled by Georg Cantor in the late nineteenth century. The details of his analysis are beyond the scope of this book because what Cantor did was to show that the real numbers $\mathbb{R}$ are a *continuous* structure, fundamentally different from $\mathbb{N}$, $\mathbb{Z}$, and $\mathbb{Q}$. A full analysis of real numbers must be approached on its own terms. We will not deal with it here in detail, but we do give a very short outline of the construction process. The process itself it not essential for our main purposes.

Outline of the Extension from the Rationals to the Reals. The intuitive approach to the definition of the reals is based (as in previous cases) on what we could do *if we knew* the reals: we could take all the rationals that are larger than any given real, and the rationals that are smaller than or equal to the given real. For instance, we know that $3.14 \leq \pi < 3.15$, $3.14159 \leq \pi < 3.14160$, $3.1415926 \leq \pi < 3.1415927$, and so on. Note that all the numbers we wrote explicitly are rational.

We define a **cut** in $\mathbb{Q}$ as a special partition $\{a, A\}$ of $\mathbb{Q}$ into precisely two blocks a and A that satisfy the additional properties: (1) if $x \in a$ and if $y \in A$, then $x < y$; (2) set A has no minimum with respect to the order $\leq$. Two special examples are $0_{\mathbb{R}} := \{\{x \in \mathbb{Q} \mid x \leq 0\}, \{x \in \mathbb{Q} \mid x > 0\}\}$ and $1_{\mathbb{R}} := \{\{x \in \mathbb{Q} \mid x \leq 1\}, \{x \in \mathbb{Q} \mid x > 1\}\}$.

Let $\mathbb{R}$ be the set of all cuts. We intend to define addition and multiplication on $\mathbb{R}$. Let $\{a, A\}$ and $\{b, B\}$ be any two cuts. Consider the set $P := \{x + y \mid x \in A$ and $y \in B\}$. There is some difficulty in proving that $\{\mathbb{R} - P, P\}$ is a cut, which we define as the sum of the two given cuts. We so have a total binary operation $+_{\mathbb{R}}: \mathbb{R} \times \mathbb{R} \to \mathbb{R}$. A similar definition is valid for multiplication, assuming that A and B each contain only positive rationals: define $P := \{xy \mid x \in A$ and $y \in B\}$; then $\{\mathbb{R} - P, P\}$ is a cut, which we define as the product of the two given cuts. There are difficulties in extending this definition to every pair of cuts. Ultimately, we have a total binary operation $\cdot_{\mathbb{R}}: \mathbb{R} \times \mathbb{R} \to \mathbb{R}$. After that, we can introduce inverses and prove that the structure ($\mathbb{R}$, $+_{\mathbb{R}}$, $\cdot_{\mathbb{R}}$, additive inverse, $0_{\mathbb{R}}$) with multiplicative identity $1_{\mathbb{R}}$ and the defined multiplicative inverse is a field, the field of the **reals.**

Outline of the Isomorphic Injection of the Rationals into the Reals. Injecting the rationals $\mathbb{Q}$ into the reals $\mathbb{R}$ isomorphically presents no problems. We define $\mathbf{j}$: $\mathbb{Q} \to \mathbb{R}$ by $z \mapsto \{\{x \in \mathbb{Q} \mid x \leq z\}, \{x \in \mathbb{Q} \mid x > z\}\}$. Let $\mathbb{Q}^*$ be the image of $\mathbf{j}$. Then $\mathbf{j}$ is an injection, which is an isomorphism with respect to addition and multiplication in $\mathbb{Q}$ and with respect to $+_{\mathbb{R}}$ and $\cdot_{\mathbb{R}}$ in $\mathbb{Q}^*$. We then define $\mathbb{R}^+$ as the set of the cuts $\{a, A\}$ other than $0_{\mathbb{R}}$ for which $0 \in a$ and construct the respective order in $\mathbb{R}$. Then $\mathbf{j}$ is order-isomorphic.

10.11.4 Complex Numbers

Once the real numbers are established, the extension to the complex numbers $\mathbb{C}$ is simple, much simpler, in fact, than the extensions of number systems we have seen before. Note that also the structure of the complex numbers is a nondiscrete one. As we mentioned before, it does not even have an order that is consistent with addition and multiplication.

Outline of the Extension from the Reals to the Complex Numbers. Define $\mathbb{C} := \mathbb{R} \times \mathbb{R}$ (no equivalence here). When $(x, y) \in \mathbb{C}$ and $(u, v) \in \mathbb{C}$, define $(x, y) +_{\mathbb{C}} (u, v) := (x + u, y + v)$ and $(x, y) \cdot_{\mathbb{C}} (u, v) := (xu - yv, xv + yu)$. In this case $+_{\mathbb{C}}$ and $\cdot_{\mathbb{C}}$ are automatically total binary operations. Define $0_{\mathbb{C}} := (0, 0)$ and $1_{\mathbb{C}} := (1, 0)$. Define the multiplicative inverse of (x, y) for anything other than $(0, 0)$. [Hint: it will be $(x/(x^2 + y^2), -y/(x^2 + y^2))$]. Prove that the structure ($\mathbb{C}$, $+_{\mathbb{C}}$, $\cdot_{\mathbb{C}}$, additive inverse, $0_{\mathbb{C}}$) with multiplicative identity $1_{\mathbb{C}}$ and multiplicative inverse as defined above is a field, the field of the **complex numbers.**

Outline of the Isomorphic Injection of the Reals into the Complex Numbers. Define $\mathbf{j}: \mathbb{R} \to \mathbb{C}$, $x \mapsto (x, 0)$ and let $\mathbb{R}^*$ be its image. Prove that $\mathbf{j}$ is an injection, which is an isomorphism with respect to addition and multiplication in $\mathbb{R}$ and with respect to $+_{\mathbb{C}}$ and $\cdot_{\mathbb{C}}$ in $\mathbb{R}^*$. That's all as far as construction and injection are concerned.

The Number i. You may object that you've seen complex numbers before and that they involved a certain i. Ours don't. The answer is simple, and it involves only some notation and a few algebraic verifications. Let's see this in some detail. Define $i := (0, 1)$. Using the definitions of addition and multiplication in $\mathbb{C}$, verify that $(x, y) = (x, 0) +_{\mathbb{C}} (0, y) = (x, 0) +_{\mathbb{C}} (0, 1) \cdot_{\mathbb{C}} (y, 0)$. Using the injection **j**, we can rewrite this result in the form $(x, y) = (\mathbf{j}x) +_{\mathbb{C}} i \cdot_{\mathbb{C}} (\mathbf{j}y)$. Pretending that $\mathbb{R} \subseteq \mathbb{C}$ (that is, pretending that **j** is the identity on $\mathbb{R}^*$) and writing $+$ for $+_{\mathbb{C}}$ and no symbol for $\cdot_{\mathbb{C}}$, the equality above becomes $(x, y) = x + iy$.

Satisfied? "No, where did $i^2 = -1$ go?" Here: $i^2 = i \cdot_{\mathbb{C}} i = (0, 1) \cdot_{\mathbb{C}} (0, 1) = (0 \cdot 0 - 1 \cdot 1, 0 \cdot 1 + 1 \cdot 0) = (-1, 0) = -_{\mathbb{C}} (1, 0) = -_{\mathbb{C}} (\mathbf{j}1)$, which equals -1 (almost). More objections of the same kind can be parried with similar verifications. We'll see some in the exercises.

Exercises 10.11

1. Prove that the relation given in Definition 42, definition of $\mathbb{Z}$, is an equivalence relation.
2. Prove Theorems 44, 46, and 47 on the operations in $\mathbb{Z}$.
3. Prove Theorem 49 on the injection of $\mathbb{N}$ into $\mathbb{Z}$.
4. In $\mathbb{Z}$, show that $(\mathbf{j}m) -_{\mathbb{Z}} (\mathbf{j}n) = (m, n)/\sim$ for every choice of m and n in $\mathbb{N}$. This justifies the shorthand $m - n$ for $(m, n)/\sim$.
5. The structure $\mathbb{N}$ satisfied a *trichotomy axiom:* relation $<$ on $\mathbb{N}$ is such that, for all x and y in $\mathbb{N}$, precisely one of $x < y$, $x = y$, $y < x$ is true. **(a)** Prove that every equivalence class $(x, y)/\sim$ in $\mathbb{Z}$ contains precisely one pair of precisely one of the following three types: **(i)** An ordered pair of the form $(n, 0)$, **(ii)** The ordered pair $(0, 0)$, **(iii)** An ordered pair of the form $(0, n)$, where $n \in \mathbb{N} - \{0\}$. **(b)** If $a \in \mathbb{Z}$ and $b \in \mathbb{Z}$, state formal definitions for the relations $a < b$, $a \leq b$, $a > b$, $a \geq b$. Show that these relations are consistent with those on $\mathbb{N}$. **(c)** Using these definitions, show that $\mathbb{Z}$ satisfies a trichotomy axiom as a consequence of the fact that $\mathbb{N}$ does.
6. Complete the details of the outline in Section 10.11.2 for the extension of $\mathbb{Z}$ to $\mathbb{Q}$.
7. In $\mathbb{Q}$, show that $(\mathbf{j}m)(\mathbf{j}n)^{-1} = m/n$ for every choice of m in $\mathbb{Z}$ and n in $\mathbb{Z} - \{0\}$. This justifies the shorthands mn^{-1} and m/n.
8. Prove that every field is an integral domain. [(Hint: if $xy = 0$ and if $x \neq 0$, then $x^{-1}xy = \ldots$.]
9. Complete the details of the outline in Section 10.11.4 for the extension of $\mathbb{R}$ to $\mathbb{C}$.
10. Let $x + iy$ and $u + iv$ be two complex numbers written in the conventional form we adopted in the text. Prove that multiplication in $\mathbb{C}$ can be performed according to the following rule: Multiply the two expressions above as if they were real polynomials in the arguments x, y, u, v, i; then replace i^2 with -1.

11. In $\mathbb{C}$, reduce to the form $x + iy$ the result of each of the following operations: **(a)** $(2 + 3i) - (5 - 2i)$; **(b)** $(2 + 3i)(5 - 2i)$; **(c)** $(2 + 3i)/(5 - 2i)$; **(d)** i^2; **(e)** i^3; **(f)** i^4; **(g)** i^5; **(h)** i^6; **(j)** i^{-1}; **(k)** i^{-2}; **(l)** i^n for every integer n.

12. Prove that the square of each of i and $-i$ equals -1.

13. Prove that the square of each of $1 + i$ and $-1 - i$ equals $2i$.

14. Given that t is the real square root of 3, prove that the cube of each of 1, $(-1 + ti)/2$, and $(-1 - ti)/2$ equals 1.

10.12 Summary

The main goal of this chapter is the study of *order structures* (Section 10.6) and *equivalence relations* (Section 10.9).

A *partial order* comprises a set and a relation $\preccurlyeq$ that is reflexive, antisymmetric, and transitive. This means: $x \preccurlyeq x$ (reflexivity, Section 10.3), if $x \preccurlyeq y$ and if $y \preccurlyeq x$, then $x = y$ (antisymmetry, Section 10.4), if $x \preccurlyeq y$ and if $y \preccurlyeq z$, then $x \preccurlyeq z$ (transitivity, Section 10.5). Useful tools for studying partial orders are the strict relation $<$ (that is, $x \preccurlyeq y$ but $x \neq y$) and the core relation (that is, the smallest amount of information that, via reflexivity and transitivity, allows to reconstruct the whole relation). A special case of partial orders is a *total order:* any two elements are related in one of the two directions.

Structures that exploit order in detail are *lattices* (every pair of elements has a least upper bound and a greatest lower bound; see Section 10.8) and several special cases: complete, distributive, or complemented lattices and, ultimately, Boolean algebras.

An *equivalence relation* on A is a relation $\sim$ that is reflexive, symmetric (that is, if $x \sim y$, then $y \sim x$), and transitive (Section 10.9). An equivalence relation determines its *equivalence classes,* that is, sets of elements such that elements in the same set are equivalent and elements in different sets are not. An important result is that the set of the equivalence classes of an equivalence relation is a partition of A.

Equivalence relations are powerful tools for extending the natural integers $\mathbb{N}$ to (and injecting them isomorphically into) the integers $\mathbb{Z}$, the integers $\mathbb{Z}$ to the rationals $\mathbb{Q}$, the rationals $\mathbb{Q}$ to the reals $\mathbb{R}$, and the reals $\mathbb{R}$ to the complex numbers $\mathbb{C}$. The resulting structures $\mathbb{N}$, $\mathbb{Z}$, $\mathbb{Q}$, $\mathbb{R}$, and $\mathbb{C}$ are the *number systems* (see Section 10.11).

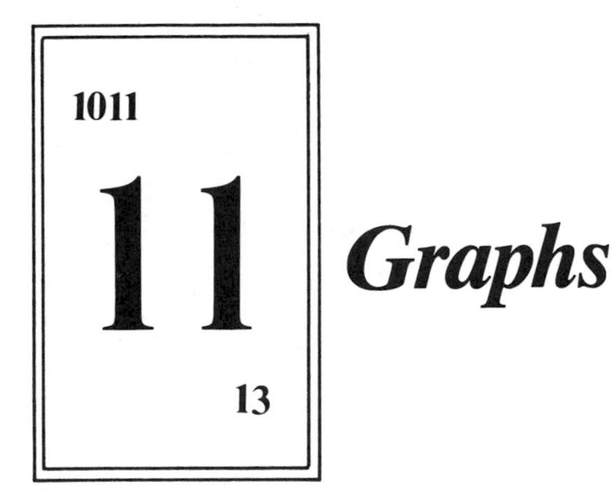

11 Graphs

The theory of graphs has to do with objects of the type shown in Figures 11.1 to 11.7. Before we go into the theory, let's spend a few words on these concrete configurations and on the problems that arise if we want to "do" anything with them.

At the time of Swiss mathematician Leonhard Euler (pronounce "oiler"; 1707–83), there were in the German city of Koenigsberg seven bridges that joined four masses of land (Figure 11.1b). A favorite pastime of the town was to try to take a walk that crossed each bridge just once. Nobody could do it. Euler wanted to show theoretically that the problem has no solution. In order to do that, he gave a schematic representation of the problem and reasoned on the representation. This representation is the drawing of a *graph*. Euler represented each mass of land with a point as in Figure 11.1a and each bridge as a segment, or arc, joining the two appropriate points. Once the idea was there, it was easy to generalize Euler's

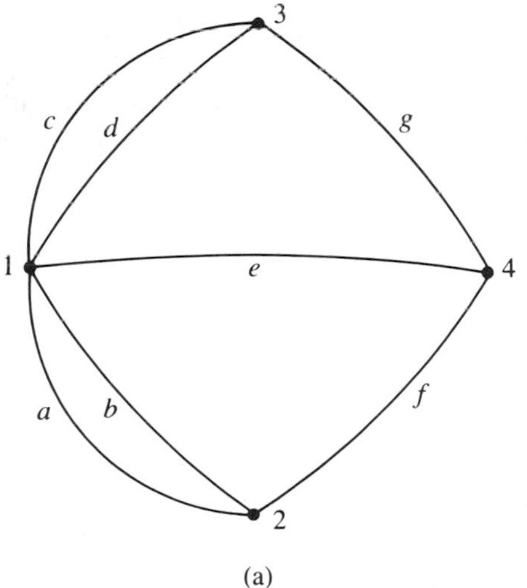

(a)

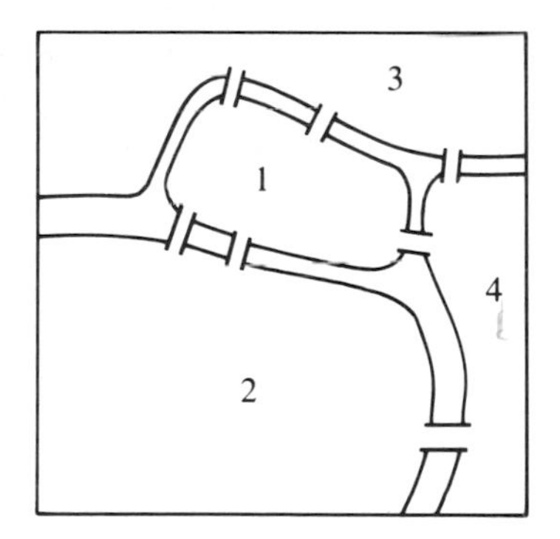

(b)

Figure 11.1 The seven bridges of Koenigsberg.

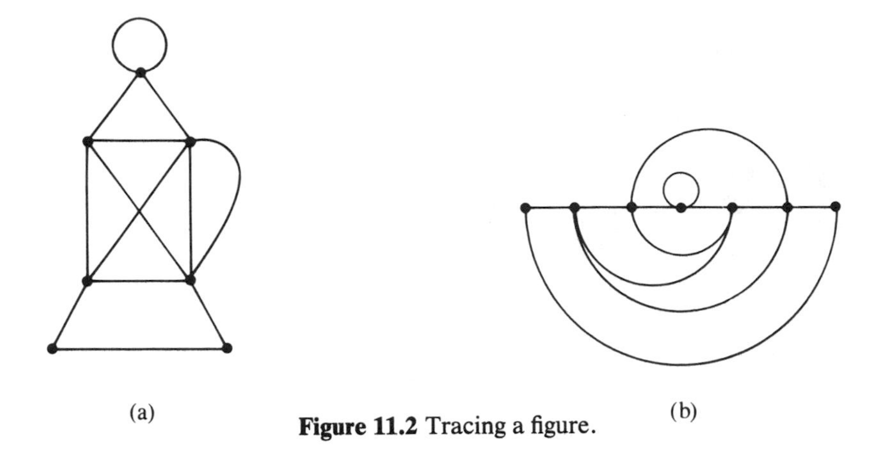

Figure 11.2 Tracing a figure.

results to similar problems. Here are some samples. Each consists of a set of points, or *nodes,* connected by some sets of *arcs.*

Can either diagram in Figure 11.2 be traced without lifting the pencil and by going over each arc exactly once? This is a situation where we must be able to make some abstraction: the central point in Figure 11.2a is to be considered as an incidental crossing, not as an intended point of the figure. Another abstraction: are the two diagrams in the figure the *same?* In appearance, certainly not, but with respect to tracing the figure, they are somewhat the same. If we grabbed the legs of Figure 11.2a and pulled, we would stretch it into Figure 11.2b (except for one arc, which we "flipped"). The relative locations of the nodes are unimportant. We only worry about which nodes are connected by arcs with which others. How do we rigorously characterize this "sameness" of graphs? How can we tell, in general, whether two graphs are "the same"?

Other problems have to do with the number of nodes or arcs possible in a graph that satisfies some restrictive conditions. Figure 11.3 gives some "small" graphs, taking the number of nodes or arcs as an indication of smallness. Part (a) "represents" the empty graph, which has no nodes and no arcs. Part (b) shows a graph with one node and no arcs. The graph is part (c) has one node and two arcs.

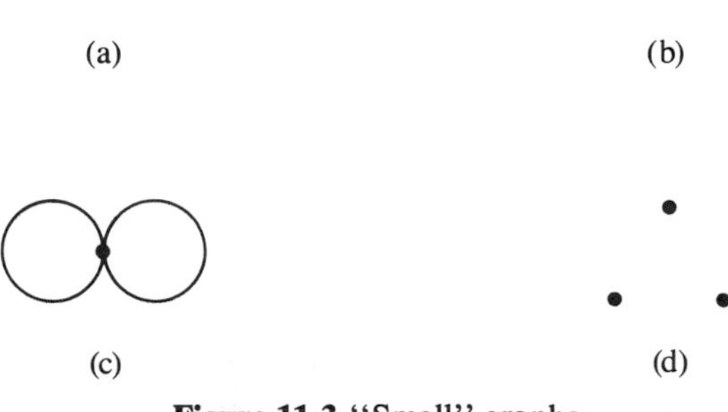

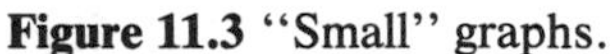

Figure 11.3 "Small" graphs.

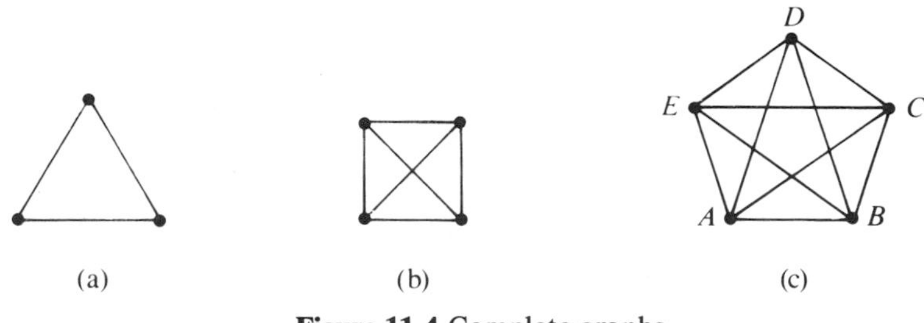

(a) (b) (c)

Figure 11.4 Complete graphs.

Note the *loops,* that is, arcs that begin and end at the same node. In part (d) there are three nodes and no arcs. On the other hand, Figure 11.4 gives "large" graphs, graphs that have the maximal number of arcs, under the restrictions that the number of nodes is given, that there should be no loops, and that any two distinct nodes cannot be joined by more than one arc. Parts (a), (b), and (c) represent the complete graphs on 3, 4, and 5 nodes, respectively. Another type of interesting graph is provided by the *complete bipartite* graphs (Figure 11.5). Each has two sets of nodes. No node is joined by any arc to any other node within the same set. But each node of each set is joined by a single arc to each node of the other set. In particular, the graph of Figure 11.5a is known as the *three-utilities graph.* Think of providing electricity, water, and gas from three sources (the three nodes on the left) to three houses (the nodes on the right). The incidental crossings would represent pipes that must be buried at different depths. This is an undesirable complication. Could another drawing be found with no incidental crossings? Is the graph in Figure 11.5b the "same" as that in Figure 11.5a? Why? The last example presents another problem, which has important applications, for instance, in the design of circuit boards: can an *equivalent* graph be drawn without incidental crossings? The answer is negative for the graph in Figure 11.5a and for the graph in Figure 11.4c. However, the graph in Figure 11.4b can be redrawn without incidental crossings. (How?)

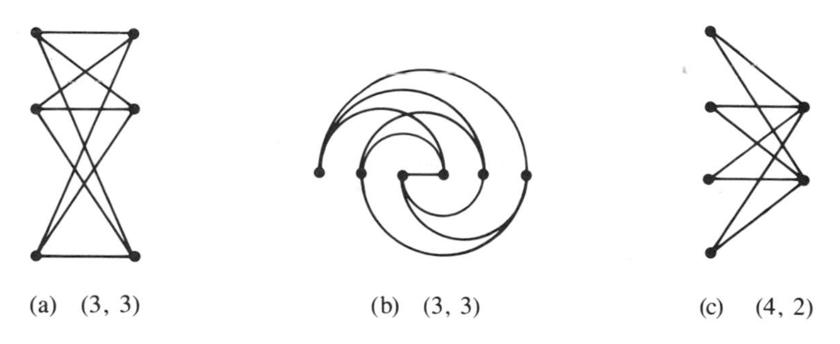

(a) (3, 3) (b) (3, 3) (c) (4, 2)

Figure 11.5 Bipartite graphs.

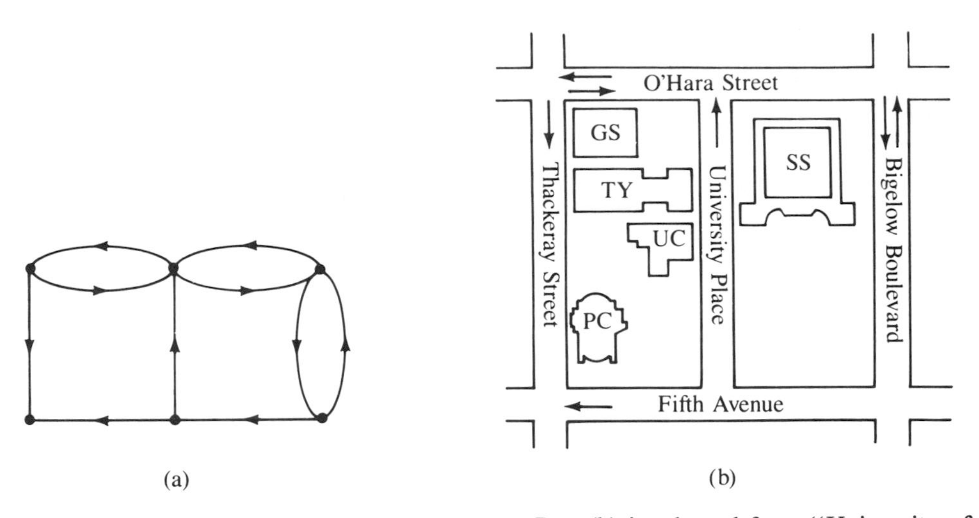

Figure 11.6 One-way and two-way streets. Part (b) is adapted from "University of Pittsburgh, College of Arts and Sciences, Course Descriptions, Fall Term 1984."

Still other problems arise in graphs for which the arcs are *one-way*. Figure 11.6 shows the street map of a small portion of a campus, including some one-way streets, in the usual sense of the phrase, and a schematization, which is the drawing of a *directed* graph. Figure 11.7 has to do with the processes that place persons into the top offices of the government of the United States, or remove them from office. This figure is limited *exclusively* to such processes. It does not consider the functions of the offices. The nodes are the offices, in a simplified form. There are, of course, many senators, many representatives, many electoral districts, and so on. Such offices are labelled with capitalized names. The arcs are the actions that influence placement into office or removal from office. The last two examples stress the problem of deciding whether or how you can "get there from here," or who can influence whom and how. In particular, Figure 11.7 shows several checks and balances. In this case, every node is the end of some directed arc; that is, every office is subject to some check.

All of these examples should make one fact evident: In extremely simple circumstances we need no theory, of course, to solve any of these problems. As an instance, there are rather few walks that cross each bridge in Figure 11.1 no more than once (how many? see Chapter 4) and it's easy to see that none crosses all the bridges. In Figure 11.6 it's again quite easy to see how to drive from one point to another and what happens if a street is closed or if a one-way street is reversed. In fact, many practical problems can be expressed quite as nicely as graphs: communication and transportation networks, energy distribution grids, oil and gas pipelines, even the interconnections of nerve cells in the brain. For our purposes we can ignore some details such as the real-world properties of the nodes or the capacities of the arcs, and consider simply nodes and arcs. Even so, these problems might easily involve hundreds or thousands, or occasionally even millions, of nodes and just as many arcs. A quick glance at a schematic will not give us any insight into such complex graphs. For such situations, we need theory.

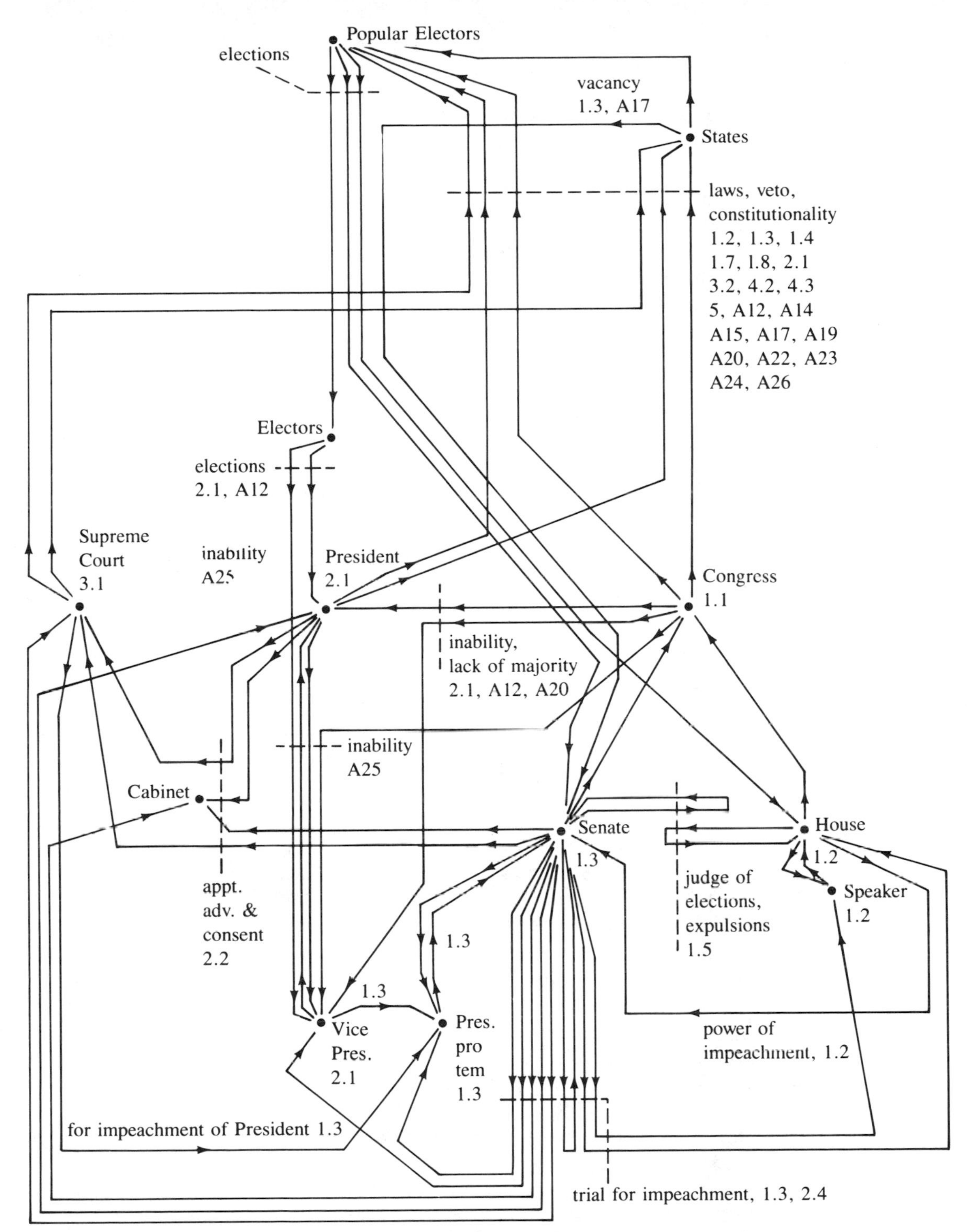

Figure 11.7 Top offices in the Government of the U.S.

Notes: a.s or a denotes Article a, Section s of the Constitution of the United States; An.s or An denotes the *n*th Amendment, Section s; Capitalized names are names of Offices (nodes), the other names are names of processes that place persons into offices or remove them (arcs).

11.1 Directed and Undirected Graphs

In order to characterize a graph, we must clearly state which arcs have which nodes as their endpoints. Specifically, we must have nodes and arcs (and be able to tell nodes from arcs) and a rule that gives us, for each pair of nodes, the list of those arcs that have those two nodes as endpoints. This is what the following formal definition is about:

Definition 1 A **graph** is a structure (V, A, b), where V and A are disjoint sets and $b: V \times V \to A$ is a surjective correspondence that satisfies one of the following two conditions: (1) If $(v, w) \mapsto x$, then $(w, v) \mapsto x$ and no other pair has image x under b. In this case, the structure is called an **undirected graph.** (2) b^{-1} is a function. In this case, the structure is called a **directed graph** or a **digraph.**

The elements of V are called **nodes,** or **vertices,** or **points.** The elements of A are called **arcs,** or **sides,** or **segments** (not necessarily straight-line segments). If $(v, w) \mapsto x$ then v and w are the **endpoints** or the **boundary points** of arc x. For a directed graph, node v is the **first endpoint,** or **origin,** and w is the **second endpoint,** or **end,** of x. If $v = w$, then arc x is called a **loop.** The correspondence b is called the **boundary correspondence** of the graph. A node that is the endpoint of no arc is called an **isolated** node.

Example Let the following be given, with the agreement that different letters represent different objects. $V = \{1, 2, 3, 4\}$; $A = \{a, b, c, d, e, f, g\}$; **b** is defined by the assignments $(1, 2) \mapsto a$, $(2, 1) \mapsto a$, $(1, 2) \mapsto b$, $(2, 1) \mapsto b$, $(1, 3) \mapsto c$, $(3, 1) \mapsto c$, $(1, 3) \mapsto d$, $(3, 1) \mapsto d$, $(1, 4) \mapsto e$, $(4, 1) \mapsto e$, $(2, 4) \mapsto f$, $(4, 2) \mapsto f$, $(3, 4) \mapsto g$, $(4, 3) \mapsto g$. Then we have an undirected graph with no loops. In fact, we have a representation of Figure 11.1a. In the exercises you'll be asked to find similar theoretical representations of graphs presented by means of a drawing.

From an intuitive point of view, the nodes of the graph are the heavily marked points of the drawing, the arcs of the graph are the arcs of the drawing, the images of a pair of nodes under **b** are the arcs that join those two nodes, not excluding the possibility that the two nodes coincide. The correspondence **b** translates the **incidence** between nodes and arcs: what node is (which) end of what arc. That's why graphs are special cases of structures that go under the general term of *incidence structures*.

Note that A and V can be both empty, for instance in Figure 11.3a; A can be empty and V nonempty, for instance Figure 11.3b or d; but, if A is nonempty, then V must be nonempty, as a consequence of the surjectivity of **b.** After all, an arc cannot exist without nodes. Finally, nodes are not necessarily points in the intuitive sense (the President is not a point) and arcs are not necessarily geometrical segments, straight or not (a bridge is usually straight, but never a segment).

Transitions Between an Undirected Graph and a Directed Graph. Clearly, an undirected graph can easily be changed to an equivalent directed graph by assigning each arc a two-way status or, more correctly, by replacing each nonloop arc by two arcs with opposite directions. Formally, if $v \neq w$ and if $(v, w) \mapsto x$ and $(w, v) \mapsto x$ in the undirected graph, then $(v, w) \mapsto x$ and $(w, v) \mapsto x^*$ in the directed graph, where x^* is a new arc. The inverse process is not possible unless the nonloop arcs of the directed graph are present in pairs, each pair joining the same nodes in opposite directions.

It is however natural to associate with a directed graph the undirected graph obtained by ignoring the orientations of the arcs. More precisely, whenever $(v, w) \mapsto x$ in the directed graph, define $(v, w) \mapsto x$ and $(w, v) \mapsto x$ in the undirected graph. What we thus create is an entirely new graph.

Exercises 11.1

1. For each graph in Figures 11.1–6, number each node and letter each arc (if any), then give (by enumeration) V, A, and b as in the definition of graph.
2. A **planar** representation of a graph is a redrawing that preserves the relation "endpoint of" (not necessarily the physical positions of nodes and arcs) and has *no incidental crossings*, that is, no intersections of arcs except at the nodes.

   Find a planar representation of each of the graphs in Figures 11.2a and 11.4b. Note that not every graph, not even every finite graph, admits a planar representation. Note also that ordinary road maps (with no tunnels or underpasses) are planar representations of the graph of streets and intersections of streets.
3. Consider each of the graphs in Figures 11.2, 11.3c, 11.4a and c, 11.5c. Trace all of its arcs without lifting the pencil and without going over each arc more than once. Note that not every finite graph admits a solution to this problem.
4. Find other examples of your own. Start from a simple line-drawing and place nodes to make it a graph. See, for instance, Figure 10.3. In each example, consider several of the problems we have mentioned.

Complements 11.1

1. Prove the following: Let g be any relation on a set V, that is, suppose $g \subseteq V \times V$ (see Section 7.2). Then g can be interpreted as a directed graph by setting $A := g$ and b as the inverse of the injection of A into $V \times V$. Similarly, any symmetric relation on V can be interpreted as an undirected graph. The converse of neither statement is true. Counterexamples are given by graphs in which there are many arcs joining the same two nodes. Find conditions under which a directed or an undirected graph is a relation on V.

11.2 Adjacency Matrices

The representation of a graph according to the definition in Section 11.1 gives all the incidence information about the graph, of course. But even this representation is somewhat impractical in concrete cases. An equivalent form of the same representation is provided, in the case of finite graphs, by *incidence matrices*. There are several forms of incidence matrices for a graph. We'll consider here only one type, in three versions. This type of incidence matrix is called *adjacency matrix*. We'll see another type in Chapter 12.

11.2.1 Path Adjacency Matrix

Intuitively, an adjacency matrix is obtained by representing the set $V \times V$ as a double-entry table (see Section 5.9). Then each box stands for an ordered pair of nodes. Fill that box with the set of the arcs that have those two nodes as endpoints, in that order. This explains the name: an entry is not empty only if the corresponding nodes are *adjacent* to each other, that is, are joined by at least one arc. Formally:

Definition 2 Let $G := (V, A, b)$ be a nonempty finite graph as given in Section 11.1. The **1-path adjacency matrix** of G is defined as the matrix W whose rows and columns are indexed by the elements of V and whose entry (v, w) $(v \in V, w \in V)$ is the set $\{x \in A \mid (v, w) \mapsto x \text{ under } b\}$. Usually, the elements of V are numbered from 1 to $\#V$. Then the definition of W is $W(i, j) := \{x \in A \mid (i, j) \mapsto x \text{ under } b\}$.

Example For the graph in Figure 11.1 W is the following matrix. The symbols in brackets are row/column labels.

$$\begin{array}{c} \\ [1] \\ [2] \\ [3] \\ [4] \end{array} \begin{array}{c} \begin{array}{cccc} [1] & [2] & [3] & [4] \end{array} \\ \begin{bmatrix} \varnothing & \{a, b\} & \{c, d\} & \{e\} \\ \{a, b\} & \varnothing & \varnothing & \{f\} \\ \{c, d\} & \varnothing & \varnothing & \{g\} \\ \{e\} & \{f\} & \{g\} & \varnothing \end{bmatrix} \end{array}$$

An intuitive interpretation of matrix W is to consider an arc as a one-step path that joins the first endpoint to the second endpoint (in both orders if the graph is undirected). Then entry (i, j) is the set of all one-step paths that join node i to node j. This interpretation will lead to the concept of *path* in general (see Section 11.3).

11.2.2 Numerical and Boolean Adjacency Matrices

The one-path adjacency matrix defined above gives all of the *incidence* information about its graph. It tells which arcs are incident to (that is, have as endpoints) which nodes. In many instances we are willing to give up some of the information in order to gain some simplicity in the matrix and in the calculations. We are going to achieve such a goal in two different ways.

Definition 3 Let $G := (V, A, b)$ be a nonempty finite graph as given in Section 11.1. Let the vertices be numbered 1 to $\#V$. Let W be the adjacency matrix defined in Definition 2. The **numerical adjacency matrix** of G is the matrix K for which entry (i, j) is the number of arcs from i to j, that is, $K(i, j) := \#W(i, j)$. The **Boolean adjacency matrix** of G is the matrix B for which entry (i, j) is *true* or *false* according as there are some arcs or there are no arcs from i to j, that is, $B(i, j) := (W(i, j) \neq \varnothing) = (K(i, j) > 0)$. (The second equality requires proof; prove it as an exercise.)

Example In the example that follows Definition 2, the two matrices K and B are, respectively,

$$K = \begin{bmatrix} 0 & 2 & 2 & 1 \\ 2 & 0 & 0 & 1 \\ 2 & 0 & 0 & 1 \\ 1 & 1 & 1 & 0 \end{bmatrix} \qquad B = \begin{bmatrix} 0 & 1 & 1 & 1 \\ 1 & 0 & 0 & 1 \\ 1 & 0 & 0 & 1 \\ 1 & 1 & 1 & 0 \end{bmatrix}$$

Clearly, all that is lost in matrix K is the names of the individual arcs. In matrix B we lose much more, but we still know what vertices are or are not joined by arcs.

Exercises 11.2

1. For several of the graphs given as examples, number the nodes, letter the arcs, and write down one or more of the matrices W, K, and B.

2. For each of the following matrices, draw a graph for which the given matrix is an adjacency matrix. State whether the graph is a directed or an undirected graph. A few of the matrices are adjacency matrices of *no* graph; identify them.

(a) $W = \begin{bmatrix} \varnothing & \varnothing \\ \varnothing & \varnothing \end{bmatrix}$

(b) $W = \begin{bmatrix} \{x\} & \varnothing \\ \varnothing & \{y\} \end{bmatrix}$

(c) $W = \begin{bmatrix} \varnothing & \{x, y\} \\ \{x, y\} & \varnothing \end{bmatrix}$

(d) $W = \begin{bmatrix} \{u\} & \{x, y\} \\ \{x\} & \{v\} \end{bmatrix}$

(e) $W = \begin{bmatrix} \{u\} & \{x, y\} \\ \{z, t\} & \{v\} \end{bmatrix}$

(f) $W = \begin{bmatrix} \varnothing & \{u, v\} & \{w, x\} \\ \{u, v\} & \varnothing & \{y, z\} \\ \{w, x\} & \{y, z\} & \varnothing \end{bmatrix}$

(g) $W = \begin{bmatrix} \{x\} & \varnothing & \varnothing \\ \varnothing & \{y\} & \varnothing \\ \varnothing & \varnothing & \{x\} \end{bmatrix}$

3. For each of the following matrices, draw a directed graph and an undirected graph that has the given matrix as its numerical adjacency matrix.

(a) $K = \begin{bmatrix} 0 & 0 \\ 0 & 0 \end{bmatrix}$ **(b)** $K = \begin{bmatrix} 1 & 0 \\ 0 & 1 \end{bmatrix}$

(c) $K = \begin{bmatrix} 0 & 2 \\ 2 & 0 \end{bmatrix}$ **(d)** $K = \begin{bmatrix} 1 & 2 \\ 2 & 1 \end{bmatrix}$

(e) $K = \begin{bmatrix} 0 & 2 & 2 \\ 2 & 0 & 2 \\ 2 & 2 & 0 \end{bmatrix}$ **(f)** $K = \begin{bmatrix} 1 & 0 & 0 \\ 0 & 1 & 0 \\ 0 & 0 & 1 \end{bmatrix}$

4. A graph is assigned by means of the set $V := \{1, 2, 3, 4\}$ of its vertices, the set $A := \{x, y, z, t\}$ of its arcs, and the correspondence b given below. In each case, **(i)** verify whether the graph is directed or undirected; **(ii)** write all three adjacency matrices (path, numerical, Boolean); **(iii)** draw an image of the graph. **(a)** $b := \{((1, 2), x), ((2, 3), y), ((3, 4), z), ((3, 4), t)\}$; **(b)** $b := \{((1, 2), x), ((2, 1), x), ((2, 3), y), ((3, 2), y), ((3, 4), z), ((4, 3), z), ((3, 4), t), ((4,3), t)\}$. **(c)** $b := \{((1, 2), x), ((2, 1), x), ((2, 4), y), ((4, 2), y), ((4, 4), z), ((4, 4), t)\}$.

5. Given an $n \times n$ matrix W whose elements are sets, give necessary and sufficient conditions for W to be the one-path adjacency matrix of a directed or of an undirected graph.

6. Prove the statement at the end of Definition 3 on the equivalence of two definitions of Boolean adjacency matrix.

11.3 Paths and Cycles

Intuitively, we can think of a *path* as a walk we take through a graph, starting at any node and proceeding—when possible and if we want to—along an arc to another node, and so on. If the graph is directed, we must respect the one-way signs. A *cycle* is a path that ends back at the starting point.

Examples In the graph in Figure 11.1, a possible path is: from 1, through d to 3, through g to 4, through e to 1, through a to 2. In Figure 11.3d there are no nondegenerate paths. We can, however, start from any one node and stay there.

A formal definition would therefore involve a sequence that alternates nodes and arcs, starting and ending with a node, and respecting the condition that each arc joins the previous node to the next one. We observe that in the partial sequence (node1, **arc,** node2), node2 is uniquely determined by node1 and **arc,** because node2 is the only other endpoint of **arc.** This remark will simplify the formal definition of path as a string, in which the first symbol identifies a node, and all remaining identify arcs.

Definition 4 Let $G := (V, A, b)$ be a nonempty graph as given in Section 11.1. Let k be a natural number. Let j be a node of G. A **k-path** in G (or path of **length k**) with **origin j** is defined, recursively as a string w of length $k + 1$, as follows.

- If $k = 0$, set $w := (j)$ and define j as the **end** of w. In this case, the path is also called a **null** path or a **null cycle.**
- For any k, let w be a k-path with origin j. Let m be its end. Let x be an arc (if any) with endpoint m, provided that, if G is directed, m be the first endpoint of x. Let m^* be the other endpoint of x. Then define $w^* := w:(x)$ as a $(k + 1)$-*path* (where : is concatenation of strings) and let m^* be defined as the **end** of w^*. Note that, whenever $k > 0$, the end of w^* is a node m^* which is not specifically written in the string w^*. It is unnecessary to specify m^* because it is uniquely determined by node m and arc x.

A **cycle** is a path whose end coincides with its origin. A path or a cycle is said to be **node-simple** if all of its nodes are distinct except, in the case of a cycle, origin and end, which coincide by definition; it's said to be **arc-simple** if all of its arcs are distinct; it's said to be **simple** if both conditions are satisfied.

Examples In the example given above, the path we considered would be symbolized as (1*dgea*) or simply 1*dgea*; its origin is 1, its end is 2. In the same graph, (1), (2), (3), and (4) are the null cycles, and other examples of cycles are (2*ab*), (2*aa*), (2*feaba*), (1*dge*), and (3*ged*).

Note the last two cycles: each is obtained from the other by "going around" the same cycle, in the same direction, but starting from a different origin. This operation is called a **cyclic permutation** of a cycle.

An evident problem is, given any two nodes, is there a path of given length k from the first node to the second node? How many such paths are there? How can we enumerate them? (Note that these are the same basic questions we asked in Chapter 4; this is a form of *combinatorics* too). What about paths of any length? The answers will all be given by the next theorem and its consequences. But let's see a few examples first.

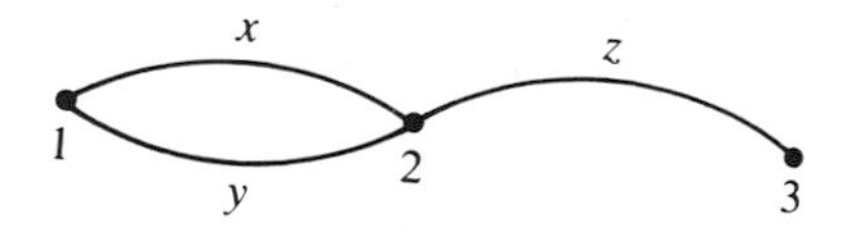

Figure 11.8

Consider the undirected graph in Figure 11.8. Its nodes are 1, 2, and 3. Its arcs are x, y, and z. The boundary function is $(1, 2) \mapsto x$, $(2, 1) \mapsto x$, $(1, 2) \mapsto y$, $(2, 1) \mapsto y$, $(2, 3) \mapsto z$, $(3, 2) \mapsto z$. The adjacency matrices are

$$W = \begin{bmatrix} \varnothing & \{x, y\} & \varnothing \\ \{x, y\} & \varnothing & \{z\} \\ \varnothing & \{z\} & \varnothing \end{bmatrix}$$

$$K = \begin{bmatrix} 0 & 2 & 0 \\ 2 & 0 & 1 \\ 0 & 1 & 0 \end{bmatrix} \qquad B = \begin{bmatrix} 0 & 1 & 0 \\ 1 & 0 & 1 \\ 0 & 1 & 0 \end{bmatrix}$$

Clearly, these three matrices answer the first three questions (in inverse order) for the 1-paths.

An an experiment, calculate W^2, K^2, and B^2. They are

$$W^2 = \begin{bmatrix} \{xx, xy, yx, yy\} & \varnothing & \{xz, yz\} \\ \varnothing & \{xx, xy, yx, yy, zz\} & \varnothing \\ \{zx, zy\} & \varnothing & \{zz\} \end{bmatrix}$$

$$K^2 = \begin{bmatrix} 4 & 0 & 2 \\ 0 & 5 & 0 \\ 2 & 0 & 1 \end{bmatrix} \qquad B^2 = \begin{bmatrix} 1 & 0 & 1 \\ 0 & 1 & 0 \\ 1 & 0 & 1 \end{bmatrix}$$

Consider, for instance, nodes 1 and 1. By inspection, the four 2-paths from 1 to 1 are $(1xx)$, $(1xy)$, $(1yx)$, $(1yy)$. This is obtained by concatenating 1 with each of the four elements of entry (1, 1) in W^2. Similarly, there are no 2-paths from 1 to 2. Again, there are no elements in entry (1, 2) of W^2. Also, entry (1, 1) of K^2 is 4: there are *four* 2-paths from 1 to 1. Entry (1, 2) of K^2 is 0: there are *no* 2-paths from 1 to 2. Entry (1, 1) of B^2 is 1: it's *true* that there are 2-paths from 1 to 1. Entry (1, 2) of B^2 is 0: it's *false* that there are 2-paths from 1 to 2. These facts are no accident (as proved below). In all cases, entry (i, j) of B^k tells us whether there are k-paths from i to j. Entry (i, j) of K^k tells us how many k-paths go from i to j. If we concatenate i with the elements of entry (i, j) of W^k we obtain an enumeration of the k-paths from i to j.

Theorem 5 Let $G := (V, A, b)$ be a finite nonempty graph as given in Section 11.1. Let $V = \{1, \ldots, \#V\}$. Let W, K, and B be the adjacency matrices of G as in Section 11.2. Let k be a natural number. Let the k-th power of a matrix be defined as in Section 8.3. Let $\{\ldots\}:\{\ldots\}$ denote concatenation of sets of strings as in Section 6.1. [Recall that $\{t, \ldots\}:\{\Lambda\} = \{t, \ldots\}$ and $(t, \ldots\}:\varnothing = \varnothing$.] Then: (1) The set of the k-paths in G with origin i and end j is $\{(i)\}:W^k(i, j)$. (2) The number of such paths is $K^k(i, j)$. (3) The truth value of the proposition "there are k-paths with origin i and end j" is $B^k(i, j)$.

Proof: We proceed by induction on k.

Initial step ($k = 0$). Recall that the entries of W^0, K^0, B^0 are $\{\Lambda\}$, 1, 1 respectively when $i = j$, and $\varnothing$, 0, 0 when $i \neq j$. When $k = 0$, the only 0-path with origin i is the null path (i), whose end is necessarily i. Consequently, when $i = j$, the answers to the three questions are $\{(i)\}$, 1, and 1 respectively. When $i \neq j$, the answers are $\varnothing$, 0, and 0. Consequently, all three statements are proved when $k = 0$.

Induction step. Let now k be any natural number, and assume that the conclusions have been proved for that value. In order to complete the proof, we have to prove that the three conclusions are valid for the value $k + 1$. (In the remainder of the proof we include in brackets [ ] those parts that refer to a directed graph.) Let w be any $(k + 1)$-path with origin i and end j, if any. By the definition of $(k + 1)$-path, $w = v{:}(x)$, where v is some k-path with origin i and x is some arc with [second] endpoint j. Let l be the other endpoint of x. Then l must be the end of v. Conversely, if v is any k-path with origin i and end at any node l, and if x is any arc with [first] endpoint l and [second] endpoint j, then $v{:}(x)$ is one of the desired paths. Consequently, the set of the desired paths is precisely

$$\bigcup_{l \in V} \{k\text{-paths with origin } i \text{ and end } l\} : \{\text{arcs from } l \text{ to } j\}. \tag{1}$$

By the hypothesis of induction, the first set $\{\ldots\}$ above is $\{(i)\}{:}W^k(i, l)$. By the definition of W, the second set $\{\ldots\}$ above is $W(l, j)$. Consequently, formula 1 reduces to

$$\bigcup_{l \in V} \{(i)\} : W^k(i, l) : W(l, j).$$

Since the concatenation operation distributes over $\cup$, we have a further reduction to

$$\{(i)\} : \bigcup_{l \in V} W^k(i, l) : W(l, j).$$

By the definition of product of path matrices, this reduces to

$$\{(i)\}{:}(W^kW)(i, j), \text{ that is, } \{(i)\} : W^{k+1}(i, j).$$

This completes the proof of part 1.

For part 2, note two facts. First, within each "addend" of expression 1, the cardinalities of the two "factors" are $K^k(i, l)$ and $K(l, j)$ respectively. The addend has therefore cardinality $K^k(i, l) \cdot K(l, j)$. Second, the union in formula 1 is a disjoint union, because the node l is different from addend to addend. Therefore, the cardinality of the set in formula 1 is

$$\sum_{l \in V} K^k(i, l) \cdot K(l, j),$$

which is precisely $K^{k+1}(i, j)$ by the definition of product of numerical matrices.

Finally, set 1 is nonempty if and only if, for at least one l, both "factors" are nonempty, that is, if and only if, for at least one l, $B^k(i, l) = 1$ and $B(l, j) = 1$. But this is precisely the condition for

$$\bigvee_{l \in V} B^k(i, l) \wedge B(l, j) = 1,$$

that is, $B^{k+1}(i, j) = 1$. This proves part 3. Q.E.D.

(See again the examples before the theorem.)

A few calculational remarks are in order. When calculating powers of matrices, it's often sufficient to calculate only a few rows or columns, or even just a few elements, depending on the goals to be reached. For instance, if we have W and want to enumerate all 3-paths from 1 to 2, then we need W^3 (1, 2). In order to calculate that, we need only row 1 of W^2. So, calculate only row 1 of W^2 and element (1, 2) of W^2W. When calculating any power of B, you seldom need to go through all l's in the expression $\bigvee_{l \in V} B^h(i, l) \wedge B(l, j)$. As soon as you hit an l for which $B^h(i, l) = 1$ and $B(l, j) = 1$, you know that the result is 1 without proceeding any further. See also the Roy–Warshall algorithm below. On occasions, it may be convenient to invert the order in which you multiply matrices. For instance, $K^5 = K^4K = KK^4 = K^2K^3$, and so on. See which matrices have more $\varnothing$'s or zeroes.

The problem of deciding whether there are paths, of any length, from any node to any node is totally manageable for finite graphs. For nodes i, j, the answer seems to be $\bigvee_{k \in N} B^k(i, j)$, but we have a problem: this may be an infinite "or," which we have not defined. There is a way around, though. If we have a path of length k (other than a cycle) and if $k \geq \#V$, then some node must be repeated in the path, because the path has more nodes than exist. Then we can "cut that path short." For instance, if the path $ixytuvwz$ is such that the second endpoint of y is the same as the second endpoint of v, then $ixywz$ is a path that has the same origin and the same end as the original path. (That's why a smart animal that is being pursued never crosses its own path.) Formalize this argument to prove the following theorem:

Theorem 6 Let G be given as in Theorem 5. Then the truth value of the proposition "there is at least one path with origin i and end j" is $\bigvee_{k=0}^{\#V-1} B^k(i, j)$. Consequently, given graph $G := (V, A, b)$, with Boolean adjacency matrix B, consider on V the relation g defined by the proposition "there is an arc with origin i and end j" if G is directed, or "there is an arc with endpoints i and j" if G is undirected. Then B is precisely the matrix of relation g, according to problems 1 and 7 in Exercises 10.9. Matrix B^k is the matrix of relation g^k, for each k in $\mathbb{N}$. When V is finite, $B^{[c]} := \bigvee_{k=0}^{\#V-1} B^k(i, j)$ is the matrix of the reflexive-transitive closure of g and, when G is undirected, also the reflexive-symmetric-transitive closure of g.

Proof: Exercise.

Example Consider the undirected graph in Figure 11.9. It has the following Boolean adjacency matrix (the numbers in brackets are the labels of nodes).

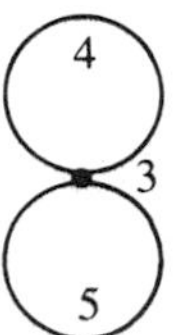

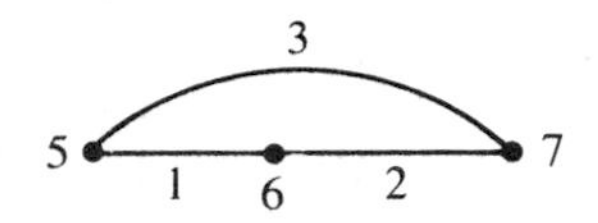

Figure 11.9

	[1]	[2]	[3]	[4]	[5]	[6]	[7]
[1]	0	1	0	0	0	0	0
[2]	1	0	0	0	0	0	0
[3]	0	0	1	0	0	0	0
[4]	0	0	0	0	0	0	0
[5]	0	0	0	0	0	1	1
[6]	0	0	0	0	1	0	1
[7]	0	0	0	0	1	1	0

Since the number of nodes is 7, we need to take the OR of the powers of B from 0 to 6. An easy calculation gives the following matrix $B^{[c]}$ (actually, in this simple case, all you have to do is look at the picture; but we want to be able to do this kind of calculation even when there is no picture, and all we know is matrix W or B).

	[1]	[2]	[3]	[4]	[5]	[6]	[7]
[1]	1	1	0	0	0	0	0
[2]	1	1	0	0	0	0	0
[3]	0	0	1	0	0	0	0
[4]	0	0	0	1	0	0	0
[5]	0	0	0	0	1	1	1
[6]	0	0	0	0	1	1	1
[7]	0	0	0	0	1	1	1

Roy–Warshall Algorithm. There is an algorithm for calculating $B^{[c]}$ efficiently. The **Roy–Warshall algorithm** calculates $B^{\#} := \bigvee_{k=1}^{\#V} B^k$. (1) Initially define B^* to be equal to B. (2) Scan the elements of B^* in the order (1, 1), (2, 1), . . ., $(n, 1)$, (1, 2), (2, 2), . . ., $(n, 2)$, . . ., (n, n) $(n := \#A)$. For each pair (i, j) for which $B^*(i, j) = 1$, consider the entries $B^*(j, k)$ $(1 \leq k \leq n)$ that are equal to 1. For each such entry, redefine the corresponding $B^*(i, k)$ to be 1. (3) At the end, B^* is $B^{\#}$. Then $B^{[c]}$ is obtained by placing 1s in positions (i, i) $(1 \leq i \leq n)$ of $B^{\#}$. The proof of correctness of this algorithm is a rather difficult exercise.

Exercises 11.3

1. Find the formal expression of several paths in several of the example graphs. Use directed and undirected graphs. Assign two nodes of the graph and, by inspection, try to find any one path from one to the other.
2. Continuing with either example of this section, find all 3-paths and all 4-paths (if any) from 1 to 1, from 1 to 2, from 1 to 3.
3. In each of the digraphs represented by the figures below **(1)** find four distinct simple cycles; **(2)** find all paths of length 4 for which the first arc is b and the last arc is a; **(3)** find all paths of length 3 or 4 (if any) from node 1 to node 3.

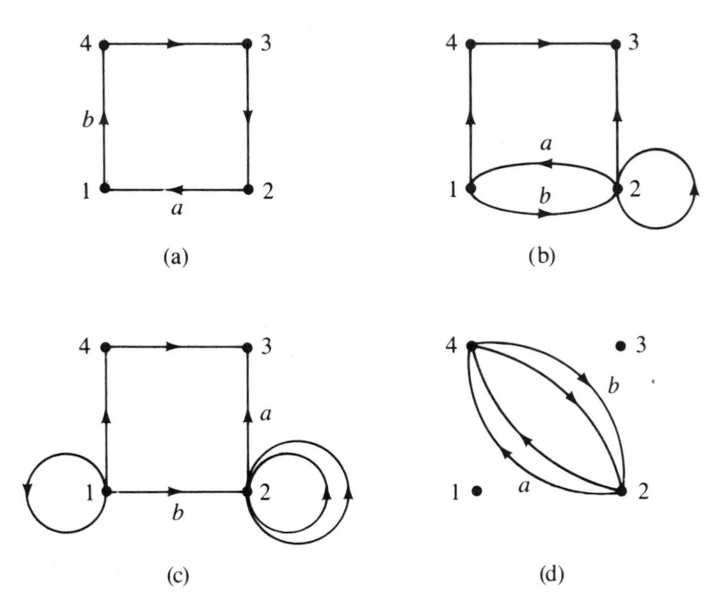

4. In several of the graphs we gave as examples, solve some of these problems: **(a)** Given concretely two nodes and a length k, find whether there are k-paths from the first node to the second node. **(b)** Find how many there are. **(c)** Enumerate them.

   (This is a suitable problem for computer solution. The crux of the program is how you want to represent a set or a matrix within the computer. Input and output may create ancillary difficulties for path matrices.)

Complements 11.3

1. Finding all the paths, of any length, from node i to node j is a rather unmanageable problem, even for finite graphs. Verify that the answer is $\{(i)\}:\bigcup_{k\in N} W^k(i, j)$, which is often an infinite set. Similar remarks apply, of course, to the number of such paths.

2. Prove Theorem 6.

3. Verify the correctness of the Roy–Warshall algorithm.

11.4 Degree of a Node

Intuitively, the degree of a node is the number of times the node is the end of an arc. A loop counts twice. Formally,

Definition 7 Let $G := (A, V, b)$ be a finite graph. Set $n := \#A$. Let K be the numerical adjacency matrix of G. If the graph is undirected and if j is any node, then the **degree** of j is defined as $K(j, j) + \Sigma_{l \in V} K(j, l)$. A node is said to be **odd** or **even** according as its degree is odd or even. If the graph is directed, then the **indegree** of j is defined as $\Sigma_{l \in V} K(l, j)$ and the **outdegree** as $\Sigma_{l \in V} K(j, l)$.

Examples In the graph in Figure 11.1, nodes 2, 3, and 4 are odd nodes of degree 3, while 1 is an odd node of degree 5. In the graph in Figure 11.4c, all nodes are even nodes of degree 4.

Theorem 8 In a finite, undirected graph, the number of odd nodes is even.

(Verify the statement on several of the undirected graphs we've seen so far.)

Proof: Let $G := (V, A, b)$ be the given graph, and let K be its numerical adjacency matrix. Let $d(i)$ denote the degree of node i. Let m be the number of odd nodes. Calculate the sum $s := \Sigma_{l \in V} d(l)$ in two different ways.

First, the definition of $d(l)$ gives

$$s = \sum_{l \in V} [K(l, l) + \sum_{p \in V} K(l, p)].$$

Moving the term $K(l, l)$ from the second summation into the first term gives

$$s = \sum_{l \in V} [2K(l, l) + \sum_{p \in V - \{l\}} K(l, p)] = 2 \sum_{l \in V} K(l, l) + \sum_{l \in V} \sum_{p \in V - \{l\}} K(l, p).$$

In the last double summation we are adding all $K(l, p)$ for which $l \neq p$. Each such addend appears exactly twice, once as $K(l, p)$ and once as $K(p, l)$. Recall that, in an undirected graph, $K(p, l) = K(l, p)$ for every pair of nodes p, l. Consequently, the double sum is an even number, and so is s.

Second,

$$s = \sum_{l \in \{\text{odd nodes}\}} d(l) + \sum_{l \in \{\text{even nodes}\}} d(l).$$

Conduct the calculation mod 2. Since each $d(l)$ is odd in the first summation, each $d(l)$ reduces to 1, and the first summation is then the number m of odd nodes. Since each $d(l)$ is even in the second summation, the second summation reduces to 0. Consequently, $s = m \bmod 2$.

Since s was proved to be even, so is m. Q.E.D.

Exercises 11.4

1. Find the degrees, indegrees, and outdegrees of several nodes of several graphs we have seen in this chapter.
2. Try to construct **(a)** an undirected graph with four nodes of degrees 3, 4, 5, and 6 respectively; **(b)** a digraph (directed graph) with four nodes of in- and out-degrees all equal to 3; **(c)** an undirected graph with three nodes all of degree 5. The last attempt will not succeed; why not?

Complements 11.4

1. **Isomorphic graphs.** Given two graphs $G := (V, A, b)$ and $G^* := (V^*, A^*, b^*)$, an **isomorphism** from G to G^* is a pair of bijections $f: V \to V^*$, $g: A \to A^*$, such that $((j, l), w) \in b$ if and only if $((fj, fl), gw) \in b^*$, for every choice of j, l in V and w in A. Two such graphs are said to be **isomorphic** to each other. Figure 11.2 provides an example of isomorphic graphs. Why? Find other examples among those we've seen. Prove that isomorphic graphs have the same number of nodes and the same number of arcs. Find other properties that are necessarily common to any two isomorphic graphs. See complement 3 below for hints.
2. **Homological graphs.** The following two operations on graphs are useful to simplify calculations: Let $G := (V, A, b)$ be a finite graph.
   1. **Free chain insertion.** Let x be an arc of G, $(j, m) \mapsto x$. Replace arc x with two arcs, u and v, by introducing a new node l so that $(j, l) \mapsto u$ and $(l, m) \mapsto v$; if the graph is undirected, we need also $(l, j) \mapsto u$ and $(m, l) \mapsto v$.
   2. **Free chain reduction.** Let l be a node that is the endpoint of no loops and that has degree 2 if the graph is undirected and in- and outdegree 1 if the graph is directed. Suppose $(j, l) \mapsto u$, $(l, m) \mapsto v$. Then remove l and replace u, v with one single arc.

   In each case, the object comprising u, l, and v is called a **free chain.**

   The two operations are inverse of each other and are illustrated in the transitions from left to right or right to left in the graphs in Figure 11.10.

An advantage of free chain reduction (operation 2) is to remove "unnecessary" nodes, as in the first row of Figure 11.10. An advantage of free chain insertion is to make the graph "loopless," as in the third row, or to create a situation in which each pair of nodes is joined by at most one arc, as in the second row.

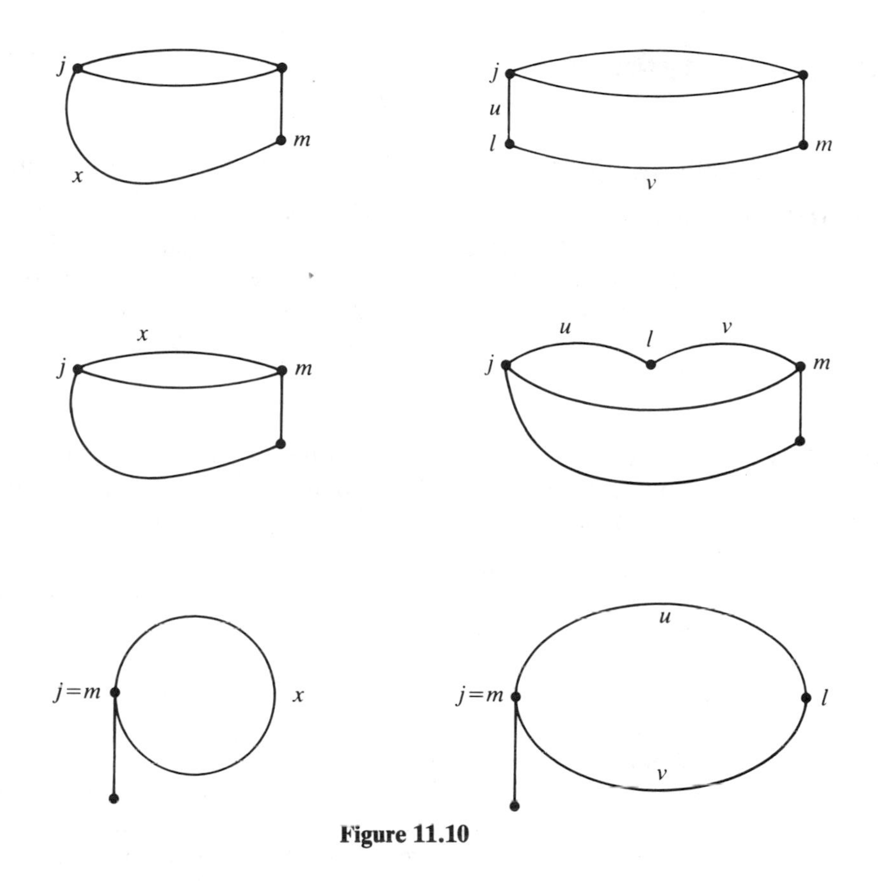

Figure 11.10

3. Let U be a given universal set of finite graphs, all directed or all undirected. Define a binary relation h on U that comprises those pairs (G, G^*) such that G^* is obtained from G by applying free chain insertion or reduction successively for a finite number of times, possibly none. Prove that, subject to a relabelling of the altered nodes or arcs, h is an equivalence relation. It is called **homology** [Greek *homo-* = same, *log-* = knowledge; so *homolog-* = same knowledge, even if not isomorphic, which would imply *identical form*]. Prove that, in two homological undirected graphs: **(a)** The nodes of degree other than 2 are the same in the two graphs. **(b)** The numbers of arc-simple paths that begin and end at any two nodes of degree other than 2 is the same in the two graphs. **(c)** If there are no loops, then the number of k-cycles that contain no repeated arcs is the same in the two graphs.

Try to find some more properties that remain unaltered under homology.

4. Prove that any finite graph is homologic to a loopless graph in which each pair of nodes is joined by at most one arc. Practice such a reduction on several of the examples we saw in this chapter and on each of the 26 "graphs" in Figure 10.3 (in these, you must place the nodes suitably). In the set of the 26 resulting graphs, find the equivalence classes with respect to homology.

11.5 Connectedness

Intuitively, a graph is connected if it's "all in one piece." Formally,

Definition 9 An undirected graph is said to be **connected** if, for every pair of distinct nodes, there is a path with origin at one node and end at the other. A directed graph is said to be connected if the undirected graph obtained from it according to the construction at the end of Section 11.1 is connected. In a directed or undirected graph, node j is said to be **accessible** from node i if there is a path with origin i and end j.

Theorem 10 Accessibility is a reflexive and transitive relation on the set of the nodes of a graph. If the graph is undirected, then accessibility is an equivalence relation.

Proof: Exercise.

The next construction will locate the "pieces" of an undirected graph. Refer to the example after Theorem 6, in which the pieces are graphically evident (they wouldn't be if the graph were given via a table). The formal construction is the following.

Let $G := (V, A, b)$ be any finite, connected, nonempty, undirected graph. Let V^* be any one of the equivalence classes given by the equivalence relation of accessibility. Define A^* as the set of those arcs of G for which at least one endpoint is in V^*. Clearly, such an arc has both endpoints in V^* because the two endpoints of an arc are in the same equivalence class. For each pair of sets V^* and A^* as above, the incidences node-arc are the same as in the given graph. Formally, define b^* by $((j, l), x) \in b^*$ if and only if $(j, l) \in V^* \times V^*$, $x \in A^*$, and $((j, l), x) \in b$. Then (V^*, A^*, b^*) is a connected graph.

The adjacency matrix of the new graph G^* is then obtained by deleting from the adjacency matrix of G the rows and columns that do not correspond to nodes in V^*. Each of the graphs (V^*, A^*, b) is called a **connection component** of G.

In the example, the connection components are

nodes 1, 2; arc 6; matrix

	[1]	[2]
[1]	$\varnothing$	{6}
[2]	{6}	$\varnothing$

node 3; arcs 4 and 5; matrix

	[3]
[3]	{4,5}

node 4; no arcs; matrix

	[4]
[4]	$\varnothing$

nodes 5, 6, and 7; arcs 1, 2, and 3; matrix

	[5]	[6]	[7]
[5]	$\varnothing$	{1}	{3}
[6]	{1}	$\varnothing$	{2}
[7]	{3}	{2}	$\varnothing$

In general, in order to construct the components abstractly (and efficiently), we need a device that translates "accessibility." In Theorem 6 we constructed such a device, the matrix $B^{[c]}$, whose entry (i, j) tells us whether there is a path from node i to node j. Consequently, each row (or column) of $B^{[c]}$ will contain 1s in the positions that correspond to the nodes of one component. Of course, we'll have the same listing in each row (or column) that belongs to the same component.

In the example, $B^{[c]}$ is

	[1]	[2]	[3]	[4]	[5]	[6]	[7]
[1]	1	1	0	0	0	0	0
[2]	1	1	0	0	0	0	0
[3]	0	0	1	0	0	0	0
[4]	0	0	0	1	0	0	0
[5]	0	0	0	0	1	1	1
[6]	0	0	0	0	1	1	1
[7]	0	0	0	0	1	1	1

The nodes of the connection components then are 1 and 2, 3 only, 4 only, and 5, 6 and 7. But do not expect such neat, compact blocks in every case: the blocks are always there, but they may not be compact. They can be made compact by reordering the sequence of the nodes.

The next theorem is not particularly important on its own, but we'll need it to prove another theorem. That's why it's called *lemma* rather than theorem.

Lemma 11 Let $G := (V, A, b)$ be a finite, connected, undirected graph. Let C be a proper, nonempty subset of A. Let E be the set of the endpoints of the elements of C. Then there exists an arc t such that $t \notin C$ and an endpoint of t is in E.

Example Let the graph be as in Figure 11.11.

Let C be the set of the arcs drawn with a heavy line. Then E is the set of the nodes marked with open circles. Any of the dashed lines is one of the desired arcs.

Proof: Start a path with a single arc not in C. This can be done, because $\varnothing \subset C \subset A$. Let the end of our path be j. Since G is connected, we can extend our path with a path from j to a node l in E. The first arc of the extended path certainly is an arc not in C. A few of the first steps may involve arcs with no endpoints in E but, at a certain moment, we must hit E. The first arc with an endpoint at a node of E can be taken as t. Q.E.D.

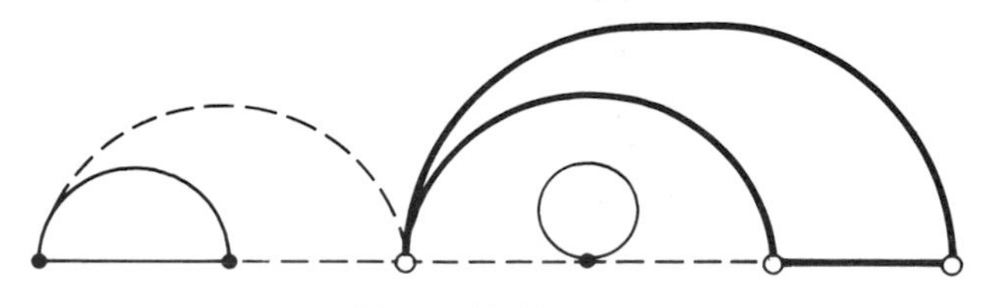

Figure 11.11

Exercises 11.5

1. In the examples of graphs given throughout this chapter, identify those that are connected and those that are not. (This is a suitable problem for computer solution. You may want to use larger graphs.)

2. Give examples of other graphs that are not connected.

3. For each of the following graphs **(a)-(f)**, find its connection components and, for each component, find: **(i)** the numbers of nodes and arcs and **(ii)** the number of nodes for each degree d ($d \in \mathbb{N} - \{2\}$). Note that **(e)** is the three-utilities graph (Figure 11.5a) and **(f)** is the complete graph on five nodes (Figure 11.4c).

4. Prove Theorem 10 on accessibility in a graph. Note that the transitivity of the relation "you can get there from here" is the feature that gives a meaning to the joke, "You can't get there from here; you've got to go somewhere else first."

11.6 An Application: PERT

The acronym PERT denotes "Program Evaluation and Review Technique," which was developed by the U.S. Navy Special Projects Office and the firm Booz, Allen & Hamilton. The primary goal of PERT is to coordinate the phases of a complex project. Our present knowledge of graph theory is sufficient to study how to reach this primary goal. Additional goals are minimization of time and cost of a project, changes to the plan in order to realize further savings, and so on. We'll just mention some of the mathematical tools that are needed for these secondary goals.

The two basic concepts in PERT are **activity** (that is, the performance of one or more specific tasks) and **event** (that is, the accomplishment of one or more concurrent activities). See Section 10.7 for a few details and a concrete example, the toy wagon construction. For each given project, the nature of the project provides the basic relation "event x must occur before event y." We denote this briefly by $x \rightarrow y$. Generally, an activity is involved in the transition from x to y. For practical reasons, it is desirable that only one activity be involved in the transition. If that is not the case, then we can easily modify the form (but not the substance) of the project in a way illustrated by the diagram below. Note that we have performed two free chain insertions, using the new nodes $y1$, $y2$.

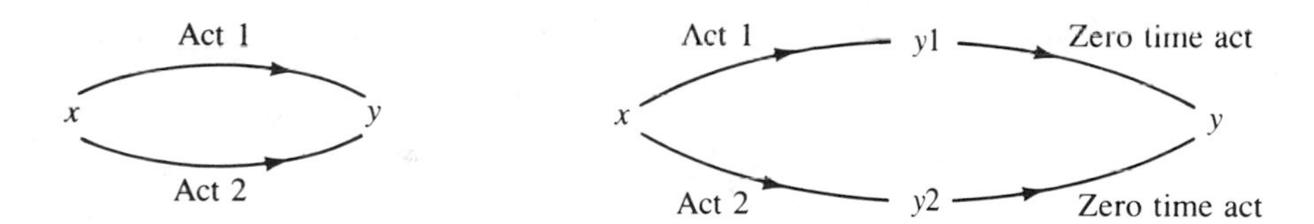

After such changes are made, we have at most one activity from one event to another. In the toy wagon example of Section 10.7, we've listed explicitly those pairs of events between which an activity occurs. In general, we represent events with the nodes of a directed graph and activities with its arcs. In the toy wagon example, a pictorial representation of the resulting digraph is given in Figure 11.12.

A few preliminary steps are necessary before we draw any conclusions. First, we must decide whether the resulting graph is connected. This can be determined by constructing the Boolean matrix $B^{[c]}$ of the undirected graph associated with the digraph (see Theorem 6). If any zero appears in the resulting matrix, then the digraph is not connected. If the digraph is not connected, the project can be split into distinct projects that can be done separately.

Assuming that the digraph is connected (as it is in the toy wagon example), we must decide whether we have a partial order. (If that were not the case, that is, if we had feedback, then PERT would not apply.) We can use now the Boolean adjacency matrix B of the digraph. If a 1 appears in any position (i, i) in any one of the *positive* powers of B, then there is a nonnull cycle and we have feedback. If no 1 appears, then the transitive closure of our relation is antisymmetric and we have a partial order.

Assuming that we have a partial order (as we do in the example), we have to determine initial and final events. These are, respectively, the nodes of indegree 0 and the nodes of outdegree 0. Although not essential for the method, it is convenient in practice to have only one initial event and only one final event. If that is not the case, we can always introduce an extra, fictitious node (the beginning or the end) and zero time activity arcs leading from or to that node to or from the previous initial or final events.

Assuming that we have just one initial node a and just one final node z (as we have in the example), the really crucial step is to determine all paths from a to z. This can be done, for instance, by considering the successive powers of the 1-path adjacency matrix W and, within each, the entry (a, z). These paths represent the "flows of activity" that are needed to complete the project. In our example, we can easily make a list of such paths (we'll indicate only their nodes):

$$\begin{array}{l} 0\ 1\ 2\ 8\ 10\ 11, \\ 0\ 1\ 2\ 4\ 6\ 7\ 9\ 10\ 11, \\ 0\ 1\ 3\ 4\ 6\ 7\ 9\ 10\ 11, \\ 0\ 1\ 5\ 7\ 9\ 10\ 11. \end{array}$$

Although again not essential for the method, we may try to split the one large project into several separate and nearly independent smaller projects. Suppose the paths we just found can be neatly partitioned into subsets sharing only initial and final nodes. If these initial and final events are merely notational conveniences (as is usually the case), they could simply be eliminated, separating the project.

Finally, we must bear in mind that each activity (arc) in a PERT graph might represent a full and complex task in its own right. In Figure 11.12, for example, each of the three activities 1 2, 1 3, and 1 5 represents the procurement of materials. This is simple in a large city, but on a ship at sea, or an Antarctic base, or a spacecraft, procuring supplies is itself a difficult and complicated operation.

The analyses we've done to this point accomplish the primary goal of PERT and are extremely useful in practical applications such as the design of assembly lines or transportation routes. We've established all the "flows of activity" within the project. In order to obtain secondary goals, we may need more mathematical tools. One of these additional goals can be reached with a methed called CPM (critical *p*ath *m*ethod), which requires a mathematical tool we have not introduced yet, a network. A **network** is a graph in which each arc has a **label** (formally, a total function from the set of the arcs to a set of symbols, the labels). In most practical cases, the labels are numerical values, such as time, capacity, or stress, and, as such, can be compared to each other, accumulated, or used in some other way. In our example, suppose that the activities of our project require the following numbers of hours to be performed:

0 1 Approximate design, 6 hours
1 2 Get hardware, 5 hours
1 3 Get wood and glue, 2 hours
1 5 Get varnish, 1 hour
2 4 0 hours
2 8 Put together wheel assembly, 2 hours
3 4 0 hours
4 6 Measure, cut, and glue wood; wait for the glue to dry, 8 hours
5 7 0 hours
6 7 0 hours
7 9 Varnish body, wait until dry, 6 hours
8 10 0 hours
9 10 0 hours
10 11 Mount wheel assembly, 1 hour

This information provides the network in Figure 11.12. The number written near each arc is the number of hours needed to perform the activity represented by that arc.

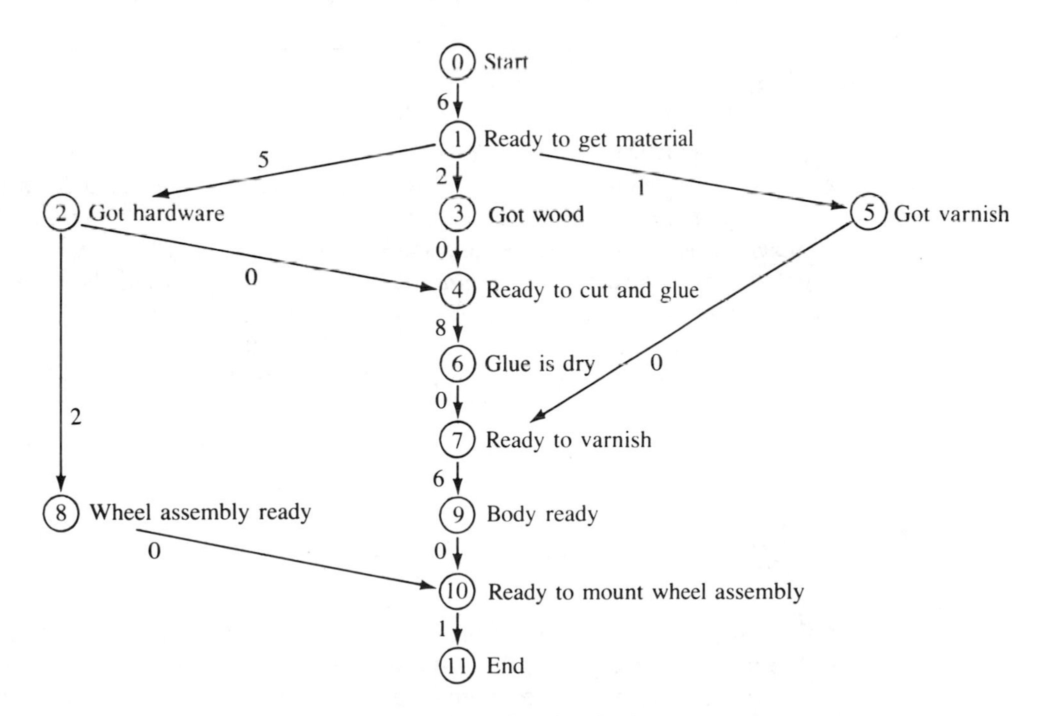

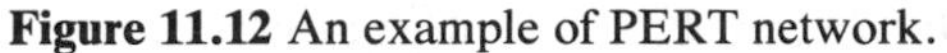
Figure 11.12 An example of PERT network.

The four paths we saw above involve, respectively,

$$\begin{aligned}
&6 + 5 + 2 + 0 + 1 = 14 \text{ hours},\\
&6 + 5 + 0 + 8 + 0 + 6 + 0 + 1 = 26 \text{ hours},\\
&6 + 2 + 0 + 8 + 0 + 6 + 0 + 1 = 23 \text{ hours},\\
&6 + 1 + 0 + 6 + 0 + 1 = 14 \text{ hours}.
\end{aligned}$$

The whole project requires therefore at least 26 hours to complete, as indicated by the second path above, which is the *critical* path. This path is critical because any delays in this path must necessarily delay the entire project. The first path is not critical, as delays of up to 12 hours ($= 26 - 14$) will have no effect whatever on the completion time of the project as a whole. Such a permissible delay is called *slack.* The critical path is the path with zero slack. The management in charge of the project must then concentrate effort on the critical path, to ensure that each critical event is attained on schedule, as well as to try to shorten the critical path so as to shorten total project time.

In general, the CPM consists in locating all the paths from initial to final event, calculating the total time for each (or another pertinent physical measurement), and selecting one path for which the time is minimal (or some other pertinent measurement is maximal or minimal). This path is the **critical path.** Elaborations on this method can provide additional information such as stumbling blocks (what should be changed to obtain a better solution) and slack time activities (in which activities would a delay *not* compromise the total time for the project?).

Exercises 11.6

1. In the wagon construction example, find the changes that occur if the following changes are made: **(a)** The time to purchase hardware is reduced to 1 hour. **(b)** The exact measurements of the hardware are known in advance. **(c)** The exact measurements of the hardware are known in advance, but it takes 12 hours to get the hardware. **(d)** The same person must purchase all the material in one single trip, which takes 8 hours. **(e)** The same person does the glueing and puts together the wheel asssembly while the glue is drying.

Find more changes of your own. Find more examples of your own.

11.7 Traceability

In this section we are going to consider Euler's theorem, which has to do with the problem of *tracing* a graph. What we seek is a path through a graph that traverses each arc precisely once. No "jumps" are allowed from one node to another except by following an arc. Tracing a graph is not by itself an especially important problem in the applications of graph theory. Our reason for studying it is that it is a simple yet elegant introduction to the kind of rigorous analysis needed to solve complex graph problems.

We need a few definitions first.

Definition 12 Let $G := (V, A, b)$ be an undirected, finite graph. Let $n := \#A$. An **Euler path** of G is defined, if existent, as an n-path that traverses every arc of G, only once each.

Example In the graph in Figure 11.13 an Euler path is $2cbaed$.

Theorem 13 **Euler's traceability theorem.** (1) A finite, connected, nonempty, undirected graph admits an Euler path if and only if there are either exactly two odd nodes or none. (2) In the first instance, any Euler path necessarily begins at one of the odd nodes and ends at the other. In the second instance, every Euler path is a cycle.

(In the case of the seven bridges of Koenigsberg all four nodes are odd, and there is therefore no Euler path. Review several other cases of undirected graphs to decide whether there is an Euler path. Try to find one by inspection.)

Remark. Let's consider first the case of a nonnull path P in a graph G. Nodes and arcs of P can be considered to form a graph Q on their own. Each node x of P will then have a degree in Q. Node x may appear in several positions in path P. Whenever x appears in a position other than origin and end, then there are two arcs of P that have x as an endpoint, and there is a contribution of two units to the degree of x in Q. The same situation occurs if P is a cycle and if x is the origin (and the end) of P. If P is not a cycle and if x is the origin or the end of P, then there is a contribution of one to the degree of x in Q. Consequently, the degree in Q is even if Q is a cycle or if the node is not the origin or the end of P, and it is odd if P is not a cycle and the node is the origin or the end of P.

Proof: For the proof of part 1 of the theorem, we have to prove an iff statement, that is, two if/then statements in opposite directions. Furthermore, we have to consider the two cases in which the number of odd vertices is two or zero. Let $G(V, A, b)$ be the given graph.

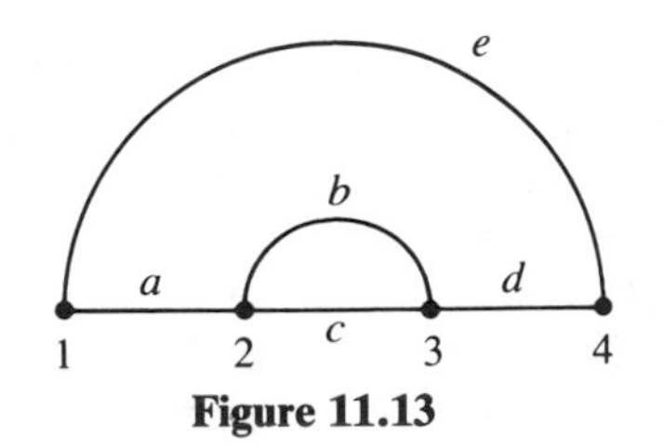

Figure 11.13

Proof of the "only if" part of part 1. Let P be an Euler path. Since P comprises all the arcs of G, once each, the graph Q provided by P is the same as graph G. Consequently, if P is a cycle, then all nodes are even (in Q and in G). If P is not a cycle, then all nodes other than origin and end are even, and the origin and end are odd. We have so proved part 2 and the "only if" part of part 1.

Proof of the "if" part of part 1. Assuming that the sufficient condition is satisfied, we'll give an algorithm for constructing an Euler path and, at each step, we'll prove the correctness of the algorithm (that is, we'll prove that the step can be done and that it accomplishes what it's supposed to accomplish). Consequently, this algorithm will provide a method of constructing an Euler path, which implies that such a path exists. Q.E.D.

Note 10 on Theorem Proving. The proof of Theorem 13 is a *constructive proof of existence*. In general, one way to prove that some object exists is to produce it, that is, to provide an explicit construction of the object. We must, of course, also be able to prove that the construction is correct.

Algorithm

1. If G has no arcs, then stop. The result is the null path with origin at the only node of G.

Proof: Since G has no arcs and is nonempty, it must have only isolated nodes. Since G is connected, there is exactly one node.

2. If G has two odd nodes, adjoin to G a new arc x with endpoints at the two odd nodes.

Proof: Consequently, all nodes of the new G are even.

3. Let u be any one arc of G. Let j be either of the endpoints of u. Define the path $z := ju$.

Proof: Since $A \neq \emptyset$ as a consequence of step 1, this step is possible and gives a 1-path.

4. Let $z = ju_1 \cdots u_h$ be the path just constructed. Let m be its end. If there is no arc of G that has m as an endpoint and does *not* appear in z, then continue at step 6. Else, let u_{h+1} be any such arc, and continue at step 5.

5. Redefine path z as $z := ju_1 \cdots u_h u_{h+1}$. Then continue at step 4.

Proof: Since an endpoint of u_{h+1} is the same as the end of u_h, the new z is a path. Since u_{h+1} does not appear in the original z, the new z is arc-simple.

6. If all arcs of G appear in z, then continue at step 8. Else, continue at step 7.

7. Let u_{h+1} be an arc of G that does *not* appear in z *and* that has an endpoint l in common with some arc that appears in z. Permute cyclically the cycle z so that the new origin (and end) is l. The new z will have a representation $ju_1 \cdots u_h$ (the new j is actually l). Continue at step 5.

Proof: Data being as at the beginning of step 7, we intend to prove first that z is a cycle. Proceed by contradiction: Suppose that the origin j and the end m of z are distinct. According to the remark before this proof, consider the graph Q that comprises nodes and arcs of z. Then m has odd degree in Q. Since m has even degree in G, there is an arc with end at m, belonging to G, but not to Q. This is an arc of the type required in step 4 and the process would not have reached step 7 at this point. By contradiction, $j = m$. Next, we can prove that an arc like u_{h+1} exists as required in step 7. This fact is a consequence of lemma 11.

8. If all the nodes of the originally given graph G were even, then stop: cycle z is the result. Else, continue at step 9.

Proof: By the proof after step 7, z is a cycle. We have already proved that z is arc-simple. Therefore, cycle z comprises all the arcs of the original G. Consequently z is an Euler cycle of G under the hypothesis of step 8.

9. Let x be the arc adjoined to the original graph. Permute z cyclically so that x is the last arc of the new cycle. Then delete x from the cycle so obtained. The final result is the result of the algorithm; stop.

Proof: We know that z is an Euler cycle of the augmented graph and therefore includes x and all the arcs of the original graph. Then the permutation can be performed. Deleting x from the augmented graph and from z clearly provides an Euler path for the original graph. Also, the endpoints of x are distinct, so that the new z is not a cycle.

Example We need no theory, of course, to construct an Euler path for the graph in Figure 11.14, but it's instructive to see how the algorithm works in a specific case.

Step 1 does not apply. Step 2: Adjoin arc x from 2 to 3. Step 3: Select e and set $z := 4e$. Step 4 does not apply. Step 5: Select a and set $z := 4ea$. Step 4 does not apply. Step 5: Select x and set $z := 4eax$. Step 4 does not apply. Step 5: Select d and set $z := 4eaxd$. Step 4: to step 6. Step 6 does not apply. Step 7: Select c. Permute: $z := 2xdea$. Step 5: Set $z := 2xdeac$. Step 4 does not apply. Step 5: Select b and set $z := 2xdeacb$. Step 4 to step 6 to step 8 to step 9. Step 9: Permute: $x := 3deacbx$; delete x. The result is $z := 3deacb$.

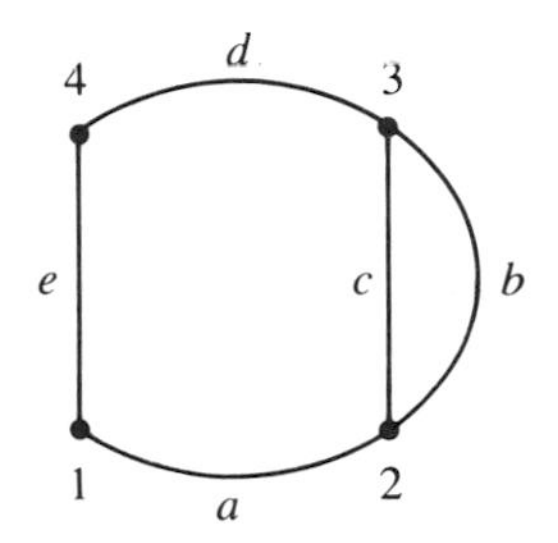

Figure 11.14

Exercises 11.7

1. Construct an Euler path, when possible, in several of the undirected graphs given in this chapter.
2. Construct an Euler path, when possible, for each of the graphs below. You may proceed by inspection, but use the algorithm for at least some of them. (This is an exercise also for the computer; you may want to use larger graphs.)

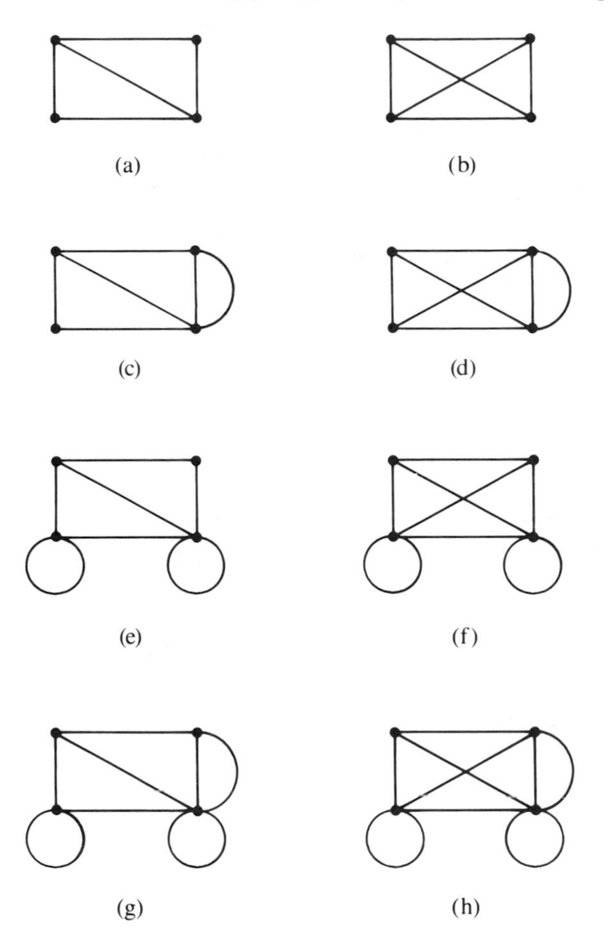

3. Find several connected, undirected graphs that have precisely ten arcs and no Euler path. What's the least number of arcs for an undirected connected graph that has no Euler path?
4. Consider a set of 14 tiles, each painted on both sides. The colors are R, O, Y, G, B, I, V. The two faces of the individual tiles are painted as follows: RO, RO, RI, RI, OY, YG, YB, YV, YV, GB, BB, BB, II, II. The task is to place the tiles in a single pile so that faces that are in contact with each other have the

same color. Is there a solution? if so, find one; if not, why not? [Hint: let the tiles be arcs, the colors nodes.]

5. The usual set of domino tiles are 28 rectangular tiles marked on the face with all 2-combinations with repetitions of the numbers 0 to 6. The task is to put all of them in a single row, head to head, face up, so that the two numbers next to each other in contiguous tiles are the same (see figure). Is there a solution? if so, find one; if not, why not?

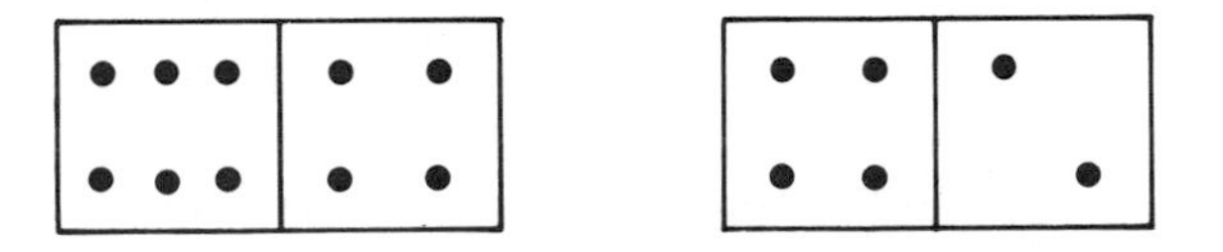

Is there a solution if the numbers are 1 to 6? If they are 2 to 6?

6. An architect's client wants the ground floor of a building to consist of precisely five rooms. Each room is to have one door to the outside and three doors that each lead from that room to an adjacent room. Is there a solution? Why or why not?

Suppose the requirements are the same except that there are 16 rooms. In this case there *is* a solution to the problem. Is there a path that starts from the outside and goes through each door exactly once? Why or why not?

7. In a usual 8×8 checkerboard, is there within the board a path that starts from inside one of the squares, avoids the corners of the squares, and crosses all partitions between two squares exactly once each? Why or why not? Does such a path exist for a square or rectangular checkerboard with a different number of squares? Give all the checkerboard sizes that admit a solution.

8. There are four boys and eight girls. Can a single piece of mail be circulated so that each boy mails it once to each girl, except the one(s) from whom he has already received it, and each girl mails it once to each boy, except the one(s) from whom she has already received it? why or why not? What if there are two boys and nine girls? three boys and eight girls?

Complements 11.7

1. Prove the following theorem:

Theorem 14 Let G be a finite, connected undirected graph with $2m$ odd nodes ($m > 0$). Then there exist precisely m paths such that each arc of G appears precisely once in precisely one of the paths, and the origins and ends of the paths are all distinct from each other.

Construct an algorithm that produces these m paths. Apply the algorithm to find the paths in those examples of undirected graphs in this section for which there is no Euler path.

11.8 Summary

Graphs are structures made up of *nodes* and of *arcs* that connect pairs of nodes. Such arcs may be *directed* or *undirected* (that is, they go from one of the nodes to another, or just join two nodes). In order to solve practical problems in large graphs (communication and transportation networks, energy distribution grids, and the like), the intuitive idea of graph must be formalized into an abstract structure, which translates the relation "a node is an endpoint (or the first endpoint, or the second endpoint) of an arc." One such formalization is obtained by means of a *boundary correspondence* (Section 11.1), another by means of *adjacency matrices*. For each ordered pair of nodes, we consider the set of all arcs that have those nodes as endpoints (Section 11.2).

In order to study the structure of a graph, we considers *paths* in the graph (that is, sequences of arcs, each beginning where the previous one ended) and, in particular, *cycles* (that is, paths that end at the same node where they begin; Section 11.3).

Several problems arise in graph theory. Among them are the following: (1) Counting or enumerating the paths of given length that join two given nodes (the answer is given by a power of an adjacency matrix; Section 11.2). (2) Determining whether a graph is *connected*, that is, whether any two nodes are joined by some path; Section 11.5). (3) Determining whether a graph can be *traced,* that is, whether there is a path that contains all arcs of the graph, precisely once each (the answer is given by Euler's Theorem; Section 11.7). (4) In Chapter 12 we'll consider the problem of *planarity*, that is, whether a graph can be pictured in a plane in such a way that distinct arcs never intersect, except at those endpoints they have in common as nodes in the given graph.

In Section 11.6 we considered PERT, a practical application of graph theory in the management of the activities that comprise a complex project.

Expressions, Trees, and Planarity

Consider the arithmetical expression $2 + 3 \times 4$. What is its value? Very likely you answered 14, but if you answered 20, you're not really wrong. To make things more complicated, try pushing the buttons on a calculator in the order shown in the expression. On some calculators, the result will be 14. On some others—even others by the same manufacturer—the result will be 20. On still others, 4 (for reasons we'll explain later). Confused? The problem is, the given expression is ambiguous. To evaluate it we need more information, or a more careful analysis. (At this point, you may want to review Definition 9.5.) The result 14 is obtained by *parsing* the expression according to the formal convention in algebra: multiplication has precedence over addition:

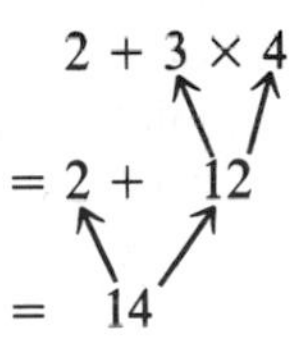

The result 20 is obtained by parsing left to right:

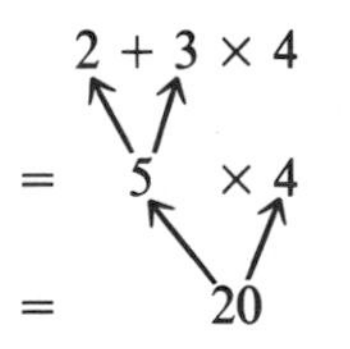

The "usual" algebraic convention applies to operations $+$, $-$, $\times$, $/$, and $^-$ on numbers: in the absence of parentheses, the unary operator $^-$ has precedence over operators $+$, $-$, $\times$, and /. Operations $\times$ and / have precedence over operations $+$ and $-$. Operations $\times$ and / have the "same" precedence; that is, in the absence of other limitations, precedence is left-to-right. Operations $+$ and $-$ have the same precedence. These are *conventions*, that is, a matter of agreement or definition. There is no mathematical or logical reason for or against these rules, for or against an overall left-to-right agreement, for or against any other consistent convention. In Boolean algebra, no such firm agreement exists. Usually (but not always), $'$ is understood to have the highest precedence, and $\wedge$ has precedence over $\vee$. When in doubt, use parentheses or look up *very* carefully the conventions of the book you're reading or of the computer language you're using.

When we presented Definition 9.5, we were careful to point out that it would always produce syntactically valid expressions but not necessarily unambiguous ones. In Chapter 9, this was all right, because syntactic validity was really all we were after. If we use Definition 9.5 (and allow integers as arguments), then $2 + 3 \times 4$ would be a syntactically correct arithmetic expression (or, in the terminology of Definition 9.5, a WFF), even though it's ambiguous. That is, $2 + 3 \times 4$ must have been produced by the rules of Definition 9.5, applied in some order. But now we see that a given expression—a WFF—might have been produced by two different sequences of the rules. Starting from the fact that 2, 3, and 4 are individually WFF's all by themselves, we could have built up $2 + 3$ as a WFF, then $2 + 3 \times 4$, or we could have built up 3×4 as a WFF, then $2 + 3 \times 4$. Both sequences produce the string $2 + 3 \times 4$, which is in any case a WFF according to Definition 9.5. The two different sequences correspond to the two different interpretations of the string, and if we knew which sequence had been used, we would have the unambiguous interpretation of $2 + 3 \times 4$. But that information has been lost in the final result. Clearly, this is unacceptable. If we are to make any use of WFF's (say, as arithmetic expressions, as here), it is not enough to know that *some* sequence of rules produced the WFF; we really want to know *which* sequence of rules did so.

Is it possible to formulate a definition similar to Definition 9.5, such that any expression can be produced by *only one* sequence of rules, and that that sequence can be uniquely reconstructed from the expression after the fact? This chapter examines how these things can be done. Following the conventions of Chapter 9, we consider a set $Y := \{a, b, c, \ldots, z\}$ of variables, a set $S := \{+, -, \times, /\}$ of operators, and the set $P := \{\ ),\ (\ \}$ of parentheses. (In the examples we give, we will also allow Y to contain natural number constants, such as 3 and 27, because the expressions are so natural to deal with. But there is a slight problem in doing this formally, so we'll hold off including them officially in Y until that problem can be discussed in Section 12.1.3.) Our alphabet is the usual $A := Y \cup S \cup P$, A^* is the free semigroup over A, that is, the set of all finite strings of elements of A, with the operation : of concatenation and the identity Λ, the null string (see Definition 6.3). Our goal is to specify a language L ($L \subseteq A^*$) whose WFF's are valid, unambiguous arithmetic expressions. We'll add some operators to S as we go, and we'll also discuss the details of including constants in Y.

12.1 Expressions and Expression Trees

12.1.1 Infix Expressions

We can easily remedy the ambiguity of the WFF's produced by Definition 9.5. That's why the parentheses are there in the first place. All we have to do is require their use. The expressions defined this way are known as *infix language* expressions. The reason for the name will become clear later.

Definition 1 Given sets Y, S, P, A, and A^* defined as usual. The **infix language** L ($L \subseteq A^*$) is defined as follows:

(1) For every variable y in Y, the expression y is in L.
(2) If $v \in L$ and $\{+, -\}$ are *unary operators* in S, then $(+v)$ and $(-v)$ are in L. In general, if $v \in L$ and $\oplus$ is a unary operator in S, then $(\oplus v)$ is in L. Expression v is the **argument** of operator $\oplus$.
(3) If $v \in L$, $w \in L$, and $\{+, -, \times, /\}$ are *binary operators* in S, then $(v + w)$, $(v - w)$, $(v \times w)$, and (v/w) are in L. In general, if $v \in L$, $w \in L$, and $\oplus$ is a binary operator in S, then $(v \oplus w)$ is in L. Expressions v and w are the **arguments** of operator $\oplus$.
(4) There are no WFF's in L except as defined above. This is **exclusion.**

Now we have what we want (we think). The string $x + y \times z$ is no longer a WFF. Not only is it ambiguous, it's also now syntactically invalid. (In fact, it's invalid *because* it's ambiguous.) Strings like $(x + (y \times z))$ are the only thing that our new definition can produce. To prevent ambiguity, parentheses are required for all operators. The expressions thus produced are said to be **fully parenthesized** and are rigorous both syntactically and arithmetically. Even operators like exponentiation $\uparrow$ lose their ambiguity when fully parenthesized like $((x \uparrow y) \uparrow z)$. (We have to use such a symbol because the familiar superscript—as is x^2— isn't in our alphabet.)

One thing that Definition 1 does not contain is *enclosure*, that is, the ability to overparenthesize an expression. For example, $(1 + 2)$ is a valid expression, but $((1 + (2)))$ is not, because it contains parentheses over and above those required by the definition. We could easily include enclosure in the definition by means of a rule

$$\text{If } v \in L, \text{ then } (v) \in L,$$

but we have good reasons for not doing so. The reasons will become apparent later.

How do we know that a fully parenthesized WFF is unambiguous? Evaluating an ambiguous expression involves some choice or judgment. In evaluating $2 + 3 \times 4$, we don't know whether to do the addition or the multiplication first and are forced either to choose between them or resort to some higher judgment, such as operator conventions. If the evaluation of an expression involves no choice or judgment whatever, it must be unambiguous. This notion is not yet rigorous enough to be a formal definition, but it will serve to present the idea.

Theorem 2 Let v be an expression in L, as defined in Definition 1. Then one and only one of the following is true.

1. Expression v consists in its entirety of a single variable v in Y.
2. There exists a unique v_1 in L and a unary operator $\oplus$ in S such that $v = (\oplus v_1)$.
3. There exists unique v_1 and v_2 in L and a binary operator $\oplus$ in S such that $v = (v_1 \oplus v_2)$.

Expressions v_1 and v_2 are referred to as **subexpressions** of v. Note that subexpressions are themselves independent expressions in L. By degeneracy, v is always a subexpression of itself, and any single variable is always a subexpression of any expression in which it occurs.

Proof: The proof follows, of course, directly from Definition 1, and is almost self-evident. But two small points should be emphasized. First, the rules of Definition 1 called for *concatenation* of expressions with operators and parentheses, but not *alteration*. For example, the addition of $(x + (y \times z))$ did not alter the argument expressions x and $(y \times z)$, hence they survive intact in $v := (x + (y \times z))$ as the subexpressions we seek. Second, we now see the importance of exclusion. We can't really prove that $xz\times+($ is not in L unless we had specifically said so in the definition. A small point, but an important one.

Let's consider as our first example the string $v := ((3 \times 2) + (+3))$. Applying Theorem 2, the subexpressions are $v_1 = (3 \times 2)$ and $v_2 = (+3)$, and the binary operator is addition. Applying the theorem again to v_1 individually yields sub-subexpressions 3 and 2 and binary multiplication, and to v_2 yields 3 and unary positive (identity).

In a general case, Theorem 2 can be applied again and again until all of the sub-sub-sub- . . . -expressions are single variables in Y.

The second example is just a little more complicated. Let $v := ((2 + 1) \times (4 - 1))$. We write

$$v = (v_1 \times v_2)$$

$$v_1 = (2 + 1) \qquad v_1 = (v_{11} + v_{12})$$

$$v_{11} = 2$$

$$v_{12} = 1$$

$$v_2 = (4 - 1) \qquad v_2 = (v_{21} - v_{22})$$

$$v_{21} = 4$$

$$v_{22} = 1$$

Why couldn't we apply a similar algorithm to the non-fully-parenthesized regime of Definition 9.5? Consider our old friend $v := 2 + 3 \times 4$, a valid expression under Definition 9.5. Can we apply a variant of Theorem 2, looking for a string of the form $v_1 + v_2$ or $v_1 \times v_2$, where v_1 and v_2 are in L? The problem is that the subexpressions would not be unique. We would choose $v_1 = 2$, $v_2 = 3 \times 4$ or $v_1 = 2 + 3$, $v_2 = 4$, and we wouldn't know which of these to use without applying some external criterion. That's why $2 + 3 \times 4$ is considered ambiguous.

The procedure is always to apply Theorem 2 to a fully parenthesized expression sufficiently many times that all that remain are single variables. This is called **parsing** the expression. Any expression constructed according to Definition 1 can be parsed by Theorem 2 *one and only one way*. Does parsing satisfy our need to know how an expression was constructed, that is, the sequence of rules that were applied? No, it doesn't; it gives us something even better! In our first example above, we still don't know whether the binary expression (3×2) or the unary expression $(+3)$ was built first. Nor do we really care. All we absolutely require is that both be fully formed *before* the final addition $((3 \times 2) + (+3))$ is formed. We could proceed on the left-to-right assumption that (3×2) was done first, and we would not really be wrong. The results would be exactly the same as if $(+3)$ had been done first. These two interpretations are thus equivalent, and either could be chosen. Something quite fundamental is going on here. We wrote down the rules of Definitions 9.5 and 1 as if they had to be executed sequentially, in some strict time-dependent order. They don't. The correct view is not that (3×2) or $(+3)$ was necessarily done first, but rather that they were constructed *independently*. They can thus be parsed independently and, most importantly, evaluated independently. This observation is the main idea behind evaluating several subexpressions simultaneously, as for example when multiple computer processors are operating in parallel. Here we would have to know, not just one possible reconstructed sequence, but all of the possibilities, in order to select the one that makes most efficient use of the computers.

Closer to our own studies, though, we now realize that we have been asking the wrong question in the first place. We started out wanting to reconstruct all the possible sequences of rules by which an expression was built up. But expressions are not built by *strict sequences* of rules, one after another. They are built up by independent, parallel operations, interconnected at the correct points. We don't yet have a systematic notation for so complex a procedure. But we do know this: Whatever process built the expression, we have to precisely reverse it in order to parse (and evaluate) the expression.

Exercises 12.1.1

Consider the expression

$$v := = (30 - (2 + 3 \times 4) + 10) + (30 - (2 + 3 \times 4) - 5)$$

1. Write one possible sequence of steps whereby the expression could have been constructed via Definition 9.5.

2. This expression is ambiguous. Before it can be parsed or evaluated, three judgment decisions must be made. They are:

(1) The subexpression $(2 + 3 \times 4)$ occurs twice in v. Of the addition and the multiplication, which is to be done first?

(2) The left-hand parenthesized subexpression contains a subtraction and an addition. Which is to be done first?

(3) The right-hand parenthesized subexpression contains two subtractions. Which one is to be done first?

a. Evaluate v using every possible interpretation, and verify that the possible final values are 5, 15, 17, 25, 27, 35, 37, and 47.

b. For each possible interpretation, write the equivalent fully parenthesized expression.

c. For each possible interpretation, write one possible sequence of steps whereby the fully parenthesized expression could have been constructed via Definition 1.

All of the remarks we have made about arithmetic algebra apply equally well to other algebras.

3. In the algebra of subsets of a set U, what sets correspond to the sets Y and S? Formulate a definition of set expressions analogous to Definition 9.5. Construct an example to show that an expression constructed by this definition can be ambiguous. Formulate a definition of infix set expressions analogous to Definition 1.

4. In the albegra of logic expressions of propositional calculus, what sets correspond to the sets Y and S? Formulate a definition of logic expressions analogous to Definition 9.5. Construct an example to show that an expression constructed by this definition can be ambiguous. Formulate a definition of infix logic expressions analogous to Definition 1.

12.1.2 Expression Trees

From the introduction to this chapter, you should have already figured out that what we really need to do is to present an expression as a graph, in particular a directed graph. This way, every possible reconstruction of the expression will be readily apparent. The particular type of graph we will use is known as an *expression tree* and is constructed as follows:

Definition 3 Let sets Y, S, and P be given. Let L be an infix language over $A := Y \cup S \cup P$, and let v be an expression in L. An **expression tree** for v is a directed graph $T := (N, R)$, where N is the node set and R the arc set, having the following properties.

(1) Every subexpression of v corresponds to a node in N, and vice versa,

(2) If v_1 is a subexpression of v of the form $v_1 := (\oplus v_2)$, where $\oplus$ is a unary operator in S and v_2 is a subexpression of v_1, then arc $(v_1, v_2) \in R$. We denote this graphically as:

$$v_1 \rightarrow v_2 \qquad \text{or} \qquad \bigoplus \rightarrow v_2$$

(3) If v_1 is a subexpression of v of the form $v_1 := (v_2 \oplus v_3)$, where $\oplus$ is a binary operator in S and v_2 and v_3 are subexpressions of v_1, then arc $(v_1, v_2) \in R$ and arc $(v_1, v_3) \in R$. We denote this graphically in one of the following ways:

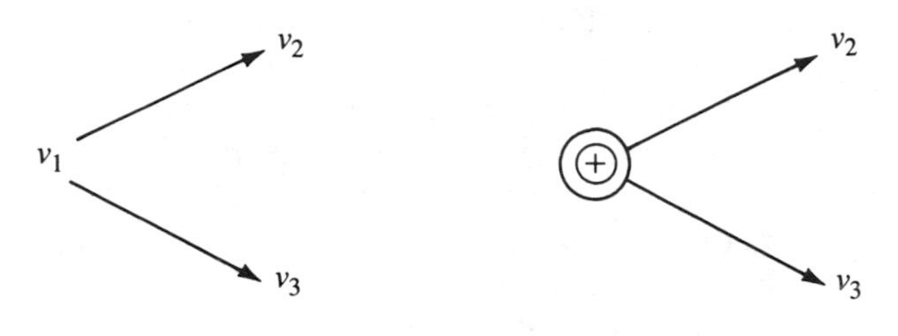

Finally, define W as the set of all such expression trees over A.

Definition 4 Suppose v is an expression in L, and T is the expression tree for v $(T \in W)$. If v_1 is a subexpression of v $(v_1 \in L)$, then there is a tree T_1 corresponding to v_1 $(T_1 \in W)$. In fact, T_1 makes up a portion of T, and is said to be a **subtree** of T. Every tree is degenerately a subtree of itself.

Note, however, that the parentheses are not given their own nodes on the tree. This is unnecessary. We'll show why after we have presented some examples of expression trees.

The first example you've already seen in the introduction. The expression $2 + 3 \times 4$ can be represented by either of two expression trees, which is why it's ambiguous. The fully parenthesized versions, however, leave no room for ambiguity, as you can see in Figure 12.1.

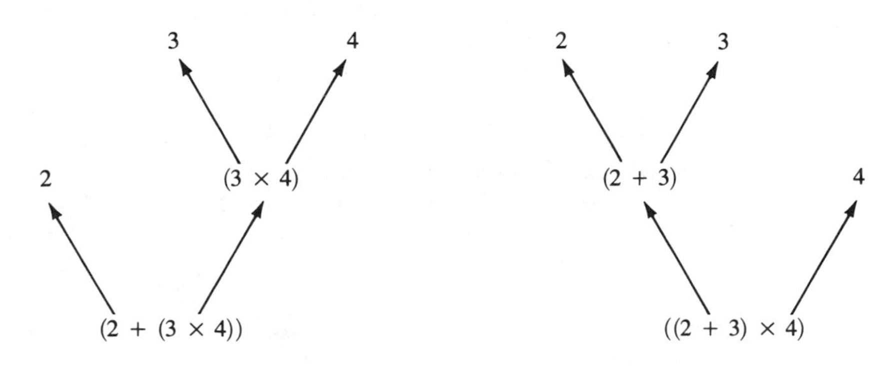

Figure 12.1

The three examples in Figure 12.2 are simply the example expressions from the previous section. Note how an expression tree can be a subtree of some larger tree.

Some conclusions and conventions may be stated immediately from Definition 3. Trees are a very rich structure, with many subtleties and nuances, and these are only a few of the conclusions that may be drawn. Just take them slowly, one at a time, and look back at the examples as you read through them.

1. In the terminology of trees, if $(a, b) \in R$ (that is, if there is an arc from node a to node b), it will be most convenient to say that a is a **predecessor** of b, and that b is a **successor** of a. Graphically, this just means that the directional arrow in the digraph points from a to b.
2. Since v is degenerately a subexpression of itself, there will be precisely one node corresponding to v itself. This node, called the **root** in tree parlance, has no predecessor. Every expression tree has precisely one root node. Each subexpression other than v is an argument for exactly one other subexpression, so every node other than the root will have exactly one predecessor.
3. Every variable y ($y \in Y$) that occurs in v is a valid expression in L all by itself. So each is a subexpression of v, and each will correspond to a node in N. But since a single variable can have no subexpressions of its own, so the corresponding nodes can have no successors. A single variable viewed as an independent expression is called a **monad** or **atom.** In tree parlance, a node with no successors is called a **leaf.** Thus each monad in v is a leaf in T.
4. Every expression other than a monad has at least one subexpression. So if v_1 is a subexpression of v and is not a monad, v_1 must itself have at least one subexpression v_{11}. Hence v_1's node has at least one successor and is not a leaf. Thus, each leaf in T is a monad in v. That is (combining this note and the last one), every monad in v is a leaf in T, and vice versa.
5. Although every node in T has at most one predecessor, some nodes may have more than one successor. The number of successors of a node is called the **arity** of the node. (So far, this can be at most two, since we've considered only up to binary operators, corresponding to nodes of arity 0, 1, or 2. In the next section, we'll see that operators could have more than two arguments. See also Section 7.2.3.)
6. Every node q in N can be viewed as the root of some subtree T_1 of T. Thus, a tree has precisely as many subtrees as it has nodes. If a tree consists only of a single leaf node, that node must be the root (corresponding to a single monadic expression). On the other hand, if a node q has a successor s, we can even speak of the **successor sub-tree** of q, that is, the subtree with root s.
7. Multiple successors must be kept in order. While addition and multiplication are commutative ($x + y = y + x$, $x \times y = y \times x$), subtraction and division are not ($x - y \neq y - x$, $x/y \neq y/x$). Even given commutativity, remember, we said that $x + y$ is not the same *expression* as $y + x$, even if it gives the same *function*. So we adopt the following convention: The successors of a

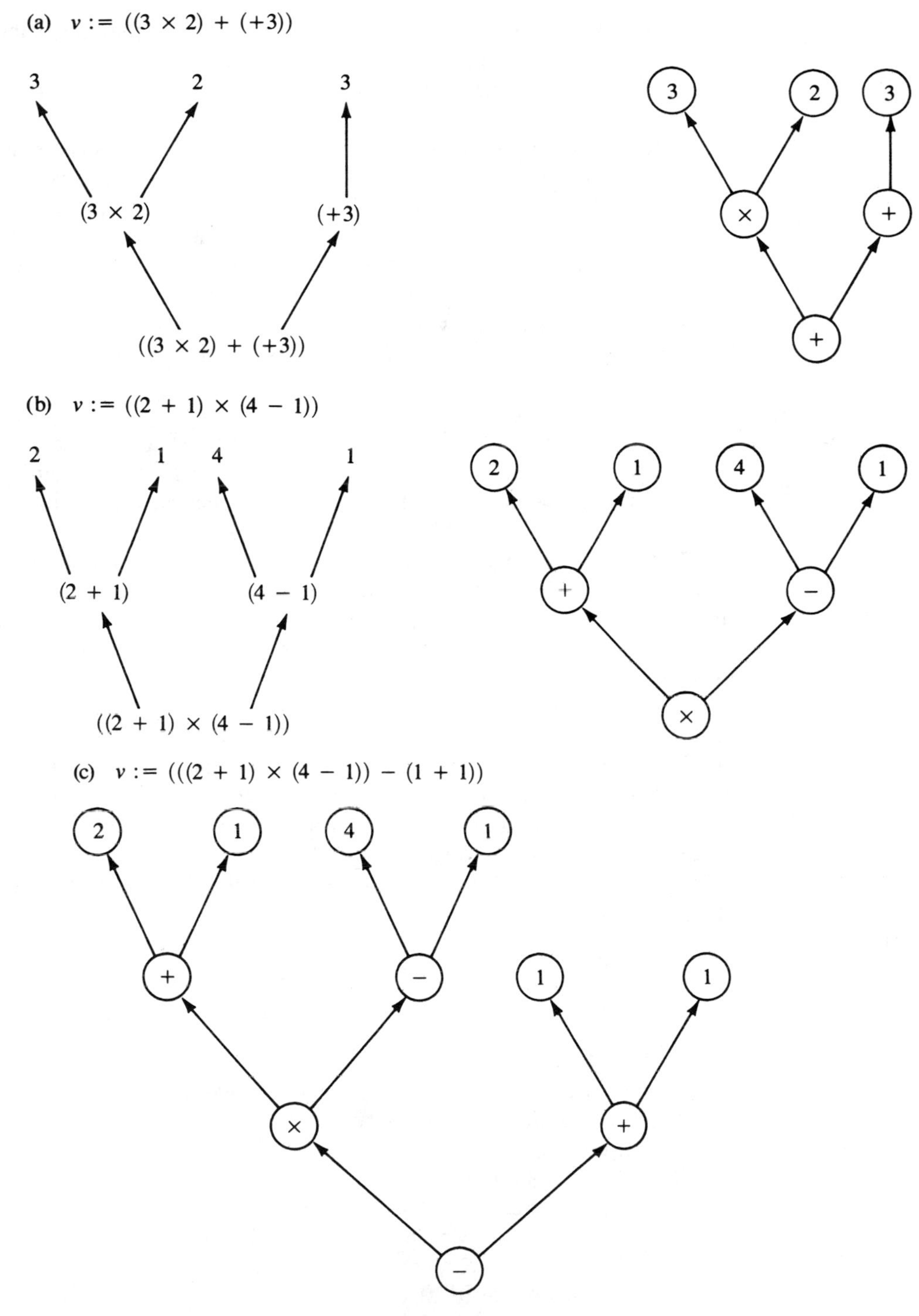

(a) v := ((3 × 2) + (+3))
3
2
3
(3 × 2)
(+3)
((3 × 2) + (+3))
3
2
3
×
+
+
(b) v := ((2 + 1) × (4 − 1))
2
1
4
1
(2 + 1)
(4 − 1)
((2 + 1) × (4 − 1))
2
1
4
1
+
−
×
(c) v := (((2 + 1) × (4 − 1)) − (1 + 1))
2
1
4
1
+
−
1
1
×
+
−

Figure 12.2

node are strictly ordered in the same sequence that the subexpressions occur left-to-right in the expression. In the *function,* it might not matter; in the *expression* (and hence the tree), we will always assume that it does matter.

8. Every node in T thus falls into one of two classes: It is either a leaf (corresponding to a single-variable monad in v), or it corresponds to a subexpression with precisely one operator from S. In the examples, you'll note that we began to associate a one-symbol *label* with each node. For a monad, we simply used a monadic symbol y in Y; for a nonmonad, we used an operator $\oplus$ in S. These labels are convenient and, as we'll see below, even necessary for correlating strings with trees. But they must not confuse the issue. Each node of T represents an entire subexpression of v. The one-symbol labels are simply a convenience.

The next thing we look at is how to construct expression trees from expressions and vice versa. To construct an expression tree $T = (N, R)$ from an expression v, we use the following algorithm. The algorithm is recursive in that it uses a procedure, BUILD_TREE_FROM_INFIX, which can invoke itself. There is also an initial portion that serves to start off the recursion. This is all similar to the recursive definitions we've used previously, starting in Chapter 9. The algorithm also provides the single-symbol labels to the nodes in N. We emphasize again that these labels are merely a convenience; the nodes really corrrespond to entire subexpressions of v.

In general, a recursive procedure could conceivably take forever, if it gets "stuck" in a never-ending process of self-invocation. To guarantee that this won't happen here, we note that for any expression v given to it, BUILD_TREE FROM_INFIX invokes itself only for subexpressions v_1 of v that are strictly shorter than v (in number of symbols). Since v is only finitely long to begin with, such a process cannot go on forever.

Following is an algorithm to build an expression tree for a fully parenthesized expression v in L.

Initial portion:
1. Set $N := \{v\}$, $R := \varnothing$.
2. Invoke BUILD_TREE_FROM_INFIX for v.
3. Halt.

Recursive portion BUILD_TREE_FROM_INFIX for v.
1. Apply Theorem 2 to expression v.
   (1.1) If $v = y$, where $y \in Y$, then **label**(v) := y.
   (1.2) If $v = (\oplus\, v_1)$, where $v_1 \in L$ and $\oplus$ ($\oplus \in S$) is a unary operator, then
      (1.2.1) **label** (v) := $\oplus$.
      (1.2.2) Add to N a node corresponding to v_1.
      (1.2.3) Add to R an arc (v, v_1).
      (1.2.4) Invoke BUILD_TREE_FROM_INFIX for v_1.
   (1.3) If $v = (v_1 \oplus v_2)$ where $v_1 \in L$, $v_2 \in L$, and $\oplus$ ($\oplus \in S$) is a binary operator, then

(1.3.1) **label** $(v) := \oplus$,
(1.3.2) Add to N a node corresponding to v_1.
(1.3.3) Add to R an arc (v, v_1).
(1.3.4) Invoke BUILD_TREE_FROM_INFIX for v_1.
(1.3.5) Add to N a node corresponding to v_2.
(1.3.6) Add to R an arc (v, v_2).
(1.3.7) Invoke BUILD_TREE_FROM_INFIX for v_2.

2. Exit.

(If you aren't familiar with recursive algorithms, you should take special note of the role played by the dummy variable v in BUILD_TREE_FROM_INFIX. Each time BUILD_TREE_FROM_INFIX is invoked "for" a concrete expression (such as v_1 or v_2), v becomes that expression and is to be interpreted as such as long as the invoked procedure lasts. In turn, the invoked procedure may invoke itself again (at which point v will change role), but when it continues, v will resume its original role.)

If we apply the algorithm to our first example above, we get the sequence of trees shown in Figure 12.3. The small box is simply to show which expression node is about to be processed by BUILD_TREE_FOR_INFIX. For clarity, as each node is assigned a label, we draw a circle with the label inside.

The exercises will ask you to apply the algorithm to the other two example trees we've shown above and also to build trees like this from other expressions. Keep in mind that such trees exist also for algebras other than the ordinary one. For instance, if we are dealing with a Boolean algebra and $C := \{0, 1\}$, $S := \{\vee, \wedge, ', \Rightarrow\}$, $Y := \{p, q, r\}$, then some trees are

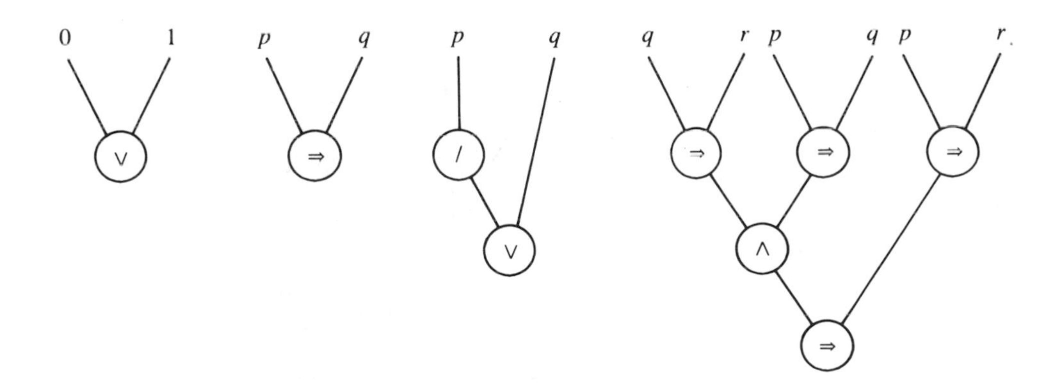

The usual Boolean expressions would be, respectively, $0 \vee 1$, $p \Rightarrow q$, $p' \vee q$, (which is *equivalent* to, but not *the same as* the previous one, and $((q \Rightarrow r) \wedge (p \Rightarrow q)) \Rightarrow (p \Rightarrow r)$ which is syllogism BARBARA.

Before we leave Algorithm BUILD_TREE_FROM_INFIX, we want to emphasize again the role played by Theorem 2. The theorem parses an expression into one of three possible cases: a single variable or an expression involving a unary or binary operator. The real power of the theorem is just how well it works.

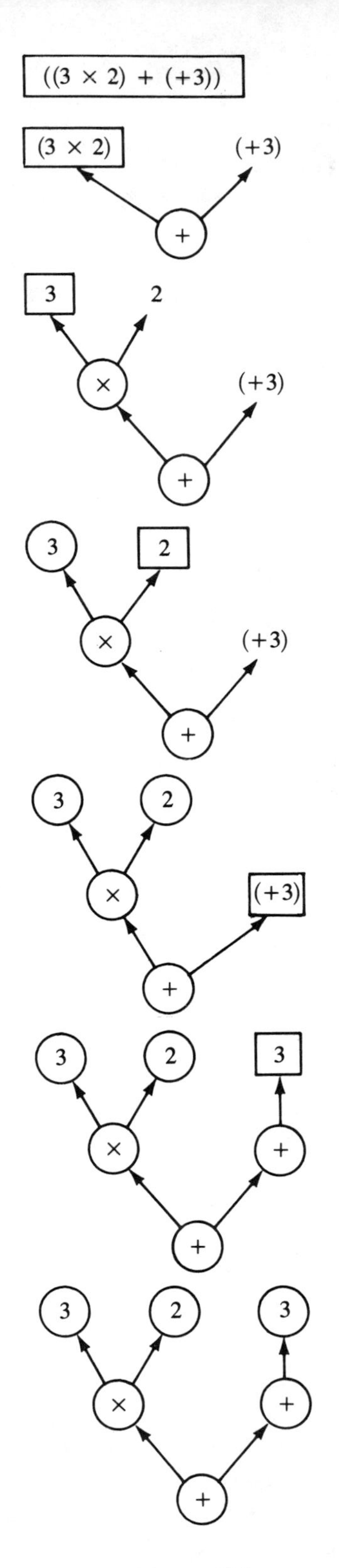

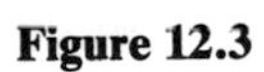

Figure 12.3

Every expression in L can be parsed and can be parsed one and only one way. No ambiguity can arise; no external judgment is called for. The parsing of fully parenthesized expressions is as mechanical as addition.

The obvious next step is to go the other way, that is, to start from an expression tree and to construct the unique fully parenthesized infix expression to which that tree corresponds. Before we do that, though, we're going to explore some different types of expressions (other than fully parenthesized ones) and their own corresponding expression trees. When we do come back to the problem of constructing expressions from trees, you'll see why we did things this way.

Exercises 12.1.2

1. Apply Algorithm BUILD_TREE_FROM_INFIX to the remaining two example expressions given in the text:
a. $v := ((2 + 1) \times (4 - 1))$
b. $v := (((2 + 1) \times (4 - 1)) - (1 + 1))$
In each case, write down all the intermediate steps as was done for the first example.

2. In Exercises 12.2.1, you constructed eight possible fully parenthesized infix interpretations for the ambiguous expression

$$v := (30 - (2 + 3 \times 4) + 10) + (30 - (2 + 3 \times 4) - 5).$$

a. Construct an expression tree for each of the eight interpretations.
b. One of the infix expressions yields the evaluation $v = 37$. For that expression, write down all intermediate steps in applying Algorithm BUILD_TREE_FROM_INFIX.

3. In the algebra of sets of a subset v, formulate a definition of expression trees analogous to Definition 3 and an algorithm for constructing them analogous to BUILD_TREE_FROM_INFIX.

4. In the algebra of logic statements of propositional calculus, formulate a definition of expression trees analogous to Definition 3 and an algorithm for constructing them analogous to BUILD_TREE_FROM_INFIX.

12.1.3 Expression Strings and Polish Notation

In order to consider only unambiguous expressions, we took Definition 9.5 and forced full parenthesization of all operators. This culminated in Definition 1. But is this really what we wanted? The strings $x + (y \times z)$ and $(x + y) \times z$ are unambiguous, yet now they are deemed syntactically invalid. An even more egregious example is $1 + 2 + 3 + 4 + 5$, which would have to be fully parenthesized as something like $((((1 + 2) + 3) + 4) + 5)$. We seem to be caught on the horns of a dilemma. How can we devise a definition of arithmetic expressions such that syntactic validity will precisely coincide with nonambiguity? Requiring parentheses is too strict, but not requiring them is too lax. Can we do nothing better? One clue lies in the observation that unparenthesized unary operators were unambiguous in the first place. Such expressions as $(x + -y)$ and even $(- - -x)$ were never ambiguous, just a bit unusual. Requiring full parenthesization of unary operators offered no real advantage. We could just as well revert back to the original rule (3) of Definition 9.5: If $v \in L$ and $\{+, -\}$ are *unary operators* in S, then $+v$ and $-v$ are in L, with a similar rule for a general unary operator $\oplus$. A second clue is found when we consider operators with more than two arguments. Suppose, for example, we would like to incorporate into our language the familiar discriminant function from the quadratic equation:

$$Dabc := b \uparrow 2 - 4 \times a \times c = ((b \uparrow 2) - ((4 \times a) \times c)).$$

We could add D to S in our language as a *ternary operator* (that is, having three arguments). The rule would look like: If v_1, v_2, and v_3 are in L, and if D is a **ternary operator** in S, then $Dv_1v_2v_3 \in L$. (The notation $Dv_1v_2v_3$ is preferable to $D(v_1, v_2, v_3)$ because it's simpler and doesn't require still more parentheses. Besides, we don't have and we don't want commas in our alphabet.) For instance (using single-digit numbers) $D973$ is easily parsed and evaluated as $9 \uparrow 2 - 4 \times 7 \times 3$ (which is -3 in usual notation). Actually, there's no limit to the number of arguments an operator could have: 3, 4, 5, or a hundred. In Section 7.2.3 we mentioned ternary, quaternary operations. Here we will not need more than the usual unary and binary and the occasional ternary or zerary. More useful to us is the general term ***n*-ary** for an operator requiring n arguments; in general, n will be positive, because a zerary operator is a constant (see Section 7.2.3; more on this later).

The need for parenthesization, we see, arises only for binary operators, where the very fact that the operator is written *between* its arguments is precisely what causes the ambiguity. It appears that writing the operator first (or, as we'll see, last) eliminates the need for parentheses, yet sacrifices nothing in rigor. It's possible, then, to construct three complete systems of expressions, that is, three complete languages L ($L \subseteq A^*$), from the sets Y (variables) and S (operators). The difference lies in whether the operator is written before, between, or after its arguments. The parentheses set $P = \{\ ),\ (\ \}$ is required for only one of them. As always, $A := Y \cup S \cup P$. We give below three complete language definitions. Compare them carefully.

Definition 5 Given sets Y, S, P, A, and A^* defined as usual.

The **prefix language** E ($E \subseteq A^*$) is defined as follows:

1. For every variable y in Y the expression y is in E.
2. If $v \in E$ and $\oplus$ is a *unary operator* in S, then $\oplus v$ is in E.
3. If v_1 and v_2 are in E and $\oplus$ is a *binary operator* in S, then $\oplus v_1v_2$ is in E.
4. There are no WFF's in E except as defined above.

The **infix language** I ($I \subseteq A^*$) is defined as follows:

1. For every variable y in Y the expression y is in I.
2. If $v \in I$ and $\oplus$ is a *unary operator* in S, then $(\oplus v)$ is in I.
3. If v_1 and v_2 are in I and $\oplus$ is a *binary operator* in S, then $(v_1 \oplus v_2)$ is in I.
4. There are no WFF's in I except as defined above.

The **postfix language** O ($O \subseteq A^*$) is defined as follows:

1. For every variable y in Y the expression y is in O.
2. If $v \in O$ and $\oplus$ is a *unary operator* in S, then $v\oplus$ is in O.
3. If v_1 and v_2 are in O and $\oplus$ is a *binary operator* in S, then $v_1v_2\oplus$ is in O.
4. There are no WFF's in O except as defined above.

Prefix notation is also commonly known as **Polish notation,** after the man who invented and investigated it, the Polish mathematician Jan Lukasiewicz (1878–1956). Postfix notation is similarly known as **reverse Polish notation**, or RPN for short. (Unlike George Boole and Bertrand Russell, whose creations bear their names, Lukasiewicz must settle for being remembered only by nationality.)

The table below compares some examples of expressions in prefix, infix, and postfix form. These are the same examples we've seen previously. We use the symbol for the unary *plus* because prefix and postfix forms require that each operator have a specific arity.

Prefix	Infix	Postfix
$+xy$	$(x + y)$	$xy+$
$+2\times 34$	$(2 + (3 \times 4))$	$234\times +$
$\times +234$	$((2 + 3) \times 4)$	$23+4\times$
$+\times 32^+3$	$((3 \times 2) + (+3))$	$32\times 3^+ +$
$\times +21-41$	$((2 + 1) \times (4 - 1))$	$21+41-\times$
$-\uparrow b2\times\times 4ac$	$((b \uparrow 2) - ((4 \times a) \times c))$	$b2\uparrow 4a\times c\times -$

In Definition 5, we have also required parentheses on unary operators in infix expressions, to keep to the spirit of full parenthesization. We also see that unary operators in infix form look exactly like the prefix form: The operator is written first. We could just as easily have defined unary infix expressions like postfix, that is, $(v\oplus)$. (This is the way that unary operators are implemented on most hand calculators. To take the square root of 9, for instance, the usual key sequence is $9\surd$.) The bottom line is that unary operators aren't the problem; the problem is binary operators.

Some hand calculators, however, actually do work in postfix notation (or Reverse Polish Notation, RPN). Recall that in the introduction to the chapter, we said that the ambiguous expression $2 + 3 \times 4$ would result in 4 on some calculators. To see why, let's examine what would happen if you key $2 + 3 \times 4$ into a calculator expecting RPN. Remember, the calculator is expecting to see the operator last, so it will assume that all arguments have already been entered. Also, as arguments are entered, they are simply stored (and the current one displayed), awaiting entry of the operator.

When you key in:	The calculator will:
2	Store and display 2
+	Add 2 to whatever was in the calculator before you started. The result, likely gibberish, is displayed.
3	Store and display 3.
×	Multiply the previous gibbberish by 3, and display the result.
4	Store and display 4.

Obviously, the calculator has accomplished nothing. The expression must be entered as proper postfix form. Keying in $234 \times +$ (with a dummy delimiter between the numbers) yields 14, while $23 + 4 \times$ yields 20.

This problem illustrates one major ambiguity that can arise with prefix and postfix forms, and we've been procrastinating in coming to grips with it. Since the beginning of the chapter, we've been hesitant to include constants in our set Y of variables. Even though many of our examples used natural number constants and the whole process seems as natural as addition, in fact there's a small problem. Consider the prefix expression $+xy$. It's perfectly clear that the two arguments are x and y, and that the infix form would be $(x + y)$. But what are we to make of $+123$? Is this (in infix) $(12 + 3)$ or $(1 + 23)$? In other words, prefix and infix form expressions are unambiguous *once we know where one argument stops and the next one begins*. In RPN calculators, there's a special key just for that purpose.

We can already hear you crying "Foul! You said that prefix and postfix expressions were unambiguous. Now you say that they're just as ambiguous as infix expressions but in a different way." It is true that parentheses would solve the problem (clumsily): $+(12)(3)$ is definitely unambiguous. But the real source of this ambiguity lies in the representation of the constants, not the expressions. If two multidigit natural numbers are expressed in base-positional notation—such as 12 and 3, or 1 and 23—their concatenation—123—will always be ambiguous. Even in infix notation, we sometimes use concatenation as a shorthand for multiplication: xy is interpreted as $x \times y$. But such a trick would never work with 12×3 or 1×23. So the ambiguity we speak of here is not in the prefix or postfix forms but in the base-positional notation of natural number constants. Constants are ambiguous in *any* expression scheme. It's just that confusion does not arise often in infix notation but arises constantly in prefix and postfix. The prefix and postfix forms themselves are unambiguous.

Most applications of prefix and postfix forms occur inside computer programs. Here, numbers are usually of a specified fixed number of digits, so that concatenation of arguments poses no problems. In fact, the arguments are usually not the numbers themselves but memory addresses, so that again no confusion can arise. But computer scientists who design programming language compilers go to great trouble and care in dealing with multidigit, base-positional natural numbers and integers. For that matter, they must also deal with multicharacter variable names, where the same concatenation problem could occur.

To summarize, we are free to incorporate constants, function names (such as roots, sines, and cosines) and anything else we like into our alphabet A. Constants are usually handled as a separate entity, a set C, so that $A := C \cup Y \cup S \cup P$. The only restriction is that if any two elements of A are concatenated, there must be no confusion later as to where the concatenation occurred. The simplest way to ensure that, as we have done, is to restrict our theoretical discussion to single-symbol variables. More complicated objects can be handled, but it takes care and effort. In contrast, the ambiguity of Definition 9.5 is genuine, and nothing short of full parenthesization or operator conventions can remedy it.

Returning to our main discussion, if we compare the prefix and postfix languages, we can see that they are simply mirror images of one another. In prefix form, the operator is always written first, while in postfix form, it's always written last. Infix form—the one we've dealt with in the previous sections—requires parentheses on binary operators to avoid ambiguity. The advantage of the prefix and postfix forms is that *neither requires any parentheses*. These expressions are unambiguous just as they are.

This claim requires justification. It's one thing to construct examples and observe that they don't need parentheses. It's quite another to make a blanket claim of unambiguity. But since we don't yet even have a formal *definition* of unambiguity, we can prove nothing yet. Both definition and proof await the next section.

At this point, you may be confused by the strange "look" of the prefix and postfix forms. They seem inconvenient and difficult to decipher, more like a code than a language. Partly this is due to convention. If you had grown up with prefix notation, it might well seem today to be perfectly natural, and infix notation would seem bizarre and confusing. You would also consider the need for the parentheses in infix to be blatantly wasteful. We are not seriously recommending that people adopt prefix or postfix notation in normal algebraic calculations, even if some hand calculators do. Prefix and postfix forms also play a pivotal role in the way some programming language compilers implement and optimize expressions in computer programs. But that discussion is beyond our current scope, and even though it's an excellent practical application of this theory, it's still not our central reason for studying it.

Our reason for studying prefix and postfix forms is that our study of expressions is clearly incomplete without them. Recall the approach we took to infix notation. We first gave a formal definition (Definition 1) of the language whose WFF's were to be considered valid, unambiguous expressions. (It took a little work to make them unambiguous, remember?) We then stated a theorem (Theorem 2) that every expression in the language could be consistently and uniquely parsed. We next recursively defined a set W of expression trees (Definition 3). Finally, using the theorem, we constructed an algorithm by which an expression tree (according to Definition 3) could be constructed for every infix expression. (Going the other way was left in abeyance.)

Precisely the same procedure can be followed for prefix and postfix notation. We have already given the formal definitions above (along with the corrected infix definition). The statement of the corresponding parsing theorems and expression tree algorithms are left as exercises.

Exercises 12.1.3

1. In Exercises 12.1.1, you constructed eight possible fully parenthesized infix expressions for the ambiguous expression

$$v := (30 - (2 + 3 \times 4) + 10) + (30 - (2 + 3 \times 4) - 5).$$

For each infix expression, formulate the corresponding prefix and postfix expressions.

2. Consider the prefix part of Definition 5.

a. State a result for the prefix language analogous to Theorem 2, permitting unique parsing of prefix form expressions.

b. Define a set of expression trees, analogous to Definition 3, for prefix form expressions.

c. State an algorithm analogous to BUILD_TREE_FROM_INFIX, which will build such an expression tree from any prefix form expression.

3. Consider the postfix part of Definition 5.

a. State a result for the postfix language analogous to Theorem 2, permitting unique parsing of postfix form expressions.

b. Define a set of expression trees, analogous to Definition 3, for postfix form expressions.

c. State an algorithm analogous to BUILD_TREE_FROM_INFIX, which will build such an expression tree from any postfix form expression.

4. The pre- and postfix form of Definition 5 can easily be extended to include operators of more than two arguments. Recall that the number of arguments required by an operator is called the *arity* of the operator. That is, unary operators have arity 1, binary operators have arity 2, and so forth.

a. Reformulate the prefix and the postfix part of Definition 5 to include the possibility of operators of arity 1, 2, 3, 4, and 5.

b. Exercise 4.a requires a separate rule for each arity. One clever way to reduce the number of rules is simply to notice that each operator in S has an arity $n > 0$. A single rule can then replace the separate arity rules. This "super rule" would have the form: If $\oplus$ is an operator of arity n in S and if v_1, v_2, . . ., v_n are in E or in O, then $\oplus v_1 v_2 \ldots v_n \in E$ (or, respectively, $v_1 v_2 \ldots v_n \oplus \in O$). How many rules are now required for the prefix and the postfix part of Definition 5?

c. Definition 5 can be made even more succinct by one final step. Suppose we combine sets C, Y, and S into one set S set of operators and declare that what used to be constants or variables are now **operators of arity** 0, or **zerary**

operators. This is perfectly consistent, because variables take no arguments. So now everything is an operator in S. The prefix and postfix part of Definition 5 can now be stated in only two rules each. What are they?

12.1.4 Traversing Expression Trees

So far, we have discussed three forms of unambiguous expressions: prefix, postfix, and infix (fully parenthesized). For each form, we have specified an algorithm to unambiguously construct an expression tree for any expression. But two major questions are unresolved. As it happens, both questions can be answered in the same stroke. The first one is simply the process we've been postponing throughout the chapter: Given an expression tree, how do we reconstruct the expression itself? Obviously, we have to answer this question three ways, for prefix, infix, and postfix expressions.

But at the same time, we should address a second and deeper question. When we considered the three expression forms, we defined a set of expression trees for each. That is, we spoke of a set of trees built from prefix expressions, a set built from infix expressions, and a set built from postfix expressions. Are there three different sets of expression trees?

No, there is only one set W of expression trees. The reason we have put off the discussion of building expressions from trees is precisely so that we could see this. Given any expression tree, no matter how it was constructed, we will show that valid prefix, infix, and postfix expressions can be constructed from it. The algorithm itself will constitute the proof that all expression trees belong to the same set.

The algorithm is widely applicable because we only make use of those properties that are common to all expression trees. We'll recall these common properties here as a review for you. Given sets Y (variables) and S (operators), an expression tree $T \in W$ was always defined as $T = (N, R)$, such that

1. One node, the root r ($r \in N$) has no predecessor. All other nodes have exactly one predecessor.

2. With each node q ($q \in N$) is associated a **label**(q), a symbol taken from Y or S.

   2.1 **Label**(q) $\in Y$ if and only if q is a leaf and has no successors.

   2.2 **Label**(q) $\in S$ if and only if q has precisely **arity**(**label**(q)) successors.

Recall from the exercises that if we incorporate the variables y ($y \in Y$) into S as zerary operators with arity 0, then statement 2.2 subsumes 2.1. Pay attention to the notation **arity**(**label**(q)). For node q, **label**(q) is either a variable in Y or an operator in S. Then **arity**(**label**(q)) is a natural number, possibly zero. An expression is a string of symbols, and the label of each node is a symbol. What we will do is to construct the expression from the tree's node labels. The process we will use is called *traversing the tree*.

Definition 6 Let Y, S, $A := Y \cup S$, and A^* be defined as usual. Let T ($T \in W$, $T = (N, R)$) be an expression tree with root node r ($r \in N$). If $q \in N$, a **full traversal** of node q is a string $t(q)$ ($t(q) \in A^*$) defined recursively as follows:

1. If **arity**(**label**(q)) $= 0$, then q is a leaf and $t(q) :=$ **label**(q).
2. If **arity**(**label**(q)) $= n > 0$, then **label**(q) $= \oplus$ (an operator in S) and q must have n successors $s_1, s_2, \ldots, s_n$. Then

$$t(q) := \oplus : t(s_1) : \oplus : t(s_2) : \oplus : \ldots : \oplus : t(s_n) : \oplus$$

Finally, the full traversal of an entire tree T is identically the traversal of its root node r, that is, $t(T) := t(r)$.

The basic idea is simply to start at the root and walk up and down the tree, encountering each node in succession. Every time a node is encountered, its label is concatenated to the traversal string. The example in Figure 12.4 will illustrate the idea of a full traversal. Don't try to make an expression out of the traversal yet; we're not finished.

A full traversal includes the node label every time the node is encountered. In an expression, whether prefix, postfix, or infix, every node label occurs exactly once. By now, we recognize the choices and can immediately write modified definitions to produce prefix and postfix expressions. If $q \in N$, a **prefix traversal** of node q is a string $t(q)$ ($t(q) \in A^*$) defined by

1. If **arity**(**label**(q)) $= 0$, then $t(q) :=$ **label**(q).
2. If **arity**(**label**(q)) $= n > 0$, **label**(q) $= \oplus$, and $s_1, s_2, \ldots, s_n$ are the n successors of q, then

$$t(q) := \oplus : t(s_1) : t(s_2) : \ldots : t(s_n)$$

as before, $t(T) := t(r)$.

The prefix traversal of our above example is $+ \times 7 3 - 2$, a valid prefix expression.

You can now immediately see how to define a postfix traversal. Do so now, and verify that the example yields $7 3 \times 2 - +$. Constructing infix expressions takes

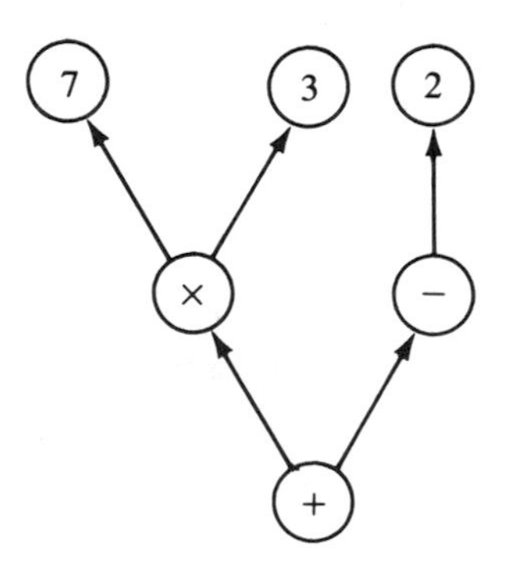

Figure 12.4 $t(T) = + \times 7 \times 3 \times + - 2 - +$

a little more thought. Remember that the problem, as always, is how to handle binary operators. The basic idea here is simple: We concatenate the binary operator *between* its arguments. But if we left it at that, we would once again produce the ambiguous expressions of Definition 9.5. Our example above would yield $7\times3+-2$, which could be either 19 (right) or 7 (wrong), in spite of the fact that the unary minus is unambiguous. Our definition will include the parentheses we need. In fact, it will fully parenthesize all operators of positive arity, even though only the binary ones need it. We will also treat nonbinary operators like prefix form.

More precisely, if $q \in N$, a **fully parenthesized infix traversal** of node q is a string $t(q)$ ($t(q) \in A^*$) defined by

1. If **arity(label(**q**))** $=0$, then $t(q) :=$ **label(**q**)**
2. If **arity(label(**q**))** $= n$, where $n > 0$ and $n \neq 2$, and if **label(**q**)** $= \oplus$, and $s_1, s_2, \ldots, s_n$ are the n successors of a, then

$$t(q) := (\oplus : t(s_1) : t(s_2) : \ldots : t(s_n))$$

3. If **arity(label(**q**))** $= 2$, **label(**q**)** $= \oplus$, and s_1 and s_2 are the successors of q, then

$$t(q) := (t(s_1) : \oplus : t(s_2))$$

As before, $t(T) := t(r)$.

Our example now yields $((7 \times 3) + (-2))$, the correct, unambiguous expression.

Note very carefully that the parentheses are provided by the traversal process; they are not part of the tree. The process provides exactly the parentheses that are needed according to Definition 1, no more, no less. We can now explain what happens to overparenthesized expressions, that is, expressions built as if enclosure had been part of Definition 1, Theorem 2, and Definition 3. Suppose we start with the overparenthesized expression $((x + (y)))$. If we first build an expression tree for it (with enclosure included in our methods), we get simply

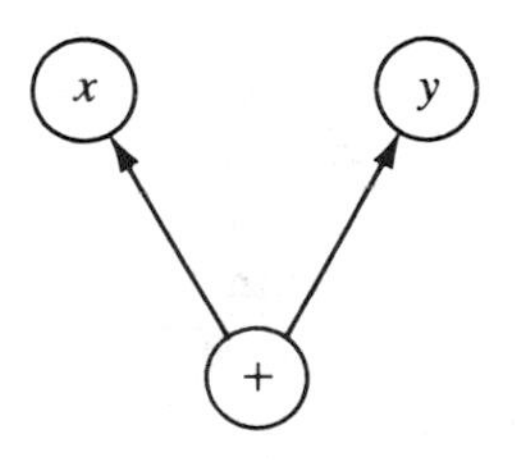

because the extra parentheses do not occur on the tree. If we now use infix traversing to traverse it, the infix traversal is simply $(x + y)$. The superfluous parentheses have been removed. This is the one case where a traversal will not faithfully reproduce the original expression. In the exercises, you'll be asked to remedy this problem. It's time now to take formal stock of what we have done. For each of the three forms of expression, we can build trees from expressions and recover the expressions from the trees. While the process of building a tree from an expression sometimes called for judgment (in other words, it was ambiguous), traversing a tree never involves any judgment and cannot be ambiguous at any point. Now, at long last, we are finally in a position to reinforce the most important concept.

Let v be an element of A^*, that is, a string over alphabet $A := Y \cup S \cup P$. Then

- v is said to be an **unambiguous prefix expression** if and only if there exists exactly one tree T ($T \in W$) such that v is the prefix traversal of T.
- v is said to be an **unambiguous infix expression** if and only if there exists exactly one tree T ($T \in W$) such that v is the fully parenthesized infix traversal of T.
- v is said to be an **unambiguous postfix expression** if and only if there exists exactly one tree T ($T \in W$) such that v is the postfix traversal of T.

Is it possible for one string v to be both, say, an unambiguous prefix expression and an unambiguous infix expression? Yes, because we always consider single variables y ($y \in Y$) to be valid expressions in any form. Also, we insisted above that the infix traversal parenthesize nonbinary operators as well as binary ones. If we relax that rule, then $+2$ is acceptable either as a prefix or infix expression. For the most part, though, the three types of expressions do not overlap very much.

Before we leave this section, let's see how the full power of these notions comes into play. Suppose we let E, I, and O refer respectively to the prefix, infix, and postfix languages over alphabet A. Our intuition tells us that these three sets are isomorphic. For every infix expression, there ought to correspond one unique prefix and one unique postfix expression, and vice versa in all combinations. The question is, how do we go about proving this? It's bad enough contemplating complex and error-prone proofs with symbol strings. We also recall that expressions are inherently misleading, in that they suggest a strict time-dependent construction process, which simply is not the case.

Expression trees change all that. We have demonstrated that for every fully parenthesized infix expression v there is a unique expression tree $T \in W$ corresponding to it, and for every $T \in W$, there is a unique corresponding v. That is, the sets W and I are isomorphic. Similarly, we have shown that W and E are isomorphic, and that W and O are isomorphic.

But set isomorphism is transitive. E is isomorphic to W, and W is isomorphic to I, so E is isomorphic to I. Sets E, I, and O are immediately seen to be pairwise isomorphic (that is, any two of them are isomorphic to each other). The proof of our intuitive conjecture is as simple as that. Moreover, we are now able to justify our claim that prefix and postfix expressions are unambiguous, even though they lack parentheses. Once again, the proof is simplicity itself. Since the trees in W are unambiguous, and since E and O are isomorphic to W, it follows that E and O contain unambiguous expressions. In other words, the following sets are pairwise isomorphic.

- The set W of expression trees over $A := Y \cup S$,
- The set E of prefix expressions over the alphabet $A := Y \cup S$,
- The set O of postfix expressions over the alphabet $A := Y \cup S$,
- The set I of fully parenthesized infix expressions over the alphabet $A := Y \cup S \cup P$.

Thus the last three sets contain unambiguous expressions.

The set W of expression trees serves not only as the most complete notational form for characterizing expressions, but it is also the connecting link among all the other forms. It is the hub to which everything else is attached. (From our pulpit now, we remind you that this is neither surprising nor unprecedented. We've learned long ago that when one spot is arrived at from several different starting points, that one spot is the one to study. Recall we had a similar experience with the combinatoric function $C(n, r)$.)

Exercises 12.1.4

1. In Exercises 12.1.1, you constructed eight possible fully parenthesized infix interpretations for the ambiguous expression

$$v := (30 - (2 + 3 \times 4) + 10) + (30 - (2 + 3 \times 4) - 5)$$

and in Exercises 12.1.2, you constructed the corresponding expression trees. In Exercises 12.1.3, you obtained the corresponding prefix and postfix expressions. For each of the eight trees, execute the algorithms given, obtain the prefix, infix, and postfix tree traversals, and verify that they reproduce the expressions exactly.

2. Change each of the following postfix expressions to the equivalent infix form.

When $C := \mathbb{Z}$, $S := \{+, -\}$, and digit symbols denote single-digit numbers, **(a)** $234+-$; **(b)** $234-+$; **(c)** $23-4+$; **(d)** $23+4-$; **(e)** $234--$; **(f)** $23-4-$.

When $C := \{0, 1\}$, $S := \{\vee, \wedge, \sim\}$ (Boolean), and $Y := \{x, y\}$, **(g)** $xy1\vee\wedge$; **(h)** $xy\vee1\wedge$; **(i)** $xy1\vee\vee$; **(j)** $xy1\wedge\wedge$; **(k)** $xy{\sim}1\vee\wedge$; **(l)** $xy{\sim}\vee1\wedge$; **(m)** $xy\vee{\sim}1\wedge$; **(n)** $xy\vee1\wedge{\sim}$. (This is a suitable exercise for computer solution.)

3. Change each of the following infix expressions to the equivalent postfix form.

For $C := \mathbb{Z}$ and $S := \{+, -\}$, **(a)** $((2 - 3) - 4)$; **(b)** $(2 - (3 - 4))$; **(c)** $(2 - (3 + 4))$; **(d)** $((2 - 3) + 4)$.

For $C := \{0, 1\}$, $S := \{\vee, \wedge, \sim\}$ (Boolean), and $Y := \{x, y\}$.

(e) $((\sim (x \vee y)) \wedge 1)$; **(f)** $((x \vee \sim(y)) \wedge 1)$; **(g)** $(\sim((x \vee y) \wedge 1))$.

4. Draw an expression tree for each of the following expressions:

For $C := \mathbb{Z}$, $S := \{+, -\}$ and digit symbols representing single-digit numbers, **(a)** $234+-$; **(b)** $23-4+$; **(c)** $234--$; **(d)** $((2 - 3) - 4)$; **(e)** $(2 - (3 - 4))$; **(f)** $(2 - (3 + 4))$; **(g)** $((2 - 3) + 4)$.

For $C := \{0, 1\}$, $S := \{\vee, \wedge, \sim\}$ (Boolean), and $Y := \{x, y\}$; **(h)** $xy1\vee\wedge$; **(i)** $xy1\vee\vee$; **(j)** $xy1\wedge\wedge$; **(k)** $xy{\sim}1\vee\wedge$; **(l)** $xy{\sim}\vee1\wedge$; **(m)** $xy\vee{\sim}1\wedge$; **(n)** $xy\vee1\wedge{\sim}$; **(o)** $(x \vee y) \wedge 1$; **(p)** $(x \wedge y) \wedge z$; **(q)** $x \wedge (1 \wedge z)$; **(r)** $\sim(x \vee y)$; **(s)** $(\sim x) \wedge (\sim y)$; **(t)** $1 \vee \sim x$; **(u)** $\sim(1 \wedge x)$. (For prefix and postfix expressions, this is a suitable problem for computer solution.)

5. Write the postfix and the infix expression of each of the following expression trees. **(a)** In elementary arithmetic:

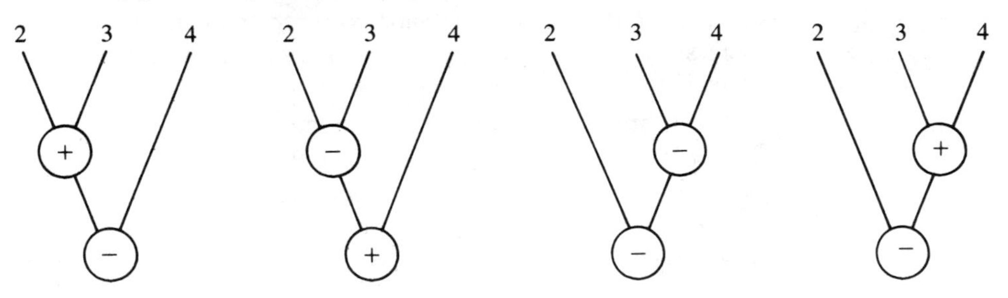

(b) $C := \{0, 1\}$, $S := \{\vee, \wedge, \sim\}$ (Boolean), $Y := \{x, y\}$

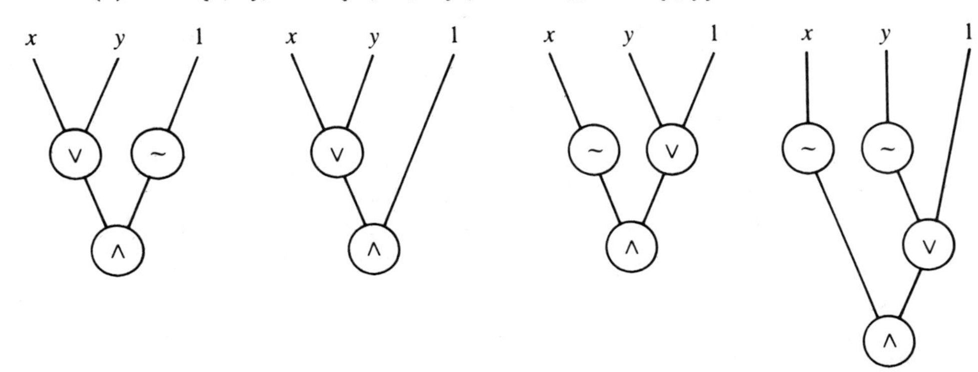

(c) For any set C, $S := \{*, \oplus, /\}$ ($*$ and $\oplus$ are binary, and / is unary), and $Y := \{x, y, z\}$

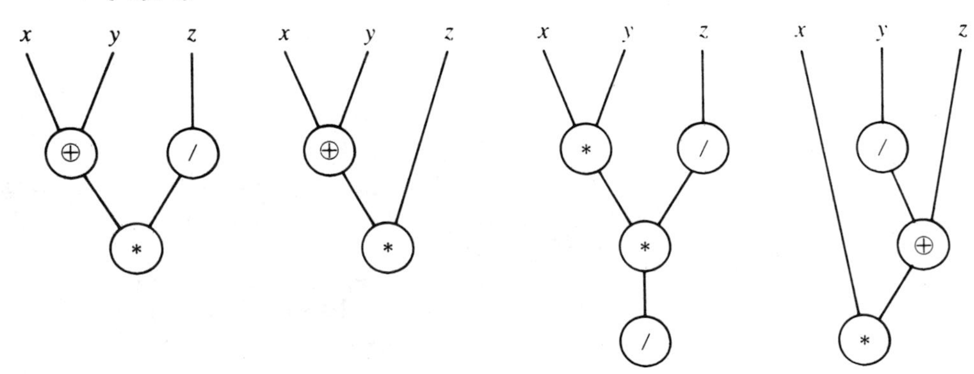

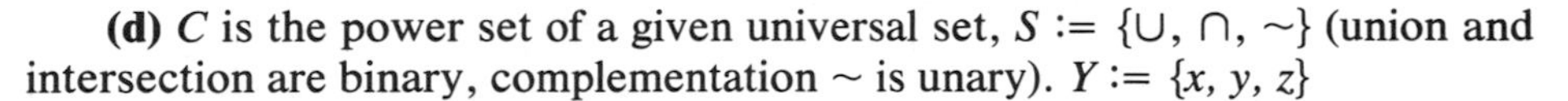

(d) C is the power set of a given universal set, $S := \{\cup, \cap, \sim\}$ (union and intersection are binary, complementation $\sim$ is unary). $Y := \{x, y, z\}$

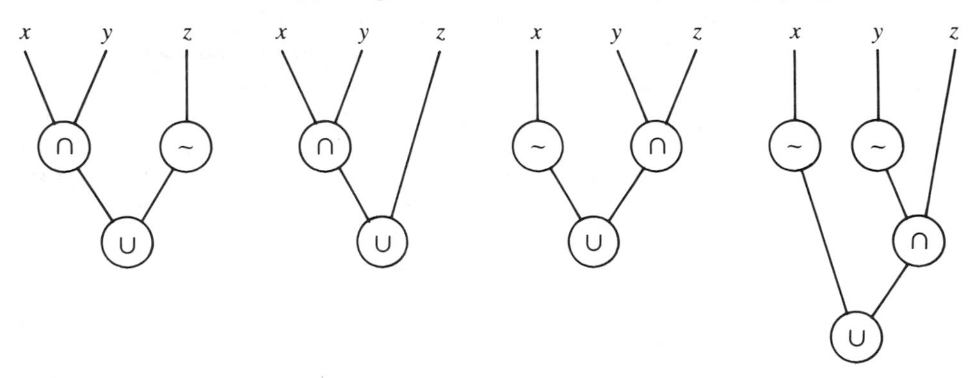

For postfix expressions, this is a suitable exercise for computer solution. See the algorithm in Complement 2.

6. Verify that the following process checks whether a given string is the string of an RPN expression.

1. Start a "counter" k with the value 0.
2. Scan the string left to right.
3. For each element of the string, let h be the arity of the respective operation (including 0 for an element of $C \cup Y$) and add to the counter the value $1 - h$.
4. The string is *well formed* if k is never less than 1 at any step (after the initial step) and is equal to 1 at the end.

7. Definition 1, Theorem 2, and Definition 3 do not allow enclosure, that is the insertion of extra parentheses. Suppose, however, that we include enclosure in S as a unary operation *having no written symbol* other than the enclosing parentheses themselves. As an operator, enclosure now occupies a node on the expression tree.

a. Modify Definition 1, Theorem 2, and Definition 3 to allow enclosure as an operator, and show that the infix part of Definition 6 now faithfully reproduces overparenthesized infix expressions.
b. How can you modify the definition of prefix and postfix traversing in order to correctly traverse trees containing enclosure? Discuss two methods:
(i) enclosure is simply treated as an identity operator, and ignored in traversals, and
(ii) enclosure now requires a written symbol in prefix and postfix traversals.
c. If enclosure is allowed in infix expressions, can we still say that the sets E, I, and O are isomorphic? (part **(b)** above provides the clue). If isomorphism turns out to be too strict a criterion, what less stringent one might be used?

Complements 12.1.4

1. Generalize the definition of expression and expression tree to the case of several sets $C_1, C_2 \ldots$, when we have operations within each set or from one or more of those sets to another. An example of the latter is the mapping $\mathbb{N} \times \mathbb{N} \rightarrow \{0, 1\}$ defined by the fact that, for each pair (x, y) $(x \in \mathbb{N}, y \in \mathbb{N})$ the statement $x > y$ is true or false. Now $C_1 = \mathbb{N}$, $C_2 = \{0, 1\}$, and we have defined a binary operation $\mathbb{N} \times \mathbb{N} \rightarrow \{0, 1\}$. In this case, examples of infix expressions are **(a)** $\sim((3 + 4) > 5)$, **(b)** $(x > y) \Rightarrow ((x + z) > (y + z))$, **(c)** $1 \Leftrightarrow ((x > 0) \Rightarrow ((x + y) > 0))$. Equivalent statements are, respectively: **(a)** $3 + 4 \leq 5$, **(b)** if $x < y$, then $x + z < y + z$, **(c)** if $x > 0$, then $x + y > 0$. In **(c)**, note that $T \Rightarrow p$ is equivalent to just p.
2. The following algorithm produces the full traversal for a rooted tree, such as the expression trees $T = (N, R)$ in W described in the text. The traversal is the same as defined in Definition 6. Recall that we have already provided the definition of one unique node as **root**(T), and for each node, $q \in N$, a one-symbol **label**(q), a natural number **arity**(q), a node **predecessor**(q) (except for the root), and a consistently ordered list of nodes **successor**(q, i), $i = 1, \ldots,$ **arity**(q) for those nodes with positive arity. Our algorithm also requires some "scratch pad" variables, which good programming practice requires we account for:

   - The node-valued variable **current-node** represents the node currently under consideration at any moment,
   - The natural number-valued variable **current-distance** represents the distance in arc traversals from **root**(T) to **current-node.** Clearly, then, **current-distance** (**root**(T)) $= 0$. Alternatively, we could simply define **current-distance**(q) for each $q \in N$ as an inherent feature of the node, like **label** or **arity,** and then ask for **current-distance**(**current-node**). We choose to count distances ourselves within the algorithm, on the grounds that it is probably cheaper in some sense, and is definitely more educational.
   - For each $q \in N$, we will need a **successor-index**(q), a natural number-valued variable whose permissible range is $0, 1, \ldots,$ **arity**(q), and whose purpose is to index the successors of the node.
   - The string valued variable **traversal** will, upon completion, contain the full traversal of T.

   The algorithm presented below is intentionally not written in any specific programming language. The style is sometimes called simply **pseudocode;** the intention is to present a rigorous statement of an algorithm without relying on any particular language's features.

   Algorithm
   PROCEDURE FULL_TRAVERSAL (T)
   current-node ← **root**(T)
   traversal ← **label**(**current-node**)
   current-distance ← 0
   successor-index(**current-node**) ← 0

```
While current-distance >= 0 repeat
  Add 1 to successor-index(current-node)
  If      successor-index(current-node) ≤ arity(current-node)
          current-node ← successor(current-node, successor-index(current-node))
          traversal ← traversal : label(current-node)
          Add 1 to current-distance
          successor-index (current-node) ← 0
  Else    successor-index(current-node) ← 0
          Subtract 1 from current-distance
          If current-node ≠ root(T)
            current-node ← predecessor(current-node)
          End if
            traversal ← traversal : label(current-node)
  End If
End repeat
Output traversal
Halt
```

a. Execute the algorithm manually on the tree in Figure 12.5 (which is not necessarily an expression tree). (Begin by identifying the root, and listing the label, arity, predecessor, and all successors for each node.) Verify that the traversal is correct by manually constructing the full traversal of T according to Definition 6, and comparing the two.

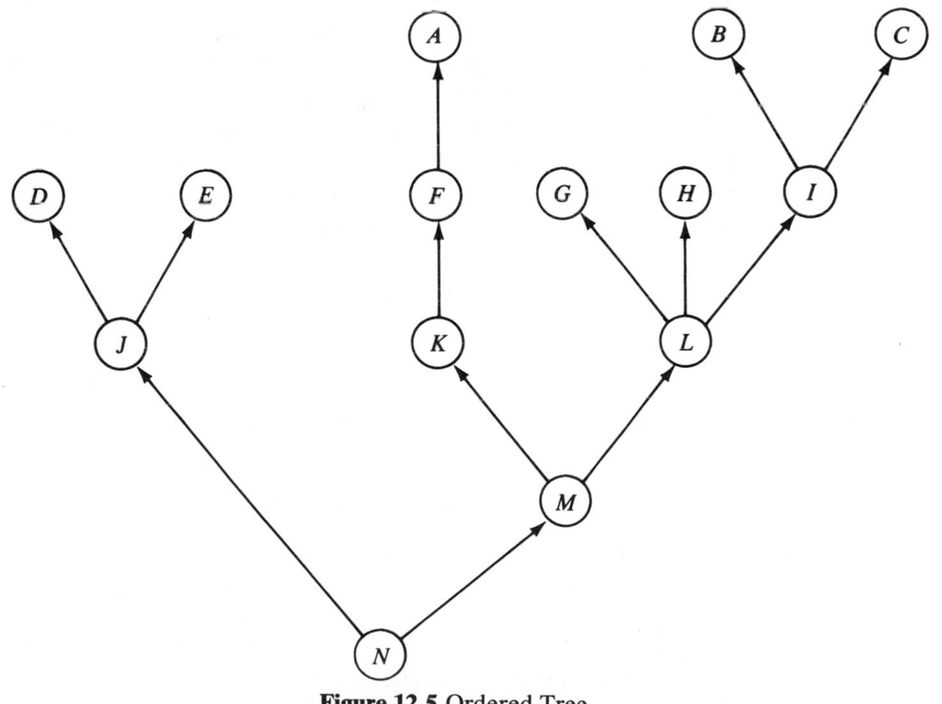

Figure 12.5 Ordered Tree

b. The algorithm as specified requires a distinct successor-index for each node of T. Explain why a single index shared by all nodes would be insufficient. Alter the procedure by replacing the six occurrences of **successor-index(current-node)** with **successor-index(current-distance)**. Execute the modified algorithm on the tree above, and verify that it still works. How many indices are required now? Explain the discrepancy. For a general tree T, how many indices would be required?

c. Modify the algorithm to produce prefix traversals. Test it on the tree in Exercise **(a)**.

d. Modify the algorithm to produce postfix traversals. Test it on the tree in Exercise **(a)**.

e. Modify the algorithm to produce infix traversals. Do not forget to provide full parenthesization of all operators. As in Definition 6, you will have to handle nodes of arity 2 as a separate case. Test it on the tree in Exercise **(a)**.

f. Modify the algorithm to produce infix traversals, but this time do not provide any parentheses anywhere. That is, the traversal contains only node labels, just as prefix and postfix traversals. Test it on the tree in Exercise **(a)**.

g. Translate at least one of these alghorithms into a working computer program and run it on several examples.

12.1.5 Evaluating Expressions

Expression trees also figure prominently in the ways in which expressions are evaluated. When an expression is **evaluated,** each of the operators in the expression is actually executed upon its appropriate number of arguments. Whether we have arithmetic operators with numeric arguments, or set theoretic operators with sets as arguments, or Boolean operators on logic statements makes little difference. If an operator $\oplus$ in S is defined as requiring fifteen arguments, five of which are numbers, five of which are sets, and five being logic statements, an expression tree can still represent the structure unambiguously. Fortunately such pathological structures don't come up very often.

In evaluating an expression, though, we must now distinguish between variables and constants. A variable can serve as an argument while an expression is being parsed, but operators are defined only on constants (addition is defined on numbers, such as 3, not on letters, such as x). So before an expression can be evaluated, all variables must be replaced by constants. The notations for doing this you have seen many times before. For instance, if we define the expression

$$f(x, y, z) := ((x \times y) + (-z))$$

then the evaluation of f for three numbers looks like

$$f(7, 3, 2) = ((7 \times 3) + (-2)) = (21 + (-2)) = 19$$

and it is understood that the constant 7 has been substituted for variable x, 3 for y, and 2 for z. Programming language compilers have many interesting techniques for making these substitutions, many of which use expression tree methods, but they are beyond the scope of this book.

In the present discussion, we have to assume that this has already been done. The effect on our notation is that now the set Y has been entirely replaced by elements of the set of constants, C. Our alphabet is now $A := C \cup S \cup P$ rather than $A := Y \cup S \cup P$ or even $A := Y \cup C \cup S \cup P$.

One approach to evaluating an expression tree uses essentially a postfix traversal of the tree. For each subtree, the operator is listed last in a postfix traversal; in an evaluation, the operator is simply evaluated. We have to use a postfix traversal, because an operator's arguments are available as evaluated constants only after all of its successor subtrees have been traversed. For instance, the function above has postfix traversal $73\times2-+$. The addition cannot be done until both the multiplication and the negation have been completed. Once an operator has been evaluated, its entire expression subtree can be discarded and replaced by the constant. This continues until the entire tree is reduced to one constant. Study the succession of trees in Figure 12.6 to see how this is done for $73\times2-+$. But this postfix-type evaluation is only one possibility for evaluating an expression. Consider the infix expression $(((1 \times 2) + (3 \times 4)) + (5 \times 6))$, with its tree, shown in Figure 12.7.

A postfix evaluation would require us to evaluate in order (in postfix now) $12\times$, $34\times$, $(2)(12)+$ (there's our multisymbol problem again!), $56\times$, and finally $(14)(30)+$ before obtaining the correct result: 44.

We could just as well do $12\times$, $34\times$, and $56\times$ before doing any additions whatever. These multiplications are independent, and there is no requirement whatever that they be done in any particular order. How do we know this? Look at the tree again, and isolate the subtrees corresponding to $12\times$, $34\times$, and $56\times$. They are disjoint! They do not come together until the additions are to be done. The three multiplications are each independent calculations in their own right.

Does this really make a difference? Why would anyone care in what order independent operations are performed? Even for computer evaluation, they all have to be done eventually, right? Right, *but not by the same computer*. If we had

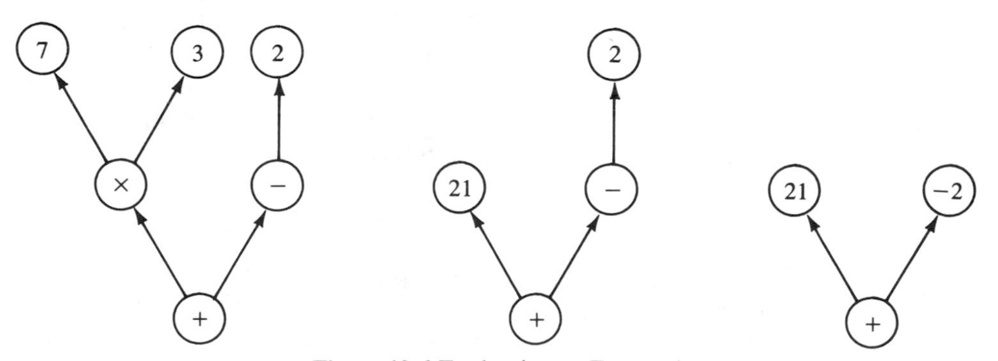

Figure 12.6 Evaluating an Expression Tree

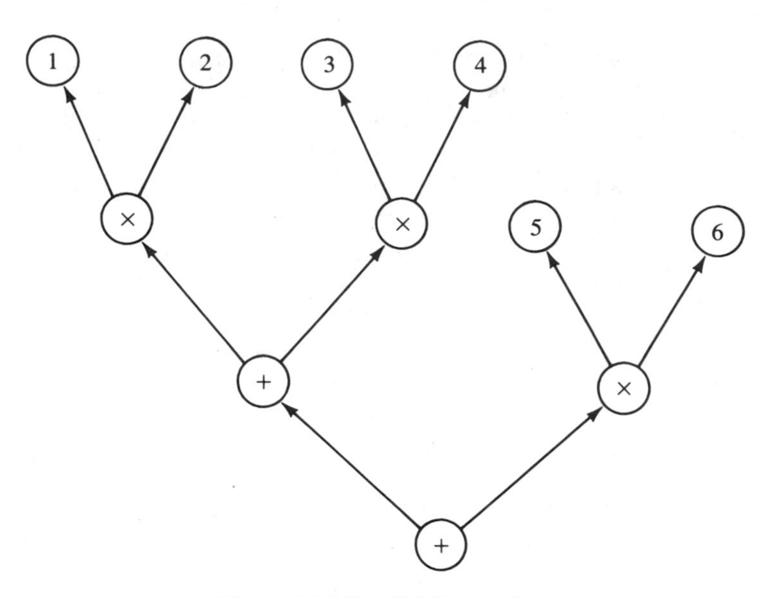

Figure 12.7 Parallel Processing

three separate computers available that could operate independently and simultaneously, one computer could be assigned to each multiplication. If each machine could perform one multiplication in one time unit (say, some small fraction of a second), the three together could perform all three multiplications in just that one time unit.

This idea is called **parallel processing**. In recent years, extremely powerful computer processors have been built, not just individually, but in groups and clusters. Sometimes the processors operate independently, sometimes they operate in lockstep parallel, sometimes some of them are busy while others are idle. To see some of the possibilities, consider the expression $12 \times 34 \times + 56 \times 78 \times - /$. (We leave it to you to draw the tree. By now, you should find that pretty simple.) Assume we have four identical processors. In the first time unit, all four are busy doing multiplication. In the second time unit, one processor is doing addition, one is doing subtraction, the other two are idle (or assigned to other duties). In the third time step, one is doing the final division, the other three are idle.

As these expressions become more complex, it is no easy task to lay out the order of calculation to make the most efficient use of multiple processors. For instance, suppose we have two lists of twenty-five numbers each, $x_1, x_2, \ldots, x_{25}$, and $y_1, y_2, \ldots, y_{25}$, and we want to calculate

$$\sum_{i=1}^{25} (x_i \times y_i) = x_1 \times y_1 + x_2 \times y_2 + \cdots + x_{25} \times y_{25}.$$

If we have seven identical processors available, and each can do either one multiplication or one addition in one time unit, what is the minimum number of time units for the whole problem? We claim that it can be done in as few as nine time units. What do you say?

Truly gigantic problems, involving millions and billions of calculations, may take months or years just to set up efficiently for calculation on parallel computers. The calculations themselves may still require hours to complete, but at least this way they are possible. Of course, these calculations don't all come in one colossal expression. In fact, the specification techniques may differ according to the source of the problem, say engineering versus weather prediction. What is certain is that expression trees form the basis for all such designs. No expression format, neither prefix, postfix, infix, nor any other, can correctly strike the balance between order that is essential and order that is arbitrary.

There is a second reason, too, that we might want to have some explicit control over the order of operation evaluation. It, too, concerns computers. Recall that computers store numbers by digit (hence, "digital" computer), and that any computer is limited in the number of digits it can store. Often numbers are stored in a *floating point* scheme, in which the location of the decimal point does not limit the size of the number as stored. For example, suppose we have a floating point computer that can store two digits per number (this is ridiculously small; most machines can store eight or more. But using two will make our example simpler). Then it doesn't matter whether we store 45, 45000000, or 0.0000045. The location of the decimal point is stored separately, so that the zeroes are implied rather than stored explicitly. Of course, storing 0.00405 is another matter. The zero between 4 and 5 cannot be inferred from the decimal point, so 0.00405 cannot be stored on our machine. But it can be *rounded* to 0.0041, which can be stored. Conversely, 0.00404 would be rounded to 0.0040. This error, called *round off error,* is what we want to consider.

We will further assume that addition on our computer produces a *three*-digit sum, rather than two. The third digit is called a *guard digit* and is present only during the addition process. As soon as the addition is done, the guard digit is used to round the three digit sum to two digits, so that it can be stored. Carefully study the two additions shown below. We use the arrow ($\rightarrow$) to mean "is rounded to".

$$\begin{array}{r} 0.12 \\ +0.045 \\ \hline 0.165 \end{array} \rightarrow 0.17 \qquad\qquad \begin{array}{r} 0.12 \\ +0.0045 \\ \hline 0.124 \end{array} \rightarrow 0.12$$

Notice that there is only one guard digit. In the second example, we do not get $0.1245 \rightarrow 0.13$, nor would we want that.

Now consider the addition $0.0031 + 0.014 + 0.11 = 0.1271$. What we want to have is that $0.1271 \rightarrow 0.13$, the closest we can get to 0.1271 on our crude machine. But it now makes a difference in what order we do the additions. Study these examples carefully.

$$\begin{array}{r} 0.0031 \\ +0.014 \\ \hline 0.0171 \end{array} \rightarrow \begin{array}{r} 0.017 \\ +0.11 \\ \hline 0.127 \end{array} \rightarrow 0.13 \neq 0.12 \leftarrow \begin{array}{r} 0.12 \\ +0.0031 \\ \hline 0.123 \end{array} \leftarrow \begin{array}{r} 0.014 \\ +0.11 \\ \hline 0.124 \end{array}$$

Discrepancies like this, accumulated over millions of calculation, may become so large that the final result is seriously in error. If the round-off error becomes larger than the result itself, the result is meaningless.

Some rules can be developed to minimize round-off error. One such rule might be stated as follows: Given a list of numbers to add up, first sort the numbers into ascending order by size (actually absolute value). Then add the numbers in order, smallest to largest. This rule would ensure that we do the more accurate calculations, as on the left above. Other rules deal with other possibilities, and as with parallel processing, some are specific to application.

To specify a rule is one thing, however. We must still communicate that rule to a computer program (or even have one computer program communicate it to another), to ensure that the calculations are done in the right order. The basic tool for this is still unambiguous expressions and expression trees. The next time you see something like

$$((0.0031 + 0.014) + 0.11)$$

don't automatically assume that it's the same thing as

$$(0.0031 + (0.014 + 0.11))$$

Theoretically, yes, but in practice, and in a computer, the two may be very different indeed. In other words, round-off addition is not associative.

To summarize, then, we have seen that ambiguity has many faces. In parsing expressions, it may cause outright confusion or error when the very meaning of the expression is called into question. In evaluating expressions, ambiguity may cause us to make decisions that we innocently believe to be arbitrary, and in the process rob ourselves of the accuracy and power of modern computers. In all cases, the cure is to provide expressions that are both efficient in form and absolutely clear in intent. The one structure that does this better than any other is expression trees.

Exercises 12.1.5

1. In each of the following cases, evaluate the given expression for the given values of the arguments.

$C := Z$, $S := \{+, -\}$, $Y := (X_1, X_2)$, $E := (X_1 - (X_1 + X_2))$. Evaluate **(a)** $E(5, 7)$, **(b)** $E(3, 5)$, **(c)** $E(1, 0)$.

$C := \{0, 1\}$, $S := \{\vee, \wedge, \sim\}$ (Boolean operations), $Y := \{x, y, z\}$, $E := ((x \vee y) \wedge \sim z)$. Evaluate **(d)** $E(0, 0, 0)$, **(e)** $E(0, 1, 1)$, **(f)** $E(1, 1, 1)$.

$C := \{00, 01, 10, 11\}$, $S := \{\vee, \wedge, \sim\}$ (bitwise operations), $Y := \{x, y, z)$, $E := ((x \vee y) \wedge \sim z)$. Evaluate **(g)** $E(00, 00, 00)$, **(h)** $E(00, 01, 10)$, **(i)** $E(11, 10, 00)$.

12.2 Trees

Trees are special cases of graphs. They are used in many applications, especially in computer science and management. Some examples are

- expression, or parsing, trees, which represent an expression or formula in any given algebraic structure. See Section 12.1.2.
- decision trees, in which each node represents a set of given circumstances, each arc originating at that node represents possible courses of action that can be taken under those circumstances.
- ordering, or sorting, trees, which provide algorithms to enumerate, in a given order, the elements of a given, finite, totally ordered set. Rewriting a given list of names in alphabetical order is an example, which we'll explain in a moment (see Figure 12.10).

Intuitively, a *tree* is defined as a structure of the type represented by either diagram in Figure 12.8.

For some strange reason, computer scientists call a tree like that in Figure 12.8a an *inverted* tree and consider Figure 12.8b a "proper" tree. The node marked with an asterisk is the *root*. We'll consider only finite trees.

Definition 7 An **undirected tree** is a connected graph in which there exist no arc-simple cycles (that is, no cycles with no repeated arcs.) A finite **directed tree,** or **rooted tree,** is a finite directed graph such that: (1) there are no loops; (2) one node, called the **root,** has indegree 0; (3) all other nodes have indegree 1.

A collection of trees, none of which has any nodes or arcs in common with another, is called, appropriately enough, a **forest.**

Rooted trees have a special terminology. The nodes with outdegree 0 are called **leaves.** The tree is said to be ***k*-ary** if each node has outdegree not exceeding k ($k \in N$). See Figure 12.9 for some examples of **binary** trees. The tree is said to be **ordered** if, at each node other than a leaf, the set of the arcs with origin at that node has a total order. Figure 12.9c represents an ordered binary tree. In a pictorial representation, the arcs are usually ordered left to

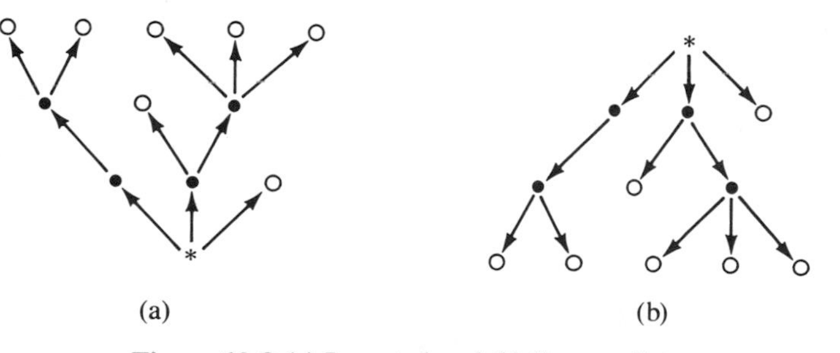

Figure 12.8 (a) Inverted and (b) "proper" trees.

right. The **height** of a node is the maximal length of the paths with origin at that node and end at a leaf. The **height** of the tree is the height of its root. The **depth** of a node is the length of the path with origin at the root and end at that node. (There is precisely one such path; see Exercises 12.2). For any two distinct nodes j, l, we call node j a **predecessor** of l if there is a path with origin j and end l; that is, if l is accessible from j. If there is an arc with origin j and end l, then j is an **immediate predecessor** of l. **Successor** and **immediate successor** are defined similarly. A finite k-ary tree is said to be **balanced** if every node of height h greater than 1 (if any) has outdegree k and all paths from that node to any successor leaf have length h or $h - 1$. See examples in Figure 12.9.

Be careful about some unfortunate terminology. A *directed* tree is a special kind of directed graph. Basically, you go *from* any node *to* its immediate successors. An *ordered* tree is a directed tree in which the arcs originating from a node are given a total order (first, second, etc.). An *ordering* tree is one that serves the function of ordering a list; the description has nothing to do with the nature of the tree itself. Note also that Definitions 4 and 6 are valid for any ordered tree. There is a close relationship between undirected and directed trees, of course. This is what we want to explore next.

Let $T := (V, A, b)$ be a given undirected tree (see the definition of a graph, Definition 11.1). Let x and y be two distinct nodes (if any). Let p and q be arc-simple paths from x to y. We intend to show that $p = q$. Suppose that this is not the case. Then the first few arcs of path p may be the same as the first arcs of path q (possibly none). Let z be the first node (possibly x itself) where this coincidence stops. Then the arc a of p and the arc b of q that have origin z are different from each other. Let w be the first node of p (after z) that is also on q. Then the part of path p that goes from z to w and the part of path q that goes "back" from w to z would form, together, a nontrivial cycle of T. Since such a cycle does not exist, we have proved this theorem:

Theorem 8 For any two nodes x and y of an undirected tree there exists precisely one arc-simple path that has origin x and end y.

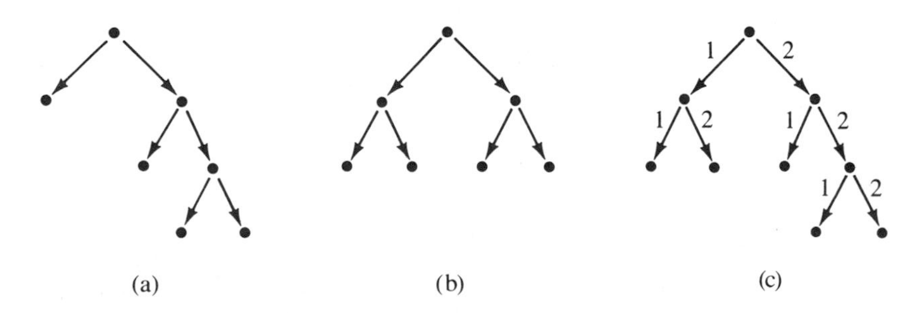

Figure 12.9 Examples of binary trees that are (a) not balanced, (b) balanced, and (c) balanced and ordered.

Suppose now that we assign one node r of T. Let x be any node other than r. Then there is precisely one arc-simple path from r to x, and this path has a last arc a, ending at x. We can give arc a the orientation that makes x the end of a. This assignment changes T to a directed graph. The next step is to verify that the directed graph so obtained is actually a directed tree with root r. This is a simple exercise. The essential steps are to verify that all three properties of a directed tree are satisfied. There is practically nothing to prove for property 1. As for properties 2 and 3, if the property were not true, then what would happen? See exercise 3 in Exercises 12.2.

The idea of a tree as a directed graph thus yields *more* freedom, not less. Since every undirected tree determines uniquely (up to the choice of r) the corresponding directed tree, we need only one theory of trees, namely that for rooted trees. Then undirected trees could be considered special cases. Returning to the original definition of a directed tree (Definition 7), we can easily verify that an expression tree (as in Definition 3) is actually an ordered directed tree. First, we observe that the initial step introduces only an isolated root. Second, the recursion step introduces a new root of indegree 0 and assigns an indegree of 1 to previously existing roots. Therefore the indegree of all nodes other than the root is 1. Finally, the recursion never introduces loops.

Exercises 12.2

1. In each of the following undirected trees select each node (one at a time) as the root. Then change the tree to a directed tree (by placing arrows on the arcs).

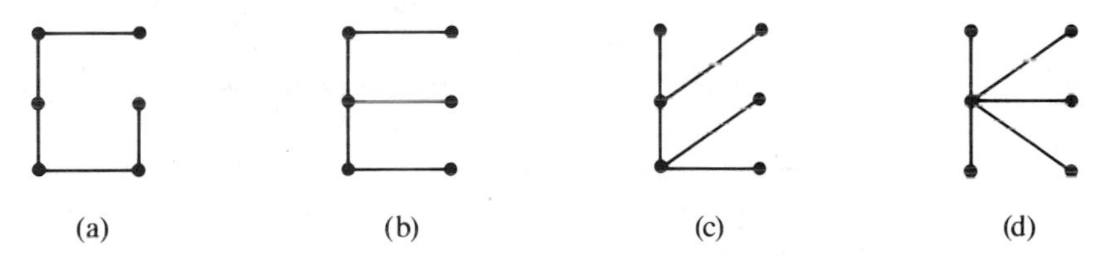

Find the depth and height of several nodes of several of the directed trees you obtained. Locate the leaves of the tree.

In each of the undirected trees given in a–d, select any two distinct nodes and find a simple path from one to the other. Repeat the process for several pairs of nodes.

2. Among the undirected graphs we saw in this chapter and in Chapter 11, consider those that are connected. In each of them, remove as few arcs as possible so that the remaining graph is still connected and has no arc-simple cycles. Then assign any one node as r and change the graph to a directed tree with root r. Find all leaves of this tree.

3. Prove the statement after Theorem 8, the alternative definition of directed tree, by showing that all three properties of Definition 7 are valid.

4. Prove that, in every tree, $\#V = 1 + \#A$.

Complements 12.2

1. Let V be a finite, nonempty set. Let r be a given element of V. Let $s:V \to V$ be a correspondence that is not many-to-one and whose image is $V - \{r\}$. Let G be the graph $(V, s, \{(t, t) \in s \times s \mid t \in s\})$. Prove that G is a finite tree with root r. To clarify the horrible set-theoretical jargon: s assigns to each node of G its immediate successors (if any); s^{-1} assigns to each node other than r its immediate precedessor; the arcs of G are the pairs (node, one of its immediate successors) if an immediate successor exists. This alternative definition of tree is practical in handling trees with a computer program. Prove also the converse of the statement above; that is, given a tree, construct correspondence s and verify its properties.

12.3 Sorting Trees

A *sorting tree* is a tree used to sort lists. For example, we might have a list of integers that we would like to sort into ascending (or descending) order, by means of the usual relation $\leq$. Or, we might have a list of strings (say, names) that we want to sort by alphabetical order, which is more correctly called *lexicographical order* (or simply *lex order;* see Section 10.6.4). (The operator $\preccurlyeq$ is called *lexicographically less than or equal to*. Other lex operators, $\prec$, $\succ$, $\succcurlyeq$, and $=$ are defined similarly.) The lists, as given, are assumed to be in some order other than what we want, possibly in random order.

But they just might already be in the order we want. In that case, whatever algorithm we derive should not change that order, of course. But intuition also tells us that in such a case, any algorithm ought to finish very quickly, since there was nothing for it to do.

The general problem of sorting lists is an extremely difficult one to deal with formally. Many algorithms exist, ranging from the extremely naive to the extremely sophisticated. Moreover, much of the complexity of sorting lies, not in the algorithms themselves, but in the measurement of their *overall* (or *global*) *efficiency*. The basic measure is always the same, though. For a list of n items, how many comparisons need to be performed to ensure (or verify) the desired order?

A second criterion for a good sorting algorithm is called *incremental efficiency*. Suppose we have a fully sorted list and wish to add one more item (that is, an *increment*). We would prefer not to have to sort the full list again from scratch every time. To what extent can we take advantage of the sorting that has already been done? After all, intuition tells us, all but the new item are already in order. The measure is still the number of one-to-one comparisons required for the new item.

Unfortunately, overall efficiency and incremental efficiency do not always go hand in hand. Some algorithms have excellent overall efficiency but poor incremental efficiency, others just the reverse. Sorting trees are rather in the last category. They're not the best for overall sorting, but their incremental properties are very attractive. This is what we want to explore. The tree itself is a rooted ordered binary tree (every node has at most two successors, designated *left* and *right*). As always, we associate with each node of the tree a *label*, which for our purposes will be one of the items in the list we wish to sort.

The algorithm is best described simply by presentation and example. We have broken the algorithm into two parts. INSERT_NODE is the main algorithm, which controls the entire sorting process. LOCATE_NODE is a subalgorithm, which is totally passive. Its function is simply to locate a particular node label in an existing tree. This accomplishes two things: First, if the label we wish to add already exists, we have a problem, and our algorithm must say so. Second, if the new item does not already exist, LOCATE_NODE will tell us where to add it.

Our data collection consists of:

- The sorting tree itself, *T*, which is created by INSERT_NODE for the first list item.
- The **desired-label** is a string-valued variable of the same type as the items in the list, be they names, numbers, or whatever. In fact, we will suppose that the list items are supplied to INSERT_NODE sequentially, through the variable **desired-label**.
- The node-valued variable **current-node** represents the node of *T* currently under consideration.
- The variable **current-successor** can take on only three values, "left," "right," or "null." Its purpose is to indicate which successor to **current-node** we intend to use. We will still use the notation **successor (current-node, current-successor)**, but we must be careful never to ask for the null-th successor of a node, since that is undefined.

As with our other algorithms, we have tried to avoid imitating any particular existing programming language. So our terminology (such as creating nodes or asking whether successors exist) should be viewed as notational rather than semantically rigorous. The algorithm itself, however, is quite rigorous.

Algorithm
  INSERT_NODE (*T*, **desired-label**)
  If tree *T* already exists then
    Invoke LOCATE_NODE (*T*, **desired-label**, **current-node**, **current-successor**)
    If **current-successor** = 'null' then
      Output "Oops! *T* already has a node with label '**desired-label**' "
    Else
      Create new node in *T* as **successor** (**current-node**, **current-successor**) with label '**desired-label**'
    End if
  Else
    Create tree *T* as single node with label '**desired-label**'
  End If
  Halt
  LOCATE_NODE (*T*, **desired-label**, **current-node**, **current-successor**)

current-node ← **root** (*T*)
current-successor ← 'null'
While **current-successor** = 'null' repeat

$$\textbf{current-successor} \leftarrow \begin{cases} \text{'left'} & \text{if } \textbf{desired-label} < \textbf{label (current-node)} \\ \text{'null'} & \text{if } \textbf{desired-label} = \textbf{label (current-node)} \\ \text{'right'} & \text{if } \textbf{desired-label} > \textbf{label (current-node)} \end{cases}$$

  If **current-successor** ≠ 'null' and **successor (current-node, current-successor)** exists then
    current-node ← **successor (current-node, current-sucessor)**
    current-successor ← 'null'
  End If
End repeat
Return to INVOKE command in INSERT_NODE.

Let's exercise the algorithm with a list of some mathematicians' names. Our list, in original order, looks like:

Euler
Russell
Pascal
Euclid
Stone
Kleene
Abel

and is hardly in alphabetical order. In the list below, we show each name as a new **desired-label**. Since our concern is with efficiency, we would also like to know how many item-to-item comparisons are required to add each new node to *T*. So we include an enumeration of the existing node labels each {desired-label} must be compared to. Finally, we list the action taken for the new node.

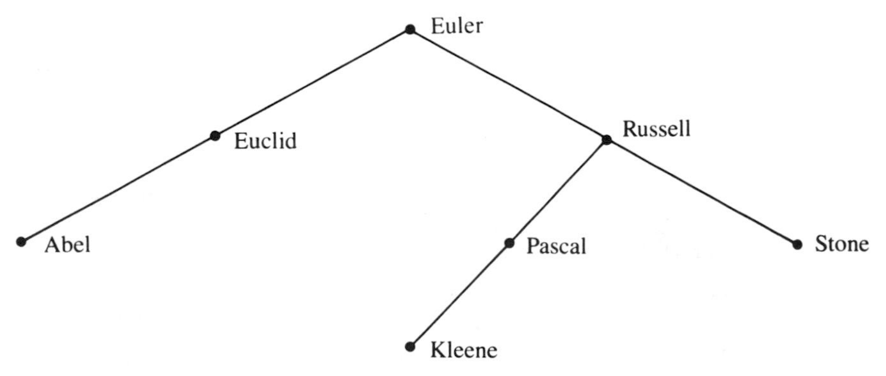

Figure 12.10 A sorting tree.

New {desired-label}	Node label compared	Action taken
Euler	—	Create T with root label 'Euler'
Russell	Euler	Add to T as right successor of 'Euler'
Pascal	Euler	Right to . . .
	Russell	Add to T as left successor of 'Russell'
Euclid	Euler	Add to T as left successor of 'Euler'
Stone	Euler	Right to . . .
	Russell	Add to T as right successor of 'Russell'
Kleene	Euler	Right to . . .
	Russell	Left to . . .
	Pascal	Add to T as left successor of 'Pascal'
Abel	Euler	Left to . . .
	Euclid	Add to T as left successor of 'Euclid'

The resulting tree is shown in Figure 12.10. If we now perform an infix order traversal of T, we obtain the correctly ordered list:

Abel
Euclid
Euler
Kleene
Pascal
Russell
Stone

For that matter, an infix order traversal of T at any point in its construction would yield a correctly ordered list of all labels that had been added up to that point.

For any given list of items, the sorting tree constructed from them is unique. The converse, however, is not true. The sorting tree we have above would also have resulted from an original list like

Euler
Russell
Pascal
Stone
Euclid
Abel
Kleene

or many others. So even if we saw the completed tree, we could not generally tell for certain what the original list looked like.

Exercises 12.3

1. Show more examples of sorting with tree. Pick a dozen or so names from the telephone book or from among classmates. Sort the names alphabetically. Then sort the same list again, this time by telephone number or social security number.
2. Lexicographical order is an extension of the usual alphabetical order, but includes more. For instance, in lex order, the name SMITH < SMITHSON (see Section 10.6.4). Also, A1 < AA in some lex order systems, because numerals < letters. On computers, we must provide a complete ordered list of all characters, including letters, numerals, the space, and all punctuation marks. Such a list is called the computer's *collating sequence*. Find the collating sequence for the computer your institution uses.
3. **Worst Case Sorting**. One important limitation of this algorithm is the peculiar results it yields on lists that are already correctly sorted. Execute the algorithm again with the sorted list above. You will notice how inefficient it becomes. Elsewhere in this book you will find references to "balanced binary trees." We are not going to discuss explicitly their use here, but you should look at them to get an idea of how to solve this problem. You might even want to try modifying the algorithm.
4. Implement the algorithm into a computer program and run it on several lists.
5. The following two lists consist of the same terms but contain them in different orders: **(a)** IN, OUT, ALL, ABOUT, UP, DOWN; **(b)** DOWN, OUT, ALL, ABOUT, UP, IN. Construct a sorting tree for each of them, according to the algorithm in the text. Observe that: **(1)** The resulting trees are different, although they both sort the list in alphabetical order; **(2)** each of the two trees is balanced at every stage of its construction.

12.4 Homological Properties of Graphs

One goal of this section is to explore the most basic properties of a graph. A few examples will illustrate what we mean by "most basic" properties. Consider the three graphs in Figure 12.11. At face value, they are all different. Under the influence of cultural factors, the second and the third graph may look more similar to each other than first and second (same number of nodes, same numbers of straight and curved arcs, same meaning as letter B, and so on). On the other hand, each one has "just two" cycles, which have a short path in common, and the third graph has two "loose ends", which the first two do not have.

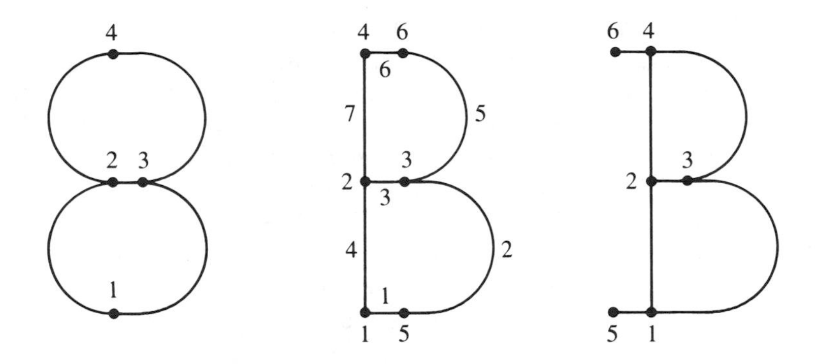

Figure 12.11 Three graphs with "two" cycles each.

Let's consider further the statement that all three graphs have just two cycles. Literally, this is not quite correct. In the first and third figure, the "two" cycles we had in mind were very likely "from 3 to 2 to 4 back to 3" and "from 1 to 2 to 3 back to 1". What about "from 1 to 2 to 4 to 3 back to 1"? That's clearly a "third" cycle. Can it be obtained by "putting together" the first two? Literally, no; informally, yes. There are three difficulties in doing that. First, we have to permute the first two cycles cyclically to make ends match. We so have 3123243 (listing nodes only). Second, we have to "cancel" arcs 23 and 32 to obtain cycle 31243. Third, we have to permute cyclically again to obtain precisely the third cycle we mentioned.

We so have a few basic properties: First, a cycle is a cycle is a cycle, no matter which way we go around it and where we start. Second, in counting the "essential" number of cycles, we can disregard those that can be obtained by putting together other cycles. Third, when putting together cycles or paths, we can disregard arcs that appear twice. Last, the process of free chain insertion or deletion (see Complements 11.4) does not alter the phenomena we just saw. In the example, this process makes first and second graph "the same", although they have different numbers of arcs and nodes.

We intend to develop a formalism that bypasses incidental details and points out some basic properties. This formalism is part of the theory called *homology theory*. The main results of this theory are the *homological properties* of the graph. A secondary goal is to handle graphs abstractly, without necessarily "seeing" the picture, and to be able to do this for graphs that are given abstractly, for instance through the correspondence b of Section 11.1 or the adjacency matrix of Section 11.2.1. This abstraction is needed in applications when, as it happens in most cases, the nodes are not points, the arcs are not segments, the incidences are given in a table or stored in a computer, and there are many nodes and arcs, so many that geometric intuition would fail even if the graph were given geometrically.

12.4.1 Definitions and Notations

Let $G := (V, A, b)$ be a finite, undirected graph as in Section 11.1. Let $n := \#V$ (the number of nodes), $m := \#A$ (the number of arcs). For convenience, let nodes and arcs be numbered, that is, suppose $V = \{1, \ldots, n\}$ and $A = \{1, \ldots, m\}$. A **0-chain** of G is defined as a total function $f: V \to \mathbb{N}$. That means that each node, say x, is "taken" a certain number of times, say fx. A **1-chain** is defined as a total function $g: A \to \mathbb{N}$ (each arc is "taken" a certain number of times). In other words, chains are multisets on V or on A (see Section 7.2.4.4).

Example In the second graph of Figure 12.11, we may decide to take node 1 once, node 2 once, node 3 twice, and node 5 twice. Then our 0-chain would be the sequence (1, 1, 2, 0, 2, 0). It will be convenient to write such sequence as a column, that is:

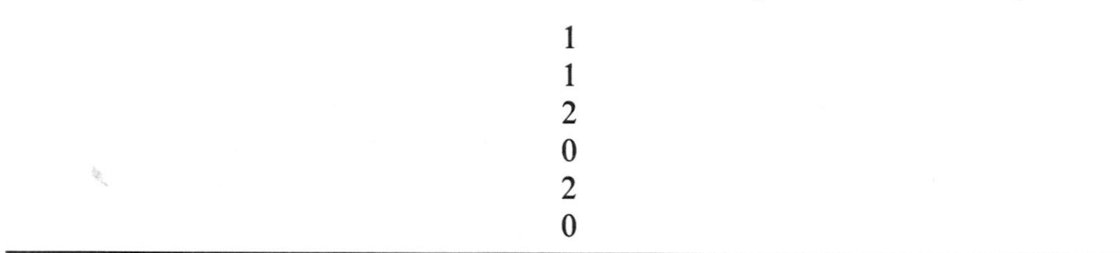

$$\begin{matrix}1\\1\\2\\0\\2\\0\end{matrix}$$

Each 0-chain will be considered as an $n \times 1$ matrix over $\mathbb{N}$, that is, as a column matrix. As such, 0-chains can be added together as matrices, a 0-chain can be multiplied by a natural number according to the definition

$$h \begin{bmatrix} f1 \\ f2 \\ \cdot \\ \cdot \\ \cdot \\ fn \end{bmatrix} = \begin{bmatrix} h\ f1 \\ h\ f2 \\ \cdot \\ \cdot \\ \cdot \\ h\ fn \end{bmatrix}.$$

Similar conventions are valid for 1-chains, of course.

We can consider chains mod 2, by reducing each fx mod 2, and so on.

The **boundary** (over $\mathbb{N}$) of an arc u is denoted δu and is defined as the 0-chain obtained by taking the endpoints of u once each if u is not a loop, or the only endpoint of u twice if u is a loop. The boundary of each arc is therefore a 0-chain in which all entries are 0 except: (1) precisely two of them, which are 1s, or (2) precisely one of them, which is a 2.

Since we have m arcs, the 0-chains representing their boundaries can be organized into an $n \times m$ matrix over $\mathbb{N}$, in which column j is the boundary of arc j. Such a matrix is called the **incidence matrix** of graph G. We'll denote it E. Clearly, matrix E carries precisely the same information as correspondence b of our previous definition of graph: it tells which nodes are *incident* to (that is, endpoints of) which arcs. Consequently, we can denote graph G as (V, A, E) instead of (V, A, b). Note that each entry of E is 0, or 1, or 2 and that in each column there is precisely one 2 (and no 1) or precisely two 1s (and no 2). This can also be expressed by saying that the sum of the entries of each column is 2.

Example In the second graph of Figure 12.11 number the arcs as follows: arc 15 := 1, arc 53 := 2, arc 32 := 3, arc 21 := 4, arc 36 := 5, arc 64 := 6, and arc 42 := 7. Then

$$E = \begin{bmatrix} 1 & 0 & 0 & 1 & 0 & 0 & 0 \\ 0 & 0 & 1 & 1 & 0 & 0 & 1 \\ 0 & 1 & 1 & 0 & 1 & 0 & 0 \\ 0 & 0 & 0 & 0 & 0 & 1 & 1 \\ 1 & 1 & 0 & 0 & 0 & 0 & 0 \\ 0 & 0 & 0 & 0 & 1 & 1 & 0 \end{bmatrix}.$$

Note that entry (i, j) of matrix E is other than zero if and only if node i is an endpoint of arc j. This justifies the symbol E for "END." Also, such an entry tells whether or not node i is an endpoint of (or incident to) arc j; this justifies the name *incidence* matrix.

There is a slight difference between a 1-chain and a path. Every path can be considered as a 1-chain, by taking each arc as many times as it appears in the path. But not every 1-chain is a path, because a path must be "organized" so that we have a sequence (not just a multiset) of arcs, and each arc of the sequence must end at the same node where the next arc (if any) begins.

Example In the second graph in Figure 12.11 the path (node 1, arc 1, arc 2, arc 3) can be considered as the 1-chain

$$\begin{matrix} 1 \\ 1 \\ 1 \\ 0 \\ 0 \\ 0 \\ 0 \\ 0 \end{matrix}$$

The boundary of this 1-chain could be constructed by putting together the boundaries of the individual arcs, and we'd obtain the 0-chain

$$\begin{matrix} 1 \\ 1 \\ 2 \\ 0 \\ 2 \\ 0 \end{matrix}$$

This wouldn't give us much information. If we reduce the 0-chain according to mod 2, then we obtain the 0-chain mod 2

$$\begin{matrix} 1 \\ 1 \\ 0 \\ 0 \\ 0 \\ 0 \end{matrix}$$

which comprises just origin and end of the original path. This is what we'd consider intuitively as the boundary of the path. For the special case in which the original path is a cycle, the boundary mod 2 would then be the *null* 0-chain, that is, the 0-chain mod 2 whose elements are all 0 mod 2. We're going to translate many of these remarks into definitions.

Let g be a 1-chain. Its **boundary δg** is defined as the 0-chain given by the matrix product Eg. The boundary mod 2 is defined as the matrix Eg mod 2. A 1-chain whose boundary mod 2 is the null 0-chain is said to be **closed**. The intuitive idea that a closed chain is something like a cycle is formalized by the following properties, whose proofs are simple verifications. The 1-chain of the arcs of a cycle of G is a closed 1-chain. Cycles obtained from each other by cyclic permutations or by inverting the order of their arcs give the same 1-chain. The arcs of any closed 1-chain mod 2 can be organized into sequences, so that (1) each of these sequences provides a simple cycle (with no repeated nodes and no repeated arcs) and (2) two distinct sequences have no arcs in common.

We'll call **circuit** a 1-chain whose arcs are the arcs of a simple cycle. Necessarily, a circuit is a closed 1-chain and every closed 1-chain mod 2 can be considered as the sum of circuits not having, pairwise, any arcs in common.

12.4.2 Block Form for the Incidence Matrix

The goal of this section and the next is to present matrix E in a form that points out some basic properties of graph G and facilitates calculations connected with such properties. We've seen in Section 11.5 the notion of *connection component* of a graph. We've also seen a process for identifying the connection components of a given undirected graph. Consequently, we can change the order of the rows of our incidence matrix E so that the rows corresponding to the nodes of one connection component are listed first, those of a second component next, and so on. Then we can perform the same operation on columns. Since no node of any one component is the endpoint of any arc of any other component, the new matrix will be in **block** form. By this we mean that the matrix is subdivided into blocks as in the following table:

$$\begin{bmatrix} E(1) & 0 & \cdots & 0 \\ 0 & E(2) & \cdots & 0 \\ \cdot & \cdot & \cdot & \cdot \\ \cdot & \cdot & \cdot & \cdot \\ \cdot & \cdot & \cdot & \cdot \\ 0 & 0 & \cdots & E(q) \end{bmatrix}$$

Namely, each of the $E(j)$ and each of the zeroes in this table denotes a submatrix. All the submatrices in the same row of the table have the same number of rows proper; all the submatrices in the same column of the table have the same number of columns proper. All blocks off the "main diagonal" are zero and those on the main diagonal are the incidence matrices of the connection components of the graph.

We may want to be able to perform both steps (locating the connection components and changing the incidence matrix to block form) under the assumption that all we have is the incidence matrix of the graph. We're going to achieve this goal via an algorithm, which contains two recursive constructions.

Algorithm E. Let graph G be given as in Section 12.4.1. Let G be nonempty. We'll illustrate our process on the following graph (which we've seen before, in Section 11.3):

The matrix of this graph is the following:

	[1]	[2]	[3]	[4]	[5]	[6]
[1]	0	0	0	0	0	1
[2]	0	0	0	0	0	1
[3]	0	0	0	2	2	0
[4]	0	0	0	0	0	0
[5]	1	0	1	0	0	0
[6]	1	1	0	0	0	0
[7]	0	1	1	0	0	0

1. *Initial step of the main recursion.* Let q be the number of rows that consist of only 0 entries ($q \geq 0$). Move those rows to the first q row positions of the matrix and let $L(q)$ be the matrix comprising the remaining rows of matrix E. In practice, labels of rows (and, later, columns) are moved together with the respective rows or columns, in order to avoid a repeated renumbering of nodes and arcs.

   In the example, we move row [4] to the first position and set $q := 1$. Matrix $L(1)$ will comprise the last six rows of the new matrix.

	[1]	[2]	[3]	[4]	[5]	[6]	
[4]	0	0	0	0	0	0	
[1]	0	0	0	0	0	1	$L(1)$
[2]	0	0	0	0	0	1	
[3]	0	0	0	2	2	0	
[5]	1	0	1	0	0	0	
[6]	1	1	0	0	0	0	
[7]	0	1	1	0	0	0	

2. *Recursion step of the main recursion.* If matrix $L(q)$ has no rows, then the recursion and the algorithm stop.
3. *Initial step of the secondary recursion.* Else, observe that matrix $L(q)$ must contain some nonzero entries. Within matrix $L(q)$, let (i, j) be a position that contains a nonzero element. Move row i to the first row position within matrix $L(q)$ and column j to the first column position. Then start a secondary recursion by letting set R comprise the first row and set C comprise the first column of the new matrix $L(q)$. (See the example that follows the algorithm.)
4. *Recursion step of the secondary recursion.* Operate within matrix $L(q)$. If any column in set C has a nonzero element outside the rows of set R, then move that row to the next row position and adjoin it to set R. If any row in set R has nonzero elements outside the columns of set C, then move that column to the next column position and adjoin it to set C. Repeat step 4, if possible. Else, the secondary recursion terminates; continue at step 5.
5. Increase q by 1, let $E(q)$ be the matrix comprising the elements that are in the rows of set R and in the columns of set C, and let the new $L(q)$ be the matrix that is made up of the elements in the rows beyond set R and in the columns beyond set C. Then resume the main recursion at step 2.

In our example, we can select entry ([3], [4]), using the original labels in square brackets, as the entry to start the secondary recursion at step 3. Then the next step implies moving such element to the new second row and first column, and then moving the only other entry equal to 2 to the next column, same row. The matrix is then

	[4]	[5]	[1]	[2]	[3]	[6]
[4]	0	0	0	0	0	0
[3]	2	2	0	0	0	0
[1]	0	0	0	0	0	1
[2]	0	0	0	0	0	1
[5]	0	0	1	0	1	0
[6]	0	0	1	1	0	0
[7]	0	0	0	1	1	0

(The last five rows and last four columns are marked $L(2)$.)

Matrix $E(2)$ is then $[2 \quad 2]$ and matrix $L(2)$ comprises the elements in the last five rows and four columns of the table above. At the next pass of the main recursion (steps 2 and 3), we select element ([1], [6]), move it to the new third row and third column and, necessarily, move the other 1 in column [6] to the next row, same column. Then the matrix becomes

	[4]	[5]	[6]	[1]	[2]	[3]
[4]	0	0	0	0	0	0
[3]	2	2	0	0	0	0
[1]	0	0	1	0	0	0
[2]	0	0	1	0	0	0
[5]	0	0	0	1	0	1
[6]	0	0	0	1	1	0
[7]	0	0	0	0	1	1

(The last three rows and last three columns are marked $L(3)$.)

Then matrix $E(3)$ is $\begin{bmatrix}1\\1\end{bmatrix}$ and matrix $L(3)$ comprises the last 3 rows and columns of the table above. The next primary recursion step leaves everything else unaltered, matrix $E(4)$ comprises the elements in the last three rows and columns of the table above, $q = 4$, and the main recursion stops. The blocks of the final matrix are: a "nominal" block with one row and "no columns" at the beginning, then the block $E(2)$, comprising one row and two columns, then block $E(3)$, with two rows and one column, and finally block $E(4)$, made up of the remaining three rows and three columns.

It's clear that block j is the incidence matrix of a connection component, say $G(j)$, of the original graph G. Consequently, the final value of q is the number of connection components of G. In particular, G is connected if and only if $q = 1$.

As a degenerate case, a row of E that contains only zeroes (one node, no arcs) provides a graph $G(j)$ that comprises only one isolated node. Another degenerate case occurs when $n > 0$, but $m = 0$. We have a graph with no arcs; matrix E has (paradoxically) several rows and *no columns*. The graph is just n isolated nodes. The really degenerate case in which $n = 0$ and $m = 0$ would occur when G is empty, but we have excluded this case from consideration.

12.4.3 Spanning Tree

The goal of the next algorithm is to put more order within each matrix $E(j)$. Specifically, we intend to start from a connected graph and see if we can delete some arcs and obtain a graph that is still connected but has no simple cycles. For instance, graph

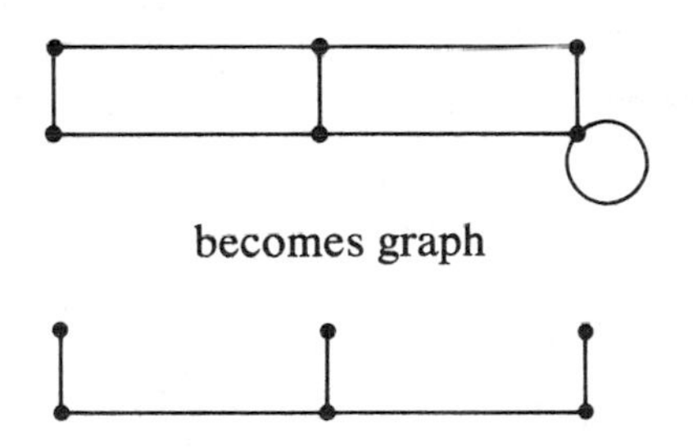

if we delete the top arcs and the loop.

As usual, to be sure that we can go through this process in general and, especially, when all we have is the incidence matrix and no picture, we need a formalism, which we shall present in the form of an algorithm. For the sake of efficiency, it would have been better to include this part of the algorithm in the main algorithm. For the sake of clarity, however, we keep it separate. Since this algorithm operates separately on each connection component of G, we may as well rename G any one of the connection components and ignore the others for a while.

We let $G := (V, A, E)$ be a finite connected graph ($A \neq \emptyset$) and adopt the notations used in the main algorithm ($q = 1$ now). Recall that each column of E contains two 1s or one 2, that each row of E contains some nonzero element, and that G is connected.

This time, we illustrate the process on the second graph of Figure 12.11, that is, the graph in Figure 12.12.

Then

$$E = \begin{array}{c|ccccccc} & [1] & [2] & [3] & [4] & [5] & [6] & [7] \\ [1] & 1 & 0 & 0 & 1 & 0 & 0 & 0 \\ [2] & 0 & 0 & 1 & 1 & 0 & 0 & 1 \\ [3] & 0 & 1 & 1 & 0 & 1 & 0 & 0 \\ [4] & 0 & 0 & 0 & 0 & 0 & 1 & 1 \\ [5] & 1 & 1 & 0 & 0 & 0 & 0 & 0 \\ [6] & 0 & 0 & 0 & 0 & 1 & 1 & 0 \end{array}$$

Algorithm

1. If no entry of E is 1, then E comprises just one row and all entries of E are 2. (*Proof:* Since no entry is 1, all arcs are loops. Since G is connected, all loops have the same node as their only endpoint, and there are no isolated nodes.) Then $n = 1$, $m > 0$, and graph G comprises one node and m loops with that node as the only endpoint. The process stops here.

2. *Initial step of a recursion.* There is an entry (i, j) of E that is a 1. Start a recursion process as follows. Locate both rows of E (say i and h) that contain a 1 in column j. Move rows i and h to the first two row positions and move column j to the first column position. Let S be the 2×1 matrix comprising the two 1s just moved, that is, the entries in the first two rows and first column of the new matrix E. Let T be the tree that comprises nodes (rows) i and h and arc (column) j. Define $v := 2 =$ number of nodes of T, $u := 1 =$ number of arcs of T.

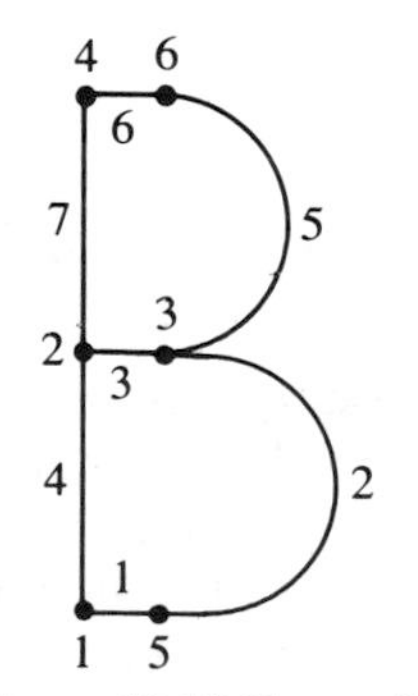

Figure 12.12 Example.

In the example, we select $j := [1]$ and, necessarily, $i := [1]$, $h := [5]$ (or the other way around). Matrix E becomes

$$E = \begin{array}{c|ccccccc} & [1] & [2] & [3] & [4] & [5] & [6] & [7] \\ [1] & 1 & 0 & 0 & 1 & 0 & 0 & 0 \\ [5] & 1 & 1 & 0 & 0 & 0 & 0 & 0 \\ [2] & 0 & 0 & 1 & 1 & 0 & 0 & 1 \\ [3] & 0 & 1 & 1 & 0 & 1 & 0 & 0 \\ [4] & 0 & 0 & 0 & 0 & 0 & 1 & 1 \\ [6] & 0 & 0 & 0 & 0 & 1 & 1 & 0 \end{array}$$

3. *Recursion step*. If $v = n$, then the recursion stops; continue at step 4. If $v < n$, then there is a node $[x]$ not in tree T. Since G is connected, there is a path from node $[x]$ to any one node of T. In this path, take the first arc, $[j]$, that has second endpoint $[h]$ in T. Let $[i]$ be its first endpoint, necessarily not in T. In E, this means that column $[j]$ (beyond the columns of matrix S) has a 1 entry $([i], [j])$ beyond the rows of matrix S and a 1 entry $([h], [j])$ within the rows of matrix S. Move column $[j]$ to make it the column in E next to the columns of S and move row $[i]$ to make it the row after the rows of S. Redefine S to include the moved row and column, increase v and u by one unit, so that S is again $v \times u$ and $v = u + 1$. Observe that T is still a tree, comprising as nodes the rows of S and as arcs the columns of S. (The reason is that node $[h]$ was already in T, while node $[i]$ and arc $[j]$ were not. The new T is therefore connected. No circuit of T can contain arc $[j]$, because $[j]$ has the loose end $[i]$, nor can it lie in the previous tree, because a tree has no nonnull circuits.) Observe also that the entries in matrix S in the new positions $(2, 1)$, $(3, 2)$, . . . are all ones, and the entries beneath these entries are all zeroes. Then continue at step 3.

Continuing with the same example, we select, for instance, column [4] and rows [1] and [2]. We obtain

$$E = \begin{array}{c|ccccccc} & [1] & [4] & [2] & [3] & [5] & [6] & [7] \\ [1] & 1 & 1 & 0 & 0 & 0 & 0 & 0 \\ [5] & 1 & 0 & 1 & 0 & 0 & 0 & 0 \\ [2] & 0 & 1 & 0 & 1 & 0 & 0 & 1 \\ [3] & 0 & 0 & 1 & 1 & 1 & 0 & 0 \\ [4] & 0 & 0 & 0 & 0 & 0 & 1 & 1 \\ [6] & 0 & 0 & 0 & 0 & 1 & 1 & 0 \end{array}$$

At the next passes of the induction steps we select column [3] and rows [2] and [3], then column [7] and rows [2] and [4], and finally column [6] and rows [4] and [6]. We obtain, successively,

$$E = \begin{array}{cccccccc} & [1] & [4] & [3] & [2] & [5] & [6] & [7] \\ [1] & 1 & 1 & 0 & 0 & 0 & 0 & 0 \\ [5] & 1 & 0 & 0 & 1 & 0 & 0 & 0 \\ [2] & 0 & 1 & 1 & 0 & 0 & 0 & 1 \\ [3] & 0 & 0 & 1 & 1 & 1 & 0 & 0 \\ [4] & 0 & 0 & 0 & 0 & 0 & 1 & 1 \\ [6] & 0 & 0 & 0 & 0 & 1 & 1 & 0 \end{array}$$

$$E = \begin{array}{cccccccc} & [1] & [4] & [3] & [7] & [2] & [5] & [6] \\ [1] & 1 & 1 & 0 & 0 & 0 & 0 & 0 \\ [5] & 1 & 0 & 0 & 0 & 1 & 0 & 0 \\ [2] & 0 & 1 & 1 & 1 & 0 & 0 & 0 \\ [3] & 0 & 0 & 1 & 0 & 1 & 1 & 0 \\ [4] & 0 & 0 & 0 & 1 & 0 & 0 & 1 \\ [6] & 0 & 0 & 0 & 0 & 0 & 1 & 1 \end{array}$$

$$E = \begin{array}{cccccccc} & [1] & [4] & [3] & [7] & [6] & [2] & [5] \\ [1] & 1 & 1 & 0 & 0 & 0 & 0 & 0 \\ [5] & 1 & 0 & 0 & 0 & 0 & 1 & 0 \\ [2] & 0 & 1 & 1 & 1 & 0 & 0 & 0 \\ [3] & 0 & 0 & 1 & 0 & 0 & 1 & 1 \\ [4] & 0 & 0 & 0 & 1 & 1 & 0 & 0 \\ [6] & 0 & 0 & 0 & 0 & 1 & 0 & 1 \end{array}$$

At this point, the recursion stops. Columns [1], [4], [3], [7], and [6] represent the left vertical stroke of our letter B and the three horizontal strokes. This set of arcs takes all the nodes of the initial graph and represents the smallest number of connected arcs that do that. In general, the situation is described in the next step.

4. The recursion stops. At this moment, the first v rows and u columns of matrix E ($v = n$, $v = u + 1$) have the form

$$\begin{bmatrix} 1 & * & * & \cdots & * \\ 1 & * & * & \cdots & * \\ 0 & 1 & * & \cdots & * \\ 0 & 0 & 1 & \cdots & * \\ \cdot & \cdot & \cdot & \cdot & \cdot \\ \cdot & \cdot & \cdot & \cdot & \cdot \\ \cdot & \cdot & \cdot & \cdot & \cdot \\ 0 & 0 & 0 & \cdots & 1 \end{bmatrix}$$

In each column, one of the symbols $*$ is 1 and the others are 0. Tree T comprises all nodes of G, but not necessarily all arcs. Such a tree is called a **spanning tree** or a **skeleton** of graph G. (If we perform this construction on each connection component of a graph, then we obtain a *spanning forest* of the graph.)

Let c be the number of those columns of E that are not included in matrix S, that is, the number of arcs not in tree T ($c \geq 0$). Clearly, $c = m - u = m - (v - 1)$

$= m-n+1$. Let $[k]$ be a column of E not in S (if any). Column $[k]$ contains one entry equal to 2 or two entries equal to 1. In each case, arc $[k]$ has its endpoints in tree T but is not in T. In the first case, the arc is a loop with its only endpoint in T. In the second case, the arc has two distinct endpoints in T, say $[i]$ and $[j]$. By Theorem 8 there is in T precisely one simple path from $[i]$ to $[j]$. Consequently, arc $[k]$ belongs to precisely one circuit of G for which all the arcs other than $[k]$ are in T. The same statement is degenerately true also in the case in which $[k]$ is a loop: the circuit comprises arc $[k]$ only. A consequence of this remark is

Theorem 9 Let G be a finite undirected connected graph with n nodes and m arcs. Define $c := m - n + 1$ as the **index of connection** of graph G. Then $c \geq 0$ and there exist c arcs of G, say a_h $(1 \leq h \leq c)$, such that each a_h belongs to precisely one circuit (mod 2) of G (say z_h) for which the other arcs (if any) are other than $a_1, \ldots, a_c$.

Furthermore, each closed 1-chain mod 2 of G is, in a unique way, the sum mod 2 of distinct circuits among $z_1, \ldots, z_c$. Consequently, graph G has 2^c closed 1-chains mod 2.

Proof: We've already proved the first part. In that proof, we denoted generically $[k]$ any one of arcs a_h.

To prove the second part, let z be any closed 1-chain mod 2 of G. Then z contains just once some of the arcs z_h (and some not at all). If we add to z mod 2 such arcs, then they will not appear any more in the resulting sum. The resulting sum is then in T. Since the resulting sum is also a closed chain mod 2 (as a sum of closed chains), it must be null (because it's in T). Consequently, the original z was a sum mod 2 of some of the z_h. To prove uniqueness of this sum, observe that, if we had two such sums, then *their* sum mod 2 would be a closed chain mod 2 in T, and therefore null, and the two sums would coincide.

The final statement is then purely combinatorial: we have a closed 1-chain mod 2 for each combination of the circuits z_h. Q.E.D.

A set of circuits z_h that satisfy the conditions of the second part of Theorem 9 is called a set of **fundamental circuits** of G.

Example In the example of Figure 12.12, a set of fundamental circuits is the following (we list only their arcs): arc [2] gives circuit [1] [2] [3] [4], arc [5] gives the circuit [3][5][6][7]. Using the original numbering of the arcs, such two circuits would be the two 1-chains

$$\begin{bmatrix}1\\1\\1\\1\\0\\0\\0\end{bmatrix} \quad \text{and} \quad \begin{bmatrix}0\\0\\1\\0\\1\\1\\1\end{bmatrix}$$

It is clear that the operation of free chain insertion or free chain deletion does not alter any of the following:

- the number of connection components of a graph;
- the value of the index of connection c for each connection component, and therefore the number of fundamental cycles;
- the spanning trees;
- the circuits.

Such properties are therefore the same for any one graph and all the graphs homological to it. These properties are therefore **homological properties** of a graph.

Exercises 12.4

1. Find the connection components of each of the following graphs, and, for each component, find: **(i)** the numbers of rows and columns, **(ii)** the index of connection, **(iii)** a spanning tree, **(iv)** a set of fundamental circuits, and **(v)** the number of nodes for each degree d ($d \in \mathbb{N} - \{2\}$).

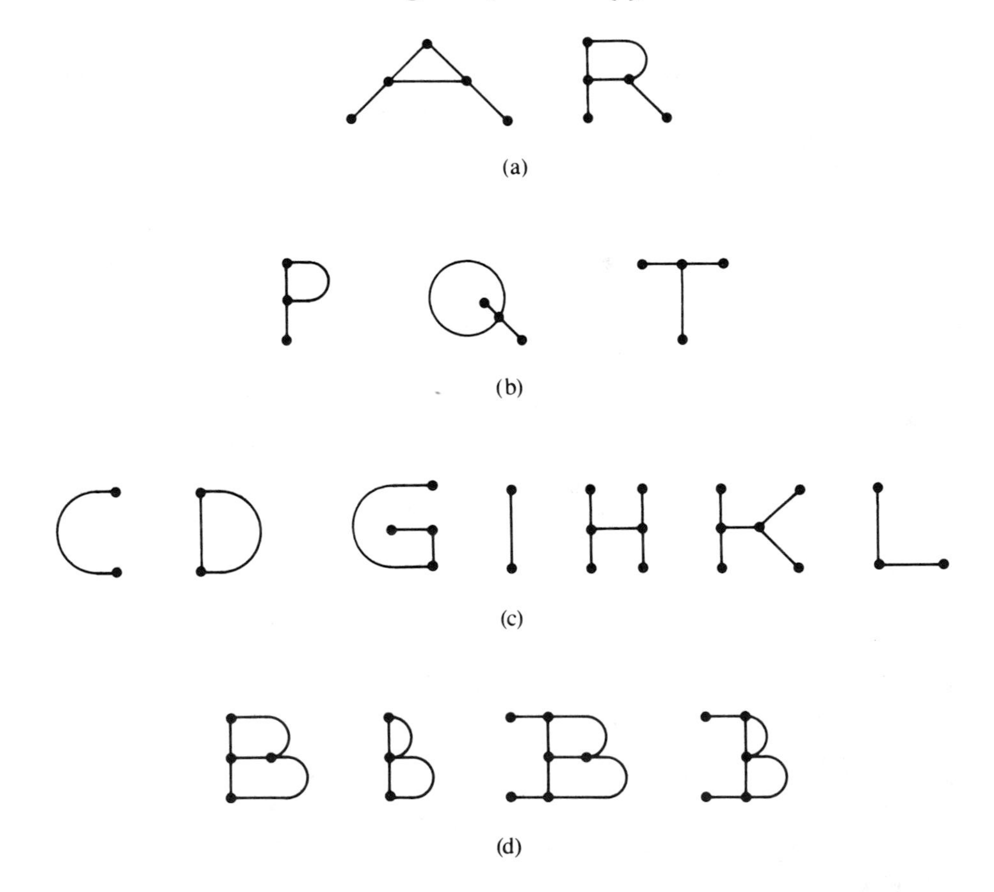

(a)

(b)

(c)

(d)

(e) The three-utilities graph (Figure 11.5a) and **(f)** the complete graph on five nodes (Figure 11.4c), that is,

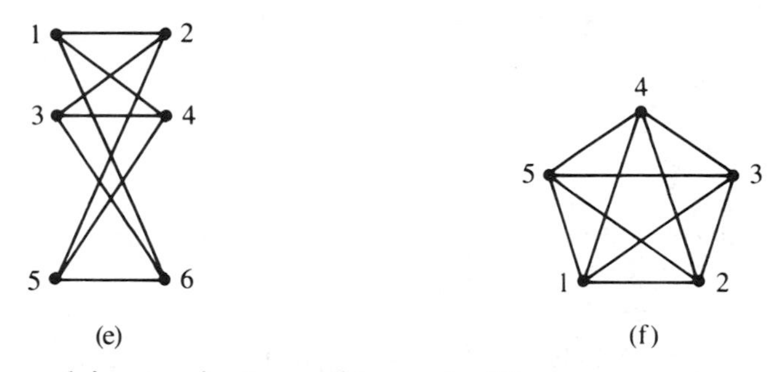

2. For each graph in exercise 1, consider each of the components as a single graph. Examine the existence of isomorphisms or homomorphisms between any two such graphs. Recall that homomorphic (and isomorphic) graphs have the same c and the same number of nodes of degrees other than 2 (this immediately excludes several possibilities).

3. Assign a graph of your choice abstractly, by assigning matrix E. Recall that the only conditions for matrix E are that each column contain precisely two 1s (the other entries being 0), or precisely one 2 (the other entries being 0). Then answer the same questions as in exercise 1. For instance:

(a)
$$\begin{bmatrix} 1 & 0 & 1 & 0 & 0 & 0 & 0 & 0 & 0 \\ 1 & 1 & 1 & 0 & 0 & 0 & 0 & 0 & 0 \\ 0 & 1 & 0 & 1 & 1 & 2 & 0 & 0 & 0 \\ 0 & 0 & 0 & 1 & 0 & 0 & 0 & 0 & 0 \\ 0 & 0 & 0 & 0 & 1 & 0 & 0 & 0 & 0 \\ 0 & 0 & 0 & 0 & 0 & 0 & 1 & 1 & 1 \\ 0 & 0 & 0 & 0 & 0 & 0 & 1 & 0 & 0 \\ 0 & 0 & 0 & 0 & 0 & 0 & 0 & 1 & 0 \\ 0 & 0 & 0 & 0 & 0 & 0 & 0 & 0 & 1 \end{bmatrix}$$

(b)
$$\begin{bmatrix} 1 & 0 & 0 & 1 & 0 & 0 & 0 & 0 & 0 & 0 & 0 & 0 & 0 & 0 & 0 \\ 1 & 1 & 0 & 0 & 0 & 0 & 0 & 0 & 0 & 0 & 0 & 0 & 0 & 0 & 0 \\ 0 & 1 & 1 & 0 & 0 & 0 & 0 & 0 & 0 & 0 & 0 & 0 & 0 & 0 & 0 \\ 0 & 0 & 1 & 1 & 0 & 0 & 0 & 0 & 0 & 0 & 0 & 0 & 0 & 0 & 0 \\ 0 & 0 & 0 & 0 & 1 & 0 & 0 & 1 & 1 & 0 & 0 & 0 & 0 & 0 & 0 \\ 0 & 0 & 0 & 0 & 1 & 1 & 0 & 0 & 0 & 0 & 0 & 0 & 0 & 0 & 0 \\ 0 & 0 & 0 & 0 & 0 & 1 & 1 & 0 & 1 & 0 & 0 & 0 & 0 & 0 & 0 \\ 0 & 0 & 0 & 0 & 0 & 0 & 1 & 1 & 0 & 0 & 0 & 0 & 0 & 0 & 0 \\ 0 & 0 & 0 & 0 & 0 & 0 & 0 & 0 & 0 & 1 & 0 & 0 & 1 & 1 & 0 \\ 0 & 0 & 0 & 0 & 0 & 0 & 0 & 0 & 0 & 1 & 1 & 0 & 0 & 0 & 1 \\ 0 & 0 & 0 & 0 & 0 & 0 & 0 & 0 & 0 & 0 & 1 & 1 & 0 & 1 & 0 \\ 0 & 0 & 0 & 0 & 0 & 0 & 0 & 0 & 0 & 0 & 0 & 1 & 1 & 0 & 1 \end{bmatrix}$$

(c)
$$\begin{bmatrix} 2 & 0 & 0 & 0 & 0 & 0 & 0 & 0 & 0 & 0 \\ 0 & 2 & 2 & 0 & 0 & 0 & 0 & 0 & 0 & 0 \\ 0 & 0 & 0 & 2 & 2 & 2 & 0 & 0 & 0 & 0 \\ 0 & 0 & 0 & 0 & 0 & 0 & 2 & 2 & 2 & 2 \end{bmatrix}$$

(d)
$$\begin{bmatrix} 0 & 0 & 0 & 0 \\ 0 & 0 & 0 & 0 \\ 2 & 0 & 0 & 0 \\ 0 & 2 & 0 & 0 \\ 0 & 0 & 1 & 1 \\ 0 & 0 & 1 & 1 \end{bmatrix}$$

This is a suitable exercise for the computer. The program must be able to accept an arbitrary integer matrix of reasonable dimensions, check whether it satisfies the conditions mentioned above, and if it does, produce the answers to the questions in the previous exercise.

4. There is only one specific type of node-simple cycle that has repeated arcs. What is it?

5. Show that every node-simple path other than a cycle is also arc-simple.

Complements 12.4

1. Describe the effect of a free chain insertion or deletion onto the incidence matrix E of a graph.

12.5 Planarity of Graphs

One of the problems in graph theory is the question of *planarity*, that is, whether a graph can be drawn in a plane so that the images of distinct arcs do not have "incidental" intersections, intersections other than actual nodes of the graph. This problem is important, for instance, in the design of circuit boards. Since many electronic components such as resistors and capacitors have two terminals, *series* connection is no problem: it is simply a free chain insertion and does not affect planarity. *Parallel* connections or multi-terminal components such as transistors and integrated circuits are a different matter. They may imply a "crossing of wires," which creates engineering problems, unless we can find a way to draw all of the circuit in a single layer, by rearranging the physical layout of the entire circuit without changing the nature of the connections.

We've already given the definition of planarity. A **planar realization** of a graph is a drawing of the graph in a plane so that nodes are represented by distinct points of the plane, each arc is represented by a segment (not necessarily a straight-line segment) that joins the geometric points that represent its endpoints, and does not intersect itself nor any other arc, except at the endpoint that it may have in common with such an arc. A graph that admits a planar realization is said to be **planar**.

Clearly, whether the graph is directed or undirected has no influence on its planarity. We may as well study only undirected graphs. Also, free chain insertions and deletions do not influence the planarity of a graph. A free chain insertion means simply to place one more node on the image of an arc (but not at either endpoint). A free chain deletion means to remove a node (provided it is the endpoint of precisely two arcs) and consider those two arcs as one. As usual, we'll limit ourselves to finite graphs only. Also, it's clear that, if each connection component of a graph is planar, then the whole graph is planar: just draw the connection components, planarly, one next to the other. We'll therefore consider only finite, undirected, connected graphs.

Problems of graph planarity have to do with the geometry of the plane. Since we have not studied any geometric properties in this book, we'll have to rely on geometric intuition. But this will not prevent us from considering some essential problems.

12.5.1 Euler's Planarity Theorem

Suppose that a finite graph G is connected. Then it has a spanning tree, as we saw in Section 12.4.3. Now suppose that G is planar. Whichever way we draw the graph in a plane, we can always draw first the spanning tree, and then the c additional arcs, c being the index of connection of the graph. When we draw the spanning tree, we do not separate any region of the plane from the rest of the plane because we are drawing a tree, which has no nonnull circuits. But, each time we draw an additional arc, this arc (together with the previous part of the drawing) encloses a region of the plane, and creates an extra **face** of the planar realization of the graph. The new face may be obtained in two ways: either the new arc cuts an existing face into two parts, or it encloses a new portion of the plane outside the existing drawing. The two cases are illustrated by the progressive drawing of Figure 12.13.

Consequently, the resulting graph has c faces. Since $c = m - n + 1$ (n and m being the numbers of nodes and arcs respectively), we have proved the following theorem:

Theorem 10 **Euler's planarity theorem**. If a finite, connected graph is realized planarly and has n nodes, m arcs, and f faces, then $n - m + f = 1$.

In the fourth drawing of Figure 12.13, we have $n = 4$, $m = 6$, and $f = 3\ (= c = m - n + 1)$.

Observe that, whenever we can draw a graph in a plane, we can also draw it on the surface of a sphere: just draw it on a small portion of the sphere. In this case, the far side of the sphere is actually an additional face that corresponds to the portion of the plane "outside" the picture of the graph. This region is often

Figure 12.13

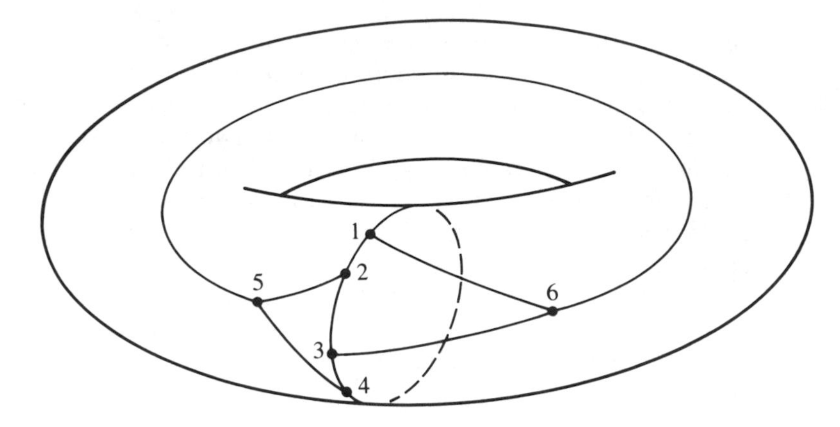

Figure 12.14 The three-utilities graph drawn on a donut.

considered an extra face, which is called the **improper face**. If we do that, or if we are on a sphere, then the conclusion of Euler's theorem is that $n - m + f = 2$.

Note that the index c of a connected graph is always $m - n + 1$, whether the graph is planar or not, whether we attempt to realize it planarly or not. But c is not always the number of faces. As a counterexample, consider again the three-utilities graph: it has six nodes (1 to 6) and each pair {even-numbered node, odd-numbered node} gives a single arc. Consequently, $n = 6$, $m = 9$, and $c = 4$ ($= m - n + 1$). As we'll see in a moment, this graph is not planar. We can, however, draw this graph on the surface of a donut, as in Figure 12.14. On the surface of the donut, there are only three faces: the "triangles" 12361 and 23452, and the rest of the surface, which is enclosed by 16521456341 or, if you cancel repetitions, by 1634521.

12.5.2 Kuratowski's Theorem

Not every finite graph is planar. Consider again the three-utilities graph (see Exercises 12.4 or Figure 12.14 above). The nodes are 1, 2, . . ., 6 and the arcs join each odd-numbered node to each even-numbered node. To show that the graph is not planar, proceed by contradiction. Suppose it is planar and try to draw it. Whichever way we draw it, we can draw first the circuit 1234561. Necessarily, we obtain a closed curve that does not intersect itself, something like Figure 12.15.

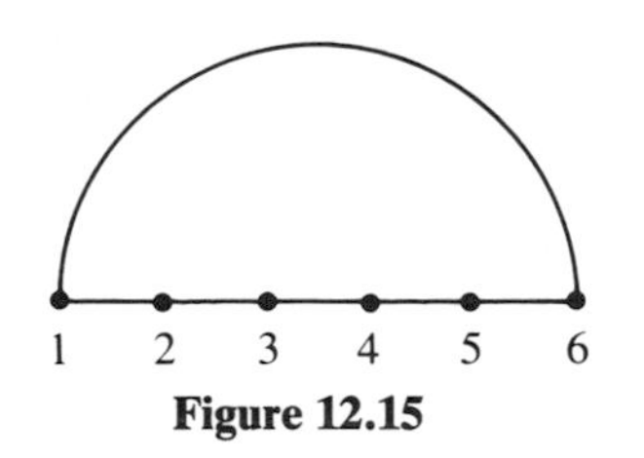

Figure 12.15

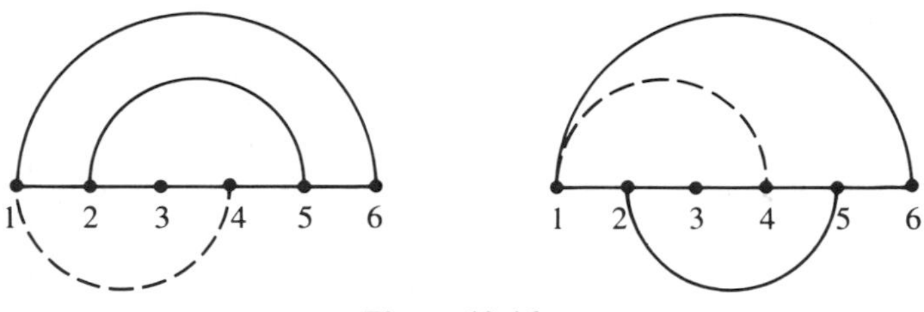

Figure 12.16

Then we have a region of the plane "inside" the curve and the rest of the plane "outside." When we draw arc 25, it must be entirely inside or entirely outside the curve. We have therefore two cases, represented in Figure 12.16. (For the time being, ignore the dotted arcs.)

After that, arc 14 must lie on the "other" side, that is, outside the first cycle if arc 25 was inside, and inside the first cycle if arc 25 was outside (see the dotted arcs in Figure 12.16). In either case, arc 36 must go from the inside of cycle 12541 to the outside of it, and there's no way to do that without crossing that cycle, that is, without introducing stray intersections. As an exercise, you'll be asked to show that the complete graph on five nodes is not planar either.

It's clear that, if a graph G *includes* a three-utilities graph or a complete graph on five nodes, then G itself is not planar, because not even a part of G can be drawn in a plane.

Example. The graph in Figure 12.17 part (a) includes the graph in part (b), which is homological to the graph in part (c), and the last graph is a three-utilities graph.

We have so proved the "if" part of Kuratowski's theorem:

Theorem 11 **Kuratowski's Theorem.** A finite undirected graph fails to be planar if and only if it *includes* a three-utilities graph or a complete graph on five nodes. That means that there is in G a subset of nodes and a subset of arcs that (with the same incidences they have in G) form a graph homological to a three-utilities graph or to a complete graph on five nodes.

Proof: A full proof of the "only if" part is involved and requires many more additional definitions. We'll not attempt it here. We can, however, see how some basic geometric ideas can provide a constructive process to realize a connected graph planarly or to recognize that it includes a three-utilities graph or a complete graph on five nodes.

1. If G has no nonnull circuits, then it is a tree, and there is no difficulty in drawing all of it at once. Otherwise, drawing a first circuit of G is again immediate.
2. Once a part of G is already on paper (call G^* such part), the next step is to add to the picture the images of additional arcs. (If there are no more arcs,

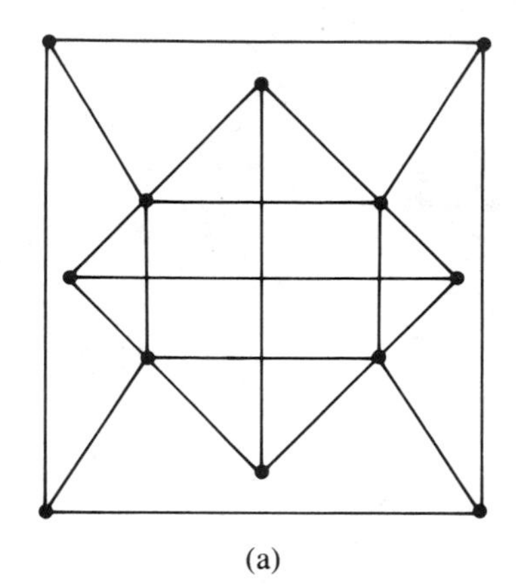

(a)

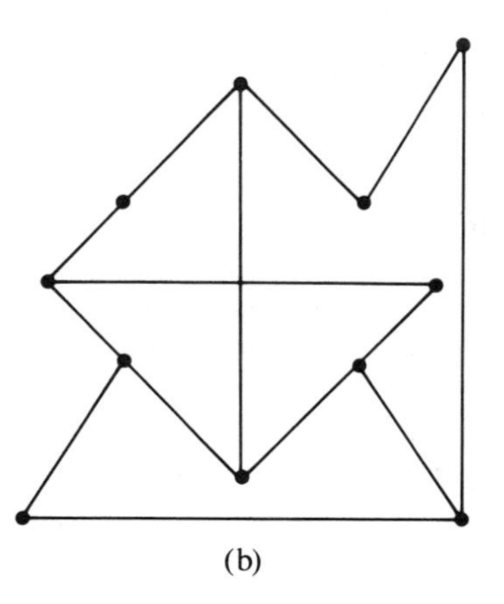

(b)

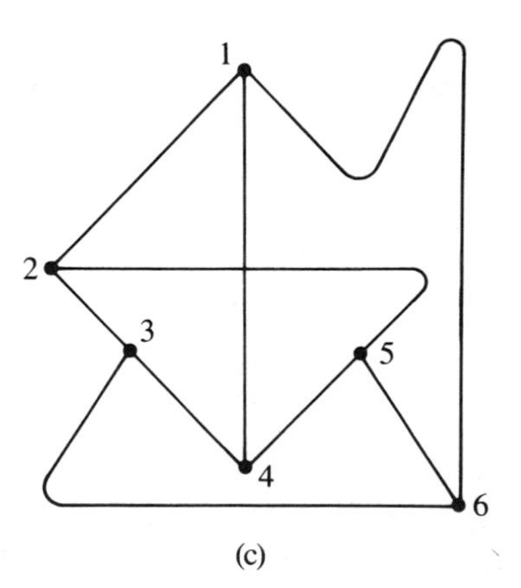

(c)

Figure 12.17

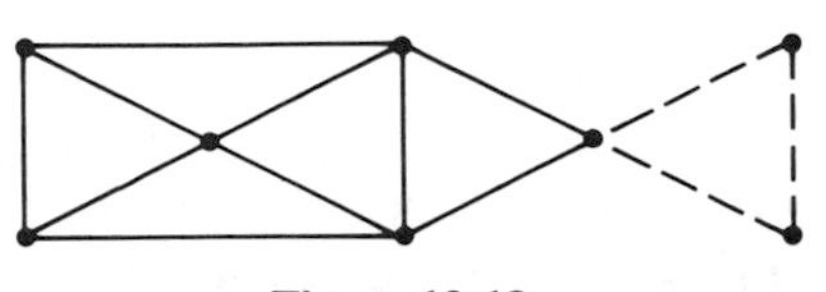

Figure 12.18

then the process stops successfully.) There may exist a simple path p that is in G but not in G^* and that joins two *distinct* nodes x and y of G^* (see part 3). There is, however, the possibility that such a path does not exist, even though G is not entirely on paper. In this second case, there exists a **node of disconnection**, that is, a node w that is the only node in common to G^* and a part of G that is not on paper. Then this second part can be drawn separately (if it's planar) and, at the end of the process, "copied" in the first picture, provided the connection through node w is respected. An example is Figure 12.18, in which the "copied" part is the dotted arcs.

3. If there is a path p as described above and if nodes x and y are on the boundary of a same face of G^* (possibly the improper face), then there is no difficulty in adding path p to the drawing (and then continue with part 2; Figure 12.19a is a typical example).

When this is *not* the case, we need a more detailed analysis. The essential step is to locate the ends, say h, of those simple paths that originate at y and end as soon as they encounter a node on the boundary of some face through x. The favorable case occurs when all such nodes h lie on the boundary of one single face F through x (see part 4). If not, then we have a situation typified by one of the graphs in Figure 12.19b (in which 1, 2, 3 are some of the nodes h). Since path p is in G and joins x and y, the six nodes 1, 2, a, b, x, y in the first drawing in Figure 12.19b provide a three-utilities graph included in G; or the five nodes 1, 2, 3, x, y in the second drawing provide a complete graph on five nodes included in G. In either case, the process has led us to the conclusion that G is not planar.

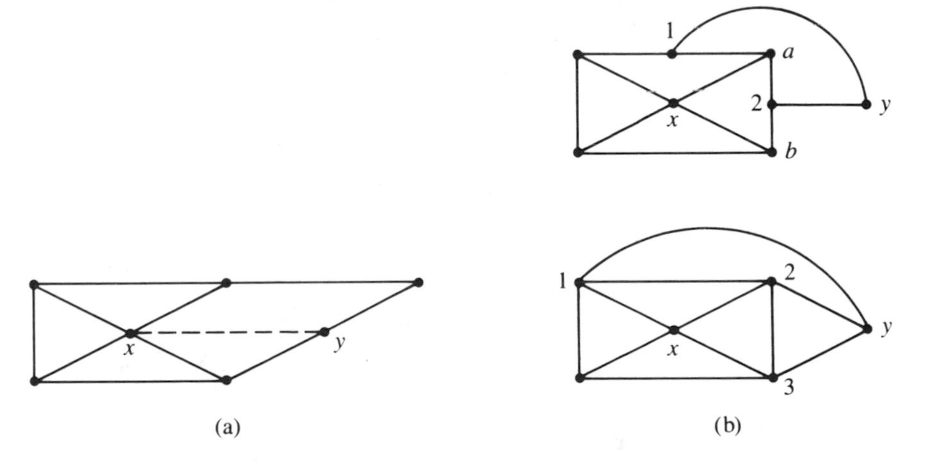

Figure 12.19

4. In the favorable case, all nodes h are on the boundary of a same face F through x. Figure 12.20a is a typical example. Then we select, among the nodes h, two "pivots," that is, two nodes u and v that (along the boundary of F) leave x in one part and all of the other nodes h (if any) in the other part. Then F is a face through both of u and v. Necessarily, there is at least one other face (such as H in the figure) through both u and v (otherwise, we'd be in the case of part 3, which we have already taken care of). Consider all faces through *both* u and v. They divide graph G^* into "slices," that is, parts that contain both u and v and have—pairwise—no other nodes in common. Of these slices, one contains x and a distinct one contains y. If needed, we can redraw the slice containing y so that it is next to the face F through x (we don't need this step in our example above). There are now three cases: (i) Nodes x and y are on the same face of the new G^*, in which case we proceed as before; that is, we draw p and continue with part 2. (ii) Node y is on an edge of its slice, but it is on the wrong edge (as in our example); in that case, we "flip" the slice containing y, that is, we redraw it "right side left" (Figure 12.20b), and then continue as before. (iii) Node y is not on the edge of the slice that contains it. An example is shown in Figure 12.20c. Then (perhaps after repositioning one pivot or both) we discover again a three-utilities graph included in G (in Figure 12.20c, observe nodes 1, 2, u, v, x, y and recall that path p is in G). Then we conclude that graph G is not planar.

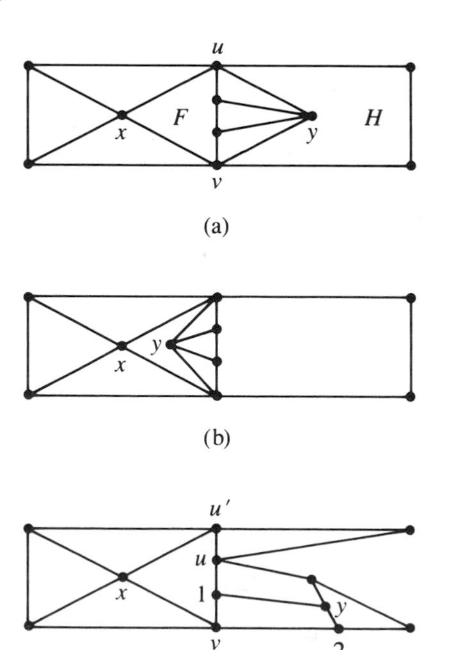

Figure 12.20

Exercises 12.5

1. Verify that the complete graph on five nodes is not planar. Verify also that it can be realized, without stray intersections, on the surface of a donut.
2. Try to realize each of the following graphs planarly or else to locate within it a three-utilities graph or a complete graph on five nodes. (In practice, much of the figure can be drawn without invoking any algorithm. Some hard steps may need guidance from parts of the algorithm.)

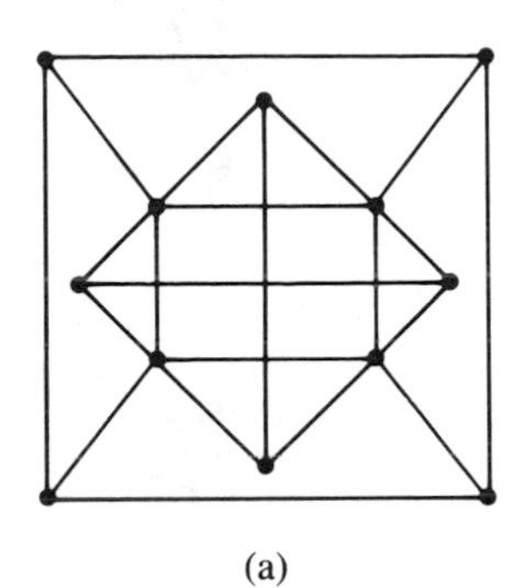

(a)

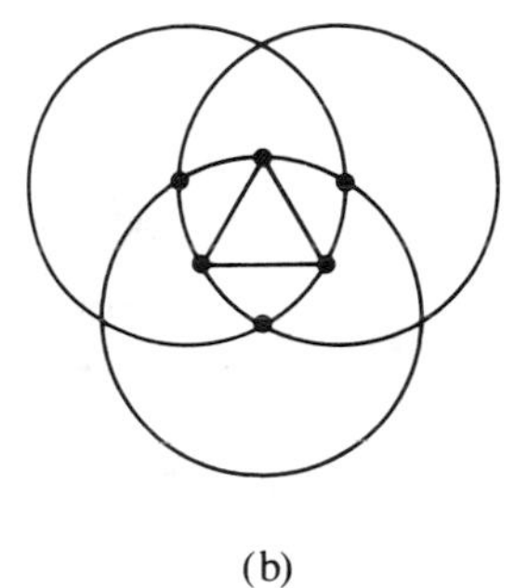

(b)

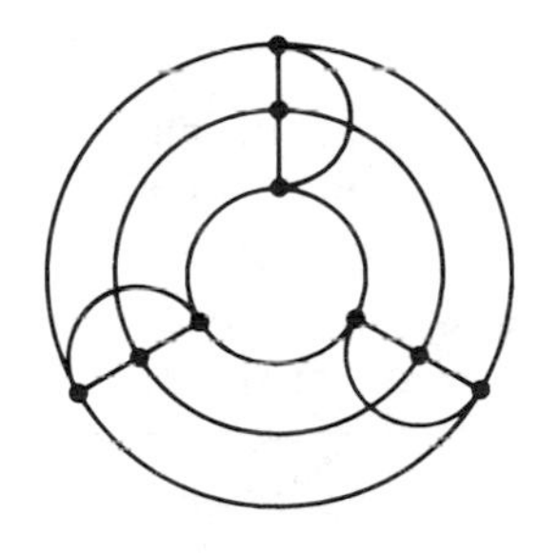

(c)

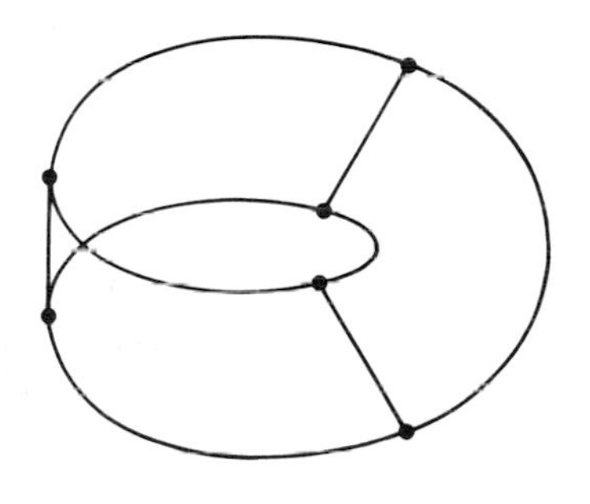

(d)

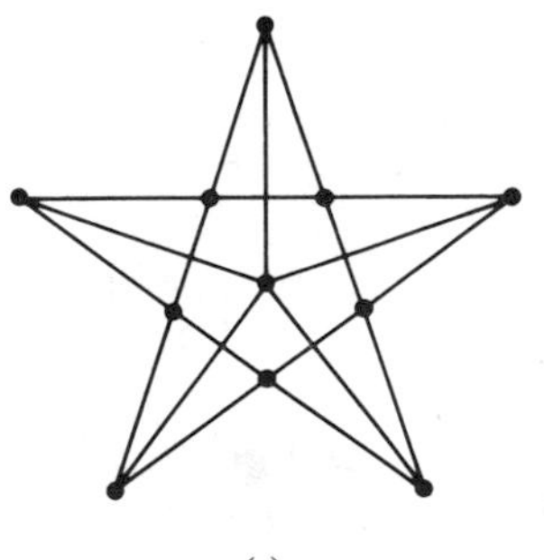

(e)

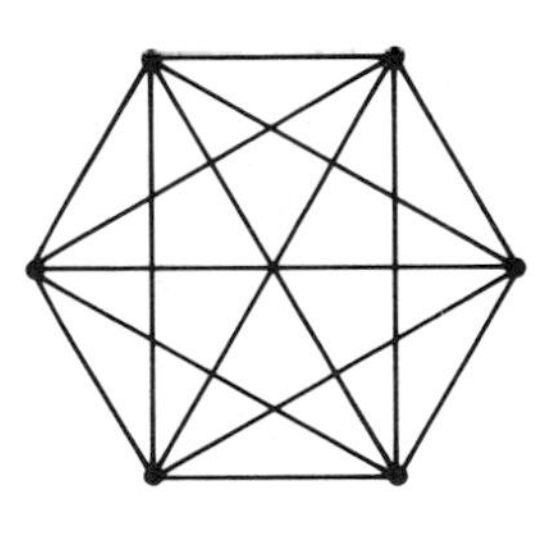

(f)

Complements 12.5

1. Consider the complete graph on five nodes 1, 2, 3, 4, and 5. Suppose we "offset" node 2; that is, we "split" it into two nodes 2 and 6 as in the figure on the right.

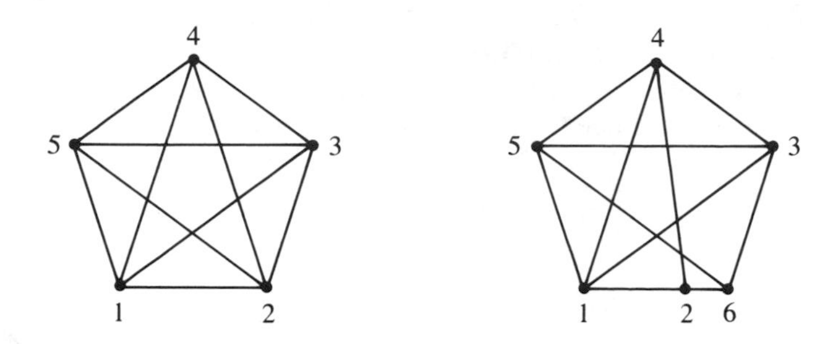

Show that the new graph includes a three-utilities graph and is therefore not planar. (Hint: ignore arcs 35 and 14.)

12.6 Summary

An *expression* is defined recursively by applying n-ary operators (for instance, addition of two numbers) to sequences of n previously defined expressions (Section 12.1.1). The initial steps of the recursion are provided by the objects of a structure (for instance, numbers) and symbols that stand for any of those objects. Two concrete representations of an expression are provided by an *expression tree*, in which a node stands for an operation and its immediate successors stand for the *arguments* of that operation (Section 12.1.2), or by an *expression string,* in which each operator is written before (or after) the string of its arguments in a *prefix* (or *postfix*) expression, or between its two arguments in *infix* expressions (Section 12.1.1 and 12.1.3). Infix expressions are those used in elementary arithmetic and algebra; additional conventions are necessary to handle nonbinary operations and precedence of operations.

The transition between expression trees and expression strings is done by means of algorithms, which are based on the concept of *traversal* of a tree (Section 12.1.4).

A given expression can be *evaluated* for any allowable sets of *values* of its arguments. This process associates a function with each expression (Section 12.1.5).

At a more general level, a *tree* is a special graph with no nontrivial cycles (Section 12.2). A tree may be *undirected* or *directed*. Directed trees are classified according to several criteria. For instance, in *ordered* trees the set of the arcs going out from each node is ordered linearly; in *binary* trees the outdegree of each node is 0, 1, or 2; and in *balanced* trees the simple paths from each node to the leaves have lengths that do not differ by more than one unit. Special uses of trees are, for instance, *parsing* trees, which are the same as expression trees, and *sorting* trees, which arrange a given list of strings in alphabetical order (Section 12.3).

The basic *homological properties* of a graph (Section 12.4) depend on concepts like that of a *spanning tree* (a tree that contains all nodes of the graph) and that of a set of *fundamental cycles* (a set such that each cycle of the graph can be obtained in a unique way by adding mod 2 some cycles of that set). A tool to calculate spanning trees and fundamental cycles is provided by the adjacency matrices presented in Chapter 11 or by *incidence matrices*, which are more convenient in the present context.

The concepts above can be used to study *planarity* of graphs (Section 12.5), that is, to investigate whether (and how) a graph can be pictured in a plane without introducing "stray" intersections of arcs, intersections at points other than nodes that are common ends of the two intersecting arcs. Two powerful theorems about planarity are Euler's planarity theorem and Kuratowski's theorem.

Coding Theory

Transmitting messages is a necessity of life. Examples in this technological era range from just ringing the doorbell to the transmission of data from the core of a computer to a magnetic tape. Any message must pass through a *transmission channel*, that is, through a composite of physical devices and processes that cause the message to be *transmitted* at one end and *received* at the other. The typical example is the transmission of digital electronic signals via some medium such as radio, cable, fiber optics, or copper wire.

It is a fact of life that any channel, good or bad, is prone to error. The wire of the doorbell may have a bad contact, the device that positions the magnetic tape may get confused when it receives too many blanks in a row. Consequently, the received message may be different from the transmitted message. An obvious goal of the entire process is to limit the frequency of errors, to detect errors that do occur, and, when possible, to correct errors and reconstruct the transmitted message. Limiting the frequency of errors is the task of the technology of the channel. We'll devote our attention to the detection and correction of errors. This is the task of *coding theory*.

In this chapter we'll see some basic ideas of coding theory and study in some detail a special class of codes, the *binary codes,* codes, that transmit messages consisting of only zeroes and ones. "But," you may say, "what if I want to send a message about something other than zeroes and ones?" "We'll address that question in Section 13.1. Coding theory is still in the making. There are many known codes, binary and otherwise, but there are also many unsolved problems. We'll mention some of them. Also, note the difference between coding theory and *cryptography*. Both theories share many common methods, but their goals are in a sense opposite: cryptography tries to *hide* the message from unauthorized receivers (that's the meaning of the Greek root *crypt*); coding theory tries to deliver the message correctly. The confusion is compounded by phrases like "breaking a code" (which belongs to cryptography) and "decoding" (which is used in both, but is replaced by "decrypting" in cryptography).

The design of a channel and the design of a code usually go together, in serious applications, with the usual "give and take" of all difficult real problems. We shall consider here only the mathematical part. And at the other end, there is the problem of decoding. An application usually allows plenty of time to design a code but, when the code is put to work, the decoding process must be quick. When bits are transmitted at a rate of thousands or millions a second, we can't dawdle at the receiving end. This implies that the *decoding algorithm* must be extremely efficient and must therefore be planned skillfully. In order to do that, the encoding and the decoding process—and therefore the structure of the code—must be understood well. This should be your main goal at this stage. As usual, we start with simple examples, play around with them, generalize, and then produce more serious codes.

13.1 Preliminary Conversions

Whatever the original message is, it must be first converted to a string of only symbols from the code's *alphabet*. For a binary code, the alphabet is {0, 1}; for a ternary code, it may be {0, 1, 2}, or {−1, 0, 1}, or {negative, no charge, positive}; for a quinary code, {0, 1, 2, 3, 4} or {on, off, -, /, #} will do; for a usual memorandum, we need several shapes of the English alphabet, digits, spaces, punctuation, and so on. There are no mathematical difficulties at this level. We give two examples that illustrate the process well enough.

13.1.1 Morse Code

A classical example of coding is the Morse code used in telegraphy. Each letter and digit and some punctuation is encoded as a sequence of dots and dashes and, of course, spaces. For instance, O is *dah-dah-dah,* N is *dah-di*, E is *di,* B is *dah-di-di-di,* I is *di-di,* T is *dah.* The message ONE BIT becomes *dah-dah-dah/dah-di/di#dah-di-di-di/di-di/dah.* We have used the hyphen - for a space within a letter, the slash / for a space between letters, and # for a space between words. Our code's alphabet is now {dah, di, -, /, #}, so the code is not binary but quinary (Latin *quin* = fivefold). We can now recall the good old rule of telegraphy, according to which a dash lasts three times as long as a dot, an intraletter space as long as a dot, an interletter space as long as two dots, an interword space as long as three (or four) dots. Then we can adopt, as usual, 1 for "on" and 0 for "off." The resulting form of the message ONE BIT is then

$$11101110111001110100100011101010100101001 11,$$

which uses only the binary alphabet {0, 1}. The last conversion is a textbook example. It would give a very impractical method if the main goal were just to convert letters into binary form. There are many codes that perform this task much better.

Binary	Octal		Binary	Octal	
0000000	000	NUL	0100000	040	SP
0000001	001	SOH	0100001	041	!
0000010	002	STX	0100010	042	”
0000011	003	ETX	0100011	043	#
0000100	004	EOT	0100100	044	$
0000101	005	ENQ	0100101	045	%
0000110	006	ACK	0100110	046	&
0000111	007	BEL	0100111	047	’
0001000	010	BS	0101000	050	(
0001001	011	HT	0101001	051	)
0001010	012	LF	0101010	052	*
0001011	013	VT	0101011	053	+
0001100	014	FF	0101100	054	,
0001101	015	CR	0101101	055	-
0001110	016	SO	0101110	056	.
0001111	017	SI	0101111	057	/
0010000	020	DLE	0110000	060	0
0010001	021	DC1	0110001	061	1
0010010	022	DC2	0110010	062	2
0010011	023	DC3	0110011	063	3
0010100	024	DC4	0110100	064	4
0010101	025	NAK	0110101	065	5
0010110	026	SYN	0110110	066	6
0010111	027	ETB	0110111	067	7
0011000	030	CAN	0111000	070	8
0011001	031	EM	0111001	071	9
0011010	032	SUB	0111010	072	:
0011011	033	ESC	0111011	073	;
0011100	034	FS	0111100	074	<
0011101	035	GS	0111101	075	=
0011110	036	RS	0111110	076	>
0011111	037	US	0111111	077	?

Figure 13.1 The 7-bit ASCII code, part 1, from the American Standard Code for Information Interchange, American National Standard Institute.

13.1.2 ASCII Code

The ASCII code is one of the most common codes used to convert letters, digits, some special signs like operation symbols and punctuation, and some machine instructions into binary strings. The 7-bit ASCII code is presented in Figures 13.1 and 2, together with the octal counterparts of the 7-bit binary strings. For instance

$$
\begin{aligned}
A &= 1\ 000\ 001 \text{ binary} = 101 \text{ octal} \\
B &= 1\ 000\ 010 \text{ binary} = 102 \text{ octal} \\
3 &= 0\ 110\ 011 \text{ binary} = 063 \text{ octal.}
\end{aligned}
$$

<table>
<tr><th>Binary</th><th>Octal</th><th></th><th>Binary</th><th>Octal</th><th></th></tr>
<tr><td>1000000</td><td>100</td><td>@</td><td>1100000</td><td>140</td><td>‘</td></tr>
<tr><td>1000001</td><td>101</td><td>A</td><td>1100001</td><td>141</td><td>a</td></tr>
<tr><td>1000010</td><td>102</td><td>B</td><td>1100010</td><td>142</td><td>b</td></tr>
<tr><td>1000011</td><td>103</td><td>C</td><td>1100011</td><td>143</td><td>c</td></tr>
<tr><td>1000100</td><td>104</td><td>D</td><td>1100100</td><td>144</td><td>d</td></tr>
<tr><td>1000101</td><td>105</td><td>E</td><td>1100101</td><td>145</td><td>e</td></tr>
<tr><td>1000110</td><td>106</td><td>F</td><td>1100110</td><td>146</td><td>f</td></tr>
<tr><td>1000111</td><td>107</td><td>G</td><td>1100111</td><td>147</td><td>g</td></tr>
<tr><td>1001000</td><td>110</td><td>H</td><td>1101001</td><td>150</td><td>h</td></tr>
<tr><td>1001001</td><td>111</td><td>I</td><td>1101000</td><td>151</td><td>i</td></tr>
<tr><td>1001010</td><td>112</td><td>J</td><td>1101010</td><td>152</td><td>j</td></tr>
<tr><td>1001011</td><td>113</td><td>K</td><td>1101011</td><td>153</td><td>k</td></tr>
<tr><td>1001100</td><td>114</td><td>L</td><td>1101100</td><td>154</td><td>l</td></tr>
<tr><td>1001101</td><td>115</td><td>M</td><td>1101101</td><td>155</td><td>m</td></tr>
<tr><td>1001110</td><td>116</td><td>N</td><td>1101110</td><td>156</td><td>n</td></tr>
<tr><td>1001111</td><td>117</td><td>O</td><td>1101111</td><td>157</td><td>o</td></tr>
<tr><td>1010000</td><td>120</td><td>P</td><td>1110000</td><td>160</td><td>p</td></tr>
<tr><td>1010001</td><td>121</td><td>Q</td><td>1110001</td><td>161</td><td>q</td></tr>
<tr><td>1010010</td><td>122</td><td>R</td><td>1110010</td><td>162</td><td>r</td></tr>
<tr><td>1010011</td><td>123</td><td>S</td><td>1110011</td><td>163</td><td>s</td></tr>
<tr><td>1010100</td><td>124</td><td>T</td><td>1110100</td><td>164</td><td>t</td></tr>
<tr><td>1010101</td><td>125</td><td>U</td><td>1110101</td><td>165</td><td>u</td></tr>
<tr><td>1010110</td><td>126</td><td>V</td><td>1110110</td><td>166</td><td>v</td></tr>
<tr><td>1010111</td><td>127</td><td>W</td><td>1110111</td><td>167</td><td>w</td></tr>
<tr><td>1011000</td><td>130</td><td>X</td><td>1111000</td><td>170</td><td>x</td></tr>
<tr><td>1011001</td><td>131</td><td>Y</td><td>1111001</td><td>171</td><td>y</td></tr>
<tr><td>1011010</td><td>132</td><td>Z</td><td>1111010</td><td>172</td><td>z</td></tr>
<tr><td>1011011</td><td>133</td><td>[</td><td>1111011</td><td>173</td><td>{</td></tr>
<tr><td>1011100</td><td>134</td><td>\</td><td>1111100</td><td>174</td><td>|</td></tr>
<tr><td>1011101</td><td>135</td><td>]</td><td>1111101</td><td>175</td><td>}</td></tr>
<tr><td>1011110</td><td>136</td><td>^</td><td>1111110</td><td>176</td><td>~</td></tr>
<tr><td>1011111</td><td>137</td><td>_</td><td>1111111</td><td>177</td><td>DEL</td></tr>
</table>

Figure 13.2 The 7-bit ASCII code, part 2.

Items 0 to 37 and 177 octal are machine instructions; item 40 octal is the usual "space" or "blank". The message ONE BIT would then be

$$1\ 001\ 111\ 1\ 001\ 110\ 1\ 000\ 101\ 0\ 100\ 000\ 1\ 000\ 010\ 1\ 001\ 001\ 1\ 010\ 100 \qquad (1)$$

The spaces in this line should not be there; we've included them to facilitate reading by us humans. The actual "space" of the original message, between the words ONE and BIT, *is* there: it's the 7-bit string 0100000.

13.1.3 Stages of the Encoding and Decoding Process

In summary, the transition from the original message to the final received message involves many stages. At the most fundamental level there is the *design stage* in which the nature of the channel is chosen, the channel built, the code chosen, and coding and decoding algorithms created. There then follow several stages for each individual message.

Stages for each message

1. Conversion of the original message to a string S over the code's alphabet is the first step. In the example we have been discussing, the alphabet is $\{0, 1\}$. In this stage the message ONE BIT is converted to string 1 using ASCII.
2. In most instances, the code is able to encode only strings of a given length, generally rather small. These strings are called *information strings*. We must then decompose string S as the concatenation of shorter binary strings which are information strings. In the example above, for a code able to encode 4-bit strings, the message would become

   $$1001\ 1111\ 0011\ 1010\ 0010\ 1010\ 0000\ 1000\ 0101\ 0010\ 0110\ 1010\ 0111 \quad (2)$$

   The last three 1s are just padding; the spaces are for human convenience only.

Stages for each information string

3. This stage is where coding theory begins to take over. Each information string, that is, each 4-bit string of the example, is *encoded* as another binary string, which will be a *code word,* usually longer than the information string.
4. In the next stage each code word is transmitted and received. Errors should be expected. The received word may or may not be the same as the transmitted word.
5. The hard stage is to decode the received word, that is, to reconstruct the transmitted code word or at least to be able to recognize whether the received word contains any errors.
6. Whenever a code word is reconstructed, it is *decoded,* that is, converted back to the original information string.

Final stages for each message

7. Decoded information strings are concatenated to form a longer string, which should be string S.
8. The resulting string is converted back (with luck) to the original message.

The channel design and construction and the transmittal of the coded message (stage 4) are technological problems and do not concern us. Choice of the code and design of the coding algorithm is a combination of mathematics, management, and technology; we're not going to examine that either. Stages 1 and 8 are inverses of each other and present no mathematical difficulties. Managerial, psychological, and sociological problems may arise: for instance, people do not like the "secrecy" of the Universal Consumer Code, the bar code that appears on boxes and cans at the supermarket. Stages 2 and 7 are also inverses of each other and are mathematically trivial; there may be some technological problems, like losing count of where the strings were concatenated. Stages 3, 5, and 6 belong to coding theory proper; stages 3 and 6, which are inverses, are mathematically simple; stage 5 will require a *decoding procedure* and will give us a hard time.

Exercises 13.1

1. Convert a few phrases or formulae to binary or octal strings using ASCII or a similar code. For instance, **(a)** $X + Y$; **(b)** $1 + 2 = 3$; **(c)** any phrase. (You can do this on a computer; read literal, print binary.)
2. Convert a few binary strings to plain alphabetical form using ASCII or a similar code. For instance, **(a)** 0110001011001001100011; **(b)** 1100001110001011000111. (You can do this on a computer.)
3. Look up in the library some other codes that perform a function similar to the ASCII code. Examples are other computer codes similar to ASCII, teletype codes, and the Braille alphabet.

13.2 Examples

There are infinitely many known binary codes, and they fall into many categories. Before we approach the problem in general, let's study a few examples. In most of them, we'll denote the *code,* that is, the set of all allowable *code words,* as C.

13.2.1 Repetition Codes

Consider the case in which the information string comprises just one bit. As usual, a longer message would be decomposed into its individual bits. An example is the doorbell: either it rings or it doesn't. What do you do if it doesn't? Most people ring it again several times.

Definition 1 Given the positive integer n, the **n-repetition code** is the code according to which each 1-bit string x ($x = 0$ or $x = 1$) is encoded as the string x^n, that is, $x \cdots x$ (n times). Usually, n is odd.

Example 1 **3-Repetition Code.** $C = \{000, 111\}$. Of the possible eight 3-bit strings, only two are code words.

The basic idea of a decoding procedure for a repetition code is, "If, in the received word, you see more ones than zeroes, then assume that the transmitted bit was 1 (repeated n times); if you see more zeroes than ones, then assume that the transmitted bit was 0 (repeated n times); if there are as many zeroes as ones, then you can't tell, but you are sure that there were exactly $n/2$ errors" (n must be even, in this case). Let's formalize this.

Decoding Algorithm A. Let y be the received binary word, that is, any element of $\{0, 1\}^n$.

1. If the number of zeroes in y is greater that $n/2$, then decide that the transmitted code word was 0^n.
2. If the number of ones in y is greater that $n/2$, then decide that the transmitted code word was 1^n.
3. If neither case occurs (in which case n is necessarily even), then decide that $n/2$ errors occurred.

Example 2 Suppose $n = 3$ and the binary message is 10110. The information strings are, in this case, the individual bits. Each 0 bit is encoded 000 and each 1 bit is encoded 111. The message is then transmitted as 111 000 111 111 000 (the spaces are for human convenience). Suppose the message is received as 110 010 111 111 100. According to Algorithm A, it is interpreted as 111 000 111 111 000 and then decoded to 10110, which is correct. As a counterexample, suppose the same message had been received as 010 010 . . . (two errors in the first code word, one in the second); then the interpretation would have been 000 000 . . ., which would have been decoded as 00 . . ., which is incorrect.

Theorem 2 Decoding Algorithm A of the n-repetition binary code makes a correct decision if the number of errors in the transmission of each code word does not exceed $n/2$. Consequently, the code is able to correct fewer than $n/2$ errors and, when n is even, to detect one more.

Proof: Nearly trivial.

Note that, in the degenerate case in which $n = 1$, the encoding process is the identity, that is, the binary message is transmitted in its original form, and the code is not able to detect, much less correct, any errors.

When $n > 2$, the same code can be used in a different way. Suppose we use this algorithm:

Decoding Algorithm B.

1. If the received word is 0^n or 1^n, then decide that the transmitted word was 0^n or 1^n, respectively.
2. Otherwise, decide that there were errors.

The decision of this algorithm is incorrect if and only if the number of errors in one code word is n (that's a really bad channel). In this form, the code is able to detect up to $n - 1$ errors but cannot correct any.

We have one more remark, which will be useful later to understand more complicated codes. The diagram in Figure 13.3 represents all possible 3-bit strings, that is, all possible received words in the 3-repetition code. The code words, the possible transmitted words, are marked with a heavy dot. The arcs in the figure join words that differ from each other at one bit only. This diagram is really a graph, with 8 nodes and 12 arcs; a *cube*, if you wish.

In the figure, the interpretation of algorithm B is to accept each received code word at face value, and to declare error if the received word is not a code word. In this case, we partition the set of the nodes of the diagram into three classes. Two of them contain one code word each; the third class contains the six strings that are not code words.

The interpretation of algorithm A is different: accept each received code word at face value, of course, but also accept, for that code word, each of the three words that are joined to it by a single arc. The nodes now fall in two classes, each containing four words: one code word and the three words that are joined to it by

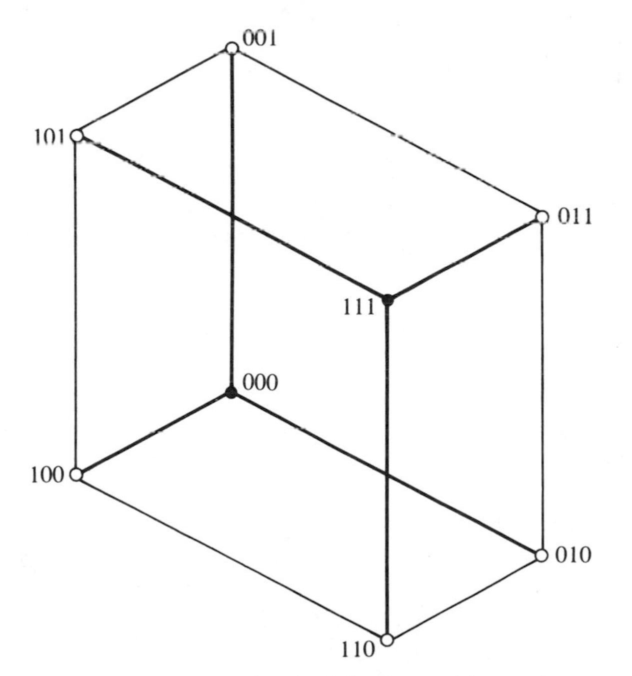

Figure 13.3 Graph of the 3-repetition code.

a single arc. In terms of graph theory, the graph includes a forest of two disjoint trees, each one with the root at a code word, three arcs, and three leaves.

As we'll see later, there are other ways to use a repetition code for which $n > 4$. For instance, when $n = 5$, we may want to be able to detect four errors, or to correct one error and to detect two more, or to correct two errors. See Exercises 13.2. Usually there are several such choices for any code. Selecting one or the other depends essentially on the quality of the channel: if it's reliable, we want to correct many errors and give up mere detection; if the channel is not reliable, we prefer to give up some correction capabilities and insist on detection.

Repetition codes are practical only in very few applications, because each code word, no matter how long, transmits only one bit of actual information. If n is "large" (3 or 4 is already large, here), then the overhead is high. Theoretically, repetition codes are perfectly valid, though, and, furthermore, provide a rather simple proof of the *fundamental theorem of information theory* (see Complements 13.2).

13.2.2 Parity Check Codes

Suppose we want to use as information strings all possible 2-bit strings, that is, 00, 01, 10, 11. We may decide to encode them by prefixing each of them with a 0 or a 1 so that the resulting 3-bit string contains an even number of ones. Then 00 is encoded 000, 01 is encoded 101, 10 is encoded 110, and 11 is encoded 011. The prefix we used is the **parity check**.

Example 3 **Parity Check Code of Length 3.** $C = \{000, 011, 101, 110\}$. The four code words each contain an even number of ones each, and the other four 3-bit strings contain an odd number of ones.

It is only natural that the decoding algorithm accept a 3-bit string as correct if it has an even number of ones and declare an error if it has an odd number of ones. The decision will clearly be correct if and only if there is no more than one error in the transmission of each code word. As a counterexample, suppose 10 is the information string; it is encoded and transmitted as 110; suppose two errors occur and the word is received as 101; then it is interpreted as 101 and decoded as 01, which is incorrect. Consequently, this code can detect one error, but cannot correct any. On the other hand, the overhead is relatively low: for every three bits we transmit, we have two bits of information, the last two in this example.

For the parity check code of length 3 again, compare Figure 13.4 with the diagram for the 3-repetition code (Figure 13.3). Now the approach is different: there are four "right" classes, each comprising one code word, and one "wrong" class, containing the other four 3-bit strings.

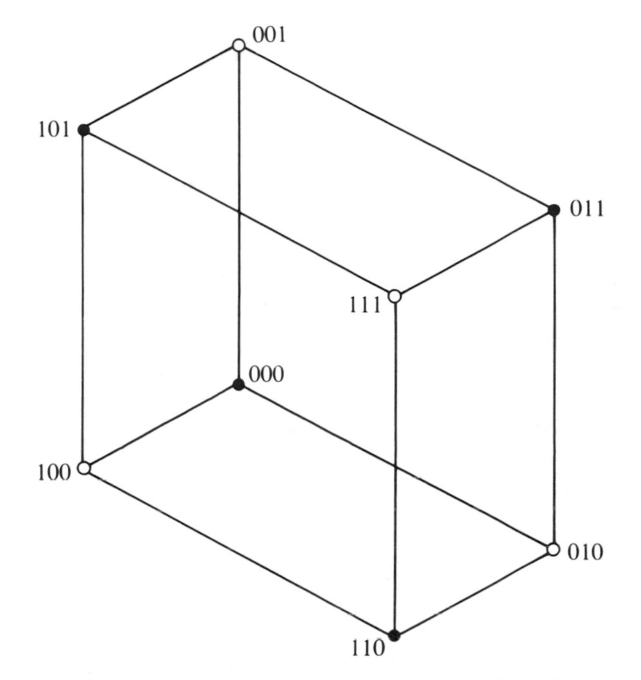

Figure 13.4 Parity check code of length 3.

The diagram points out also the fact that this code cannot be used to correct errors: if we tried to go from a code word to another word through one arc only, we would have some interference, because two different code words may point to the same wrong word. For instance, 011 and 101 both differ from 111 at one bit only.

Clearly, the idea of parity check can be extended to words of any given positive length. Do that as an exercise.

13.2.3 The Hamming Code of Length 7

Most of the codes used in applications are rather long. Usually they are so big that the code words are never enumerated, much less the possible received words. In this example, we begin to follow that practice, although we'll make some exceptions this time.

The *Hamming code of length 7* can be constructed as follows. We start from a specific set of four given code words of length 7, which we write here as four rows:

Example 4 Hamming Code of Length 7. (See also Figure 13.5 below.)

$$\begin{matrix} 0 & 1 & 1 & & 1 & 0 & 0 & 0 \\ 1 & 0 & 1 & & 0 & 1 & 0 & 0 \\ 1 & 1 & 0 & & 0 & 0 & 1 & 0 \\ 1 & 1 & 1 & & 0 & 0 & 0 & 1 \end{matrix}$$

(The special spacing is only to point out a pattern: the last four columns of the array make up the 4×4 identity matrix.) Consider each word as a 1×7 matrix with elements in $\mathbb{Z}_2$ (*not* a Boolean matrix; recall that, in $\mathbb{Z}_2$, $1 + 1 = 0$, $-1 = 1$, etc.). We construct all code words by adding together, as matrices and in all possible ways, any number of distinct words among the four given code words. By convention, adding "no words" gives the result 000 0000.

This is a way to handle the code without enumerating the words. We know how to construct any or all of them, but we don't have to "see" them. If you really want to check, Figure 13.5 lists them all.

0 0 0	0 0 0 0	Adding "no words"
0 1 1	1 0 0 0	"Adding" one at a time
1 0 1	0 1 0 0	
1 1 0	0 0 1 0	
1 1 1	0 0 0 1	
0 1 1	0 1 1 0	Adding two at a time
1 0 1	1 0 1 0	
1 1 0	1 1 0 0	
1 0 0	1 0 0 1	
0 1 0	0 1 0 1	
0 0 1	0 0 1 1	
1 0 0	0 1 1 1	Adding three at a time
0 1 0	1 0 1 1	
0 0 1	1 1 0 1	
0 0 0	1 1 1 0	
1 1 1	1 1 1 1	Adding all four

Figure 13.5 Hamming code of length 7.

There are many patterns in the list of Figure 13.5, but for now we'll comment on only one. When we add distinct words among the four given code words, no 1s in the last four columns cancel out, because the 1s in the four given words occupy different positions. As a consequence, we obtain in the last four positions all sixteen possible 4-bit strings. Then there are 16 code words, among the 128 possible 7-bit strings, and in the last four positions of the code words we have all 16 4-selections from set $\{0, 1\}$, one for each code word. So, we can use the code words to encode all sixteen 4-bit information strings: each 4-bit string is encoded as the only code word that includes that string in its last four bits.

Suppose now that, next to each of the sixteen code words, we write the seven words that are obtained from it by changing a single bit. In the interpretation of coding theory, these would be the received words if the channel made a single mistake. We obtain a total of $16 \times 8 = 128$ strings. Are they all distinct? We could verify that by writing out everything and just checking. We have done that in

Figure 13.6, but it's not really necessary. We can verify the result otherwise: take the two words obtained by changing at most *one* bit in each of two distinct code words. If the two results were the same, then the two original code words would differ at no more than *two* positions. Verifying that this is not the case is much easier, because it implies comparing, two at a time, 16 words, the code words. This implies $\binom{16}{2}$, or 120, verifications. (Later, we'll see an even faster way.) The other approach would take much more work. First, we'd have to write down all 128 words obtained by changing one bit (or none) and, second, we'd have to compare 128 words, two at a time; this implies $\binom{128}{2}$, or over 8000, comparisons, many more than before. (Do you see how counting in advance improves efficiency?)

The conclusion is that, with our one-bit change, we've obtained 128 distinct 7-bit strings. Since 128 is the number of all 7-bit strings ($= 2^7$), we've obtained them all! In other words, we've proved that every received word differs at no more than one bit from precisely one code word. It follows that we can use for this code an algorithm similar to algorithm A of the 3-repetition code: replace any received word with the only code word that differs at no more than one bit from the received word. The decision is correct if the channel makes no more than one error in each code word. In other words, this use gives a 1-error correcting code.

Here too we have a graph, this one with 128 nodes and 448 arcs that join any two nodes differing at one bit only. This gives a forest of 16 mutually disjoint trees, each one having the root at a code word, 7 arcs, and 7 leaves at the "near" words. (Perhaps you see now why we don't want to physically draw this graph.) The favorable circumstance is that, in this case too, these trees take all the nodes of the graph. Not all codes will be so "perfect."

0000000	1000000	0100000	0010000	0001000	0000100	0000010	0000001
0111000	1111000	0011000	0101000	0110000	0111100	0111010	0111001
1010100	0010100	1110100	1000100	1011100	1010000	1010110	1010101
1100010	0100010	1000010	1110010	1101010	1100110	1100000	1100011
1110001	0110001	1010001	1100001	1111001	1110101	1110011	1110000
0110110	1110110	0010110	0100110	0111110	0110010	0110100	0110111
1011010	0011010	1111010	1001010	1010010	1011110	1011000	1011011
1101100	0101100	1001100	1111100	1100100	1101000	1101110	1101101
1001001	0001001	1101001	1011001	1000001	1001101	1001011	1001000
0100101	1100101	0000101	0110101	0101101	0100001	0100111	0100100
0010011	1010011	0110011	0000011	0011011	0010111	0010001	0010010
1000111	0000111	1100111	1010111	1001111	1000011	1000101	1000110
0101011	1101011	0001011	0111011	0100011	0101111	0101001	0101010
0011101	1011101	0111101	0001101	0010101	0011001	0011111	0011100
0001110	1001110	0101110	0011110	0000110	0001010	0001100	0001111
1111111	0111111	1011111	1101111	1110111	1111011	1111101	1111110

Figure 13.6 Hamming code of length 7 and all 128 7-strings.

Another algorithm would be to accept a received word at face value if it is a code word, and to declare an error if it is not. As an exercise, show that this use of the Hamming code of length 7 gives a correct decision if no more than two errors occur in the transmission. The result is therefore a 2-error detecting code.

Exercises 13.2

1. Convert the message ONE BIT by means of the ASCII code, then encode it by means of the 3-repetition code. Do the same for other messages of your choice. Using the computer may help.
2. For each of the following received messages, use decoding algorithm A of the 3-repetition code and ASCII to reconstruct the transmitted alphabetic message:
   (a) 111 000 000 000 111 111 111 111 000 000 111 111 111 111 111 000 000 111 111 111 111 111 000 000 000 111 000 000
   (b) 111 011 000 000 001 110 000 111 000 000 000 001 001 110 111 011 000 100 011 100 001
   (c) 110 000 110 001 100 101 110 110 001 001 110 111 110 111 110 100 110 001 100 101 110 110 100 001 110 111 110 111
   Do the same, as far as possible, using algorithm B. Note the differences.
3. For the 5-repetition code, provide a decoding algorithm in each of the following cases: **(a)** The code is able to detect four errors. **(b)** The code is able to correct one error and to detect two more. **(c)** The code is able to correct two errors.
   In each case, specify under what conditions the decisions will be correct. This exercise does not require special theories: good sense will be enough. Test each algorithm on a few concrete cases. For instance, 11111 is received as 11010, 00000 is received as 01000, 00000 is received as 10001, 11111 is received as 11111.
4. **Parity check code of length n.** Let n be given ($n > 1$). Let the information strings be all the ($n - 1$)-bit strings. Describe the code obtained by adjoining a single bit, as a parity check, to each information string, to obtain a code word of length n. Explore the correction or detection capabilities of this code. Give decoding algorithms, together with a statement of the conditions under which each algorithm will make correct decisions. [Hints. Use the parity check code of length 3 as a guideline. Do first the case in which $n = 4$.]
5. Show that the parity check code for which $n = 2$ is just the 2-repetition code.
6. When $n = 8$, use the parity check code to do a few examples of encoding the 7-bit strings of the ASCII code, considered as information strings. Such encoded 8-bit strings comprise the *8-bit parity check ASCII code,* which is widely used to transmit ASCII data from one computer to another. It fits well any 8-bit, 16-bit, or 32-bit internal configuration of the computer.
7. Show that the Hamming code of length 7 can be used as a 2-error detecting code.

8. Consider the Hamming code of length 7. For each of the 4-bit strings **(a)** 0010, **(b)** 0101, **(c)** 1101, **(d)** 1111, **(e)** others of your choice, do the following:
  (i) Encode it into a code word.
  (ii) Assume that one error occurs in the transmission, at bit 2. Write the received word. Decode the received word using the 1-error correcting algorithm. Check whether the decoded word equals the transmitted word.
  (iii) Still assuming that one error occurs in the transmission, at bit 2, decode the received word using the 2-error detecting algorithm. Compare the conclusions.
  (iv) Assume that two errors occur in the transmission, at bits 2 and 6. Write the received word. Decode the received word using the 1-error correcting algorithm. Check whether the decoded word equals the transmitted word.
  (v) Still assuming that two errors occur at bits 2 and 6, decode the received word using the 2-error detecting algorithm. Compare the conclusions.

Complements 13.2

1. Find the number of arcs of the graph whose nodes are the 128 7-bit strings and whose arcs are the joins of strings that differ at one bit only. [Hints: The result is in the text. Use induction of n, the length of the code words.]

2. As we said before, repetition codes are practical only in very few applications, but we can use them to provide a rather simple proof of the following theorem:

Theorem 3 **Fundamental theorem of information theory.** If the probability that a channel transmit one bit correctly is greater than 0.5, then there exist codes for which the probability that a received word be decoded correctly is as near certainty as desired.

Proof: A full proof would require arguments from probability theory and we'll not attempt it here. We point out only one key element: although a repetition code would generally not be the most efficient code to obtain the result, a suitably long repetition code proves the theorem. If n is large enough, then the probability of receiving more correct bits than wrong bits increases to near certainty, and that's all we need to decode correctly the received word.

13.3 Canonical Linear Binary Codes

In this section, we'll consider, in general, an ample family of codes, the *canonical linear binary codes*. (We'll define the adjectives "canonical" and "linear" in a moment.) These codes are interesting theoretically, but many of them are extremely important in concrete, practical applications. Some of them, for example, have been used for the transmission of data from satellite to ground station. The efficiency of these codes is particularly good when the channel produces *random* errors, that is, when there is no reason to believe that there is a concentration of errors at any one special bit of every code word. This assumption is not true for every channel; a voice-activated microphone, for instance, tends to make more errors at the beginning of each utterance.

We'll use matrices over $\mathbb{Z}_2$, that is, matrices with elements 0 and 1 only, subject to addition and multiplication mod 2 (recall, for instance, that $1 + 1 = 0$ and $-1 = 1$). Such matrices will be rectangular or, in particular, square. A code word of length n, as well as any n-bit string, will be considered as a $1 \times n$ matrix. A given sequence of any number h of such strings can be then written as an $h \times n$ matrix. Clearly, each column of this matrix will be an h-bit string, if we want to interpret it so.

13.3.1 Definitions

A **canonical linear binary code** is defined as a code C that admits the following construction:

1. There exists a $k \times n$ matrix G with elements in $\mathbb{Z}_2$ such that:
   - $0 \leq k \leq n$.
   - The last k columns of G constitute the $k \times k$ identity matrix. Consequently, G is a matrix of the form $G = [K \quad I]$, where I is the $k \times k$ identity matrix and K is a $k \times (n - k)$ matrix.
   - In the degenerate case in which $0 < k = n$, K is empty (to be pedantic, K has k rows and no columns). This case is acceptable, theoretically and practically, but it is not very useful in practice. In the degenerate case in which $k = 0$, both K and I are empty (K has no rows and n columns and I is really empty). This case has only abstract interest.
2. All the elements of the code C are only those strings that are obtained by adding, as matrices, any number of rows of matrix G, chosen in all possible ways. The degenerate case in which we "add no rows" yields as a result the zero n-string.

The elements of C are called **code words**. They are **binary** code words because the code words are strings over the **code alphabet** $\{0, 1\}$. The code is **linear** because the sum of any two elements of C is again in C (as a consequence of the fact that a sum of sums of rows of G is still a sum of rows of G). And finally the code is **canonical** because the last k bits of the code words include *all* k-bit strings (same proof as in the case of the Hamming code of length 7 above).

The number n is called the **length** of the code; n is also the length of every code word. The number k is called the **dimension** of the code. A code of length n and dimension k is said to be an (n, k) code. Matrix G is called the **canonical generating matrix** of the code.

In example 3, the parity check code of length 3, we have $n = 3$, $k = 2$, and

$$G = \begin{bmatrix} 1 & 1 & 0 \\ 1 & 0 & 1 \end{bmatrix} \qquad K = \begin{bmatrix} 1 \\ 1 \end{bmatrix}$$

In example 4, the Hamming code of length 7, we have $n = 7$, $k = 4$, and

$$G = \begin{bmatrix} 0 & 1 & 1 & 1 & 0 & 0 & 0 \\ 1 & 0 & 1 & 0 & 1 & 0 & 0 \\ 1 & 1 & 0 & 0 & 0 & 1 & 0 \\ 1 & 1 & 1 & 0 & 0 & 0 & 1 \end{bmatrix} \qquad K = \begin{bmatrix} 0 & 1 & 1 \\ 1 & 0 & 1 \\ 1 & 1 & 0 \\ 1 & 1 & 1 \end{bmatrix}$$

Example 5 In the special cases in which $0 < k = n$, we obtain **the identity code of length n.** Here $n > 0$, $k = n$, $G = I_n = n \times n$ identity matrix. K is, nominally, the $0 \times n$ empty matrix.

13.3.2 Basic Properties of Canonical Linear Binary Codes

Theorem 4 Let data be as in Section 13.3.1. Then: (1) every word of C can be obtained uniquely as the sum of distinct rows of G, including the possibility of a "sum of no rows" for the zero word. (2) If u is an $(n-k)$-bit string and if v is a k-bit string, then $[u \ \ v] \in C$ if and only if $u = vK$ (matrix product; the symbol $[u \ \ v]$ denotes the concatenation $u{:}v$). In other words, $C = \{[vK \ \ v] \in \{0, 1\}^n \mid v \in \{0, 1\}^k\}$. (3) The function $\mathbf{E}$: $\{0, 1\}^k \to C$, $v \mapsto [vK \ \ v]$ is a bijection. (4) $\#C = 2^k$. Consequently, k is uniquely determined by C.

Proof: By definition, every word of C is a sum of rows of G. If two addends of such a sum are equal, then they cancel each other and can be omitted. Ultimately, every word is obtained as a sum of distinct addends, possibly none. This proves part 1, except for uniqueness. Observe the k-string v. Each column of I has precisely one 1, and these 1s appear in different rows. If we want to obtain the 1s of v in positions $i, j \ldots$ (within v), then the only choice is to add rows $i, j, \ldots$ of G. This proves the uniqueness conclusion of part 1.

According to the definition of product of matrices, adding rows $i, j, \ldots$ of G as above is the same as performing the matrix product vG. Consequently, $[u \ \ v] = vG = [vK \ \ v]$; that is, $u = vK$. This proves part 2, with the additional remark that the mapping $\mathbf{E}$ maps the k-string comprising the last k bits of any one code word to precisely that code word. Clearly $\mathbf{E}^{-1}$ is the function $[u \ \ v] \mapsto v$. This proves part 3. Since $\mathbf{E}$ is a bijection, $\#C = \#(\{0, 1\}^k)$, so $\#C = 2^k$. Part 4 is so proved. Q.E.D.

Example 6 Consider again example 3, the parity check code of length 3, $C = \{000, 011, 101, 110\}$. Then $n = 3$, $k = 2$, and

$$G = \begin{bmatrix} 1 & 1 & 0 \\ 1 & 0 & 1 \end{bmatrix}.$$

Figure 13.7 shows how we can obtain the entire code. Observe that, if $0 < k \leq n$, then all the information about the code is contained in matrix K. Namely,

- since K is $k \times (n - k)$, the number of rows of K is k;
- the sum of the numbers of rows and columns is n;
- $G = [K \quad I]$, where I is uniquely determined by the fact that it is the $k \times k$ identity matrix; and
- C is obtained by adding distinct rows of G in all possible ways.

It is advisable at this point to do the exercises that ask you to verify some of these properties on concretely given codes.

13.3.3 Encoding Algorithm

In Section 13.1 we've commented on all steps of the encoding process except one, how to encode an information string as a code word. This is what an encoding algorithm does. In the present case of a canonical linear binary code, the information strings are the k-bit strings. Encoding one of them means simply applying the function **E** to it. More explicitly, if the code is given as in Section 13.3.1 and if v is an information string (that is, if v is a k-bit string), then v is encoded as the word $\mathbf{E}v := [vK \quad v]$. Equivalently, $v \mapsto vG$.

As an example, see the columns headed "When v" and "and obtain" in Figure 13.7. In the exercises, we've already asked you to practice this process (problem 8 in Exercises 13.2). You may want to practice some more encoding.

13.3.4 Criteria for a Decoding Process

If the channel were so good as to introduce no errors, then the decoding process would consist in just applying the function $\mathbf{E}^{-1}$ to the received word (see Section 13.3.3). But, in that case, we wouldn't need anything more complicated than a (1, 1) code or an identity code (example 5). But channels are not that good, and

When $v =$	we're adding	and obtain	which is
11	rows 1 and 2	011	$[11]\begin{bmatrix} 1 & 10 \\ 1 & 01 \end{bmatrix}$
01	row 2	101	$[01]\begin{bmatrix} 1 & 10 \\ 1 & 01 \end{bmatrix}$
10	row 1	110	$[10]\begin{bmatrix} 1 & 10 \\ 1 & 01 \end{bmatrix}$
00	no row	000	$[00]\begin{bmatrix} 1 & 10 \\ 1 & 01 \end{bmatrix}$

Figure 13.7 Parity check code of length 3.

we must make some decisions on how to handle errors. We could, of course, give you a list of decoding algorithms and the proofs that they are correct, but our goal is not to provide you with a manual. What we want to do here is to help you understand where the difficulties are, how to overcome them, how to construct a procedure and, ultimately, how to produce a decoding algorithm.

Any decoding procedure must follow some good-sense guidelines:

1. If the received word y is a code word, then we'd like to assume that the transmitted word was the same as y. This conclusion may be correct or not.
2. If the received word is not a code word, then we can draw the certain, but not very profound, conclusion that there was an error. Stopping at this conclusion could be expensive, because we'd have to either accept the error or ask for a retransmission. When we cannot, or do not want to, do anything better, this process goes under the technical name of **error detecting**.
3. Otherwise, we may want to try to guess what the transmitted word was. This process is called **error correcting**.
4. Whatever decision we make, we have to have criteria to know whether we stand a chance of having made a correct decision.
5. Efficiency is of paramount importance. Keep in mind that, in most real-life codes, n and k are big, while $n - k$, although not minuscule, is comparatively small.

13.3.5 More Properties of Canonical Codes

The first question within guideline 1 is how to recognize whether an n-bit string y is a code word. One method would be simple, but extremely inefficient: construct all 2^k code words and compare y with each of them. The following theorem gives a much more efficient criterion.

Theorem 5 Let C be an (n, k) code as in Section 13.3.1. Let H be the $n \times (n - k)$ matrix defined by

$$H := \begin{bmatrix} I \\ K \end{bmatrix}$$

where I is the $(n - k) \times (n - k)$ identity matrix and K is defined as in Section 13.3.1. Matrix H is called the **check matrix** of the code. Occasionally it's called the *parity check matrix,* although it may not check parity proper. Let y be any n-bit string. Then $y \in C$ if and only if $yH = 0$.

Proof: By part 2 of Theorem 4 we know that $y \in C$ if and only if $y = [vK \;\; v]$ for some v in $\{0, 1\}^k$. Suppose y has such form. Then $yH = [vK \;\; v]\begin{bmatrix} I \\ K \end{bmatrix} = vK + vK = 0$ (remember that $1 + 1 = 0$ in $\mathbb{Z}_2$). The "only if" part is so proved. For the "if" part, suppose $yH = 0$ and split y in the form $[u \;\; v]$, where u is $1 \times (n - k)$ and v is $1 \times k$. Then $yH = 0$ gives $[u \;\; v]\begin{bmatrix} I \\ K \end{bmatrix} = 0$, that is, $u + vK = 0$, which implies $u = vK$, $y = [vK \;\; v]$, and $y \in C$. Q.E.D.

Example 7 For code {000, 011, 101, 110} of example 3, we have $K = \begin{bmatrix}1\\1\end{bmatrix}$, $H = \begin{bmatrix}1\\1\\1\end{bmatrix}$. If y is written as $[a\ b\ c]$, then $yH = 0$ gives $[a\ b\ c]\begin{bmatrix}1\\1\\1\end{bmatrix} = 0$; that is, $a + b + c = 0$, which means precisely that the number of ones among a, b, and c is even.

This method is much more efficient than the brute force comparison, because the calculation of yH involves no more than $n(n - k)$ additions of bits, and checking whether the result is zero involves just one comparison on a string of $n - k$ bits. The first method would imply the construction of the code (which could be done once for all and stored, but that's still a long process) and, potentially, 2^k comparisons at n bits for each y.

Putting together guidelines 1 and 4, and assuming that the received word y is a code word, how safe is the conclusion that the transmitted word was y? The conclusion might be wrong only if there were two code words, say x and y, that can both be received as y. For example, the parity check code {000, 011, 101, 110} contains $x = 011$ and $y = 101$. If y is transmitted and received with no errors, and if x is transmitted and received with two errors in positions 1 and 2, then y is received again. The crucial point is to establish at how many positions two code words can differ.

Definition 6 The **weight** of a binary string y is denoted **w**y and defined as the number of ones that appear in y. The **Hamming distance** of two strings x and y of equal lengths is defined as $\mathbf{w}(x - y)$. Clearly, $\mathbf{w}(w - y)$ is the number of positions at which x and y differ. The **distance** of a binary code whose words have all the same length is defined as the minimal value of the Hamming distance of pairs of distinct code words. The usual symbol is d. For the degenerate case in which the only code word is zero, we set $d := 0$. For a linear binary code as in Section 13.3.1, the correspondence **syn**: $\{0, 1\} \to \{0, 1\}^{n-k}$ is defined as $y \mapsto yH$. For each y, **syn** y is called the **syndrome** of y.

Later we'll see the reason for this name. Clearly, **syn** $y = 0$ if and only if $y \in C$.

Example 8 For code {000, 011, 101, 110}, we have $\mathbf{w}(000) = 0$, $\mathbf{w}(011) = 2$, $\mathbf{w}(101) = 2$, $\mathbf{w}(110) = 2$; also, $\mathbf{w}(000 - 011) = \mathbf{w}(011) = 2$, $\mathbf{w}(000 - 101) = \mathbf{w}(101) = 2$, $\mathbf{w}(000 - 110) = \mathbf{w}(110) = 2$, $\mathbf{w}(011 - 101) = \mathbf{w}(110) = 2$, $\mathbf{w}(011 - 110) = \mathbf{w}(101) = 2$, $\mathbf{w}(101 - 110) = \mathbf{w}(011) = 2$; therefore $d = 2$. Since $H = \begin{bmatrix}1\\1\\1\end{bmatrix}$, we have, for instance, $\mathbf{syn}(011) = 0$, $\mathbf{syn}(111) = 1$.

Theorem 7 Let n be a given positive integer. For every choice of t, u, v in $\{0, 1\}^n$,

1. $\mathbf{w}t = 0$ if and only if $t = 0$; otherwise, $\mathbf{w}t > 0$;

2. $\mathbf{w}(t - u) = 0$ if and only if $t = u$; otherwise, $\mathbf{w}(t - u) > 0$;
3. $\mathbf{w}(t - u) = \mathbf{w}(u - t)$ **(symmetry of the Hamming distance)**;
4. $\mathbf{w}(t - v) \leq \mathbf{w}(t - u) + \mathbf{w}(u - v)$ **(triangularity of the Hamming distance)**.

Proof: Parts 1, 2 and 3 are obvious if you recall that $t - u = t + u$. For part 4, observe that, if t and v differ at any one position, then necessarily t and u, or u and v (but not both pairs) differ at that same position. Therefore {positions at which t and v differ} = {positions at which t and u differ} ∪ {positions at which u and v differ} − (intersection of these two sets). Calculating the cardinalities of these sets implies the conclusion. Q.E.D.

Theorem 8 Let C be an (n, k) code with matrices G, H and K as defined in Section 13.3.1 and Theorem 5.

1. The minimal weight s of the nonzero code words equals the distance d of the code. The integer s is to be interpreted as 0 if the only code word is zero.

2. Let r be the only positive integer such that
- (a) either $r > n$, or
  (b) there exist r distinct rows in H whose sum is zero;
- *and* the sum of any positive number, less than r, of distinct rows of H is other than zero.

Then $r = d$.

3. The correspondence **syn**: $\{0, 1\}^n \to \{0, 1\}^{n-k}$, $y \mapsto yH$ is a total surjective function.

4. For any natural number p, define
- $A_p := \{z \in \{0, 1\}^n \mid \mathbf{w}z \leq p\}$,
- $V_p := \{zH \in \{0, 1\}^{n-k} \mid z \in A_p\}$,
- $L: A_p \to V_p$, $Lz := zH$ for every z in A_p.

If $2p < d$, then L is a bijection.

Proof: The degenerate case in which $d = 0$ is trivial. When $d > 0$, let t and u be any code words such that $\mathbf{w}(t - u) = d$. Then $t \neq u$ and $v := t - u$ is a nonzero code word. The minimality of s implies that $s \leq \mathbf{w}v = \mathbf{w}(t - u) = d$. Conversely, let v be any nonzero code word of weight s. Since zero is a code word distinct from v, the minimality of d implies $d \leq \mathbf{w}(t - \text{zero}) = \mathbf{w}t = s$. Since $s \leq d$ and $d \leq s$, part 1 is proved.

We know that there exists a nonzero code word, say t, whose weight is d. This means that there are precisely d ones in t and, therefore, the string tH is the sum of d distinct rows of H, and this sum is zero. Also, there are no strings u in C with positive weight less than d. That is the same as saying that there is no combination of rows of H (in a number less than d, but positive) whose sum is zero. By the definition of r, we have proved that $d = r$. Part 2 is so proved.

The fact that the mapping $t \mapsto \mathbf{syn}\ t := tH$ is a total function is obvious. In order to prove that it is surjective, we have to prove that for each $(n - k)$-string z there is an n-string y such that $yH = z$. Observe that the first $n - k$ rows of H comprise the identity matrix. Set $y := [z \ \ 0]$, where 0 denotes the $1 \times k$ matrix with all zero entries. Then $yH = z * \mathrm{I} + 0 * K = z$, as we wanted to prove. Part 3 is so proved.

For part 4, observe that L is the result of applying the function **syn** just to the elements of A_p. Consequently, and independently of the value of p, L is total. By the definition of V_p, L is surjective. All we need to prove is that L is injective. Let's analyze under what circumstances L may fail to be injective. There must be two distinct n-strings t and u such that $\mathbf{w}t \le p$, $\mathbf{w}u \le p$, and $tH = uH$. This implies $t - u \ne 0$, $(t - u)H = 0$, and $t - u \in C$. By the definition of d, we must have $d \le \mathbf{w}(t - u)$. By triangularity, $d \le \mathbf{w}(t - u) \le \mathbf{w}(t - \mathrm{zero}) + \mathbf{w}(\mathrm{zero} - u) = \mathbf{w}t + \mathbf{w}u \le 2p$. We have so proved: if L is not injective, then $d \le 2p$. Then we have also proved the contrapositive statement: if $2p < d$, then L is injective. Part 4 is proved, and so is the theorem. Q.E.D.

Now we can clarify guideline 1: if $0 \le 2p < d$, if no more than p errors occur in the transmission of one code word, and if **syn** $y = 0$, then y was the transmitted word.

This line of reasoning will clarify also guideline 3. Suppose the received word y is not a code word. Then a reasonable guess is that the transmitted word x is the code word that is "closest" to y, because that would be the code word that presupposes the least number of errors. (We cannot *assume* this; it's a guess. We have to analyze it.) Consider the possibility that there are two distinct code words, x and t, that fit the bill, that is: $\mathbf{w}(x - y) = a$, $\mathbf{w}(t - y) = a$, and $a \le \mathbf{w}(u - y)$ for every u in C. If that is the case, then $2a = \mathbf{w}(x - y) + \mathbf{w}(t - y) \ge \mathbf{w}(x - t) \ge d$. It can happen, but only if $2a \ge d$. If $2a < d$, then it can't.

Combining guidelines 3 and 4, we have another firm conclusion: Let p be such that $0 \le 2p < d$ and assume that no more than p errors occur in the transmission of one code word. Then, for each received word y, there is at most one code word x such that $\mathbf{w}(x - y) \le p$. Two questions arise immediately: Is there such an x? How can we find it? Set $t := x - y$; then $x = y + t$ and t is called the **correction**. If we can calculate t, then we can calculate x as $x := y + t$. Since we assume that no more than p errors occur, we know that $\mathbf{w}t \le p$, that is, $t \in A_p$. By the definition of mapping L, we know that **syn** $t = Lt \in V_p$. Since L is a bijection, $t = L^{-1}Lt = L^{-1}$ **syn** t. Also, **syn** t = **syn**$(x + y) = (x + y)H = xH + yH$. Since $x \in C$, we have that $xH = 0$ and **syn** $t = yH =$ **syn** y. Calculating **syn** y is immediate. Then $t = L^{-1}$ **syn** y and $x = y + L^{-1}$ **syn** y. This answers both questions: if no more than p errors occur, then a unique x exists and we have a method to calculate it. This method is certainly more efficient than comparing y with all code words and selecting the closest one. We are now following also guideline 5.

In order to see how efficient this method is, observe that the calculation of **syn** y is very quick. Calculating L^{-1} is somewhat more time consuming. We need a table lookup over V_p (remember that a function is a set of pairs, that is, a table). To estimate the efficiency of the lookup operation, we need the cardinality of V_p. Since L is a bijection, $\#V_p = \#A_p$. Using the methods of Chapter 4, the elements of A_p can be obtained, once each, by changing the zero n-string at no position, at one position, at two positions and so on up to p positions. There are $\binom{n}{p}$ ways to change p positions, . . ., $\binom{n}{2}$ ways to change two positions, $\binom{n}{1}$ ways to change one position, $\binom{n}{0}$ ways to change 0 positions. Therefore

$$\#V_p = \#A_p = \sum_{i=0}^{p} \binom{n}{i}. \tag{3}$$

As we'll see in Complements 13.3, $d \leq n - k + 1$. Then the expression in equation 3 does not exceed $\Sigma_{0 \leq i < (n-k+1)/2} \binom{n}{i}$. Since $n - k$ is reasonably small, the process, although not extremely quick, is reasonably efficient.

We have so found a fail-safe error correcting procedure: let p be given so that $0 \leq 2p < d$. Assume that no more that p errors occur in the transmission of one code word. Then, if y is the received word and if $\mathbf{syn}\, y = 0$, then $y + L^{-1}\, \mathbf{syn}\, y$ is the transmitted word. We'll be able, in a moment, to refine this conclusion slightly, for any linear binary code. (Because of the uncertainty in the efficiency of the table lookup, one of the goals of research in coding theory is to find special codes for which that part can be made even more efficient or replaced with another procedure.)

Since $0 \leq 2p < d$, we can select another natural number q $(q \geq 0)$ such that $2p + q < d$. We can draw some more conclusions under the more relaxed assumption that the transmission introduces no more than $p + q$ errors in one code word. In other words, let x be the transmitted word, y the received word, and assume $\mathbf{w}(x - y) \leq p + q$. If $\mathbf{syn}\, y \notin V_p$, then we know that more than p errors occurred. But if $\mathbf{syn}\, y \in V_p$ (remember, we're working under the assumption of no more than $p + q$ errors), can we conclude that there weren't actually more than p errors? The answer is affirmative, for the following reason. We know that there is a unique correction t such that $\mathbf{syn}\, t = \mathbf{syn}\, y \in V_p$ and $0 \leq \mathbf{w}t < p$. If there were another correction u such that $p < \mathbf{w}u \leq p + q$, then $u \neq t$ (because $\mathbf{w}u \neq \mathbf{w}t$) and $\mathbf{w}(u - t) \leq \mathbf{w}(u - 0) + \mathbf{w}(0 - t) = \mathbf{w}u + \mathbf{w}t \leq p + q + p < d$. Since $\mathbf{syn}\, u - \mathbf{syn}\, t = 0$, we know that $u - t$ is a code word. The fact that $\mathbf{w}(u - t) < d$ would imply, by the definition of d, that $u = t$, which is a contradiction.

With the given choice of p and q we have therefore a stronger conclusion: Let p and q be given so that $p \geq 0$, $q \geq 0$, and $2p + q < d$, and assume that there are no more than $p + q$ transmission errors in the received word y. Then,

- if $\mathbf{syn}\, y \in V_p$, no more than p errors occurred and $y + L^{-1}\, \mathbf{syn}\, y$ was the transmitted word;
- if $\mathbf{syn}\, y \notin V_p$, more than p errors occurred.

We are so able to *correct* up to p errors and to *detect* q more errors (that is, no pattern of up to $p + q$ errors goes undetected).

This is not the ultimate decoding procedure. Special types of codes may have better and faster decoding procedures. But the one we've presented is efficient, and it's general in the sense that it applies to all canonical linear binary codes.

Furthermore, it's rather easily extended to all linear codes, binary or not, canonical or not.

There's one more, final catch. Mathematics is ideal, but our knowledge of it is not. It's a fact of life that, although each code has its own uniquely determined code distance d, we don't have today a general formula to calculate d in all cases. In principle, the calculation of d is simple: use part 1 of Theorem 8, scan all nonzero words of C, and take the weight of one with the smallest number of ones. In the example of the parity check code of length 3, the nonzero code words are 011, 101, 110, and the result is 2. Easy. But, for large codes, this method is time consuming and does not give a formula. In specific cases, there may be other methods. When d cannot be calculated, the next best is to find a *lower bound* for d, that is, a number b such that, although we don't know d, we know that $b \leq d$.

We now have the criteria, in principle, to construct a decoding procedure. We'll actually do this in Section 13.5.

Exercises 13.3

1. For each of the following codes, determine the values of n and k, write matrix G, calculate $\#C$, and compare it with $\#(\{0, 1\}^n)$. **(a)** The code in example 1. **(b)** The code in example 3. **(c)** The code in example 4. **(d)** $C := \{00, 11\}$. **(e)** $C := \{0000, 1111\}$. **(f)** $C := \{00, 01, 10, 11\}$. **(g)** $C := \{0000, 0011, 0101, 0110\}$. **(h)** $C := \{0, 1\}^4$.
2. For several of the codes in exercise 1, verify some of the properties stated in Theorems 4 and 8.
3. Observe that **syn** is a total surjective function from $\{0, 1\}^n$ onto $\{0, 1\}^{n-k}$. According to the methods of Section 10.10, function **syn** determines on set $\{0, 1\}^n$ an equivalence relation, "having the same syndrome". The code itself is clearly the class of zero, that is, the set of those words whose syndrome is zero. Count the elements of each class; count the number of classes; check that you've obtained all elements of $\{0, 1\}^n$.

Complements 13.3

1. **Upper bounds for *d*.** Prove that, in any canonical linear binary (n, k) code, $n - k + 1 \geq d$. [Hint. Use part 2 of Theorem 8.] Prove that, when $n - k$ is given, there are precisely two canonical linear binary codes for which $d = n - k + 1$. Which ones? [Hint. First do some examples, for instance, $n - k = 2$ or 3 or 4. Look at H.]

   Another upper bound for d is found as follows: For each code word, consider the set of those n-strings that have distance less than $d/2$ from that code word. Find the cardinality of such a set. These sets, one for each code word, are pairwise disjoint. Find the cardinality of their union. This union is a subset of the set of all n-strings. Formalize this argument to prove that $\Sigma_{0 \le j < d/2} C(n, j) \le 2^{n-k}$. Verify that, for given n and k, this inequality provides an upper bound for d.

13.4 Summary of Notations

Our system of notations is proliferating. Since the symbols we have used are all rather commonly accepted, let's summarize them here.

For any *canonical linear binary code C:*

- $n :=$ length of C; $k :=$ dimension of C, $0 \le k \le n$.
- Canonical generating matrix of C, $G := [K \quad I]$ where K is $k \times (n - k)$, and I is the $k \times k$ identity matrix; G is $n \times k$.
- Canonical check matrix, $H := \begin{bmatrix} I \\ K \end{bmatrix}$, where I is the $(n - k) \times (n - k)$ identity matrix; H is $n \times (n - k)$.
- Weight of string y, $\mathbf{w}y :=$ number of ones in y. String y is said to be *odd* or *even* according as $\mathbf{w}y$ is an odd or even integer. Hamming distance of strings x and $y := \mathbf{w}(x - y)$.
- Distance of code C, $d :=$ minimum value of $\mathbf{w}(x - y)$ when $x \in C$, $y \in C$, $x \neq y$. Also, $d =$ minimum value of $\mathbf{w}x$ when $x \in C - \{\text{zero}\}$; also, some d distinct rows of H have zero sum *and* every choice of fewer than d distinct rows of H gives a nonzero sum.
- $(x \in C) \Leftrightarrow (xH = 0) \Leftrightarrow (x = [vK \quad v]$ for some v in $\{0, 1\}^k)$.
- The encoding bijection is $\mathbf{E}$: $\{0, 1\}^k \to C$, $v \mapsto [vK \quad v] = vG$.
- The syndrome is the surjection **syn**: $\{0, 1\}^n \to \{0, 1\}^{n-k}$, $y \mapsto yH$.
- Bijection L: $A_p \to V_p$, $y \mapsto yH$, where $A_p := \{y \in \{0, 1\}^n \mid \mathbf{w}y \le p\}$, $V_p := \{\mathbf{syn}\ y \in \{0, 1\}^{n-k} \mid y \in A_p\}$, $0 \le p < d/2$. Each element of A_p is called a **class leader.** (This refers to the equivalence classes of the relation $\mathbf{syn}\ x_1 = \mathbf{syn}\ x_2$ on $\{0, 1\}^n$. Another common term is **coset leader.**)

13.5 A Decoding Algorithm

Putting together the results above and adding a few remarks, we can now give a decoding algorithm, whose correctness we've already proved. Among the objects needed in the algorithm, we should distinguish between those that can be calculated in advance, once and for all, and those that have to be recalculated for each received word. Extreme efficiency is needed for the latter, moderate efficiency is advisable for the former.

Theorem 9 Let a canonical linear binary (n, k) code C be given. Let a decoding algorithm be constructed as follows. Let no more than $p + q$ errors occur in the transmission of a code word. Then the decision of the algorithm is correct and the code can correct up to p errors and detect up to q additional errors.

Preliminary Constructions. Calculate d, the code distance, for instance as the smallest value of $\mathbf{wt}\, t$ when $t \in C - \{\text{zero}\}$. Alternatively, calculate a lower bound for d.

Select two nonnegative integers p, q such that $2p + q < d$. The closer $2p + q$ is to d, the higher the efficiency with which C is used.

Construct the set $A := \{t \in \{0, 1\}^n \mid \mathbf{wt}\, t \leq p\}$.

Construct the set $V := \{\mathbf{syn}\, t \in \{0, 1\}^{n-k} \mid t \in A\}$.

Tabulate the inverse of the bijection $L: A \to V$ defined by $Lz := \mathbf{syn}\, z$ $(z \in A)$; that is, store the set of pairs $\{(\mathbf{syn}\, t, t) \mid t \in A\}$.

Decoding Algorithm C. For each received word y, that is, for each n-bit string y, calculate $\mathbf{syn}\, y$. (Greek *syn-* = together, *drom-* = run. In medicine, a *syndrome* is a concurrence of symptoms, characteristic of a specific disease; in coding theory, $\mathbf{syn}\, y$ tells how "sick" the word y is.)

1. If $\mathbf{syn}\, y \in V$, then decide that the transmitted word was $y + L^{-1}\, \mathbf{syn}\, y$.
2. If $\mathbf{syn}\, y \notin V$, then decide that there were more than p errors in the transmission.

This use of the code can correct up to p errors and detect up to q additional errors.

In the following sections we'll see more examples, more general binary codes, and transformations of codes, which provide new codes starting from known ones.

Exercises 13.5

1. Reword algorithm C for the case in which $q = 0$. This will provide a p-error correcting code. Also, reword the algorithm when $p = 0$, to obtain a q-error detecting code.
2. For the codes in examples 1, 3, and 4, construct the ingredients that are necessary for algorithm C and test a few concrete cases. Namely, for each of those codes, **(i)** recall n, k, d, G, H, K; **(ii)** select, in several possible ways, nonnegative integers p and q such that $2p + q = d - 1$; do parts (iii) to (viii) for each of these choices; **(iii)** when $p > 0$, construct sets A and V, and the bijection L. **(iv)** Assign yourself two or three k-bit strings. Do parts (v) to (vii) for each of these strings, considered as information strings. **(v)** Encode the string according to our encoding algorithm. **(vi)** Alter one or two bits in the resulting encoded string, to simulate transmission errors. Assume that the result is the received

word. **(vii)** Apply algorithm C to the "received" word; state the conclusions. Verify the correctness of such conclusions (this time you *know* the transmitted word). **(viii)** Assign yourself a few n-bit strings to be considered as received words, apply algorithm C to them, and state the conclusions (this time you do *not* know the transmitted word).

3. Construct the matrix G of a canonical linear binary code of your choice (remember that K is arbitrary, the rest is standard). Calculate n, k, and H. Try to calculate d; failing that, calculate a lower bound for d. Then, for this code, do exercise 2. Repeat this problem for a few nontrivial codes we haven't seen before. Consider also the identity code of length n. Stop when you understand what's going on.

A computer may help, but do by hand at least one nontrivial case. For computer purposes, the only data for the code are n, k, and K. For the calculations, the data are the words to be encoded or decoded.

13.6 Hamming Codes

Definition 10 Given a positive integer m, the canonical *Hamming code of length* $2^m - 1$ is defined as the canonical linear binary $(2^m - 1, 2^m - m - 1)$ code for which the rows of matrix K are all the m-bit strings that have more than one 1 each, arranged in any one given order.

Example 9 When $m = 2$, then $K = [1\ 1]$ and we obtain the repetition code of length 3; see example 1. When $m = 3$, then we obtain the Hamming code of length 7 (see example 4).

Theorem 11 The Hamming code given by Definition 10 has length $n := 2^m - 1$ and dimension $k := n - m$. If $m > 1$, then the code distance is 3. This code can therefore be used as a 1-error correcting code or as a 2-error detecting code.

Proof: The first m rows of H comprise all the m-strings with one 1 each. Consequently, the rows of H comprise all nonnull m-bit strings, that is, $2^m - 1$ strings. This gives immediately the values of n and k. Assume $m > 1$. By construction, no two distinct rows of H are equal, and therefore the sum of any two distinct rows is never zero. By part 2 of Theorem 8, this implies that $d \geq 3$. On the other hand, rows 100 . . ., 010 . . ., and 110 . . . add up to zero. Therefore $d = 3$. The rest of the theorem is a consequence of Theorem 9. Q.E.D.

The exercises ask you to verify the conclusions above on concrete examples and the following conclusions in general.

Select $p := 1$. Then (1) the elements of set A (as given in Theorem 9) are zero and the rows of the $n \times n$ identity matrix; (2) the elements of set V are zero and the rows of H; (3) the function L^{-1} maps row j of H to the n-string with one single 1 in position j; (4) the 1-error correcting decoding algorithm reduces to: if y is the received word, then necessarily **syn** y (if not zero) is a row of H (say row j) and the transmitted word is obtained by changing bit j of word y.

Exercises 13.6

1. Construct the Hamming codes for which m is 1 or 2; note that they are codes we've seen before. Calculate d.
2. When $m = 4$, construct the matrices G and H of the (15, 11) Hamming code. Find the value of d. When $p = 1$, construct the bijection L (see Section 13.4), and enumerate the pairs of L^{-1}.
3. Do exercise 2 on a computer for an arbitrary (but reasonable) value of m. That is, write a program that accepts any reasonable positive value of m and calculates n, k, G, H, K, d. (See also exercise 2 in Exercises 13.5.) Test the program for the values $m = 1, 2, 3$. Check the results towards the Hamming codes we've seen. Run the program when $m = 4$ and for a few higher values. When $m = 4$, check the results towards your answer to exercise 2.
4. Verify in general, the conclusions given in the text after Theorem 11.

13.7 Parity Check

Any canonical linear binary code can be modified slightly by increasing each word by one parity check bit, which is defined to be 0 or 1 so that the augmented word is always even. The adjoined bit is called an **overall parity check** on the given code. For instance, the left-hand column below lists all the code words of a code. The right-hand column lists the corresponding augmented words with an additional parity check bit as a prefix.

000 000	0 000 000
011 100	1 011 100
101 010	1 101 010
110 001	1 110 001
011 011	0 011 011
101 101	0 101 101
110 110	0 110 110
000 111	1 000 111

We might expect the new code to be linear and to carry the same information as the previous code (and no more), but with one extra margin of error. With one exception, this conclusion proves correct.

Theorem 12 Let G^* and H^* be, respectively, the generating matrix and the check matrix of a canonical linear binary $(n - 1, k - 1)$ code C^* $(0 < k \leq n)$. If $k = 1$, then we have a degenerate case, which is included in our considerations. Let G be the $k \times n$ matrix obtained by adjoining to G^* a first column, calculated as the sum of all columns of G^*. Let H be the $n \times (n - k)$ matrix obtained by adjoining to H^* first a first row of all zeroes and then a first column which is calculated as the bitwise complement of the sum of all columns of the increased matrix. Then, (1) matrices G and H are the generating and check matrices, respectively, of a canonical linear binary (n, k) code C; (2) the code words of C are precisely those strings that are obtained by augmenting each code word of C^* by one initial parity check bit; (3) the distance d of code C is the same as the distance d^* of code C^* if d^* is even, and $d = d^* + 1$ if d^* is odd; (4) consequently, if d^* is odd and if C^* can be used as a p-error correcting, q-error detecting code, then C can be used as a p-error correcting, $(q + 1)$-error detecting code, or, if $q > 0$, also as a $(p + 1)$-error correcting, $(q - 1)$-error detecting code.

Proof: Because of the definition of G, any addition of distinct rows of G produces a word of C and, (by deletion of the first element) a word of C^*; and conversely. The sum of all columns of G is zero, because it is actually twice the sum of all columns of G^*. Consequently, each row of G is an even word. Since addition of rows may delete ones only in pairs, the parity of any code word will be the same as the overall parity of the added rows. Since each of the rows of G is even, so is every code word of C. This proves part 1 and the first half of part 2. For the second half of part 2, number the columns of H^* from 2 to $n - k$ and those of H from 1 to $n - k$. Recall that the conditions for x (or x^*) to be in C (or in C^*) are

$$x \text{ (column } l \text{ of } H) = 0 \quad (1 \leq l \leq n - k)$$

or for x^*

$$x^* \text{ (column } l \text{ of } H) = 0 \quad (2 \leq l \leq n - k).$$

Because of the definition of H, those conditions are the same in both cases when $l > 1$. We have to analyze the additional condition obtained when $l = 1$. Given the other equations, the first equation is equivalent to the sum of all $n - k$ equations, which happens to be

$$\sum_{1 \leq i \leq n} \sum_{l=1}^{n-k} h_{il} x_l = 0. \tag{4}$$

Because of the definition of H the first element of each row of H is the complement of the sum of the others. When the sum of the others is added again, then a one is obtained. Then equation 4 is $x_1 + x_2 + \cdots = 1$, which is just the parity check. Part 2 is so proved. For part 3, consider the distances calculated as in part 1 of Theorem 8. If d^* is even, then no 1 is adjoined to any of those words of C^* that contain d^* ones. The other words of C^* have more ones, and another may still have to be adjoined to form a word of C. This implies that $d = d^*$. If d^* is odd, then a one is adjoined to each word of C^* that contains d^* ones. The other code words have more than d^* ones anyhow. This implies that $d = d^* + 1$, and part 3 is proved. The rest of the theorem is a direct consequence of general properties of linear binary codes. Q.E.D.

Note that, according to decoding algorithm C, there is no gain in an overall parity check when the original code has an even distance. This applies, in particular, when all words of the original code are even, including the special case in which a parity check is superimposed to a previous parity check. A (bad) example of this type is the following:

Example 10 **Parity check over the parity check code of length 3.** $C := \{0000, 0011, 0101, 0110\}$. Only four of the eight 4-bit even strings are code words. Then there are eight more 4-bit strings, which are odd and are not code words.

Exercises 13.7

1. For each of the codes in examples 1, 3, and 4: **(a)**: Adjoin an overall parity check; **(b)** For the resulting code, do exercise 2 of Exercises 13.5.
   Do the same for some canonical linear binary codes of your choice.
2. Verify that the parity check code of length n can be obtained by adjoining an overall parity check to the identity code of length $n - 1$.

13.8 Dual Codes

Recall the definition of **transpose** of a matrix. Intuitively, it's the matrix obtained by interchanging the roles of rows and columns of a given matrix. For instance, the transpose of matrix

$$\begin{bmatrix} 1 & 0 & 1 \\ 1 & 1 & 0 \end{bmatrix} \quad \text{is} \quad \begin{bmatrix} 1 & 1 \\ 0 & 1 \\ 1 & 0 \end{bmatrix}.$$

Formally, if $a(i, j)$ is a given $m \times n$ matrix $(1 \leq i \leq m,\ 1 \leq j \leq n)$, then the transposed matrix is defined by $(k, l) \mapsto a(l, k)$ $(1 \leq k \leq n,\ 1 \leq l \leq m)$. The transpose of matrix a is denoted a^{T}.

Suppose that

$$G := [K \;\; I] \quad \text{and} \quad H := \begin{bmatrix} I \\ K \end{bmatrix}$$

are, respectively, the generating and check matrix of a canonical linear binary code C. It's then evident that

$$G^* := [K^{\mathrm{T}} \;\; I] \quad \text{and} \quad H^* := \begin{bmatrix} I \\ K^{\mathrm{T}} \end{bmatrix}$$

are the generating matrix and check matrix of a canonical linear binary code C^*. Code C^* is called the **dual code** of C. Clearly, the dual of an (n, k) code is an $(n, n - k)$ code. Also, the dual of the dual of a code is the original code.

Example 11 The parity check code of length 3 (example 3) has matrices

$$K = \begin{bmatrix} 1 \\ 1 \end{bmatrix} \qquad G = \begin{bmatrix} 1 & 1 & 0 \\ 1 & 0 & 1 \end{bmatrix} \qquad H = \begin{bmatrix} 1 \\ 1 \\ 1 \end{bmatrix}.$$

The dual code has matrices

$$K^* = [1 \quad 1] \qquad G^* = [1 \quad 1 \quad 1] \qquad H^* = \begin{bmatrix} 1 & 0 \\ 0 & 1 \\ 1 & 1 \end{bmatrix};$$

and is the 3-repetition code (example 1).

The operation of dualizing a code is simple and sounds trivial but, in certain instances, it is a powerful tool for producing useful codes. See, for instance, Reed–Mueller codes in Section 13.10 below.

Exercises 13.8

1. For each of the examples 4, 5, and 10, find generating and check matrices of the dual code. Find length, dimension, and distance of the resulting code.

Complements 13.8

1. Let C be any nontrivial canonical linear binary code. For each code word $z_1 z_2 \ldots z_n$ consider the equation $\Sigma_{i=1}^{n} x_i z_i = 0$, in which the zs are given and the xs are unspecified elements of $\mathbb{Z}_2$.

   Define $C^* := \{x_{n-k+1} \ldots x_n\, x_1 \ldots x_{n-k} \in \{0, 1\}^n \mid \Sigma_{i=1}^{n} x_i z_i = 0$ for every $z_1 z_2 \ldots z_n$ in $C\}$. Prove that C^* is the dual code of C.

13.9 Reduction of a Code

It may happen that all the words of a canonical code contain a zero in a given position, say at position j. Clearly, $j \leq n - k$, because the code words in G contain ones in positions $n - k + 1$ through n. Also, all code words have a zero in position j if and only if column j of G is zero. Clearly, in this case position j does not carry any information. We'll see in a moment that, according to decoding algorithm C, position j does not contribute anything to error-handling capabilities either. We may therefore want to eliminate position j altogether. This will shorten transmission time without lessening the power of the code in any form. It's a matter of easy verification that the suppression of bit j in every code word is formally equivalent to deleting column j in G (and in K), and deleting column j *and* row j in H. The result is an $(n - 1, k)$ code C^*. As an exercise, show that the distance of C^* is the same as the distance of C. Under these conditions, the operation of deleting position j is called **reduction** or **expurgation** of the code. There are more general processes to reduce a code but, since we do not need them, we can disregard them.

Exercises 13.9

1. For the codes in examples 4, 5, and 10, see whether a reduction is possible and, if it is, perform it. Do the same for each of the codes obtained in the first problem of Exercises 13.8.
2. Do all possible reductions on the code with generating matrix

$$\begin{bmatrix} 0 & 0 & 1 & 1 & 1 & 1 & 1 & 0 & 0 & 0 \\ 1 & 0 & 0 & 0 & 1 & 0 & 0 & 1 & 0 & 0 \\ 1 & 0 & 1 & 1 & 0 & 0 & 0 & 0 & 1 & 0 \\ 1 & 0 & 1 & 0 & 1 & 0 & 0 & 0 & 0 & 1 \end{bmatrix}.$$

   Verify that the resulting code has the same distance as the original code.
3. Prove that reducing a code does not alter the code's distance nor its dimension.
4. Prove that, if an overall parity check is adjoined to a code for which all words are even, then the added bit can be reduced.

13.10 Reed–Mueller Codes

An ample family of codes is provided by the **Reed–Mueller codes,** or **RM codes** for short. There are RM codes of several **orders**. We'll consider only RM codes of two particular orders. They are defined thus:

- For any positive integer m, let C^{**} be the canonical Hamming code of length $2^m - 1$ (see Section 13.6).
- Let C^* be the code obtained by adjoining to C^{**} an overall parity check (see Section 13.7). Then C^* is a $(2^m, 2^m - m - 1)$ code, which is defined as the RM code of length 2^m and order $m - 2$.
- Let C be the dual of C^* (see Section 13.8). Then C is a $(2^m, m + 1)$ code, which is defined as the RM code of length 2^m and order 1.

The Reed–Mueller code of length 2^m and order 1 has distance 2^{m-1} and the Reed–Mueller code of length 2^m and order $m - 2$ has distance 4.

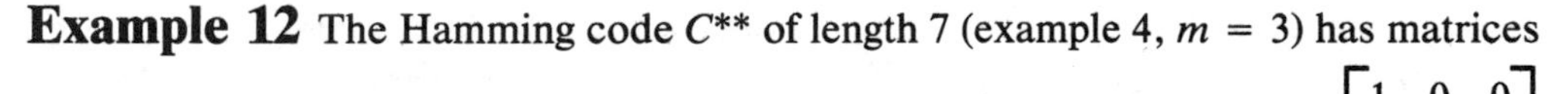

Example 12 The Hamming code C^{**} of length 7 (example 4, $m = 3$) has matrices

$$G^{**} = \begin{bmatrix} 0 & 1 & 1 & 1 & 0 & 0 & 0 \\ 1 & 0 & 1 & 0 & 1 & 0 & 0 \\ 1 & 1 & 0 & 0 & 0 & 1 & 0 \\ 1 & 1 & 1 & 0 & 0 & 0 & 1 \end{bmatrix} \quad \text{and} \quad H^{**} = \begin{bmatrix} 1 & 0 & 0 \\ 0 & 1 & 0 \\ 0 & 0 & 1 \\ 0 & 1 & 1 \\ 1 & 0 & 1 \\ 1 & 1 & 0 \\ 1 & 1 & 1 \end{bmatrix}.$$

The distance of C^{**} is 3. Adjoining an overall parity check gives the following code C^*.

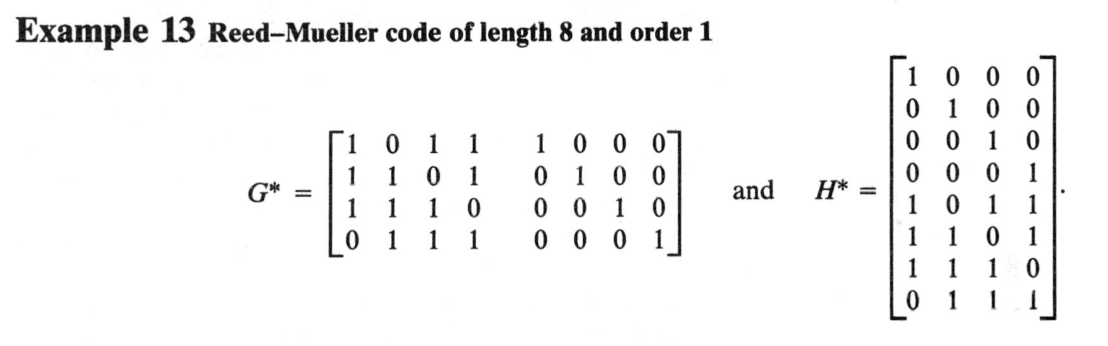

Example 13 **Reed–Mueller code of length 8 and order 1**

$$G^{*} = \begin{bmatrix} 1 & 0 & 1 & 1 & 1 & 0 & 0 & 0 \\ 1 & 1 & 0 & 1 & 0 & 1 & 0 & 0 \\ 1 & 1 & 1 & 0 & 0 & 0 & 1 & 0 \\ 0 & 1 & 1 & 1 & 0 & 0 & 0 & 1 \end{bmatrix} \quad \text{and} \quad H^{*} = \begin{bmatrix} 1 & 0 & 0 & 0 \\ 0 & 1 & 0 & 0 \\ 0 & 0 & 1 & 0 \\ 0 & 0 & 0 & 1 \\ 1 & 0 & 1 & 1 \\ 1 & 1 & 0 & 1 \\ 1 & 1 & 1 & 0 \\ 0 & 1 & 1 & 1 \end{bmatrix}.$$

The distance of C^* is 4. The dual of C^* is the following code C.

Example 14 **Reed–Mueller code of length 8 and order 1**

$$G = \begin{bmatrix} 1 & 1 & 1 & 0 & 1 & 0 & 0 & 0 \\ 0 & 1 & 1 & 1 & 0 & 1 & 0 & 0 \\ 1 & 0 & 1 & 1 & 0 & 0 & 1 & 0 \\ 1 & 1 & 0 & 1 & 0 & 0 & 0 & 1 \end{bmatrix} \quad \text{and} \quad H = \begin{bmatrix} 1 & 0 & 0 & 0 \\ 0 & 1 & 0 & 0 \\ 0 & 0 & 1 & 0 \\ 0 & 0 & 0 & 1 \\ 1 & 1 & 1 & 0 \\ 0 & 1 & 1 & 1 \\ 1 & 0 & 1 & 1 \\ 1 & 1 & 0 & 1 \end{bmatrix}.$$

The distance of C is 4. In this case ($m = 3$), the codes C^* and C are essentially the same, although G^* and G are different. These two codes can be reduced to each other by means of a simple operation which we will see in Section 13.11.

Exercises 13.10

1. Construct the RM code of length 4 (its order is necessarily 1). Recognize it as a code we've seen before.
2. What can be said about RM codes in the degenerate cases in which m is 0 or 1?
3. Construct generating and check matrices for the RM codes of length 16 and orders 1 or 2. Calculate k and d.
4. For the RM code of length 8 and order 1 (example 14), do exercise 2 of Exercises 13.5.
5. Prove that the RM code of length 2^m and order $m - 2$ has distance 4.

13.11 Permutation of Bit Positions

It's clear that, if we consistently permute the positions of the bits in every word of a canonical linear binary code, then we do not alter the substance of the code. As an example, consider $C := \{0000, 0011, 0101, 0110\}$. If we interchange bits 1 and 3 consistently, then we obtain $C^* := \{0000, 1001, 0101, 1100\}$. The code is different in form, but the essential characteristics are the same, such as code length, code distance, information power, error-handling capability, one "useless" bit, and so on. Note that we are not permuting code C as a *set* of words. First, C is a set, and the order of its elements is therefore irrelevant. Second, even if we want to place the elements of C in a *sequence* (as we do occasionally for book-keeping purposes) this is not the operation we are doing now. Also, a permutation of bit positions may destroy the fact that the code is canonical. In the example above, the original generating matrix $G := \begin{bmatrix} 0 & 1 & 1 & 0 \\ 0 & 1 & 0 & 1 \end{bmatrix}$ would become $G^* := \begin{bmatrix} 1 & 1 & 0 & 0 \\ 0 & 1 & 0 & 1 \end{bmatrix}$, which is not canonical. Finally, if we use also matrix H, then we must perform the same permutation on the rows of H in order to be consistent and preserve the property $(x \in C) \Leftrightarrow (xH = 0)$. In the example, the matrices

$$G = \begin{bmatrix} 0 & 1 & 1 & 0 \\ 0 & 1 & 0 & 1 \end{bmatrix} \quad \text{and} \quad H = \begin{bmatrix} 1 & 0 \\ 0 & 1 \\ 0 & 1 \\ 0 & 1 \end{bmatrix}$$

become

$$G^* = \begin{bmatrix} 1 & 1 & 0 & 0 \\ 0 & 1 & 0 & 1 \end{bmatrix} \quad \text{and} \quad H^* = \begin{bmatrix} 0 & 1 \\ 0 & 1 \\ 1 & 0 \\ 0 & 1 \end{bmatrix}.$$

Exercises 13.11

1. Practice several permutations of bit positions on codes we considered as examples, for instance examples 4, 5, 10, 13 and 14.
2. Prove that, in the special case in which $m = 3$, the RM codes of length 8 and order 1 (examples 13 and 14) are obtained from each other by permuting bit positions.

13.12 Summary

The major goal of mathematical code theory is to detect and possibly to correct errors that occur in a transmission channel during the transmission of a message (Section 13.1). We concentrated on binary codes, that is, codes that use the alphabet $\{0, 1\}$ for the transmission. The basic steps of the process are translating the original message into a sequence of binary strings that can be handled by the *code* (such binary strings are the *information strings*); *encoding* the information strings into *code words* (Section 13.3); transmitting and receiving the code words; and *decoding* the code words to try to reconstruct the original message. The difficulties arise from the fact that errors do occur during the transmission, that some of them can be *corrected,* some can only be *detected,* and some more may go undetected. We considered, in particular, the *canonical linear binary codes* (special cases of binary codes). They provide an opportunity to study almost all essential phenomena of coding theory and, furthermore, many useful codes are among them. Canonical linear binary codes can be described compactly by means of a *generating* matrix, which also provides an immediate encoding algorithm (Sections 13.3 and 13.5; Section 13.4 summarizes the pertinent symbols). The *check matrix* of the code is easily constructed from the generating matrix and provides a quick calculation to determine whether a received word is a code word. A *decoding algorithm* can be constructed from the check matrix, but the process is not immediate nor totally satisfactory in practice. Special codes may have better decoding algorithms.

In Section 13.2 we considered simple examples. Two advanced, typical, binary codes are Hamming codes (Section 13.6) and Reed–Mueller (RM) codes (Section 13.10).

There are techniques for constructing new codes starting from known codes, for instance adjunction of an overall parity check (Section 13.7), dualization (Section 13.8), and permutation of bit positions (Section 13.11).

Boolean Functions

In this chapter we shall study in more detail some of the properties of a Boolean algebra. We devote the first few sections to an elementary Boolean algebra, that is, to a Boolean algebra with two elements (0 and 1) and operations we've seen before ($\vee$, $\wedge$, $\sim$; see Sections 6.3, 6.4, and 8.2). In Sections 14.4 and 14.5 we'll consider a general finite Boolean algebra.

14.1 Boolean Functions and Expressions

In dealing with Boolean algebra, many students—as well as experienced practitioners who should really know better—tend to use the terms *Boolean function* and *Boolean expression* interchangeably. Of course, every Boolean expression such as $f(p, q, r) := (p \wedge q) \vee \sim r$ yields a Boolean function through the process of *evaluation,* which we saw in Section 12.1.5. Briefly, the function is obtained by calculating the expression for every selection of values (0 or 1) of the arguments p, q, and r. Since all of $\vee$, $\wedge$, $\sim$ are total operations, the resulting function will also be a total function in this case (see exercises). Whether every Boolean function can be written as a Boolean expression is less clear. As we'll see, the answer will be affirmative for an elementary Boolean algebra. So why is a sharp distinction necessary? Let's review some definitions.

A *Boolean function* is a mathematical abstraction. For some number n ($n \geq 0$), we consider n independent variables, or *arguments*. Then a Boolean function is a total function f: $C^n \rightarrow C$, where C is the set of the *constants,* that is $C := \{0, 1\}$ or $\{F, T\}$. By the methods of Chapters 4 and 7, $\#(C^n) = 2^n$. Then there are 2^{2^n} possible, distinct Boolean functions with n arguments. Note that it's perfectly possible to have zero arguments.

A *Boolean expression* is a syntactic construction, defined recursively by the methods of Sections 9.1.1 and 12.1. To make the discussion a bit quicker, let's review that definition. If we use prefix or postfix expressions, parentheses are not necessary; if we use infix expressions, as we shall in this chapter, parentheses are needed, but we'll try to use them sparingly.

To define the alphabet A, we need several components.

1. The set of the constants is C, that is, $\{0, 1\}$.
2. The set of the arguments is the set $Y := \{p, q, \ldots\}$, which we will assume finite, with n elements. We'll give Y a *total order*, by placing its elements in a sequence $X := (p, q, \ldots)$. When n is small, we'll adopt the usual conventions, such as $Y = \varnothing = \{\}$ and $X = \Lambda = (\,)$ when $n = 0$, $Y = \{p\}$ and $X = (p)$ when $n = 1$, etc.
3. The set of the *operators* is $S := \{\vee, \wedge, \sim\}$.
4. For infix expressions, the set of the *parentheses* is $P := \{\,),\,(\,\}$. If we were using prefix or postfix expressions, P would be empty.

Then we set alphabet $A := C \cup S \cup P \cup Y$, that is, $A := \{0, 1, \vee, \wedge, \sim,), (\,, p, q, \ldots\}$. The number n of arguments is unimportant, save to say that, in any application, we have enough and we decide once for all what n is in that application.

We then define a Boolean expression by the following rules:

1. Every element of $C \cup Y$ is a Boolean expression. This is the initial condition for the recursive definition.
2. If E is a Boolean expression, then (E) is a Boolean expression. This is *enclosure*.
3. If E is a Boolean expression, then $\sim E$ is a Boolean expression. This is the unary operation of *negation*.
4. If E and F are Boolean expressions, then
   - $E \vee F$ is a Boolean expression. This is the binary operation of *disjunction*, although it's usually read simply as OR.
   - $E \wedge F$ is a Boolean expression. This is the binary operation of *conjunction*, usually read simply as AND.
5. There are no other Boolean expressions.

In order to avoid ambiguities, we could require fully parenthesized expressions as in Section 12.1. Or we could set out operator precedence conventions, as is usually done in programming languages (**not** first, then **and,** then **or**). Our method will be to parenthesize only as much as necessary to remove ambiguities in any given expression. You can quickly verify that the example $(p \wedge q) \vee \sim r$ above is indeed a valid unambiguous Boolean expression.

In addition to all their other differences, one crucial distinction between Boolean functions and Boolean expressions now becomes clear. For any fixed n (the number of arguments), there are finitely many Boolean functions, but infinitely many Boolean expressions. It must be the case that some Boolean functions have more than one possible expression. The truth is that *every* function has infinitely many expressions. To see this, suppose that $f(p)$ is a Boolean expression of one variable. Then all of the expressions $f(p)$, $f(p) \wedge 1$, $(f(p) \wedge 1) \wedge 1$,

$((f(p) \wedge 1) \wedge 1) \wedge 1, \ldots$ represent exactly the same Boolean function and are *distinct* expressions. Of course, some expressions are more "efficient" than others, in that they contain fewer symbols. But all are valid expressions nonetheless. Moreover, although our example is a toy, in that it's clear where the inessential terms are, in more complicated cases this isn't so obvious.

What we have to address, then, are four questions. First, since every Boolean function can be represented by infinitely many Boolean expressions, we ask whether there is some "preferred" expression for a function, and if so, why it's preferred. Such a preferred expression is called a *normal form* or *canonical form*. Second, given two Boolean expressions, we seek a standard algorithm for determining whether or not they represent the same function. This notion is called the *equivalence* of Boolean expressions. In the next section, we'll see that these two are really the same question. Third, we consider the problem whether every Boolean function can be represented by means of a Boolean expression. Fourth, we consider very briefly the concept of *Boolean minimization;* that is, for a given Boolean function, we seek an expression that is the most "efficient" in the sense of having the fewest symbols.

Exercises 14.1

1. If the number n of arguments is odd, define the **majority function** as $M_n(p, q, \ldots) := 1$ if at least half of the arguments are 1; $M_n(p, q, \ldots) := 0$ otherwise. When $n = 3$, write an expression for $M_3(p, q, r)$. When $n = 5$, write an expression for $M_5(p, q, r, s, t)$. State an algorithm for constructing an expression for a general M_n.

2. If the number n of arguments is greater than 1, define the **parity check function** as $K_n(p, q, \ldots) := 1$ if the number of arguments that are equal to 1 is even, $K_n(p, q, \ldots) := 0$ otherwise. When $n = 3$, write an expression for $K_3(p, q, r)$. When $n = 4$, write an expression for $K_4(p, q, r, s)$. State an algorithm for constructing an expression for a general K_n.

3. Give an example of a Boolean function that is not onto $\{0, 1\}$.

4. How many possible distinct Boolean functions are there with no arguments? Enumerate them, and identify each by its usual name. How many possible distinct Boolean functions are there with one argument? Enumerate them, and identify each by its usual name. How many possible distinct Boolean functions are there with two arguments? Enumerate them, and identify each by its usual name (if there is one).

5. Show by an example that two Boolean expressions can be equivalent even if they have different numbers of arguments.

14.2 Boolean Normal Forms

We have seen that it's entirely possible and perfectly valid for two Boolean expressions to represent the same Boolean function and that, in fact, every function has infinitely many expressions. How, then, do we determine whether two expressions represent the same function? There are two primary tools for doing this, and you've already used both of them: deduction and exhaustion.

Deduction is simply the formal name given to the application of axioms, that is, to algebraic manipulations within the realm of Boolean algebra. Given two expressions, E and F, we successively apply the axioms of Boolean algebra to one of them, say E, to turn it into, or *deduce* F.

Example 1 Consider the two Boolean expressions

$$E(p, q, r) := (q \wedge r) \vee (\sim q \wedge r) \vee \sim(p \vee r) \tag{1}$$

and

$$F(p, q, r) := \sim p \vee r \tag{2}$$

Expression 1 reduces thus:

$(q \wedge r) \vee (\sim q \wedge r) \vee \sim(p \vee r)$

$$
\begin{aligned}
&= ((q \vee \sim q) \wedge r) \vee (\sim p \wedge \sim r) && \text{[distributivity, De Morgan's law]} \\
&= r \vee (\sim p \wedge \sim r) && [\text{properties } x \vee \sim x = 1,\ 1 \wedge x = x] \\
&= (r \vee \sim p) \wedge (r \vee \sim r) && \text{[distributivity]} \\
&= r \vee \sim p && [\text{properties } x \vee \sim x = 1,\ x \wedge 1 = x] \\
&= \sim p \vee r && \text{[commutativity].}
\end{aligned}
$$

Deductive proofs of equivalence may be short and clear or long and involved. Worse yet, they're quite difficult to master, since they depend more on experience and insight than on simple manipulation. Moreover, if two expressions are *not* equivalent, attempts at deduction won't make that clear. A lot of time could be wasted going up blind alleys.

Exhaustion is the technique you've come to know as the *truth-table* approach. For every possible selection of argument truth values, both expressions are evaluated, and the results compared. If the results match exactly, the expressions are deemed equivalent. If not, we can see precisely where they differ, a feature deduction lacks.

Example 2 The truth tables for E and F are the following. Since the last column shows **functions** E and F to be equal in every case, **expressions** E and F, while not identical, are equivalent.

p	q	r	$\sim p$	$\sim q$	$q \wedge r$	$\sim q \wedge r$	$p \vee r$	$\sim(p \vee r)$	E	F	expression $E \equiv F$
1	1	1	0	0	1	0	1	0	1	1	1
1	1	0	0	0	0	0	1	0	0	0	1
1	0	1	0	1	0	1	1	0	1	1	1
1	0	0	0	1	0	0	1	0	0	0	1
0	1	1	1	0	1	0	1	0	1	1	1
0	1	0	1	0	0	0	0	1	1	1	1
0	0	1	1	1	0	1	1	0	1	1	1
0	0	0	1	1	0	0	0	1	1	1	1

Exhaustion is simple and mechanical, but also tedious and error prone. The *normal forms* we're going to study are a formalization of the exhaustive technique. For a function f, we evaluate f for every possible selection of arguments. From this tabulation, we form a new expression for f, the normal form of f whose purpose is to represent the tabulation conveniently. We then determine the equivalence of two expressions simply by comparing their normal forms. The idea is simple, but requires a bit of definitional machinery. Read carefully.

14.2.1 Disjunctive Normal Form

Let's consider an informal definition first. Let n be the given number of arguments. A *minimal term* or *minterm* in n arguments is a Boolean expression in which

- every variable appears precisely once, either simple or negated, and
- the variables are related by conjunction.

For instance, consider two arguments p and q. There are four minterms:

$$\mathbf{min}_0 := \sim p \wedge \sim q,$$

$$\mathbf{min}_1 := \sim p \wedge q,$$

$$\mathbf{min}_2 := p \wedge \sim q,$$

$$\mathbf{min}_3 := p \wedge q.$$

Do you see where the subscript indices of the minterms come from? If you rewrite the subscripts in binary form, then the pattern will be more evident:

$$\mathbf{min}_{00} := \sim p \wedge \sim q,$$

$$\mathbf{min}_{01} := \sim p \wedge q,$$

$$\mathbf{min}_{10} := p \wedge \sim q,$$

$$\mathbf{min}_{11} := p \wedge q.$$

Whenever there is a 0 in the first position of the subscript, p appears negated in the first position of the expression; if a 1 appears, then p is not negated. The same rule applies to q for the second positions, in the subscript and in the expression. This leads us to the following formal definition, which will be complemented by the next theorem.

Definition 1 Let notations be as in Section 14.1. For convenience, set $B := C^n$. Let J be any one element of B. Then the **minterm** J in n arguments is defined as the Boolean expression $\mathbf{min}_J$ given by the conjunction of n terms the kth of them being the kth variable if 1 appears in the kth position of J, and the negation of the kth variable if 0 appears in the kth position of J $(1 \leq k \leq n)$. Through the process of evaluation, this expression defines a function $\mathbf{min}_J$: $B \rightarrow C$. In an elementary Boolean algebra, the same name is used also for the Boolean function $\mathbf{min}_J^*$: $B \rightarrow C$ determined by the fact that $\mathbf{min}_J^*(I) = 1$ if $I = J$ and $\mathbf{min}_J^*(I) = 0$ if $I \neq J$ $(I \in B)$.

Theorem 2 Let notations be as in Definition 1. Then (1) there are 2^n possible distinct minterms; (2) in an elementary Boolean algebra, the Boolean function of the expression $\mathbf{min}_J$ is precisely $\mathbf{min}_J^*$.

Proof: Part 1 is an immediate consequence of the results of Chapter 4. For part 2, observe that a minterm is a conjunction of arguments, and its value will be 1 if and only if each argument *of the minterm* has value 1. On the other hand, the kth argument of the minterm is the negation of the kth variable or the kth variable itself according as the kth element of J is 0 or 1. Consequently, the kth variable must also be 0 or 1, respectively; that is, the sequence $pq \cdots$ must be the same as J. Q.E.D.

Now consider a total function f of n arguments, that is, f: $B \rightarrow C$. If we again use the labels $p, q \cdots$, *in order*, for these variables, then a notation like $f(01 \cdots)$ is a shorthand for evaluating f when $p = 0$ **and** $q = 1$ **and** . . . . In general, $f(J) := f$(value of first term of J, value of second term of J, . . .). That is, $f(J)$ is the evaluation of f for minterm $\mathbf{min}_J$.

For instance, consider $f(p, q) := p \wedge {\sim}q$. The four evaluations are $f(00) = 0 \wedge {\sim}0 = 0, f(01) = 0 \wedge {\sim}1 = 0, f(10) = 1 \wedge {\sim}0 = 1, f(11) = 1 \wedge {\sim}1 = 0$.

Definition 3 Let notations be as in the definitions above. Let f:$B \rightarrow C$ be a given total function in an elementary Boolean algebra. Then the **universal disjunctive form** of f is denoted $\mathbf{UDF}_f$ and is defined as the expression $\mathbf{UDF}_f := \bigvee_{J \in B} (f(J) \wedge \mathbf{min}_J)$. The **disjunctive normal form** of f is denoted $\mathbf{DNF}_f$ and is defined as the expression $\mathbf{DNF}_f := \bigvee_{J \in B \wedge f(J)=1} (\mathbf{min}_J)$. In the degenerate case in which $f(J) = 0$ for every J in B, the expression above is to be interpreted as 0. Normal forms are called also **canonical forms.** The DNF is sometimes called the **sum of products** form, from the analogy between Boolean and integer algebras. It is an inaccurate and misleading terminology, though, and is fortunately becoming less common.

Theorem 4 Let data be as in Definition 3. Then (1) $f(J) = \mathbf{UDF}_f(J)$ for every J in B; that is, function f and the function of expression $\mathbf{UDF}_f$ are the same function; (2) similarly, function f and the function of expression $\mathbf{DNF}_f$ are the same function; (3) every total Boolean function has a disjunctive normal form, which is unique except possibly for

- the labelling and order of its arguments, and
- the order in which the minterms are enumerated.

(4) Thus, every Boolean function is *completely characterized* by its universal disjunctive form or its disjunctive normal form. In other words, two expressions are equivalent if and only if their disjunctive normal forms are the same except for the reordering and relabelling mentioned above. (5) Every total Boolean function is the function of a Boolean expression.

Proof: Exercise.

In our example of $f(p, q) := p \vee {\sim}q$, Definition 3 gives us the universal disjunctive form $f(p, q) = (f(00) \wedge ({\sim}p \wedge {\sim}q)) \vee (f(01) \wedge ({\sim}p \wedge q)) \vee (f(10) \wedge (p \wedge {\sim}q)) \vee (f(11) \wedge (p \wedge q)) = (0 \wedge ({\sim}p \wedge {\sim}q)) \vee (0 \wedge ({\sim}p \wedge q)) \vee (1 \wedge (p \wedge {\sim}q)) \vee (0 \wedge (p \wedge q))$, which could be reduced to $p \wedge {\sim}q$, which, in turn, is the DNF and is clearly equivalent to the originally given f. In fact, we gave f in a form which was already normal.

Notice that the UDF has produced a good deal of redundant information. Any minterm for which the coefficient $f(J)$ is 0 need not be written at all, since the conjunction with 0 simply annihilates it without affecting the disjunction of minterms. For those minterms that have coefficient 1, the conjunction with 1 is similarly redundant, since 1 is the identity for conjunction. The definition of DNF removes much of the deadwood.

Consider again the example in formula 1. We intend now to construct the disjunctive normal form directly and to point out that this is an automatic process, that is, an algorithm.

14.2.1.1 Algorithm A

This algorithm exhaustively constructs the DNF of an expression in an elementary Boolean algebra. As we present it, we apply it to the expression

$$E(p, q, r) = (q \wedge r) \vee ({\sim}q \wedge r) \vee {\sim}(p \vee r)$$

of Example 1. (Ignore the column labelled ''$\mathbf{max}_J$''; we'll get to that in a moment.)

The algorithm is executed as follows:

1. Construct a table with one row for each J in B. In our example table below, J is expressed in both binary and decimal, 0 through 7, bottom to top.
2. For each J, write the minterm expression $\mathbf{min}_J$, and tabulate $E(\mathbf{min}_J)$, the evaluation of E for $\mathbf{min}_J$.

J	$\mathbf{min}_J$	$\mathbf{max}_J$	$E(\mathbf{min}_J)$
$111 = 7$	$p \wedge q \wedge r$	$\sim p \vee \sim q \vee \sim r$	1
$110 = 6$	$p \wedge q \wedge \sim r$	$\sim p \vee \sim q \vee r$	0
$101 = 5$	$p \wedge \sim q \wedge r$	$\sim p \vee q \vee \sim r$	1
$100 = 4$	$p \wedge \sim q \wedge \sim r$	$\sim p \vee q \vee r$	0
$011 = 3$	$\sim p \wedge q \wedge r$	$p \vee \sim q \vee \sim r$	1
$010 = 2$	$\sim p \wedge q \wedge \sim r$	$p \vee \sim q \vee r$	1
$001 = 1$	$\sim p \wedge \sim q \wedge r$	$p \vee q \vee \sim r$	1
$000 = 0$	$\sim p \wedge \sim q \wedge \sim r$	$p \vee q \vee r$	1

3. Disjoin (that is, form the disjunction of) all minterms for which $E(\mathbf{min}_J) = 1$. (In our example, $J = 0, 1, 2, 3, 5$, and 7.) This expression is the disjunctive normal form of E. The example yields:

$$\begin{aligned}\text{DNF } E(p, q, r) = {} & (\sim p \wedge \sim q \wedge \sim r) \vee \\ & (\sim p \wedge \sim q \wedge r) \vee \\ & (\sim p \wedge q \wedge \sim r) \vee \\ & (\sim p \wedge q \wedge r) \vee \\ & (\ p \wedge \sim q \wedge r) \vee \\ & (\ p \wedge q \wedge r)\end{aligned}$$

If two expressions are to be reduced to DNF for comparison, they must have the same number of arguments. In Example 1, E has three arguments, p, q, and r, while F has only two, p and r. This is easily remedied by writing $F(p, q, r)$ and observing that, since F does not depend on q, $F(p, q, r) = F(p, \sim q, r)$ for every combination of p and r. Thus each of the four evaluations of $F(p, r)$ will appear twice in the eight-row table of $F(p, q, r)$: once as $F(p, q, r)$, and once as $F(p, \sim q, r)$.

14.2.1.2 Algorithm B

Although they are neither very attractive nor widely used, there are algorithms to reduce an expression to DNF *deductively*, that is, by means of algebraic manipulations in an elementary Boolean algebra. An informal algorithm is given in Complements 14.2.

14.2.2 Conjunctive Normal Form

There is another pair of forms analogous to (in fact, the duals of) the disjunctive forms. These *conjunctive forms* have to be defined a bit more carefully. In fact, conjunctive forms are generally less intuitive, but they do arise in practical applications. The idea behind conjunctive forms is that expressing a function in a normal form, whether conjunctive or disjunctive, does not mathematically alter it, that is, does not alter the respective function. Thus, suppose we take any function f, negate it, express it in a normal form, and negate it again. The middle step is mathematically "transparent", and the two negations cancel each other. The function is mathematically unchanged. As we'll see later, this is one of the possible methods. Another method, which we are going to present first, is similar to what we just saw for disjunctive forms. We'll present it briefly.

Definition 5 Let notations be as before. Let J be any one element of B. Then the **maxterm** J in n arguments is defined as the Boolean expression $\mathbf{max}_J$ given by the disjunction of n terms, the kth one being the kth argument if 0 appears in the kth position of J, and the negation of the kth argument if 1 appears in the kth position of J ($1 \leq k \leq n$). Through the process of evaluation, this expression defines a function $\mathbf{max}_J$: $B \rightarrow C$. In an elementary Boolean algebra, the same name is used also for the Boolean function $\mathbf{max}_J^*$: $B \rightarrow C$ determined by the fact that $\mathbf{max}_J^*(I) = 0$ if $I = J$ and $\mathbf{max}_J^*(I) = 1$ if $I \neq J$ ($I \in B$).

Theorem 6 Let notations be as in Definition 5. Then (1) there are 2^n possible distinct maxterms; (2) the Boolean function of the expression $\mathbf{max}_J$ is precisely $\mathbf{max}_J^*$; (3) for every J in B, $\sim\mathbf{min}_J = \mathbf{max}_J$, provided the symbol $\sim$**min** is interpreted as the result of the application of DeMorgan's law to the minterm as an expression (unless the expression is 0, in which case ~ 0 is to be replaced by 1). Also, $\sim\mathbf{max}_J = \mathbf{min}_J$, with the analogous agreement.

Proof: Exercise.

There is one crucial difference, however, between a minterm $\mathbf{min}_J$ and a maxterm $\mathbf{max}_J$: **a Boolean function cannot be evaluated for a maxterm**! A function $f(p, q)$ can be evaluated for $\sim p \wedge q$, but not for $\sim p \vee q$. To see why, consider an analogous integer function. The function $x + y$ has the value 8 for $x = 3$ **and** $y = 5$, but it makes no sense to speak of evaluating $x + y$ for $x = 3$ **or** $y = 5$. Thus, in the following definition, $f(J)$ still refers to the evaluation of f for **minterm** $\mathbf{min}_J$.

Definition 7 In an elementary Boolean algebra, let notations be as in the definitions above. Let f: $B \rightarrow C$ be a given total function. Then the **universal conjunctive form** of f is denoted $\mathbf{UCF}_f$ and is defined as the expression $\mathbf{UCF}_f := \bigwedge_{J \in B} ((\sim f(J)) \vee \mathbf{max}_J)$. The **conjunctive normal form** of f is denoted $\mathbf{CNF}_f$ and is defined as the expression $\mathbf{CNF}_f := \bigwedge_{J \in B \wedge f(J)=0} (\mathbf{max}_J)$. In the degenerate case in which $f(J) = 1$ for every J in B, the expression above is to be interpreted as 1.

The CNF is sometimes called the *product-of-sums form,* from the analogy between Boolean and integer algebra. As with sum of products, this too is less common.

Theorem 8 Let data be as in Definition 7. (1) Then $f(J) = \mathbf{UCF}_f(J)$ for every J in B. That is, function f and the function of expression $\mathbf{UCF}_f$ are the same function. (2) Similarly, function f and the function of expression $\mathbf{CNF}_f$ are the same function. (3) Every total Boolean function has a conjunctive normal form, that is unique except possibly for the labelling and order of its arguments, and the order in which the maxterms are enumerated. (4) Thus, every Boolean function is *completely characterized* by its universal conjunctive form or its conjunctive normal form. In other words, two expressions are equivalent if and only if their conjunctive normal forms are the same except for the reordering and relabelling mentioned above. (5) For every total function $f{:}B \to C$, $\mathbf{CNF}_f = \sim\mathbf{DNF}_{\sim f}$, provided the negation $\sim\mathbf{DNF}$ is interpreted syntactically as

- if DNF is 0, then replace ~ 0 with 1;
- else, apply DeMorgan's law to the k-fold $\vee$ (if $k > 1$), then to each of the resulting n-fold $\wedge$s (if $n > 1$).

Proof: Exercise.

Algorithm *A* in Section 14.2.1.1 can easily be modified to produce the CNF of an expression rather than the DNF. The necessary changes are:

1. Same as in Algorithm *A*.
2. Tabulate also the maxterms $\mathbf{max}_J$ for each J in B. (The table already shows the $\mathbf{max}_J$ for the expression $E(p, q, r)$ in Example 1.)
3. The CNF of E is then simply the conjunction of all maxterms for which $E(\mathbf{min}_J) = 0$. (Note again that it makes no sense to speak of $E(\mathbf{max}_J)$!) The example thus yields:

$$\text{CNF } E(p, q, r) = (\sim p \vee \sim q \vee r) \wedge (\sim p \vee q \vee r)$$

To compare the CNF's of $E(p, q, r)$ and $F(p, r)$, we insert the dummy argument q into F exactly as in Algorithm *A*.

14.2.3 Comments

It's pretty obvious that the CNF or the DNF is *not* an attempt to minimize the number of symbols in a Boolean expression. The CNF or the DNF is sometimes quite inefficient in that sense. In the example of formula 1 a much more efficient expression would be the originally given one. But that's not the point. The advantage is that *every function has one and only one normal form of a given type, conjunctive or disjunctive*! Well, almost.

We've always listed functional arguments in the order $p, q, \ldots$, and you'll recall that we always enumerated the minterms in a binary counting order. These are simply bookkeeping details, inessential to the purely mathematical aspects of the function, but in a computer program, such conventions must be selected in advance and then strictly enforced. A computer program? To *do what*?

Suppose we have two different expressions for the same function. For each of them, we follow the above procedure for constructing the DNF, following strictly our conventions on labelling and order. What can we say about the two resulting versions of the DNFs? *They are identical*! Not merely mathematically equivalent, not merely similar in form, but syntactically, symbol-by-symbol identical. The DNF for a function can be constructed quickly and mechanically from any of its expressions, and the results will always be the same. Computer programs that accept as input two Boolean expressions and test them for equivalence usually adopt this technique.

Note that a total function $f{:}B \to C$ is a function to set $\{0, 1\}$ and is therefore a characteristic function. Clearly, it is the characteristic function of the set of those minterms that are present in $\mathbf{DNF}_f$. Set B ($B := \{0, 1\}^n$) has a natural order, as we saw before, obtained by interpreting its elements as binary representations of integers. Each element of B, being a sequence, is totally ordered and its order corresponds to the order we chose for the set of the arguments of the function. In this sense, each minterm is ultimately a sequence of bits, the UDF is a specific sequence of all minterms (each with a "coefficient" 0 or 1), and the DNF is a subsequence of such sequence. Similar remarks apply to UCF and CNF, except that $\sim f$ is now the characteristic function in question. This presentation of normal forms is what would ultimately be considered inside the computer or, as we'll see in a moment, in a chip.

From a purely mathematical standpoint, the UDF and UCF are primarily technical tools. They come into play only to develop and prove results about the more important normal forms, the CNF and DNF. Seldom will the universal forms enter into calculations or program designs. When computer hardware is considered, however, the universal forms become much more important, as we'll see.

Exercises 14.2

1. Verify that the following expressions are all equivalent: **(a)** $(\sim p) \vee q$; **(b)** $\sim(p \wedge \sim q)$; **(c)** $(\sim p) \vee (p \wedge q)$; **(d)** $p \Rightarrow q$.
2. For each of the following pairs, check whether the two expressions are equivalent or not: **(a)** $((p \vee q) \wedge r) \vee \sim((p \vee r) \wedge \sim r)$ and $(p \wedge r) \vee (\sim p \wedge r)$; **(b)** $(p \wedge q) \vee ((\sim(p \wedge q)) \wedge r)$, and $(q \wedge r) \vee (p \wedge q \wedge \sim r) \vee ((\sim q) \wedge r)$; **(c)** $(p \wedge q \wedge r) \vee \sim r$ and $(p \wedge q) \vee (\sim(p \wedge q) \wedge \sim r)$.
3. Verify that each of the following expressions is equivalent to 1: **(a)** $[(q \Rightarrow r) \wedge (p \Rightarrow q)] \Rightarrow (p \Rightarrow r)$, **(b)** $((p \Rightarrow q) \wedge p) \Rightarrow q$. An expression that is equivalent to 1 is called a **tautology.** [Greek *taut-* = same, *log-* = say, *tautology* = saying the same thing.]
4. State both the conjunctive and disjunctive normal forms (CNF and DNF) for each of the following functions. **(a)** $f(p, q) := p \vee q$; **(b)** $f(p, q, r) := p \wedge r$; **(c)** $f(p, q, r) := (p \vee r) \wedge (q \vee \sim r)$; **(d)** $f(p, q, r) := (p \text{ XOR } q) \text{ XOR } r$, where XOR is the "exclusive or," defined by $x \text{XOR} y := (x \vee y) \wedge \sim(x \wedge y)$; **(e)** several of the Boolean expressions or functions given in Exercises 12.1.4; consider also some of those given by expression trees.
5. The following truth table defines three functions f, g, h $\{0, 1\}^3 \to \{0, 1\}$. For each of them, find an expression in terms of $\vee$, $\wedge$, $\sim$.

			f	g	h
1	1	1	1	0	1
1	1	0	1	1	0
1	0	1	0	1	0
1	0	0	1	1	0
0	1	1	0	0	0
0	1	0	0	0	0
0	0	1	0	1	1
0	0	0	0	1	0

6. The following truth table defines three functions f, g, h $\{0, 1\}^3 \to \{0, 1\}$. For each of them, find an expression in terms of $\vee$, $\wedge$, $\sim$.

			f	g	h
1	1	1	1	0	1
1	1	0	1	1	0
1	0	1	0	1	1
1	0	0	1	0	0
0	1	1	0	1	0
0	1	0	1	0	0
0	0	1	1	0	0
0	0	0	0	1	0

7. For each of the following expressions, find a DNF and a CNF **(i)** in two variables, **(ii)** in three variables, and **(iii)** in four variables. **(a)** $p \wedge \sim q$; **(b)** $(p \vee \sim q) \wedge \sim p$; **(c)** $p \Rightarrow q$; **(d)** $p|q$ (Recall the definition of Sheffer stroke; Exercise 6.2).

8. Construct the DNF of expression 2 in three arguments p, q, and r.

9. Prove Theorems 4 and 8, about the existence and uniqueness of a normal form for a function.

10. Prove Theorem 6, about duality of minterms and maxterms.

Complements 14.2

1. Algorithm B. The following informal algorithm reduces an expression to its DNF deductively, that is, by means of algebraic manipulations in an elementary Boolean algebra.

1. If the expression is other than just 0 or 1, and if there is, in the expression, a 0 or a 1, then resolve it by using one of the properties $\sim 0 = 1$, $\sim 1 = 0$, $u \vee 0 = u$, $u \vee 1 = 1$, $u \wedge 0 = 0$, $u \wedge 1 = u$ (that is, replace the left-hand member of these formulae with the right-hand member). After such a change, resume from step 1.

2. If there is an operator $\sim$ whose argument is *not* a single variable, then resolve it by using one of the properties $\sim(\sim u) = u$, $\sim(u \vee v) = \sim u \wedge \sim v$, $\sim(u \wedge v) = \sim u \vee \sim v$. After such a change, resume from step 2.

3. If there is a sequence of consecutive $\vee$s or of consecutive $\wedge$s, then replace it with a k-fold $\vee$ or $\wedge$ respectively. Example: $((u \vee v) \vee w) = (u \vee v \vee w)$. After such a change, resume from step 3.

4. If there is a k-fold $\vee$ or a k-fold $\wedge$ that contains two equal arguments, then delete all but one of the equal arguments; that is, use commutativity, associativity, and idempotency of $\vee$ or $\wedge$. Example: $u \vee v \vee u = u \vee v$. After such a change, resume from step 3.

5. If there is as k-fold $\wedge$ operator that contains two terms that are the negation of each other, then replace it with 0 and continue at step 1. Example: $u \wedge v \wedge \sim u = 0$.

6. If there is a k-fold $\vee$ operator that contains two terms that are the negation of each other and *neither of which is a single variable,* then replace it with 1 and continue at step 1. Example: $u \vee v \vee \sim u = 1$; but *do not* perform a replacement like $p \vee \sim p = 1$.

7. If there is an $\wedge$ operator that can be distributed over an $\vee$ operator, then distribute and continue at step 3. Do *not* distribute $\vee$ over $\wedge$. Example: $(u \vee v) \wedge w = (u \wedge w) \vee (v \wedge w)$.

8. If there is an h-fold $\wedge$ whose arguments are only variables, negated or not, then order the variables according to their order in sequence X. After such a change, resume from step 4.

9. If the entire expression is just 0, then stop. This is the normal form. If the entire expression is just 1, then replace it with $p \vee \sim p$ and continue at step 10.

10. At this point, the entire expression is a k-fold $\vee$ ($k \geq 1$) for which all arguments are distinct from each other and each argument u is an h-fold $\wedge$ ($h \geq 1$) whose arguments are distinct single variables, negated or not, which appear in the same order given by sequence X. If, in each argument u of the k-fold $\vee$, the number h equals the total number n of variables we want for the normal form, then this is the normal form; stop. If, for some argument u of the k-fold $\vee$, the number of variables actually present is less than n, then replace u with $[(\bigwedge_{x \in \{\text{missing variables}\}} (x \vee \sim x)] \wedge u$ and continue at step. 7.

The original expression (1) is changed thus:

Step 1 does not apply.
Step 2, De Morgan's law, yields $(q \wedge r) \vee (\sim q \wedge r) \vee (\sim p \wedge \sim r)$.
Steps 2–9 do not apply.
Step 10 yields $((p \vee \sim p) \wedge q \wedge r) \vee ((p \vee \sim p) \wedge \sim q \wedge r) \vee (\sim p \wedge (q \vee \sim q) \wedge \sim r)$.
Step 7 yields $(p \wedge q \wedge r) \vee (\sim p \wedge q \wedge r) \vee (p \wedge \sim q \wedge r) \vee (\sim p \wedge \sim q \wedge r) \vee (\sim p \wedge q \wedge \sim r) \vee (\sim p \wedge \sim q \wedge \sim r)$.
Steps 8 and 9 do not apply.
Step 10 makes no changes and yields the expression above, which, except for reordering the terms, is the normal form.

2. Give a formal algorithm to replace algorithm B above. A strict form for the definition of an expression is necessary; postfix form is of some help here. Test your algorithm on a few of the examples from the text. Prove the correctness of the algorithm; in particular, prove the statement that appears in step 10.

14.3 Functional Completeness and Minimization

14.3.1 Functional Completeness

In the previous section we showed that every Boolean function in an elementary Boolean algebra has conjunctive and disjunctive normal forms. Of course, this makes the CNF and DNF powerful tools, but it says something even more fundamental. Our original alphabet for Boolean expressions included the arguments themselves, of course, plus the self-defining terms 0 and 1, and parentheses. These things would be needed regardless of which Boolean operations we used. Our alphabet, however, included also the three operations $\vee$, $\wedge$ and $\sim$; that is, the set of the operations was $S := \{\vee, \wedge, \sim\}$. What Theorems 4 and 8 say is that *any* Boolean function can be expressed using only these three operations. These three operations thus form a set which is *functionally complete.*

Definition 9 A set of Boolean operations $\{O_1, O_2, \ldots\}$ is said to be **functionally complete** if every Boolean function can be expressed solely in terms of the set, the function's arguments, parentheses, and the terms 0 and 1. Equivalently, in the definition of expression given in Section 14.1, replace set S with $\{O_1, O_2, \ldots\}$. If the set of the functions of the expressions so obtained is all of C^B (that is, the set of all Boolean functions in n arguments), for every value of n in $\mathbb{N}$, then set $\{O_1, O_2, \ldots\}$ is said to be **functionally complete.** The set is called **minimally functionally complete** if the removal of any one operation would render the set not functionally complete.

Notice that the definition does not prescribe the number of arguments. The amazing thing is that one unary operation and two binary operations are enough to express *any* function having *any* number of arguments. In fact, the set $\{\vee, \wedge, \sim\}$ is functionally complete, but not *minimally* so. By DeMorgan's laws, $p \vee q = \sim(\sim p \wedge \sim q)$ and $p \wedge q = \sim(\sim p \vee \sim q)$, so that operation $\vee$ can be expressed in terms of operations $\{\wedge, \sim\}$ and operation $\wedge$ can be expressed in terms of operations $\{\vee, \sim\}$. Consequently, anything that can be expressed in terms of $\{\vee, \wedge, \sim\}$ can also be expressed only in terms of $\{\wedge, \sim\}$ or only in terms of $\{\vee, \sim\}$. In other words, each set $\{\wedge, \sim\}$, $\{\vee, \sim\}$ is functionally complete. As an exercise, prove that each is minimally functionally complete. Furthermore, each set $\{\vee, \wedge, \sim\}$, $\{\vee, \sim\}$, and $\{\wedge, \sim\}$ is functionally complete in a stronger sense than Definition 9 would suggest; the alphabet needed to construct expressions for *all* functions in each set is only $\{\vee, \wedge, \sim, p, q, \ldots\}$, $\{\vee, \sim, p, q, \ldots\}$, and $\{\wedge, \sim, p,$

$q, \ldots\}$, respectively. In other words, even the zerary functions (that is, the constants), can be expressed by means of one of these alphabets; for instance, $0 = p \wedge {\sim}p$.

The fact that $\{\wedge, \sim\}$ and $\{\vee, \sim\}$, and thus $\{\vee, \wedge, \sim\}$, are all functionally complete suggests a method for testing a set of operations for functional completeness.

Theorem 10 A set of Boolean operations $\{O_1, O_2, \ldots\}$ is functionally complete if and only if it can express either of the Boolean operation sets $\{\wedge, \sim\}$ and $\{\vee, \sim\}$.

Proof: If $\{O_1, O_2, \ldots\}$ can express either $\{\wedge, \sim\}$ or $\{\vee, \sim\}$, then it can express $\{\vee, \wedge, \sim\}$, because each of $\vee$, $\wedge$ can be expressed in terms of the other two, as we saw before. Also, every Boolean function has a CNF, which in turn requires only the operation set $\{\vee, \wedge, \sim\}$. Consequently, the set can express any function's CNF. A parallel argument applies using the DNF. Q.E.D.

There are two operations, however, each of which is functionally complete all by itself: the *stroke* and the *dagger* operations.

Definition 11 The **stroke** operation is called also **Sheffer stroke** or NAND and is defined by $p|q := {\sim}(p \wedge q)$. The term NAND (not AND) is more frequently used where computer hardware is concerned. Similarly, the **dagger** operation is called also NOR (not OR) and is defined by $p \dagger q := {\sim}(p \vee q)$.

As an exercise, prove that each is functionally complete by itself in a strong sense; that is, each alphabet $\{|, p, q, \ldots\}$ and $\{\dagger, p, q, \ldots\}$, can express all total functions.

14.3.2 Minimization

We close this part on elementary Boolean algebras with a brief discussion of minimization. The term *minimization of Boolean expressions* refers to algorithms applied to Boolean expressions to construct equivalent expressions using fewer symbols. But the phrase "fewer symbols" requires comment. For example, consider the expression $p \vee {\sim}(q \wedge r)$. By De Morgan's laws, it's equivalent to $p \vee (({\sim}q) \vee ({\sim}r))$. Have we used "fewer" symbols? If you count the symbols written on paper, no; we now have 13 versus 8 in the original. If you count the types of symbols, yes; we have only p, q, r, $\vee$, $\sim$,) and (versus p, q, r, $\vee$, $\wedge$, $\sim$,) and (. Which minimization do we want? There are two immediate aspects of the question to consider. First, we've already seen examples of expressions which are blatantly inefficient. Constructions like $(\ldots((((f(p)) \wedge 1) \wedge 1) \wedge 1) \ldots)$ cry out for minimization to remove absolutely useless excess. More subtle though is the choice of a functionally complete operation set. Are we to use only $\{\vee, \wedge, \sim\}$? Can we include the stroke or dagger? At the one extreme, we could demand a minimally functionally complete set; at the other we could define 16 single symbols, one for each possible binary operation. If the answer is that "it depends on the application," then what is the application?

Boolean minimization began with George Boole himself in the 1800s, and continued with a succession of mathematicians into the present day. But these people were concerned with developing a new branch of mathematics, not with pursuing any particular application. Being so general, their work is highly theoretical, and most of it is beyond the scope of this book. Besides, relatively little of it is of practical application in its general form. With the coming of computers in the 1940s and 50s, Boolean minimization took on an entire new urgency. The inventors recognized very early that Boolean operations had to be implemented in electronic circuitry, so the choice of the operation set was simple: what circuits can we build? With each transistor costing a dollar or more, and each gate employing several transistors, the need for minimal expressions was clear. Graphical techniques such as Karnaugh's "maps" were very successful, but difficult to learn and cumbersome to apply. The early computers themselves were used to implement these techniques, sparking the first awe-struck whispers about computers designing their own successors. But technology has had the last laugh—so far. As circuits became smaller, they also became cheaper to build. Intricate designs to squeeze the last bit out of each gate became not merely unnecessary, but costly. "Brute force" techniques, a term not usually associated with microchips, became cheaper.

Consider for example a 1024×1-bit *R*ead-*O*nly *M*emory (ROM). For each of 1024 addresses is stored one bit of data. But the 1024 addresses are themselves just the possible selections of 10 individual address signals. If the address signals are the arguments and the stored data the output, what we have here is a function $f\colon \{0, 1\}^{10} \rightarrow \{0, 1\}$, and a circuit to implement that function. The memory is read-only, so the value of the function will never change. After the function has been tabulated on paper, it's time actually to build the chip. This is done in two stages.

- First, we build a generic "blank" chip which, for the ten input signals, generates all 1024 possible minterms. These signals are made available at tiny connection areas on the top surface of the unfinished chip.
- Second, to implement a specific function (memory contents), we OR together (OR is a verb here, something like "put together by means of an $\vee$ operation") all minterm signals for which we want $f = 1$. This one signal is now the output of the chip. The chip is now completed and sealed, ready for use.

The advantage of this technique is that the "blank" chips of stage one are all the same for any function. Thus, they can be cheaply mass produced. The only step that is specific to any function is the final customization of selecting which minterms to OR into the output. More sophisticated chips may have more inputs, and even multiple outputs, but the basic idea remains the same. From a mathematical viewpoint, what this process does is to implement a Boolean function by implementing its universal disjunctive form (UDF). All minterms are evaluated, with each being either selected or ignored. The UDF is syntactically one of the least efficient expressions for any function, so its use here is a bit paradoxical. We've postponed discussion of efficient expressions until last, only to find that we already have the most efficient method anyway. We don't mean to suggest that efficient circuit design has become totally unnecessary. Most computer circuits do more than just implement Boolean functions, and here efficiency is critical. Moreover, engineering constraints such as speed, size, power requirements, and heat dissipation are still very much with us, and just as difficult to solve now as ever. But the syntactic minimization of Boolean expressions has become, in the computer age, a victim of its own success.

Exercises 14.3

1. Prove that {stroke} and {dagger} are both minimally functionally complete sets of Boolean operations. The fact that the set is minimal is obvious. The proof that it is functionally complete is based on some formulae we saw before, namely $p \vee q = (p|p) \mid (q|q)$, $p \wedge q = (p|q) \mid (p|q)$, $p|p = \sim p$, and the analogues for the dagger operation. Observe also that stroke and dagger are functionally complete in a strong sense because each of the alphabets $\{|, p, q, \ldots\}$ and $\{\dagger, p, q, \ldots\}$ is able to produce expressions that represent all functions (including the zerary ones).
2. Prove that the sets $\{\wedge, \sim\}$ and $\{\vee, \sim\}$ are both minimally functionally complete sets of Boolean operations.
3. Find the set of all unary and binary functions that can be expressed by means of the operation set $\{\sim\}$. Do the same for each set $\{\vee\}$, $\{\wedge\}$ and $\{\vee, \wedge\}$.
4. Prove that neither $\{\vee\}$ nor $\{\wedge\}$ nor even $\{\vee, \wedge\}$ is a functionally complete set. [Hint: $\vee$ and $\wedge$ are both idempotent.]

Complements 14.3

1. Show that the set $\{\Rightarrow\}$ is minimally functionally complete. (Note, however, that this completeness is valid in a weak sense. In this case, we need the alphabet $\{\Rightarrow, 0, p, q, \ldots\}$ to express all functions.) [Hint: calculate $x \Rightarrow 0$ and $(y \Rightarrow x) \Rightarrow x$.] Then find all binary total Boolean operations that are individually functionally complete (in a strong or in a weak sense). Classify them according to the following criteria: **(i)** only the alphabet {operation, $p, q, \ldots$} is needed, or an alphabet {operation, constant, $p, q, \ldots$} is necessary; **(ii)** the operation is commutative or not; **(iii)** negation can or cannot be obtained in one step, that is, in the form $\sim x = (x$ **operation** $x)$.

14.4 Boolean Forms in General

So far in this chapter we have considered an *elementary* Boolean algebra, that is, a Boolean algebra with only two elements (see Section 6.3.3). We want to extend the notion of normal form and the properties that go with it to the case of any finite Boolean algebra. There will be analogies with the elementary case, and there will be differences.

Let a finite Boolean algebra S be given. Let A be the set of its elements and define $m := \#A$. Let $\vee$, $\wedge$ and $\sim$ be the Boolean operations. In symbols, $S = (A, \vee, \wedge, \sim, 0, 1)$. Occasionally, we'll refer to operation $\sim$ as the *negation,* although *complementation* would be more correct. For each positive integer n, we'll keep the definition of *minterm* and *maxterm* as the conjunction and disjunction, respectively, of n distinct arguments each one negated or not. We'll also keep the notation. In two arguments, for instance, $\mathbf{min}_{00} := \sim p \wedge \sim q$, $\mathbf{min}_{01} := \sim p \wedge q$, $\mathbf{min}_{10} := p \wedge \sim q$, and $\mathbf{min}_{11} := p \wedge q$. Also, J will still denote a string of n zeroes and ones (that is, an element of $\{0, 1\}^n$), and $\mathbf{min}_J$ will denote the corresponding minterm (and similarly for the maxterms). It is still true that $\mathbf{min}_J\, J = 1$ for every J in $\{0, 1\}^n$. The first difference from the elementary case is that $\mathbf{min}_J\, I$ is not necessarily 0 when $I \neq J$. Under the present general circumstances, a sequence I of n values, one for each argument, is not necessarily made up of 0s and 1s only. Such a string is an arbitrary element of A^n and therefore comprises any elements of the set A of the m elements of our Boolean algebra. It is still true that $\mathbf{min}_J\, I = 0$ whenever $I \neq J$ and $I \in \{0, 1\}^n$. The reason is the same as in the elementary case.

Example 3 Recall from Exercises 6.3 that the following is a Boolean algebra: $A := \{00, 01, 10, 11\}$, 0 is 00, 1 is 11, each of the operations $\vee$, $\wedge$, $\sim$ is a bitwise operation. For instance, $01 \vee 10 = 11$, $01 \wedge 11 = 01$, $01 \wedge 10 = 00$, $\sim 00 = 11$, $\sim 01 = 10$.

In two variables, consider the expression $(10 \wedge p \wedge \sim q) \vee (01 \wedge p)$. Using the general rule of Boolean algebra, this expression can be changed to the following equivalent expressions.

$$(10 \wedge p \wedge \sim q) \vee (01 \wedge p \wedge 1),$$

$$(10 \wedge p \wedge \sim q) \vee (01 \wedge p \wedge (q \vee \sim q)),$$

$$(10 \wedge p \wedge \sim q) \vee (01 \wedge p \wedge q) \vee (01 \wedge p \wedge \sim q),$$

$$(01 \wedge p \wedge q) \vee ((01 \vee 10) \wedge p \wedge \sim q),$$

$$(01 \wedge p \wedge q) \vee (11 \wedge p \wedge \sim q).$$

The last expression is normal and can be rewritten as $(01 \wedge p \wedge q) \vee (11 \wedge p \wedge \sim q) \vee (00 \wedge \sim p \wedge q) \vee (00 \wedge \sim p \wedge \sim q)$.

The definitions of **UDF** and **UCF** are a little more general than in the elementary case, because the "coefficients" of the minterms and maxterms can be other than 0 or 1.

Definition 12 Given a Boolean algebra S as above, a **UDF** in n arguments is defined as the disjunction of 2^n expressions $a_J \wedge \mathbf{min}_J$, where, for each J in $\{0, 1\}^n$, the *coefficient* a_J is any assigned element in the set A. The **DNF** is then obtained by omitting those terms for which the coefficient a_J is 0. Conjunctive forms are obtained in a dual manner.

From now on, we'll focus our attention on disjunctive forms only.

If m is the number of elements of set A, then there are m^{2^n} distinct UDFs, because we have m choices for each of the 2^n coefficients a_J. As usual, let's denote f the function defined by a given UDF. As in the elementary case, it's true that $a_J = f(J)$ for every J in $\{0, 1\}^n$. This is a direct consequence of the fact that $\mathbf{min}_J\ I = 1$ or $\mathbf{min}_J\ I = 0$ according as $I = J$ or $I \neq J$, provided that $I \in \{0, 1\}^n$ and $J \in \{0, 1\}^n$. It is therefore still true that each function that has a UDF or DNF has a unique one. The most important difference is an immediate consequence of the count we just did: unless the Boolean algebra has exactly two elements ($m = 2$), not every function has a universal form. The reason is simple. Theorem 7.3 gives the number of distinct total Boolean function with n arguments as $\#(A^{A^n})$, which equals m^{m^n}. As we saw above, the number of distinct universal forms is m^{2^n}. Whenever $m > 2$ and $n > 0$, we have $m^n > 2^n$ and $m^{m^n} > m^{2^n}$, as we wanted to show.

Example 4 Consider the same Boolean algebra of Example 3. Let f be the function defined by the fact that $f(00, 00) = 1, f(00, 01) = 1$, and $f(p, q) = 0$ otherwise. If f had a UDF, then the form would be $(a \wedge p \wedge q) \vee (b \wedge p \wedge {\sim}q) \vee (c \wedge {\sim}p \wedge q) \vee (d \wedge {\sim}p \wedge {\sim}q)$, where a, b, c and d are elements in A. Then, necessarily,

$$f(00,00) = 1 = (a \wedge 00 \wedge 00) \vee (b \wedge 00 \wedge 11) \vee (c \wedge 11 \wedge 00) \vee (d \wedge 11 \wedge 11) = d;$$

$$f(00,11) = 0 = (a \wedge 00 \wedge 11) \vee (b \wedge 00 \wedge 00) \vee (c \wedge 11 \wedge 11) \vee (d \wedge 11 \wedge 00) = c;$$

$$f(11,00) = 0 = (a \wedge 11 \wedge 00) \vee (b \wedge 11 \wedge 11) \vee (c \wedge 00 \wedge 00) \vee (d \wedge 00 \wedge 11) = b;$$

$$f(11,11) = 0 = (a \wedge 11 \wedge 11) \vee (b \wedge 11 \wedge 00) \vee (c \wedge 00 \wedge 11) \vee (d \wedge 00 \wedge 00) = a.$$

Consequently, $a = 0$, $b = 0$, $c = 0$, $d = 1$, and f should be equivalent to the function of expression ${\sim}p \wedge {\sim}q$. This is not the case: when $p = 00$ and $q = 01$, we would have $f(p, q) = 1$ (by definition), and the value of the latter expression would be ${\sim}00 \wedge {\sim}01 = 11 \wedge 10 = 10 \neq 1$. Consequently, this is an example of a Boolean function that does *not* have a normal form.

The next question is, of course, which Boolean functions *do* admit a normal form? The answer in one direction is obvious: those functions that admit a normal form admit also an expression in terms of the n arguments, the elements of A, and the three operators $\vee$, $\wedge$ and ${\sim}$. Can this property be inverted? The next theorem gives the answer, which is affirmative. Before we state and prove the theorem, let's recall the definition of the term *expression* under the present circumstances (see Section 14.1). The initial steps of the recursive definition are: (1) each of the m elements of A is an expression; (2) each of the n arguments is an expression. The recursion steps are: given that x and y are expressions, each of (x) (*enclosure*), ${\sim}x$ (*negation*), $x \vee y$ (*disjunction*), $x \wedge y$ (*conjunction*) is an expression. There are no other expressions.

Theorem 13 Let S be a Boolean algebra whose set A comprises m elements. Let $f: A^n \rightarrow A$ be any total function, that is, any Boolean function in n arguments. Then f admits a normal form if and only if f is the function of an expression represented in terms of the operators $\vee, \wedge, \sim$.

Proof: We have already proved the "only if" part: If f has a normal form, then it has an expression as desired. The "if" part is proved by a process of induction that parallels the recursive definition of an expression. We have to prove first that each element of A and each argument has a normal form, and then that enclosure, negation, disjunction, and conjunction of normal forms provide functions that have normal forms. Each of these statements is proved directly by applying algorithm B of Complements 14.2. Q.E.D.

As a consequence, if a Boolean function admits an expression in terms of the operators $\vee$, $\wedge$, and $\sim$, then its normal form can be calculated almost exactly as in the elementary case because its coefficients depend only on the values of the function when each of the arguments is 0 or 1. The only difference is that the coefficients a_J ($J \in \{0, 1\}^n$) are the values of the function at J, and these values are not necessarily zero or one only.

Example 5 On the same Boolean algebra as in Example 3, consider the function f defined by $f(00, 10) = 10, f(00, 11) = 10, f(10, 10) = 10, f(10, 11) = 10, f(01, q) = q$ for every $q, f(11, q) = q$ for every q, and $f(p, q) = 0$ otherwise. Then its normal form is (if existent) $(10 \wedge \sim p \wedge q) \vee (11 \wedge p \wedge q)$. Since this expression has the same table as f, it is the normal form of f.

Minor changes are also necessary in the algorithms. In algorithm A, step 3, replace the minterm with (entry of last column of table) $\wedge$ minterm. In algorithm B, step 1, replace contexts like $\sim$constant, constant $\vee$ constant, constant $\wedge$ constant with their constant values. In step 8, the h-fold $\wedge$ may contain also a constant, which is to be placed first. In step 9, if the entire expression is just a constant c, it is to be replaced with $c \wedge (p \vee \sim p)$. In step 10, consider also $(h + 1)$-fold $\wedge$s for which one argument is a constant.

Exercises 14.4

1. Let a Boolean algebra be given as in Example 3 in the text. In each of the three cases below, the given set of conditions (each of which is valid for all p and q) defines a function f. Find a DNF in two arguments p and q for the function f. **(a)** $f(p, q) := \sim p$. **(b)** $f(p, p), := 1, f(p, \sim p) := 0, f(p, 1) := p, f(p, 0) := \sim p, f(1, q) := q, f(0, q) := \sim q$. **(c)** $f(1, q) := 0, f(0, q) := 1, f(01, q) := 10, f(10, q) := 01$.
2. In the same Boolean algebra as in Example 3 in the text, verify that the following functions do not admit a normal form. **(a)** f defined by $f(00, 01) = 1$, $f(01, 01) = 1$, and $f(p, q) = 0$ otherwise. **(b)** f defined by $f(00, 00) = 1$,

$f(01, 00) = 01$, and $f(p, q) = 0$ otherwise. **(c)** f defined by $f(p, q) = 01$ if $p = q$, $f(p, q) = 11$ otherwise. **(d)** f defined by $f(p, q) = p \vee q$ if $p = q$, $f(p, q) = 11$ otherwise.

3. Find more examples of Boolean functions that cannot be expressed only in terms of operators $\vee$, $\wedge$ and $\sim$.

4. Explain the meaning of the following statement: the set of operators $\vee$, $\wedge$, and $\sim$ is not functionally complete in a Boolean algebra with more than two elements.

14.5 Stone's Theorem

The conclusions of Section 14.4 answer several questions on a finite Boolean algebra in general, but leave also some questions unanswered. The following theorem provides a method of answering many more questions, although in a different direction.

Recall first the definition of direct product of elementary Boolean algebras that we saw in Exercises 6.3: given the positive integer k, define $A := \{0, 1\}^k$ and let $\vee$, $\wedge$ and $\sim$ be the bitwise operations OR, AND and NOT, respectively.

Example 6 When $k = 1$, set A is practically the same as $\{0, 1\}$. Assuming now $k = 3$, some elements of A are, for instance, 000, 001, 010, 011. Some operations are $011 \vee 101 = 111$, $011 \vee 001 = 011$, $011 \wedge 101 = 001$, $011 \wedge 001 = 001$, $\sim 000 = 111$, $\sim 001 = 110$, and $\sim 010 = 101$.

Stone's Theorem, named for American mathematician Marshall Harvey Stone, born in 1903, states, in essence, that this process provides *all* finite Boolean algebras. The only difference between an arbitrary finite Boolean algebra and a direct product of elementary Boolean algebras may be in the symbols used to denote elements and operations.

Theorem 14 **Stone's theorem.** Let A be the set of the elements of a finite Boolean algebra S. Then, necessarily, $\#A = 2^k$ for some positive integer k and S is isomorphic to the direct product of k elementary Boolean algebras.

Proof: We'll proceed by induction on $\#A$. The definition of a Boolean algebra implies existence and distinctness of the elements 0 and 1. Therefore $\#A \geq 2$. If $\#A = 2$, then, necessarily, the two elements must be 0 and 1 and, except possibly for the symbols used for them, our algebra is the elementary Boolean algebra (see Section 6.3.3). This proves the initial step of our induction.

Proof of the induction step. Let A comprise more than two elements. Then we can apply the construction we used in Complements 6.3.

- There are in A two special subsets X and Y.
- X and Y are both Boolean algebras with the same operations $\vee$ and $\wedge$ as in

the given Boolean algebra S. Let such operations be denoted OR_1, AND_1 within X and OR_2, AND_2 within Y. In each of X and Y there is a complementation, denoted NOT_1, NOT_2 respectively.

- X and Y each have a 0 that is the same as the 0 of S.
- Each element z of A admits a unique representation of the form $z = z_1 \vee z_2$, under the conditions that $z_1 \in X$, $z_2 \in Y$, and that the $\vee$ operation be in S.
- If z is represented as above and if an element w of A is similarly represented as $w = w_1 \vee w_2$, then $z \vee w = (z_1 \text{ OR}_1\ w_1) \vee (z_2 \text{ OR}_2\ w_2)$, $z \wedge w = (z_1 \text{ AND}_1\ w_1) \vee (z_2 \text{ AND}_2\ w_2)$, and $\sim z = (\text{NOT}_1\ z_1) \vee (\text{NOT}_2\ z_2)$.

(In other words, the Boolean algebra S on A is the direct product of the Boolean algebra on X and the Boolean algebra on Y.) By the hypothesis of induction, each of the two algebras on X and on Y is the direct product of elementary Boolean algebras, say r of them and s of them respectively. Then each z_1 is a string of r 0s and 1s and each z_2 is a string of s 0s and 1s. Then z can be represented by concatenating these two strings; the result is an $(r + s)$ string. Verifying that the operations of S are equivalent (that is, isomorphic) to the bitwise operations on such $(r + s)$-strings is then immediate. Both statements are therefore proved, with k equal to $r + s$. Q.E.D.

Exercises 14.5

1. Given $A := \{0, 1, a, b\}$ ($\#A = 4$), we want on A a structure of a Boolean algebra with the assigned 0 and the assigned 1. How many such structures are possible? [Hint: very few!] How can we find them all; that is, how must the Boolean operations be defined in each case?
2. On the Boolean algebra S obtained as the direct product of three elementary Boolean algebras ($k = 3$), consider each of the following functions. Recall that p and q are to be replaced by elements of $\{0, 1\}^3$. Does the given function admit a normal form? If so, find it. If not, why not? **(a)** $f(p) :=$ first term of string p. **(b)** $f(p) := 1$ if and only if string p contains an even number of ones; else, $f(p) = 0$. **(c)** $f(p) := 1$ if and only if string p contains fewer 0s than 1s; otherwise, $f(p) = 0$. **(d)** $f(p) := 1$ if and only if string p contains precisely one 1; otherwise, $f(p) = 0$. **(e)** $f(p, q) := 1$ if and only if the first terms of p and q are different from each other; otherwise, $f(p, q) = 0$. **(f)** $f(p, q) := 1$ if and only if string p contains more 1s than string q; otherwise, $f(p, q) = 0$. **(g)** $f(p, q) := 1$ if and only if string p contains as many 1s as string q; otherwise, $f(p, q) = 0$.
3. Repeat problem 2 for the direct product of k elementary Boolean algebras ($k > 1$).

Complements 14.5

1. Let A be a given finite set with n elements. Give a method to construct all possible structures of Boolean algebra on set A (when they exist). [Hints: If n

is not a power of 2, then what? If n *is* a power of 2, then the n elements must be (except possibly for the order and the symbolism) the binary numbers from 0 to $n - 1$.]

14.6 Summary

A systematic way to study Boolean functions and expressions is provided by *normal forms*: these are either disjunctions of distinct minterms (where a minterm is a conjunction of *all* the arguments, each negated or not) or conjunctions of distinct maxterms (where a maxterm is a disjunction of *all* the arguments, each negated or not). Normal forms on an elementary Boolean algebra (Section 14.2) are important for theoretical purposes (study of Boolean functions in a given number of variables) and for practical purposes (circuitry in a chip). They provide an insight into the problem of functional completeness (which operators are able to represent *all* total Boolean functions; Section 14.3). Each Boolean function has a unique *disjunctive normal form* and a unique *conjunctive normal form;* each can be constructed via an exhaustive algorithm (Section 14.2).

On a finite Boolean algebra in general, normal forms are *not* able to represent all total Boolean functions (that is, the set of operations $\vee$, $\wedge$ and $\sim$ is not *functionally complete;* Section 14.4). This disadvantage is somewhat compensated by the fact that, according to Stone's Theorem, every finite Boolean algebra is the direct product of elementary Boolean algebras (Section 14.5).

Finite State Acceptors

15.1 Introduction

Throughout this book, we've stressed the analogy of mathematical structures to games. We begin this chapter only slightly differently, looking at the mathematical aspects of a familiar game.

15.1.1 A Game of Tennis

The scoring method in tennis is both interesting and unique. Two players, the server S and the receiver R, can each score one point at a time. Within a *game* (as opposed to a *set* or a *match*), the score of 0 is traditionally denoted as "love", the scores of 1, 2 and 3 points are denoted 15, 30 and 40, respectively. A score of 4 or more may be denoted "game" or "advantage", depending on the situation, as we'll see in a moment. To win, one player must score at least four points total *and* score at least two points more than the other player. Moreover, if a score of "40 all" is reached (that is, "40-40"), numerical scores are not kept at all any more. Rather, if player R scores one point beyond "40 all", the score is denoted "advantage R". If R scores an additional point, the score is "Game R" and R wins. On the other hand, if the score is "advantage R" and if S scores a point beyond that, then the score goes back to "40 all". In other words, the "advantage" score merely keeps track of one player being one point ahead of the other, regardless of how many times this has happened in the past. Thus, at any point in a tennis game, either before, during, or after, there are only 20 possible scores. (Take a moment now to list them!) The initial score for any game is always "love-love", and the final score is either "game S" or "game R". We're going to refer to each of these 20 possible scores as a *state* of the game at that point. As any point is scored, the game undergoes a *transition* from one state to another, and then to still another, until it's over.

We can express this process much more easily by the diagram in Figure 15.1. Here, each node corresponds to one state, and each arc corresponds to the event of a single point being scored. The arcs are here called *state transitions,* because they move the game from state to state. A game always begins with state "love-love". When the first point is scored, Figure 15.1 moves either to state "15-Love" (if S scored) or to state "Love-15" (if R scored). As each additional point is scored, Figure 15.1 moves to the appropriate new state. The game ends when state "Game S" or state "Game R" is reached. (The twenty-first "illegal" state will be discussed in a moment.)

We can uniquely and completely characterize any tennis game as a sequence of scored points by R and S. Suppose we define an alphabet $A := \{R, S\}$ and let $(A^*, :, \Lambda)$ be the free semigroup over A as defined in Section 6.1.2.2. The elements of A^* are the finite strings of Rs and Ss. The strings we consider here are finite, although arbitrarily long. (In tennis, a game could theoretically go on indefinitely if the players alternated scoring points. In the real world, though, the referee would eventually declare a draw so that everybody could go home.)

Any tennis game, then, either still in progress or completed, whether won by R or S, can be characterized as a string $p \in A^*$. Not every string in A^* corresponds to a legal tennis game, however. Once a game has been won, no further points may be scored, and any string reflecting such an event must be deemed an "illegal" game. What we can do is to add one additional "score" state labelled "Illegal" (as we've done in Figure 15.1) to reflect this possibility. Now every string in A^* corresponds to exactly one of the following:

- A game in progress (or a draw). For example, the string $q :=$ SSRRRSRSS is a game currently in the state "Advantage S". The null string Λ is degenerately a game in progress, just about to start, that is, in state "Love-Love".
- A completed game, won by either R or S. Continuing game q above to completion might yield q:(S) (won by S) or q:(RRR) (won by R) or any one of many others.
- An illegal string, that is, a string that doesn't correspond to a legal tennis game. For example, the string q:(SS) is illegal, since game q:(S) is complete (having been won by S), and the additional S (or any string except Λ) concatenated to q:(S) would produce an illegal game.

Thus every game corresponds to a string, and every string either determines a game or is illegal. To *process* a string means to classify it definitely as either a game in progress, a game won by R, a game won by S, or as an illegal string, and to do so in a finite amount of time, that is, in a finite number of state transitions. (An algorithm that requires an infinite amount of time can hardly be said to determine anything definitely.) The graph in Figure 15.1 can be viewed as a potential algorithm for doing this. Given a string p in A^*, we begin in the initial state "Love-Love", and process string p left to right, symbol by symbol. By the definition of A^*, string p is finitely (although arbitrarily) long. So we will eventually reach the last symbol of p, and the state we wind up in at that time determines the

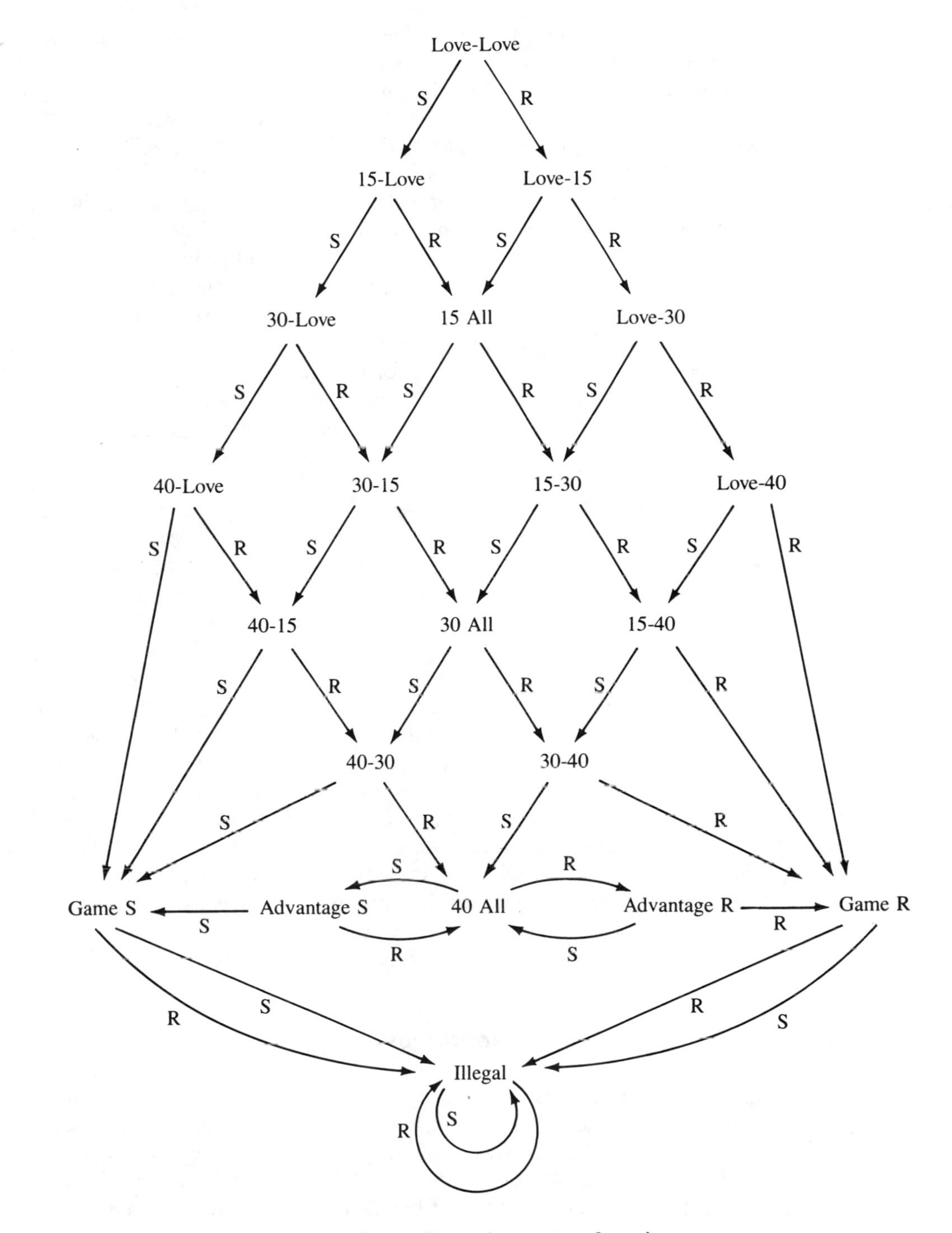

Figure 15.1 Scores in a game of tennis.

classification of p. If we find ourselves at "Illegal", then p is illegal (note that once we arrive at state "Illegal" at any point in the string, we can never leave.) States "Game R" and "Game S" classify p accordingly *if our arrival coincides with the end of the string*. Otherwise, we simply move through these two to "Illegal". If we arrive at any state other than "Game R", "Game S", or "Illegal" at the end of p, then p is a game in progress.

The procedure will work for any string in A^*, that is, any string of arbitrary but finite length. (It would even work for strings of infinite length, if we adopted the convention that, once in state "Illegal", we halt processing. What's the point of going on? *Any* string concatenated to an illegal string produces another illegal string.) The crucial idea here is that all finite strings are correctly processed in a finite number of state transitions. That's pretty impressive for a graph having only 21 states. By comparison, a graph for baseball would require an infinite number of states, since theoretically either team could score arbitrarily (but finitely) many runs *and still lose the game* if their opponents scored one more.

15.1.2 Remarks

In this chapter we shall consider the mathematics of structures such as Figure 15.1. We lay the formal groundwork in the next section, but three aspects of their purpose are already apparent. These structures, called *finite state acceptors* (or FSAs), must definitely accept as valid or reject as invalid each finite string of a free semigroup of strings. They must do so having themselves only a finite number of internal states, and must return a definite determination after only a finite number of state transitions. Each FSA so determines, within A^*, the subset of the strings it accepts. We will also look at the question itself in a somewhat inverted fashion. First, we'll attain some practice by looking at specific subsets of specific free semigroups and trying to build FSAs for them. But our real purpose will be to look at the semigroups themselves and examine what kinds of subsets we can build FSAs for. In other words, just how powerful and useful is the notion of FSAs? What sorts of information can they process, and what determinations can they make? What are their limitations? As you might guess, therefore, our work will be as much with specification of string sets as with specification of FSAs.

15.1.3 Strings and Languages

Since we'll be dealing with strings in some detail, we need some convenient notational machinery. Everything we need we've been through before; see Section 6.1 and problem 7 in Exercises 8.2. It will be useful to review those ideas.

Let A be a given finite nonempty alphabet and let n be its size, that is, $n := \#A$. A k-string (or string of length k) is a sequence of k elements of A. For instance, if $x, y, z, \ldots$ are any elements of A, then (x, y, x, z) and (x, x, x) are strings over A. For short, they are written $(xyxz)$ and (xxx) or even $xyxz$ and xxx, respectively. The second one is even denoted x^3. But such informal notations should be used only when there is no ambiguity in the interpretation of symbols. The symbol A^k denotes the set of all k-strings from alphabet A. The symbol A^* denotes the free semigroup over A, that is, the set of all finite strings over A. If p and q are two such strings, then the concatenation of p and q is denoted $p{:}q$ and is the string obtained by writing q right after p. Here too we have some informal notations. For instance, if $p \in A^*$, then p^2 denotes $p{:}p$, p^3 denotes $p{:}p{:}p$, and so on. Also, if $p \in A^*$ and if $x \in A$, then concatenating the symbol x at the end of string p produces string $p{:}(x)$, which is also denoted just $p{:}x$ or even px. Just be careful in distinguishing elements of A from elements of A^*. A language is any subset of A^*, that is, any element of $\mathcal{P}(A^*)$ where $\mathcal{P}$ denotes the *power set*. If L and M are any two languages, then the concatenation of *languages*, $L{:}M$, is obtained by concatenating each string in L with each string in M. The result is again a language. If L is a language, then we can concatenate each string in L with each string in L, any finite number of times. The result is the iteration of L, or Kleene star of L, denoted $L^\star$, which is also a language. There are a few more informal ways of denoting these operations. If $x \in A$, then (x) is a string, which we sometimes denotes just x; $\{(x)\}$ is a language (a language of just one string) and is sometimes written again just x (this is dangerous); $\{(x)\}^\star$ is the iteration of language $\{(x)\}$ and is a language, informally denoted $x^\star$. Note also that $\{x\}$ is an alphabet, and therefore we can construct $\{x\}^*$ (the free semigroup), which is also denoted informally x^*. If there's no ambiguity, informal symbolism is all right, but if there is any doubt, use parentheses ( ) and braces { } as needed.

Here are a few examples of combined operations. Work them out carefully. Square parentheses [ ] are just that: parentheses. By contrast, ( ) and { } have specific meanings.

Assuming $x \in A$ and $y \in A$: (1) $x^2 = xx$ [$\in A^*$]; (2) $x^\star = \{\Lambda, x, xx, xxx, \ldots\}$ [$\subseteq A^*$; strings made up only of any number of xs, possibly none; formal notation is $\{(x)\}^\star$]; (3) $[xy]^\star = \{\Lambda, xy, xyxy, xyxyxy, \ldots\}$ [$\subseteq A^*$; strings comprising only any number of pairs xy, possibly none; formal notation is $\{(x, y)\}^\star$]; (4) $x^\star y^2 = \{yy, xyy, xxyy, \ldots\}$; [$\subseteq A^*$; strings comprising any number of xs (possibly none), followed by precisely two ys; formal notation $\{(x)\}^\star{:}\{(y, y)\}$]; (5) $[x^\star y^2]^\star = \{\Lambda, yy, xyy, xxyy, xxxyy, \ldots, yyyy, yyxyy, yyxxyy, \ldots, xyyyy, xyyxyy, xyyxxyy, \ldots\}$ [$\subseteq A^*$; any number of xs and any number of nonoverlapping pairs yy, in any positions; same as $[x, y^2]^\star$; formal notations $[\{(x)\}^\star{:}\{(y, y)\}]^\star$ or $\{(x), (y, y)\}^\star$, respectively.

A few more words on degenerate cases are in order. The null string Λ is a string, that is, an element of A^*. The empty set $\varnothing$ is a subset of A^* and therefore a language. Concatenating a language L with the language comprising only the null string gives as a result the language L again, that is, $L{:}\{\Lambda\} = L$ and $\{\Lambda\}{:}L = L$. Concatenating language L with the empty language $\varnothing$ gives the empty language: $L{:}\varnothing = \varnothing$ and $\varnothing{:}L = \varnothing$. Recall also that languages are subsets of A^*. If we consider this set as our universal set, then we have the set-theoretical operations on languages. In particular we'll need *union* (if L and M are languages, then $L \cup M$ is a language, $L \cup \varnothing = L$, etc.) and *complementation* (if L is a language, then $\sim L := \{p \in A^* \mid p \notin L\}$ is a language).

Putting this all together, we can state that $(A^*, :, \Lambda)$ is a monoid with identity Λ and $(\mathcal{P}A^*, \cup, :, \varnothing)$ is a semiring with identity $\varnothing$ for operation $\cup$ and identity $\{\Lambda\}$ for operation :.

Exercises 15.1

1. Given $A := \{x, y, z\}$, $B := \{\Lambda, x, xy, yz\}$, and $C := \{\Lambda, x, y, z\}$, note that B and C are languages over alphabet A. Construct the following: **(a)** $B \cup C$; **(b)** $B{:}C$; **(c)** $B{:}B$; **(d)** $C{:}C$; **(e)** $\{\Lambda\}{:}B$, **(f)** $\{\Lambda\}{:}C$, **(g)** $\varnothing{:}B$; **(h)** $\varnothing{:}C$; **(i)** $\{\Lambda\}^\star$; **(j)** $\varnothing^\star$.
2. Given $A := \{a, b\}$ and $C := \{c, d\}$ as languages over the alphabet $\{a, b, c, d\}$, construct **(a)** A^2; **(b)** C^2; **(c)** C^3; **(d)** $A^2{:}C^3$; **(e)** $A{:}C{:}A$.
3. The alphabet is $A := \{x, y, z\}$. Expand each of the following string notations. If infinitely many strings are obtained, limit consideration to those obtainable in 0, 1, or 2 steps. **(a)** yx^5, **(b)** z^0y^2x, **(c)** $[xy^2]^\star$, **(d)** $[xy^\star]^2$, **(e)** $[x^\star y^\star z^\star]^0$, **(f)** $[x^0y^0z^0]^\star$.
4. The alphabet is $\{x, y, \ldots\}$. What are the relationships between $[\{(x)\}{:}\{(y)\}]^\star$ and $\{(x)\}^\star{:}\{(y)\}^\star$? Are they equal, disjoint, overlapping, or does one contain the other?
5. What's the relationship between $\{(x, y)\}^*$ and $\{(x, y)\}^\star$? [Hint: interpret the symbols carefully; recall the difference between $A^h{:}A^k$ and A^{h+k}.]

15.2 Finite State Acceptors

As in Figure 15.1, it's easiest to imagine a finite state acceptor as a directed graph, or digraph. The formal basis, though, rests as usual in sets and functions.

Definition 1 A **finite state acceptor (FSA)** over an alphabet A is a quintuple (A, V, g, u, T), where

- A is a nonempty, finite alphabet of size n;
- V is a nonempty, finite set of **states**, of size k;
- $g{:}V \times A \to V$ is a total function called the **state transition function**; since g is often given in tabular form, it can also be called the **state table**;
- u is a designated element of V, the **initial state**;
- T is a designated subset of V; the elements of T are the **final states**.

In the case of the tennis game, we set $A := \{\text{R}, \text{S}\}$, V is the set of the 20 possible scores plus the special "illegal" state, and Love-Love is the initial state. The state transition function g wasn't tabulated explicitly, but it is given via the graph in Figure 15.1. Finally, we never explicitly named a set of final states, because what we wanted then was to classify all score strings. If, however, we had wanted to isolate and identify only completed, legal games, then we should have set $T :=$ {Game R, Game S}. We'll have more to say about T in a moment.

Here is an example of an FSA in which function g is given in tabular form. The alphabet is $A := \{x, y\}$; the nodes are 1 and 2; node 1 is the initial node and, at the same time, the only final node. Function g is assigned by the following table, in which the entry in row i and column j ($i = 1$ or 2; $j = x$ or y) is the value of $g(i, j)$.

	x	y
1	2	1
2	1	2

(Verify that Figure 15.4 provides a digraph representation of this FSA.)

For any FSA as above and for each string p in A^*, the operation of M on p is completely determined. At the beginning, M is in the initial state and, at any one point in the processing of p, M can be in only one unique state, and no others. The FSA M is thus classed as a *deterministic* FSA. But what exactly do we mean by the *operation of M on p?* If we write the sequence of states that M occupies as it processes p, we can trace exactly what M did for p. Such a sequence of states is called the *run* of p on M.

Definition 2 Given the FSA M as above, let p be an element of A^*, let m be its length, and suppose $p := (x_i)_{1 \leq i \leq m}$, where $x_i \in A$ for each i. The **run** of p on M is denoted $\mathcal{M}(p)$ and is the $(m + 1)$-selection $(v_h)_{0 \leq h \leq m}$ from V defined recursively by

- $v_0 := u$ (the initial state of M);
- $v_{h+1} := g(v_h, x_{h+1})$ $(0 \leq h < m)$.

In the tennis example, we can use a completed game $p :=$ SSRRRSRSRSSS as the string. The run of p is then $\mathcal{M}(p) =$ (Love-Love, 15-Love, 30-Love, 30-15, 30 All, 30-40, 40 All, Advantage R, 40 All, Advantage R, 40 All, Advantage S, Game S).

A few things should be noticed before we go on. First, although a string p is a selection from A (that is, $p \in A^*$), the sequence $\mathcal{M}(p)$ is a selection from V (that is, $\mathcal{M}(p) \in V^*$). Also, $M(p)$ always starts with the initial state u, and thus is always precisely one symbol larger than p. Finally, and most importantly, $\mathcal{M}(p)$ always contains, as its very last symbol, the state M was in after the last symbol of p had been processed. This is the one piece of information we want, the reason for constructing M in the first place.

Definition 3 A string p ($p \in A^*$) is said to be **accepted** by FSA M if the last state in the run $\mathcal{M}(p)$ is a final state, that is, if $v_m \in T$ (using notations as above). Otherwise, p is said to be **rejected**. The set of strings which FSA M accepts is called the **language of *M*** and is denoted $\mathcal{L}(M)$. Clearly, $\mathcal{L}(M) \subseteq A^*$ (so that the terminology is consistent with the definition of language in Section 6.1).

The run $\mathcal{M}(p)$ is dependent on p (and thus on A), and on V, g, and u. But the run is not dependent on T. Specifying a different T will not alter the operation of

M at all. It will, however, affect our interpretation of the results. The reason for the name *finite state acceptor* is that M *accepts* certain strings from A^* and rejects all others. In the tennis example, we wished to accept all strings representing completed legal games and reject all others. Hence we chose $T :=$ {Game R, Game S}. Had we wished to accept a different subset of strings, our choice of T (and possibly of the entire design of M) would have been different.

The definition of *language of M* demonstrates what we alluded to in the introduction. While V, g, u, and T are important details in the construction of an FSA, the overriding importance must be given to A, and specifically to A^* and its subsets (that is, the languages over A). The question we really wish to address is this: Given A and some language L ($L \subseteq A^*$), is there an FSA M that accepts precisely L, that is, such that $\mathcal{L}(M) = L$? If so, we really do not place any other restrictions on M. In fact, if more than one FSA will work, we will consider them to be interchangeable even if their internal mechanisms (that is, their respective V, g, u, and T) are different. This idea is stated formally as follows:

Definition 4 Let M and M' be FSAs over the same alphabet A. The two FSAs M and M' are said to be **equivalent over** A if and only if $\mathcal{L}(M) = \mathcal{L}(M')$. In operation, M and M' are said to **simulate** each other over A.

Let's look now at some examples of FSAs. As in the tennis example, we shall represent the FSA as a digraph. The nodes of the digraph are precisely the states of the FSA. The arcs of the digraph represent the state transition function g. For instance, if an arc is labelled x ($x \in A$) and has first endpoint v and second endpoint w ($v \in V$, $w \in V$), then this is the same information as $g(v, x) = w$. Since the state function must be total, each node must be the origin of n arcs ($n = \#A$), each labelled with a different element of A. The notation for initial and final states is simply a visual convention: we'll use ∘ for the initial state and □ for the final states (if any).

The first three examples we consider are, in a sense, degenerate. Chosen to illustrate the lower limit of what FSAs can do, they are extremely simple but extremely important. Figure 15.2a shows an FSA that accepts all of A^*, that is, $\mathcal{L}(M) = A^*$. All strings are accepted, none are rejected. The initial and final states coincide (perfectly permissible) since there is only one state.

Note the arc labelled A in the figure. Such an arc is used for *any* symbol from alphabet A. It's simpler just to label the arc with the name of a set of symbols rather than attempting to draw many arcs with one symbol for each. In fact, here the individual symbols aren't even known. Such extensions of notation are per-

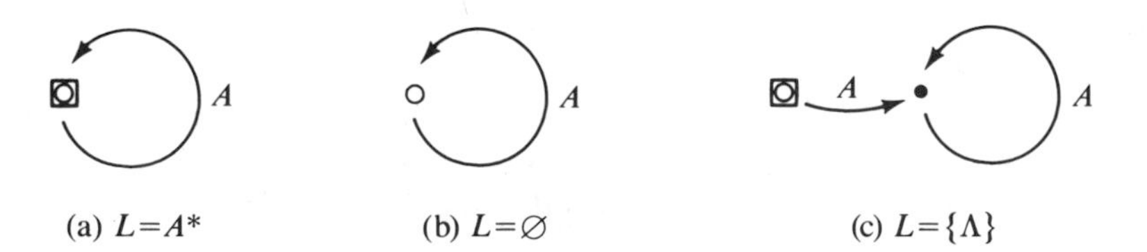

Figure 15.2 Extreme cases of FSAs.

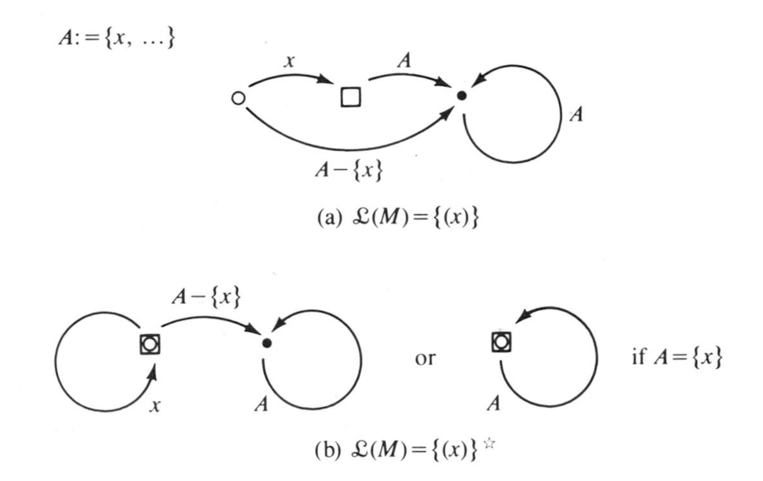

Figure 15.3 FSAs with a single letter language.

fectly permissible, even desirable and necessary, so long as they are complete, unambiguous, and consistent.

Conversely, Figure 15.2b rejects every string in A^*, that is, $\mathscr{L}(M) = \varnothing$. This FSA has no final states, that is, $T = \varnothing$. Put another way, any FSA in which $T = \varnothing$ will be equivalent to Figure 15.2b (see Exercises 15.2). Figure 15.2c shows an FSA that accepts only the null string Λ, that is, $\mathscr{L}(M) = \{\Lambda\}$. Be careful not to confuse the last two cases: Λ is a full fledged string in A^* and $\{\Lambda\}$ is *not* empty!

Figures 15.3a and b illustrate two extreme cases of the same idea. Suppose x is some specified element of alphabet A. FSA 15.3a accepts only x, that is, a single valid string consisting of a single occurrence of x. The label $A - \{x\}$ in the lower arc is intuitively self-explanatory and is permissible even if $A = \{x\}$, in which case $A - \{x\} = \varnothing$. (Having $\varnothing$ as an arc label is perhaps wasteful, but it is neither ambiguous nor inconsistent.) FSA 15.3b accepts any string in the iteration $x^\star$, which clearly contains $\{x\}$ as a proper subset. In the exercises, you'll consider other concatenations of x only.

In Figure 15.4, a specific symbol x has again been designated from alphabet A. This time, however, what we want is a **parity checker**, that accepts those strings having an even number of xs (including none) and rejects those strings having an odd number of xs. All other symbols from A (that is, all of $A - \{x\}$) are simply

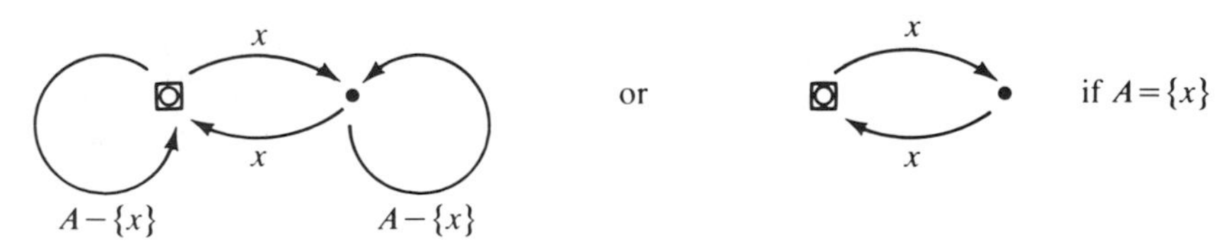

Figure 15.4 Parity checker.

ignored, even if they occur between *x*s. Thus, for instance, if $A := \{x, y\}$, then the string *yxyxy* will be accepted. In Exercises 15.2 you'll consider some variations on this idea.

Finally, let's look at Figure 15.5 Let $A := \{x, y\}$. Suppose we wish to accept only strings $x^h y^h$ for any $h = 1, 2, \ldots$. Thus, we want to accept strings like *xy*, *xxyy*, *xxxyyy*, but reject such strings as *x*, *yx*, and *xyy*. The general pattern of the figure is straightforward enough, but there is a glaring problem in the use of the ellipses in the digraph: just how many states does the FSA have? If l is some fixed natural number, then an FSA with $2l + 2$ states arranged in this pattern can correctly accept any valid string for which $h \leq l$. For instance, an FSA with eight states arranged as in the figure ($l := 3$) can recognize *xy*, *xxyy*, *xxxyyy*, but would reject *xxxxyyyy* as an error. Of course, we could design the FSA with an arbitrarily large l, but once l is chosen, the FSA can recognize and accept only strings for which $h \leq l$.

What we are exploring here, of course, is the notion of *counting*. If this problem were posed to a human being, the most obvious technique would be simply to count the *x*s, to count the *y*s, and to compare the results. Humans can count arbitrarily high, we reason, so FSAs would be able to do the same. But stop now and examine the notion of a finite counter. (The upper row in the figure can be considered as a counter of initial, consecutive *x*s.) Any counter must have one situation, or state, for every number it is expected to count. The human hand cannot count above five, the second hand of a watch cannot count above sixty, and most watches themselves cannot keep track beyond twenty-four hours. If that maximum number is exceeded, the counter loses count. Some counters, such as the mileage odometer on an automobile, simply "wrap around" to zero and begin

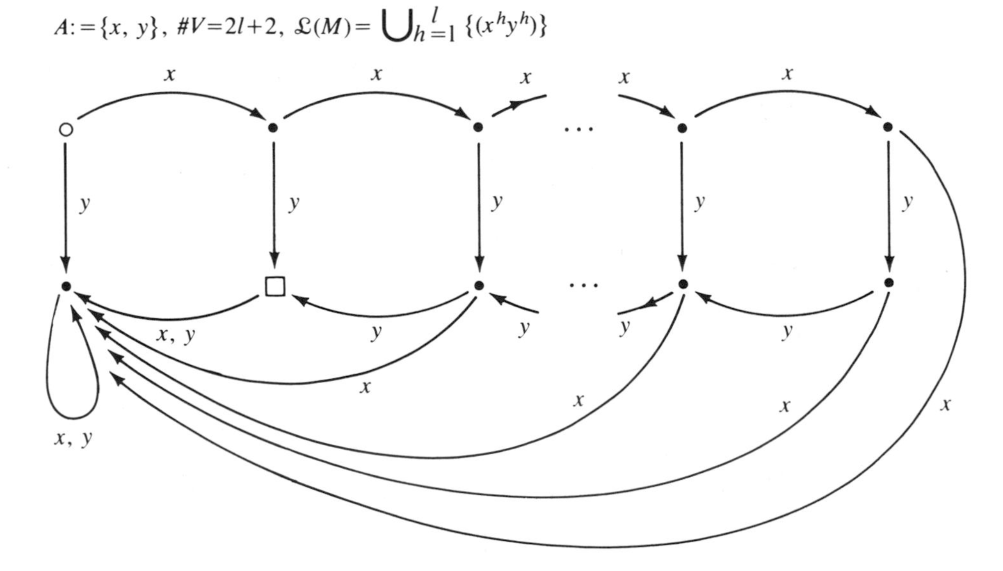

Figure 15.5 Double counter.

again. Others, like the one in Figure 15.5, break down. The important concept is that *every finite counter has an upper limit, beyond which it loses count*. Even digital computers cannot keep counting forever. Most computers store integers in binary form, allocating to each number from 8 to 64 bits. But we've seen, in Section 4.2, that a binary number with t bits can count any number from 0 to $2^t - 1$. Beyond that, the number cannot be represented; the counter loses count. In fact, we can imagine an entire large digital computer devoted in its entirety to storing a single binary number. The magnitude of such a single number would, in our human scale, be astronomically, stupendously, mind-bogglingly large. But it would be finite and, if it were exceeded, the computer would lose count. Thus the problem with the FSA in Figure 15.5 is not so surprising after all. An FSA can function as a counter, but it suffers the same limitations as any other finite counter. The maximum number we expect to count must be specified in advance and must not be exceeded in operation. In fact, this same observation applies in general to the concept of *memory*. The state set of an FSA serves as its memory, with one state allocated to each situation which the FSA can "remember." Since the number of states is finite, so is the memory capacity of the FSA. Even the largest finite computer can store only so much; all else is lost.

But suppose we take a different approach to the problem, an approach that does not involve counting at all. Some human beings would tackle the original problem in the following way. Place pointers at both ends of the string, left and right. In any valid string, the left pointer will point to an x, the right pointer to a y. Eliminate that outside pair of symbols, and then move each pointer inward toward the string's center. Repeat this process until the two pointers meet or until an error is detected (that is, the left pointer points to y or the right pointer to x). Such a method does not involve counting and would work for an arbitrarily large string, but it requires these *two* pointers. An FSA has only one "pointer," the current symbol in the string. And this pointer cannot alternate between the left and the right end of the string, cannot reverse its direction, and cannot "look ahead" to later (farther right) portions of the string. Nor can it modify the input string any way. Other structures, more advanced than FSAs, can be made to do all of these things. As you would expect, these structures can accomplish many things that FSAs cannot, and in later courses in computer science and mathematics you'll study them. We will not, however, consider them in this book. For all of these limitations, FSAs are still capable of considerable sophistication. Now that we have seen the lower (and a few of the upper) limits of their operations, we can begin to explore the power of FSAs in detail.

Exercises 15.2

1. Consider an alphabet $A := \{x, y, z, r, s, t\}$. For each case, construct an FSA M whose language is the set of finite strings over A that satisfy the given criteria: **(a)** The string contains precisely one occurrence of symbol x, in any position; **(b)** it contains precisely two occurrences of x, in any positions; **(c)** the number of occurrences of x in the string is a multiple of 3, and the xs may be in any positions; **(d)** the string contains precisely one x and two ys, in any positions and in any order; **(e)** it contains an even number of xs and a number of ys that is a multiple of 3, in any positions and in any order; **(f)** the string contains no x; **(g)** it contains neither x nor y; **(h)** it contains x (if at all) only in nonoverlapping pairs xx (e.g., xx and $xxxx$ are accepted, but xxx is not); **(i)** it contains x (if at all) only in pairs xy.

2. Let M be an FSA for which $T = \varnothing$. Show that M is equivalent to the FSA of Figure 15.2b.

3. Let M be an FSA for which $T = V$; that is, all states are final. Characterize M and demonstrate an equivalent FSA having the minimum possible number of states.

4. For any FSA M, state a necessary and sufficient condition that M accept the null string Λ, that is, $\Lambda \in \mathcal{L}(M)$. If $\Lambda \notin \mathcal{L}(M)$, then clearly state a procedure for constructing another FSA M' such that $\mathcal{L}(M') = \{\Lambda\} \cup \mathcal{L}(M)$. [Hint: the answer is in Section 15.3.3.]

5. Let $A := \{x\}$. For each of the languages L given below, first list or describe the strings in L, then construct an FSA M whose language is the given language L, that is, such that $\mathcal{L}(M) = L$. **(a)** $L := A^3$. **(b)** $L := \bigcup_{h=0}^{3} A^h$ (note that Λ is accepted). **(c)** $L := \bigcup_{h=1}^{3} A^h$ (note that Λ is not accepted). **(d)** $L := A^3 \cup A^5$. **(e)** $L := \bigcup_{h \in \mathbb{N}-\{4\}} A^h$. **(f)** $L := (\bigcup_{h=2}^{4} A^h) \cup (\bigcup_{h=7}^{9} A^h)$.

6. Construct and compare pairs of FSAs that accept languages that are unions of sets A^h for the values of h specified below. Clearly summarize your observations on these FSAs. **(a)** $h = 5$ versus $h \neq 5$. **(b)** $h \leq 5$ versus $h > 5$. **(c)** $h < 5$ versus $h \geq 5$.

7. Let M_1 and M_2 be FSAs with identical A, V, g, and u. Suppose, however, that $T_2 = V - T_1$. State clearly and prove the relationship between $\mathcal{L}(M_1)$ and $\mathcal{L}(M_2)$. [Hint: see exercise 6.].

8. Let $A := \{x, y, z\}$. For each of the following string sets from A^*, construct an FSA that accepts precisely those strings. In each case, use the minimum possible number of states. **(a)** Only the string xxx. **(b)** Any string containing xxx as a prefix (including xxx by itself); that is, strings of the form $xxx{:}p$ where $p \in A^*$ (string p might iself begin with one or more xs; it might also be null). **(c)** Any string containing xxx as a postfix (including xxx by itself); that is, strings of the form $p{:}xxx$ where $p \in A^*$. **(d)** Any string containing xxx as a substring (including xxx by itself); that is, strings of the form $p{:}xxx{:}q$; where $p \in A^*$ and $q \in A^*$. **(e)** Any string containing at least two non-overlapping occurrences of xxx; that is, strings of the form $p{:}xxx{:}q{:}xxx{:}r$ (p, q, and r are in A^*; e.g., $xxxxxx$ is accepted, but $xxxxx$ is not). **(f)** Any string containing precisely three xs, not necessarily contiguous (e.g. $yxzxxz$). **(g)** Any string containing at most three xs, not necessarily contiguous.

9. Find a tabular representation of function g for the FSAs given in **(a)** Figures 15.2 and 15.3 (assume $A := \{x, y\}$ or $A := \{x, y, z\}$), **(b)** Figure 15.5 (assume $l = 3$), **(c)** Figure 15.1.

10. Find a digraph representation of the FSA M given by each of the following tables. Describe the respective language $\mathcal{L}(M)$.

(a)

		x	$*$
initial	1	2	5
final	2	3	1
	3	4	2
	4	5	3
	5	5	4

(b)

		x	$\sim$	$*$
initial	1	2	5	5
final	2	3	2	1
	3	4	3	2
	4	5	4	3
	5	5	5	5

(c)

		Z	U	B	T	Q
initial	1	2	5	5	5	5
final	2	3	2	1	5	5
	3	4	3	2	1	5
	4	5	4	3	2	1
	5	5	5	5	5	5

(d)

		a	b	c
in. & fin.	1	2	3	5
	2	1	4	5
	3	5	5	1
	4	5	5	2
	5	5	5	5

(e)

		a	b	c
initial	1	2	1	1
	2	3	1	1
	3	4	1	1
final	4	4	4	4

(f)

		a	b	c
initial	1	2	1	1
	2	3	2	2
	3	4	3	3
final	4	4	4	4

15.3 Finite State Acceptor Languages

Our study of FSAs has so far been limited to constructing FSAs to accept specific strings or, at best, specific sets of strings defined by such operations as iteration. In the exercises you constructed FSAs one state at a time, one arc at a time, to satisfy some specific string criterion. In this section, we're going to take a quantum leap forward by constructing entire classes of FSAs at one time. The question we now wish to address fully is the one we alluded to in the introduction to this chapter: Just how powerful is the FSA concept? What kind of languages can we build FSAs for?

Definition 5 Let A be a nonempty finite alphabet and let L be any language over A (that is, $L \subseteq A^*$). Then L is called a **finite state acceptor language** if there exists an FSA, say M, such that $L = \mathcal{L}(M)$. The set of all FSA languages over A is denoted $\mathcal{F}(A)$. In formulae, $\mathcal{F}(A) := \{L \in \mathcal{P}(A^*) \mid \text{there exists an FSA } M \text{ such that } \mathcal{L}(M) = L\}$.

Clearly, $\mathcal{F}(A) \subseteq \mathcal{P}(A^*)$. There is the possibility that this inclusion is a proper one, that is, that some language is not the language of any FSA. We already have an indication in this direction: we were unable to construct an FSA over alphabet $\{x, y\}$ whose language was

$$Z := \bigcup_{h \in N} \{(x^h y^h)\}. \tag{1}$$

The fact that we were not able to construct such an FSA does not prove that it does not exist, of course, but we'll see later that, in fact, it does not.

In the examples we've already proved the existence of some FSA languages, for any given nonempty A. They are:

$$(1)\ \varnothing, \quad (2)\ \{\Lambda\}, \quad (3)\ A^*, \quad \text{and} \quad (4)\ \{(x)\} \text{ for each specific symbol } x \text{ in } A. \tag{2}$$

How can we build a *structure* of FSA languages starting from these few examples? We begin by assuming that L_1 and L_2 are two FSA languages. (This assumption is acceptable, because we already know that there are at least four of them; our goal is, of course, to construct many more.) This implies that there are two FSAs, say M_1 and M_2, whose languages are L_1 and L_2, respectively. Can we construct an FSA M whose language can be obtained from L_1 and L_2 according to some of the operations we saw in this chapter? If so, then we are able to enlarge our collection of FSA languages. Note that all of the FSAs we are considering in this context are over the same alphabet A. Our first result is one you saw in the exercises:

Theorem 6 Let L_1 be a language over A. If $L_1 \in \mathcal{L}(A)$, then $\sim L_1 \in \mathcal{L}(A)$ (the symbol $\sim$ denotes complementation within the universal set A^*).

Proof: See problem 7 in Exercises 15.2.

Our immediate goals are three: Given that L_1 and L_2 are in $\mathcal{F}(A)$, we intend to show that each of $L_1 \cup L_2$, $L_1{:}L_2$, and $L_1^\star$ is in $\mathcal{F}(A)$. We'll consider union, concatenation, and iteration individually. In each case, we'll assume that we have two FSAs, say $M_1 := (A, V_1, g_1, u_1, T_1)$ and $M_2 := (A, V_2, g_2, u_2, T_2)$ such that $\mathcal{L}(M_1) = L_1$ and $\mathcal{L}(M_2) = L_2$. In each case, we wish to construct an FSA $M := (A, V, g, u, T)$ such that $\mathcal{L}(M)$ is the desired language.

15.3.1 Union

If M is constructed correctly, then we know that M accepts any string p if and only if either M_1 or M_2 (possibly both) accepts p. Conversely, M should reject p if both M_1 *and* M_2 would reject it. An initial strategy may be the following: Define M to first simulate M_1, and let it operate on p. If M_1 would have accepted p, then M will accept it. If M_1 would have rejected p, then M will switch to a new mode of operation and reread p from the beginning, now simulating M_2. If M_2 would have accepted p, then now M will accept p. So far, so good. The problem is that an FSA cannot reread its input string. Once p has been read once, it is unavailable for further processing.

But suppose now that M is capable of simulating both M_1 and M_2 simultaneously. Then we would not have to reread the string, and that problem would be solved. On the other hand, the only memory an FSA has is its current state at any point. A state of M must therefore contain the same information as a state of M_1 *and* a state of M_2. We have a mechanism to do that, the Cartesian product $V_1 \times V_2$. Recall also that the runs $\mathcal{M}_1(p)$ and $\mathcal{M}_2(p)$ are of exactly the same length, one more than the length of p. We are thus led to the following construction of M:

$$M := (A, V, g, u, T), \quad \text{where}$$

- A is the given alphabet;
- $V := V_1 \times V_2$ (each state of M provides one state in M_1 and one state in M_2);
- $g: V \times A \to A$ is given by $g((v_1, v_2), x) := ((g(v_1, x), g(v_2, x))$; at each step, the state of M_1 is changed as it would have changed if p had been run in M_1, and similarly for the state of M_2;
- $u := (u_1, u_2)$; each simulation starts at the initial state of each FSA;
- $T := (T_1 \times V_2) \cup (V_1 \times T_2)$; either a final state from M_1 is accepted, or a final state from M_2, independently on the end position in the *other* FSA).

We have thus proved

Theorem 7 If $L_1 \in \mathcal{F}(A)$ and if $L_2 \in \mathcal{F}(A)$, then $L_1 \cup L_2 \in \mathcal{F}(A)$.

We should here point out that we have constructed one possible FSA for language $L_1 \cup L_2$. For the same given language, there may well be any number of possible constructions for M. Moreover, in terms of efficiency (minimum number of states and so forth), our construction is probably far from ideal. That's not the point. We have guaranteed the existence of *one* FSA to accept the desired language, and that's all we really wanted. We have specifically sacrificed efficiency in order to ensure success. In our next step, that sacrifice is even more pronounced.

15.3.2 Concatenation

Proving the existence of an FSA for the concatenation of two given FSA languages poses special problems, problems best demonstrated by an example.

Suppose $A := \{x, y\}$, $L_1 := \{x, xy, xyxx, xyxxy\}$, and $L_2 := \{(xx)\}^{\star}{:}\{(yy)\}$. Consider the special string $q := xyxxyy$. Clearly, $q \in L_1{:}L_2$, since $q = (xy){:}(xxyy)$. We'll use the concatenation colon as a marker, to show where a string is decomposed into its left substring (q_1) and its right substring (q_2). The fact that this example can be decomposed in either of two places ($q = (xy){:}(xxyy)$ and $q = (xyxx){:}(yy)$) should give you some preview of the difficulties we face.

Suppose we simply consider string q, without knowing how to decompose it in order to determine whether it is in $L_1{:}L_2$. How can we expect M correctly to accept or reject the left and right substrings individually, when we don't even know when the left ends and the right begins? Let's explore some possiblities.

Earliest left substring. Suppose we construct M to simulate M_1 until the earliest point at which M_1 would accept some portion of q, then switch M to simulate M_2 for the remainder. (If M_1 would never accept any left portion of q, then the switch would never be made, and q would be rejected by M.) Since $\Lambda \notin L_1$ and $(x) \in L_1$, this method would yield $q = x{:}yxxyy$. But the remainder, $yxxyy$, is not in L_2, so M_2 would reject it. Clearly, M has been too eager to halt M_1 and switch to M_2, and so it has tended to reject q.

Latest shift substring. Suppose we construct M to simulate M_1 until the latest point at which M_1 could possibly accept a left portion of q, then switch to simulate M_2. First of all, this raises problems: M would not know where that latest point would be, unless it could "look ahead" to later portions of the string. For the moment, though, we'll ignore this objection, because there are more serious difficulties anyhow. In our example, q is not in L_1, but $xyxxy$ is. Then our "latest point" decomposition would be $xyxxy{:}y$, and the right portion would not be accepted by M_2. Hence M will reject q. This time, M has waited too long to make the switch.

Every possible left substring. These examples demonstrate that M cannot use any simple single rule as to when to switch from M_1 to M_2. The next possibility is that M recognize *every* possible opportunity to accept a left substring as M_1 and switch to M_2 at each of those points. For a string of length l, there will be at most $l + 1$ such opportunities because that's the number of places at which q may be decomposed (including the possibilities $\Lambda{:}q$ and $q{:}\Lambda$). We could therefore construct M by putting together $l + 1$ FSAs, each inserting the colon at one of the possible places and simulating M_1 and then M_2 accordingly. In the example, this would give the following results (each line is devoted to one of the $l + 1$ FSAs).

h	Decomposition	M_1 accepts q_1	M_2 accepts q_2	M accepts q
0	Λ:$xyxxyy$	no		no
1	x:$yxxyy$	yes	no	no
2	xy:$xxyy$	yes	yes	yes
3	xyx:xyy	no		no
4	$xyxx$:yy	yes	yes	yes
5	$xyxxy$:y	yes	no	no
6	$xyxxyy$:Λ	no		no

Since at the end of the run we have at least one case (in fact, two) where one of the $l + 1$ FSAs would accept q, the FSA M would accept q. We seem to be on the right track at last! But, if we want to be able to process all possible strings p, the length, although finite, would be arbitrarily long and we would need infinitely many FSAs like the $l + 1$ we considered for one specific string. It seems that, once again, arbitrarily long strings will be our undoing. But that's not true! Look again carefully at the tabulated results and at the criterion for acceptance.

For an arbitrary string p of length l, consider (in general now) $l + 1$ hypothetical FSAs, each of which would simulate M_1 for the left m-substring ($0 \leq m \leq l$) and then (if the left substring is accepted by M_1) switch to simulating M_2 for the right portion (starting, of course, from the initial state u_2 of M_2). Consider a given value of h ($0 \leq h \leq l$). Each of the $l + 1$ FSAs would present one of three cases at step h of the processing:

1. If h has not passed the separation of p into two substrings, then the simulation on M_1 is in effect, and every one of these FSAs is at the simulation of a same state of M_1.
2. If h has passed the separation but the left substring has not been accepted by the simulation on M_1, then the entire string p has already been rejected by the corresponding hypothetical FSA.
3. If h has passed the separation *and* the first part of p has been accepted by the simulation on M_1, then position h in p determines a specific state in the simulation on M_2 for each one of the $l + 1$ hypothetical FSAs.

Although l is arbitrary, the number of all possible cases listed above is not. For every possible string, every possible position in the string, and every possible set of hypothetical FSAs, case 1 represents *one state* of M_1 (and the number of states of M_1 is finite and given), case 2 does not represent any possibility, and case 3 represents a *set of states* of M_2 (and the number of sets of states of M_2 is again finite and given). This suggests that the states of M should represent the pairs (one state of M_1, a set of states of M_2). We need not keep track of each of the possible hypothetical FSAs, but only of their possible states, whose number is finite and given once for all. This leads to the following construction, for which we need just one more remark. Whenever a first portion of any given string p is accepted by the simulation on M_1, we must continue the simulation on M_1 *and* attempt simulation on M_2 for the remaining part. At this point, the acceptance of M_1 places the processing also at the initial state u_2 of M_2 at the point where *we are about to begin processing* the right substring on M_2. The next step will therefore be a first step in M_2, starting from u_2.

The construction for $M := (A, V, g, u, T)$ thus looks like this:

- A is, as always, the given alphabet;
- $V := V_1 \times \mathcal{P}V_2$; that is, a state v of M is a pair (v_1, W), comprising a state v_1 of M_1 and a subset W of the set of the states of M_2;
- $g: V \times A \to V$ is given as follows: for each v in V ($v = (v_1, W)$) and for each x in A, $g(v, x)$ is defined by either of:
  if $g_1(v_1, x) \notin T_1$, then $g(v, x) := (g_1(v_1, x), \{g_2(t, x) \mid t \in W\})$,
  if $g_1(v_1, x) \in T_1$, then $g(v, x) := (g_1(v_1, x), \{g_2(t, x) \mid t \in W\} \cup \{u_2\})$;
- $u := (u_1, \varnothing)$ if $u_1, \notin, T_1$, $u := (u_1, \{u_2\})$ if $u_1 \in T_1$;
- $T := \{(v_1, W) \mid v_1 \in V_1, W \in \mathcal{P}V_2, W \cap T_2 \neq \varnothing\}$; that is, M will accept p if and only if the M_2 simulator component of T_2 contains *any* final state from the original M_2.

The FSA M will have $(\#V_1) \cdot 2^{\#V_2}$ states. This may be a very large number. Still, it's a *given, finite* number, independent of any one particular string p.

"Wait a minute!" you object, reaching to make sure that your wallet hasn't been stolen. "We've reduced an arbitrarily large FSA to one whose size can be prespecified. Surely we've lost something in the process! Even in mathematics, there's no such thing as a free lunch!" Quite right. Remember that at the beginning of this discussion we described the difficulty of deciding where to decompose p, where to place the colon? If M accepts p, *we still don't know* where the colon went. All we know is that, in at least one case, it went somewhere that worked. For each string p, a simple yes or no is what we asked for, and a simple yes or no is all we get. There's a moral here for the general process of mathematical thinking. We started our discussion on concatenation based on conclusions about union. Then we proceeded to expand that idea far beyond even specifiability, let alone efficiency, which was already shaky in advance. We were simply exploring the possibilities, "brainstorming" every conceivable hope and exploring every avenue. But then we asked the crucial question: what *precisely* do we want to know? We found that we were really demanding more than we needed and, once the superfluous information was eliminated, what was left worked. The moral is not to panic too early. Do not be afraid of thinking big, even too big, during the exploratory process. When the time comes to ask the crucial question, it's often the case that most of the troublesomely large baggage can be dispensed with.

In formal terms, we have proved this theorem:

Theorem 8 If $L_1 \in \mathcal{F}(A)$ and if $L_2 \in \mathcal{F}(A)$, then $L_1{:}L_2 \in \mathcal{F}(A)$.

15.3.3 Iteration

The final problem we face is to do a similar construction for iteration: given L_1 in $\mathcal{F}(A)$, construct an FSA M over A such that $L_1{}^\star = \mathcal{L}(M)$. We can start with the remark that iteration is multiple concatenation, but there is one detail that might give us trouble later: Λ is always in the iteration of any language and is not necessarily in L_1. In that case, concatenating L_1 with itself would not provide a language that contains Λ. We bypass this problem by replacing M_1 with an FSA M' whose language is $\{\Lambda\} \cup L_1$. In problem 4 of Exercises 15.2 we asked you to show that this could be done. You should have come up with something like the following: If $\Lambda \in L_1$, then $M' := M_1$, $g' := g$, and $s := u_1$. Otherwise, let the set of states of M' consist of the same states as M_1 plus one extra state s, to be considered both as initial and final state for M'. Then the original u_1 is not the initial state any more, of course. The arcs $g_1(v, x) = w$ of M_1 remain as arcs $g'(v, x)$ of M' for all v and x. We adjoin a new arc $g'(s, x) = w$ for each arc $g_1(u_1, x) = w$ of M_1. Then s has no loops and the arcs originating from it end precisely where the arcs originating from u_1 ended in M_1. Then $\Lambda \in \mathcal{L}(M')$ because s is initial and final. Furthermore, for every nonnull string p, the first step on M' leads to the same node as the first step on M_1. Therefore all the strings that were accepted by M_1 are accepted by M', and the only string accepted by M' but not by M_1 is Λ (see Figure 15.6). That's what we wanted.

Clearly, $L_1^\star = [\{\Lambda\} \cup L_1]^\star$. Consequently, we can construct an FSA for $[L']^\star$ ($L' := \mathcal{L}(M')$) and obtain the same result as the originally proposed one.

The construction of M is almost the same as the construction of M in the case of concatenation, and simpler because we do not have to examine the first alternative, that is, the processing of two parts of the string on two FSAs. We obtain the following:

- A is the given alphabet,
- $V := \mathcal{P}V'$, (that is, a state v of M is a set of states of M'),
- $g(v, x) := \{g'(t, x) \mid t \in v\}$ if no final state is in this set, otherwise $g(v, x) := \{g'(t, x) \mid t \in v\} \cup \{s\}$,
- $u := \{s\}$,
- $T := \{W \in V \mid W \cap T' \neq \varnothing\}$.

We have then proved Theorem 9:

Theorem 9 If $L_1 \in \mathcal{F}(A)$, then $L_1^\star \in \mathcal{F}(A)$.

$A := \{x, y\}$

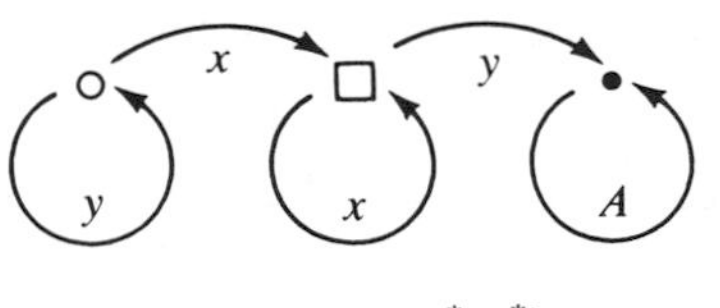

(a) $\mathcal{L}(M) = \{y^*xx^*\}$

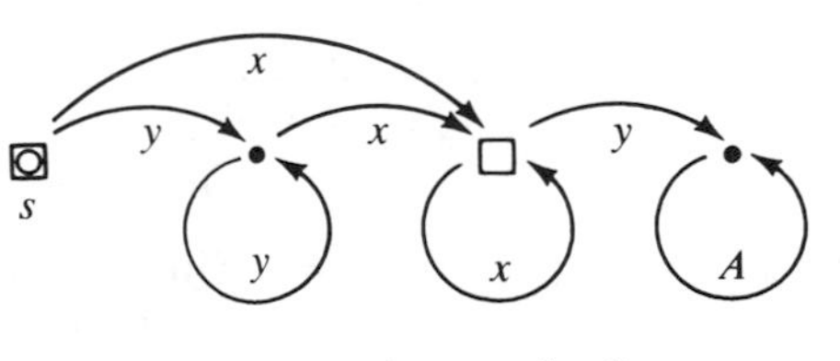

(b) $\mathcal{L}(M') = \{\Lambda\} \cup \{y^* xx^*\}$

Figure 15.6

Exercises 15.3

1. In each of the following cases, the data are alphabet A and two FSAs over A, say M_1 and M_2. Let L_1 and L_2 be their respective languages. Construct—according to the algorithms of this section—three FSAs M_3, M_4, M_5 whose respective languages are $L_1 \cup L_2$, $L_1{:}L_2$, and $L_1^\star$. Then, for each of M_3, M_4, M_5, construct—if possible—an equivalent FSA with fewer states. **(a)** $A = \{x, y\}$, M_1 as in Figure 15.2a, M_2 as in Figure 15.2b. **(b)** $A = \{x, y\}$, M_1 as in Figure 15.3a, M_2 as in the left part of Figure 15.4. **(c)** $A = \{x\}$, M_1 as in Figure 15.2c, M_2 as in the right part of Figure 15.3b. **(d)** $A = \{x, y\}$, M_1 as in the left part of Figure 15.4, M_2 as in Figure 15.3a.
2. Given $A := \{x, y\}$, $L_1 := \{x, xy, xyxx, xyxxy\}$, and $L_2 := \{[xx]^\star yy\}$. List the runs for the FSA described in the text as they determine **(a)** $xyxxy \in L_1 \cup L_2$; **(b)** $xyxxyy \in L_1{:}L_2$; **(c)** $yyxxyy \in L_2^\star$.

Complements 15.3

1. Prove

Theorem 10 If L_1 and L_2 are any languages in $\mathcal{F}(A)$, then $\mathcal{F}(A)$ also contains every language that can be expressed in terms of L_1, L_2, and the set-theoretical operations $\cup$, $\cap$, and $\sim$ (complementation in A^*).

[Hint: in Sections 6.2 and 6.3 we showed that the three operations are Boolean operations in the algebra of sets; by DeMorgan's law, $\cap$ can be expressed in terms of $\cup$ and $\sim$; consequently, $\cup$ and $\sim$ are sufficient to express what?]

2. Given two FSA languages, construct an FSA for their intersection. [Hint: use a process similar to the process we used for union; note that only T needs to be changed.]

15.4 Regular Languages and Kleene's Theorem

In Section 15.3 we defined the set $\mathcal{F}(A)$ as the set of the languages over a given alphabet A for which it is possible to construct an FSA. We then showed that certain languages were always in $\mathcal{F}(A)$ (formulae 2) and that $\mathcal{F}(A)$ was closed under union, concatenation, and iteration. We now consolidate these ideas in a structure called *regular language*.

Definition 11 For any finite alphabet A, the set of the **regular languages** over A is denoted $\mathcal{R}(A)$ and is defined recursively as follows:

- $\varnothing$, $\{\Lambda\}$, and $\{(x)\}$ (for each x in A) are regular languages.
- If L_1 and L_2 are regular languages, then each of $L_1 \cup L_2$, $L_1{:}L_2$, $L_1{}^\star$ is a regular language. (In fact, unions and concatenations of finitely many regular languages are also regular languages.)
- There are no other regular languages.

What we proved in Section 15.3 is that every regular language is an FSA language, that is, $\mathcal{R}(A) \subseteq \mathcal{F}(A)$. This is true even in the degenerate case in which $A = \varnothing$, since then $\mathcal{R}(A) = \mathcal{F}(A) = \{\varnothing, \{\Lambda\}\} = \mathcal{P}(A^*) = \mathcal{P}\{\Lambda\}$. In the exercises, you'll explore some of the properties of regular languages. Note especially that, although each of the languages in formulae 2 is always regular, not every language is regular. Regular languages were first explored by Kleene in the twentieth century, and the crucial result is the theorem that bears his name. We have already seen half of this theorem.

Theorem 12 **Kleene's Theorem.** For any finite alphabet A, $\mathcal{R}(A) = \mathcal{F}(A)$.

We've already shown that $\mathcal{R}(A) \subseteq \mathcal{F}(A)$. The astounding part of the theorem, though, is the other half: $\mathcal{F}(A) \subseteq \mathcal{R}(A)$. Regular languages give therefore a complete characterization of everything FSAs are capable of. Thus, in any application, theoretical or practical, where we suspect that FSAs might play a role, we need only ask whether that application can be characterized in terms of a regular language over some finite alphabet.

The proof of the other half of Kleene's theorem requires many ideas from theory of graphs. Although we've already presented most of these ideas, the proof would still be long and rather intricate. In later courses you'll work through it in its entirety. But for a general treatment of graphs and FSAs as discrete structures, as this book is, it's sufficient for us to recognize the importance of Kleene's theorem and to gain some experience working with regular languages. The proof we defer to another time.

Exercises 15.4

1. Let A be a finite alphabet. Show that **(a)** A^h is regular for every natural number h; **(b)** A^* is regular; **(c)** if $B \subseteq A$, then each of B^h ($h \in \mathbb{N}$) is regular over A and B^* is regular over A.
2. Let L_1 and L_2 be languages over alphabet A. Show that **(a)** $L_1 \subseteq (L_1 \cup L_2)$; **(b)** $L_1 \subseteq L_1{}^\star$; **(c)** if $\Lambda \in L_2$, then $L_1 \subseteq L_1{:}L_2$; **(d)** if $L_2 \neq \varnothing$, then every substring of a string in L_1 is a substring of a string in $L_1{:}L_2$.
3. Each of the following diagrams represents an FSA M over the alphabet $A :=$ $\{x, y\}$. In each case, list the elements of $\mathcal{L}(M)$ or describe $\mathcal{L}(M)$.

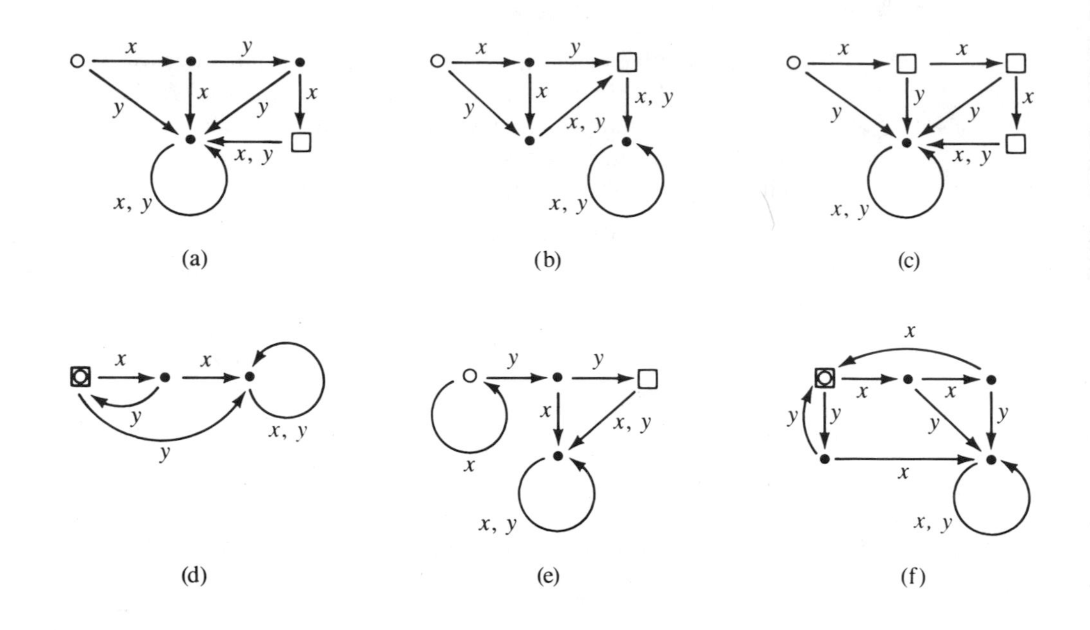

Complements 15.4

1. Let the alphabet be $A := \{x, y\}$. Consider the language Z we saw in formula 1. Show that Z is not a regular language. [Hint: Starting from $\varnothing$, $\{\Lambda\}$, $\{x\}$, and $\{y\}$, finitely many unions and concatenations alone cannot produce Z because Z is infinite. We must invoke iteration of some of the languages produced. But which one? If some set contains xy, can its iteration be in Z?] As a consequence of Kleene's theorem, there is no FSA over $\{x, y\}$ whose language is Z.

   At this point, we have the result that, if $\#A = 2$, then $\mathcal{F}(A)$ is a proper subset of $\mathcal{P}(A^*)$. Extend this result to the cases in which $\#A > 2$ and $\#A = 1$. [Hint: for the case $\#A = 1$: consider $L := \{x^{2^k} \in A^* \mid k \in \mathbb{N}\}$.]

15.5 Summary

An FSA (*finite state acceptor*) is a "device" that admits finitely many states and whose inputs are the finite strings from a given alphabet (Section 15.1). One of the states is designated as the *initial state*. Whenever the FSA starts processing a given string, its state is by definition the initial state. For each letter of the alphabet and for each state that the FSA may be in, there is a firm rule (the *transition function* or *state table*) that prescribes the next state. Consequently, each letter of the string "moves" the FSA to a new state and, at the end of the string, the FSA is at a uniquely determined state; if this state is one of the pre-assigned *final states*, then the string is accepted, otherwise it's rejected. The set of the strings that would be accepted by the given FSA is the *language* of that FSA (Section 15.2).

Among all possible languages over a given alphabet, we have therefore the FSA languages, that is, those that would be the languages of some FSA over that alphabet. The collection of the FSA languages is a structure, with three operations (*union* of languages, *concatenation* of languages, and *iteration* of a language) and some special elements: the empty language $\varnothing$, the language $\{\Lambda\}$, and, for each x in the alphabet, the language $\{(x)\}$ (Section 15.3).

Applying those three operations to the special elements generates recursively a collection of languages, which are called *regular languages* over the given alphabet. A crucial result about the collection of the FSA languages and the collection of the regular languages (over the same alphabet) is Kleene's theorem, which states that these two collections are the same (Section 15.4).

Appendix A
The Greek Alphabet

Name	Symbol Lower Case	Symbol Upper Case
alpha	α	A
beta	β	B
gamma	γ	Γ
delta	δ	Δ
epsilon	ϵ	E
zeta	ζ	Z
eta	η	H
theta	θ	Θ
iota	ι	I
kappa	κ	K
lambda	λ	Λ
mu	μ	M

Name	Symbol Lower Case	Symbol Upper Case
nu	ν	N
xi	ξ	Ξ
omicron	o	O
pi	π	Π
rho	ρ	P
sigma	σ	Σ
tau	τ	T
upsilon	υ	Υ
phi	ϕ	Φ
chi	χ	X
psi	ψ	Ψ
omega	ω	Ω

Appendix B Answers to Selected Exercises

This appendix contains answers, if compact and unique, to the odd-numbered parts of odd-numbered Exercises and Complements.

Chapter 2

Exercises 2.1

1. (a) $500 + 10 + 2$, $5 \cdot 10^2 + 1 \cdot 10^1 + 2 \cdot 10^0$; (c) $1\,000\,000 + 1$, $1 \cdot 10^6 + 1 \cdot 10^0$.

3. E.g. $3000 + 20$, $3 \cdot 10^3 + 2 \cdot 10^1$.

5. $256 \cdot 1000 = 256\,000$; $(2 \cdot 10^2 + 5 \cdot 10^1 + 6 \cdot 10^0) \times 10^3 = 2 \cdot 10^5 + 5 \cdot 10^4 + 6 \cdot 10^3 = 200\,000 + 50\,000 + 6\,000 = 256\,000$. $1234 \cdot 100 = 123\,400$; $(1 \cdot 10^3 + 2 \cdot 10^2 + 3 \cdot 10^1 + 4 \cdot 10^0) \times 10^2 = 1 \cdot 10^5 + 2 \cdot 10^4 + 3 \cdot 10^3 + 4 \cdot 10^2 = 100\,000 + 20\,000 + 3\,000 + 400 = 123\,400$. $1234 \times 10000 = 12\,340\,000$; $(1 \cdot 10^3 + 2 \cdot 10^2 + 3 \cdot 10^1 + 4 \cdot 10^0) \times 10^4 = 1 \cdot 10^7 + 2 \cdot 10^6 + 3 \cdot 10^5 + 4 \cdot 10^4 = 12\,340\,000$.

7. (a) 50; (c) 26; (e) 111 111 111 111; (g) 85.

9. (a) $3^y + 4^y + 5^y + 6^y$.

Exercises 2.2

1. E.g., addition of several addends. Hints: write in column, aligned; start from units column; add the digits, write units of the result, carry the number of tens; shift to next column on left, add carry and the digits of that column, etc.; stop if there is no carry and there are no digits in next column.

3. E.g. 62×79

```
   18
  54
  14
 42
 ----
 4898
```

5. Hints. Algorithm S1: the borrow operation at position k is equivalent to adding 10 to the minuend at position k and subtracting 1 from the minuend at position $k+1$. Algorithm S2: the carry operation at position k is equivalent to adding 10 to the minuend at position k and adding 1 to the subtrahend at position $k+1$.

Complements 2.2

1. E.g., $62 + 79$

62	79	out
31	158	
15	316	
7	632	
3	1264	
1	2528	
	4898	

Exercises 2.3

1.

binary	decimal
0	0
1	1
10	2
11	3
100	4
...	...
101 111	47
110 000	48
110 001	49
110 010	50
110 011	51

3. (a) 11 111 111; (c) 1 001 001 001 001 001.

5. $2^{30} = (2^{10})^3 \simeq 1000^3 = 1\,000\,000\,000$.

7. $(\sum_{h=0}^{n} b_h \cdot 2^h) \times 2^m = \sum_{h=0}^{n} b_h \cdot 2^{h+m}$, that is, adjoin m zeroes at the end of the given numeral.

Complements 2.3

1. Hint. $(\sum_{k=0}^{n} a_k \cdot 2^k) \cdot b = \sum_{k=0}^{n} (a \cdot (b \cdot 2^k))$ and the sum can be restricted to only those terms for which $a_k = 1$, that is, $ab = \sum_{a_k = 1} b \cdot 2^k$. The entries in row k are $\sum_{h=k}^{n} a_h \cdot 2^{h-k}$ and $b \cdot 2^k$. The first is odd when $a_k = 1$, and the corresponding terms $b \cdot 2^k$ must be added together.

Exercises 2.4

1.

E.g.

	decimal check
101	5
× 11	3
101	
101	
1111	15

3. (a)

1 011 001 ×	1001
1 011 001	
1 100 100 001	

(c) 111 111, (e) 111 100.

Exercises 2.5

1. (a)

0	1	2	5	10	20	41	83	167	335	671	1342	
	1	0	1	0	0	1	1	1	1	1	0	←

(c) $257 = 256 + 1 = 2^8 + 1 = 100\,000\,001$; (e) $2^5 + 1 = 33$; (g) $2 \cdot (2^4 - 1) = 30$.

3. 10 binary $= 2^1 = 2$, 100 binary $= 2^2 = 4$, 1000 binary $= 2^3 = 8$, 10...0 binary $= 2^n$; 11 binary $= 100 - 1 = 2^2 - 1 = 3$, 111 binary $= 2^3 - 1 = 7$, 1111 binary $= 2^4 - 1 = 15$, 11...1 binary $= 2^{n+1} - 1$.

Complements 2.5

1. Hint. Initially, the number in the top line is $h = \sum_{k=0}^{n} b_k \cdot 2^k$. At step m, the top number will be $h = \sum_{k=m}^{n} b_k \cdot 2^{k-m}$, as we'll see in a moment. Division by 2 gives remainder b_m and quotient $\sum_{k=m+1}^{n} b_k \cdot 2^{k-[m+1]}$ (the new h). The same formula is therefore valid throughout.

Exercises 2.6

1.

binary	oct	dec	hex
0	0	0	0
1	1	1	1
10	2	2	2
11	3	3	3
100	4	4	4
...	..	..	..
101 111	57	47	2F
110 000	60	48	30
110 001	61	49	31
110 010	62	50	32

2.

n	8^n	16^n
0	1	1
1	8	16
2	64	256
3	512	4 096
4	4 096	65 536
..	.....	......
8	16 777 216	4 294 967 296
9	134 217 728	68 719 476 736
10	1 073 741 824	1 099 511 627 776
11	8 589 934 592	17 592 186 044 416

3. 20, 40, 100, 200.

5.

	decimal	binary	octal	hex
(a)	1234	10 011 010 010	2 322	4D2
(c)	264	100 001 000	410	108
(e)	1025	10 000 000 001	2 001	401
(g)	1028	10 000 000 100	1 004	404

7. (a) 11 111 111; (c) 1 001 001 001 001 001; (e) 15555.

Exercises 2.7

1. E.g., refer to problem 5 of Exercises 2.6

Exercises 2.8

1. octal

+								
	0	1	2	3	4	5	6	7
	1	2	3	4	5	6	7	0
	2	3	4	5	6	7	0	1
	3	4	5	6	7	0	1	2
	4	5	6	7	0	1	2	3
	5	6	7	0	1	2	3	4
	6	7	0	1	2	3	4	5
	7	0	1	2	3	4	5	6

×							
	1	2	3	4	5	6	7
	2	4	6	10	12	14	16
	3	6	11	14	17	22	25
	4	10	14	20	24	30	34
	5	12	17	24	31	36	43
	6	14	22	30	36	44	52
	7	16	25	34	43	52	61

hex

+															
0	1	2	3	4	5	6	7	8	9	A	B	C	D	E	F
1	2	3	4	5	6	7	8	9	A	B	C	D	E	F	0
2	3	4	5	6	7	8	9	A	B	C	D	E	F	0	1
. . .															

×														
1	2	3	4	5	6	7	8	9	A	B	C	D	E	F
2	4	6	8	A	C	E	10	12	14	16	18	1A	1C	1E
3		9	C	F	12	15	18	1B	1E	21	24	27	2A	2D
4			10	14	18	1C	20	24	28	2C	30	34	38	3C
5				19	1E	23	28	2D	32	37	3C	41	46	4B
6					24	2A	30	36	3C	42	48	4E	54	5A
7						31	38	3F	46	4D	54	5B	62	69
. . .														

3. E.g.

octal	decimal check	hex	decimal check
276	190	F1A	3866
× 35	29	× 1AB	427
36		6E	
22		64	
43		A	
25		1AB	
12		A5	
6		96	
12606	5510	F	
		19305E	1650782

octal	decimal check	hex	decimal check
35	29	1AB	427
276)12607	190, 5511	F1A)19306F	3866, 1650799
1072		F1A	
1667		A166	
1666		9704	
1	1	11	17

Exercises 2.9

1. $\frac{3}{5} = .6$, $\frac{7}{20} = .35$, $\frac{11}{125} = .088$, $.03 = \frac{3}{100}$, $10.7 = \frac{107}{10}$, $.125 = \frac{1}{8}$.

3. To multiply (or divide) an r-nary number by r^m, move the radix point m positions to the right (to the left) if $m > 0$, or $-m$ positions to the left (to the right) if $m < 0$. Don't move if $m = 0$.

5. (a) C5.AB hex $\times$ 16^2 decimal = C5AB hex (197.66 796 875 $\times$ 256 = 50 603).

7. The r-nary digit values are the integers b such that $0 \leq b < r$. The r-nary rationals are the numbers $\sum_{k=-m}^{n} b_k \cdot r^k$, where the b_k are r-nary digits and m and n are integers such that $-m \leq n$.

9. Hint. Let $q = p^h$ $(p > 1, h > 0)$. Let (*) $a := \sum_{k=-m}^{n} a_k q^k$ be a q-nary rational. A p-nary representation of a_k is (**) $a_k = \sum_{j=0}^{h-1} b_{k,j} p^j$ ($b_{k,h}$, $b_{k,h+1}$, ... are 0 because $a_k < q = p^h$; $0 \leq b_{k,j} < p$). Substitute (**) into (*) and obtain a p-nary representation of a. The process is reversible. Then a is a q-nary rational if and only if a is a p-nary rational. Same conclusion for r and p, and for q and p.

Chapter 3

Exercises 3.1

1.

	−7	−2	−1	0	1	2	7	reduce to:
(a) mod 3	2	1	2	0	1	2	1	
(c) mod 6	5	4	5	0	1	2	1	

3. (a) (1) 1, (2) 4, (3) 3, (4) 3, (5) 3, (6) 4, (7) 3; (c) (1) 1, (2) 1, (3) 1, (4) 0, (5) 0, (6) 0, (7) 1, (8) 3, (9) 1.

5. Hints. $x = u + am$, $y = v + bm$, (1) $x + y = u + v + (a + b)m$; (2) $xy = uv + (av + bu + abm)m$; (3) $x^k = u^k$ + integer terms containing m as a factor (expand $[u + am]^k$ and separate u^k). In general, the definition of x^k cannot be extended when $k < 0$ (a counterexample is 3^{-1} mod 9, which would be $\frac{1}{3}$ mod 9, which is undefined); it can when m is prime and $x \neq 0 \pmod{m}$.

Exercises 3.2

1. (a) (1) F, (2) T, (3) T, (4) F, (5) T. (c) (1) T, (2) F, (3) T, (4) F, (5) T, (6) T, (7) T, (8) F, (9) T, (10) T, (11) F, (12) T, (13) T.

3. (a) (1) (1, 1), (2) (0, 0). (c) (1) (2, 2), (2) (4, 4), (3) (1, 1).

5. No: for each x, there are infinitely many formulae x^k and only finitely many results mod m.

7. Hints. Do part (b) first. (b) $xy = 0 \bmod m$. Then $xy = hm$, m divides xy, m is prime, m divides x or y, x or y is 0 mod m. (a) $x \neq 0 \bmod m$. Then (*) $0 \cdot x, 1 \cdot x, \ldots, (m - 1) \cdot x$ are m integers. Suppose $ax = bx \bmod m$. Then $(a - b)x = 0 \bmod m$, $x = 0 \bmod m$ (no) or $a - b = 0 \bmod m$ (no, because $0 < |a - b| < m$). Then (*) represent m distinct integers mod m, that is, all the integers mod m. The first is 0 mod m, one of the others must be 1 mod m, $kx = 1 \bmod m$ for precisely one k ($k \neq 0 \bmod m$), $k = x^{-1} \bmod m$.

Exercises 3.3

1. E.g. this book: $10 \cdot 0 + 9 \cdot 1 + 8 \cdot 5 + 7 \cdot 5 + 6 \cdot 1 + 5 \cdot 7 + 4 \cdot 6 + 3 \cdot 8 + 2 \cdot 3 + 1 \cdot 8 = 187 = 0 \bmod 11$.

3. From formula 1.

5. E.g.: 0-11-111002-5 is correct. Mistype one digit and obtain 0-11-111003-5 or exchange two digits and obtain 0-11-111020-5; neither checks.

7. E.g.: 0-11-111002-5 is correct. Mistype two digits and obtain 0-11-111003-3, or mistype one digit and exchange two and obtain 0-11-110102-6; both check, although different from the intended number.

Exercises 3.4

1. 

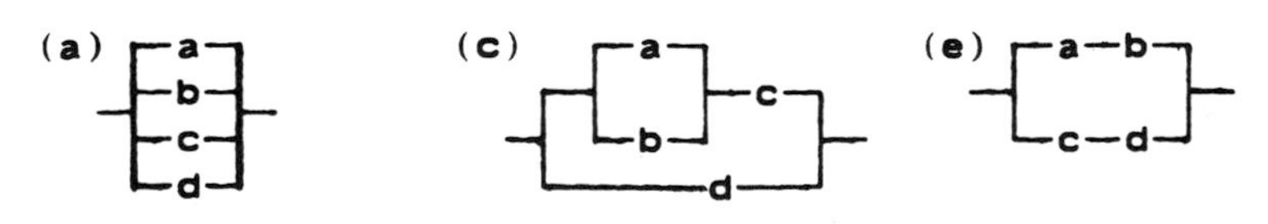

3. $a \vee a = a$, $a \wedge a = a$; definition: $a^0 := 1$, $a^{k+1} := a^k \wedge a$ ($k \geq 0$); then $a^0 = 1$, $a^k = a$ if $k > 0$.

Complements 3.4

1. Definition: $0a := 0$, $(k + 1)a := (ka) \vee a$ ($k \geq 0$); the definition cannot be extended when $k < 0$.

switching algebra

a	$0a$	$1a$	$2a$	$3a$
0	0	0	0	0
1	0	1	1	1

mod 3

a	$0a$	$1a$	$2a$	$3a$
0	0	0	0	0
1	0	1	2	0
2	0	2	1	0

mod 2

a	$0a$	$1a$	$2a$	$3a$
0	0	0	0	0
1	0	1	0	1

mod 5

a	$0a$	$1a$	$2a$	$3a$
0	0	0	0	0
1	0	1	2	3
2	0	2	4	1
3	0	3	1	4
4	0	4	3	2

integers

a	$0a$	$1a$	$2a$	$3a$
0	0	0	0	0
1	0	1	2	3
2	0	2	4	6
3	0	3	6	9
4	0	4	8	12

Exercises 3.5

1. Hint. $(n+1)(a_0+hn/2) = (n+1)(2a_0+hn)/2 = (n+1)(a_0+a_0+hn)/2 = (n+1)(a_0+a_n)/2$.

3. (a) 44 850, (c) 2450, (e) 1683, (g) 32 512, (i) 29 484.

5. (10 000 000 − 999)(10 000 000 + 1000)/2.

7. 1 111 000.

9. (a) $(x^{2[n+1]}-1)/(x^2-1)$ if $x \neq \pm 1$ or $n+1$ if $x = \pm 1$; (c) $x^m(x^{n-m+1}-1)/(x-1)$ if $x \neq 1$ or $n-m+1$ if $x = 1$.

Exercises 3.6

1. $x^9 + 9x^8y + 36x^7y^2 + 84x^6y^3 + 126x^5y^4 + 126x^4y^5 + 84x^3y^6 + 36x^2y^7 + 9xy^8 + y^9$; $a^5 + 5a^4b + 10a^3b^2 + 10a^2b^3 + 5ab^4 + b^5$; $t^6 + 6t^5 + 15t^4 + 20t^3 + 15t^2 + 6t + 1$; $u^8 - 8u^7v + 28u^6v^2 - 56u^5v^3 + 70u^4v^4 - 56u^3v^5 + 28u^2v^6 - 8uv^7 + v^8$; $u^{18} + 9u^{16} + 36u^{14} + 84u^{12} + 126u^{10} + 126u^8 + 84u^6 + 36u^4 + 9u^2 + 1$.

3, 4. Hint, cont.: Substitute into the formula of Theorem 8.

5.

1
1 1
1 2 1
1 3 3 1
1 4 1 4 1
1 0 0 0 0 1
1 1 0 0 0 1 1
1 2 1 0 0 1 2 1
1 3 3 1 0 1 3 3 1
1 4 1 4 1 1 4 1 4 1
1 0 0 0 0 2 0 0 0 0 1
1 1 0 0 0 2 2 0 0 0 1 1
1 2 1 0 0 2 4 2 0 0 1 2 1
1 3 3 1 0 2 1 1 2 0 1 3 3 1
1 4 1 4 1 2 3 2 3 2 1 4 1 4 1
1 0 0 0 0 3 0 0 0 0 3 0 0 0 0 1
1 1 0 0 0 3 3 0 0 0 3 3 0 0 0 1 1
1 2 1 0 0 3 1 3 0 0 3 1 3 0 0 1 2 1
1 3 3 1 0 3 4 4 3 0 3 4 4 3 0 1 3 3 1
1 4 1 4 1 3 2 3 2 3 3 2 3 2 3 1 4 1 4 1
1 0 0 0 0 4 0 0 0 0 1 0 0 0 0 4 0 0 0 0 1

7.

1
1 1
1 1 1
1 1 1 1
1 1 1 1 1
1 1 1 1 1 1

Chapter 4

Exercises 4.1

1. (a) aa ab ac ba bb bc ca cb cc; (c) a b; (e) aaa aab aba abb baa bab bba bbb; (g) aa.

3. E.g. 000A 1A2B FFFF. Four-digit hex numerals.

5. (a) 0; (c) 0.

7. Hints. Any one formula is obtained by starting from 0 and 1 and then applying either of the following constructions—any finite number of times and in any order—to any previously obtained formulae.

*any PREVIOUSLY CONSTRUCTED FORMULA followed by any of ∨∧ followed by any PREVIOUSLY CONSTRUCTED FORMULA;
*the symbol (followed by any PREVIOUSLY CONSTRUCTED FORMULA followed by the symbol).
Alternatively:
Each of 0 or 1 is preceded by nothing or any symbol except) and followed by nothing or any symbol except (.
*Each of ∨∧ is preceded by 0 or 1 or) and followed by 0 or 1 or (.
*Each (is preceded by nothing or one of ∨∧ or another (and followed by 0 or 1 or another (.
*Each) is preceded by a 0 or 1 or another) and followed by nothing or one of ∨∧ or another).
*Furthermore: after all symbols except) (are removed, either we are left with nothing or there must be a pair (). In the second instance, after the pair () is removed, the same conclusion must apply.
E.g. $(0 \vee 1) \wedge 1$. Start from 0 and 1. The first construction yields $0 \vee 1$, the second $(0 \vee 1)$, and the first the final formula.

Exercises 4.2

1. 16; 64; 25; 125.

3. (a) 8^{28}. (c) 254^{527}.

5. (a) AAA AAB ABA ABB BAA BAB BBA BBB, 2^3; (c) string of length 0, $3^0 = 1$, (e) none, $0^4 = 0$; (g) string of length 0, $0^0 = 1$.

Exercises 4.3

1. $5^2 \cdot 20^2$.

3. (a) $26^4 \cdot 25$; (c) $26 \cdot 25 \cdot 24 \cdot 23 \cdot 22$.

5. $7 \cdot 8^5$.

7. (a) $2000 \cdot 1500 \cdot 2000$. (c) $2000 \cdot 1300 \cdot 2000 + 2000 \cdot 1200 \cdot 1800 + 2000 \cdot 200 \cdot 199$.

9. $10 \cdot 9^3$.

11. (a) $999 \cdot 10^6$.

Exercises 4.4

1. $0! = 1$, $1! = 1$, $2! = 2$, $3! = 6$, $4! = 24$, $5! = 120$, $6! = 720$, $7! = 5040$, $8! = 40\ 320$, $9! = 362\ 880$, $10! = 3\ 628\ 800$, $11! = 39\ 916\ 800$, $12! \simeq 4.79 \times 10^8$.

3. n, 1.

5. (a) 6; (c) 120; (e) ACR ARC CAR CRA RAC RCA.

7. Hints. Consider permutations as k-digit numerals; recall the rule for $a < b$ in numeral form. Step 1 produces the lowest numeral subject to the constraint of no repetition. If the digit in the last position is less than $n - 1$, then steps 3 and 4 give—if possible—the immediately higher numeral, subject to the constraint. Otherwise, steps 6, 4, and 5 test each position, right to left, for the possibility of increasing the digit in that position by the least possible number of units and then fill the remaining positions subject to the constraint. When possible, the remaining positions are filled in a way that produces the lowest possible numeral. Clearly, the last will be $n|n - 1| \ldots |n - k + 1$, with which the process stops.

Exercises 4.5

1.

1
1 1
1 2 1
1 3 3 1
1 4 6 4 1
1 5 10 10 5 1
1 6 15 20 15 6 1
1 7 21 35 35 21 7 1
1 8 28 56 70 56 28 8 1
1 9 36 84 126 126 84 36 9 1
1 10 45 120 210 252 210 120 45 10 1

3. 1, 999, 498 501, 498 501, 999, 1.

5. 21.

7. (a) $C(25, 3) \cdot C(22, 5) = 60\,568\,200$.

9. $C(n, 2) - n = n(n - 3)/2$.

$n = 3, n_{\text{diagonals}} = 0$ $\quad$ $n = 4, n_d = 2$ $\quad$ $n = 5, n_d = 5$ $\quad$ $n = 6, n_d = 9$

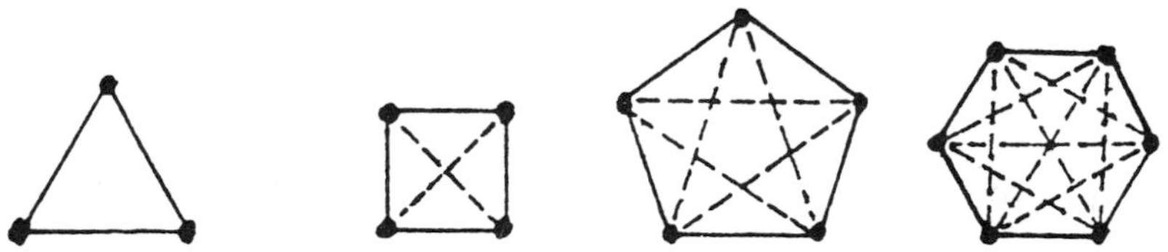

11. (a) 2600; (c) 343; (e) 720.

13. $7 + C(7, 2) = 28$.

Exercises 4.6

1. See results in problem 1 of Exercises 4.4.

3. (a) $2^3 = 8$, (c) $2^5 = 32$.

Complements 4.6

1. Hint, cont. Selecting y in k of the factors can be done in $C(n, k)$ ways. The sum of these monomials is then $C(n, k)x^{n-k}y^k$ and must coincide with $C_{n,k}x^{n-k}y^k$.

3.

	k = 1	2	3	4	5	6
n = 1	1					
2	1	1				
3	1	3	1			
4	1	7	6	1		
5	1	15	25	10	1	
6	1	31	90	65	15	1

Chapter 5

Exercises 5.1

1. (a) {0, 1, 2, 3, 4}, 5; (c) {−2, −1, 0, 1, 2}, 5; (e) e.g., {C, O, Q, S}, 4; (g) ∅, 0; (i) {{a}}, 1.

3. (a) 5, 5, equal; (c) 5, 5, equal; (e) 11, 7, unequal.

Exercises 5.2

1. (a) 2; ∅, {a}, {b}, {a,b}; (c) 3; ∅, {{∅}}, {{a}}, {{a, {a}}}, {{a}, {a, {a}}}, {∅, {a, {a}}}, {∅, {a}}, {∅, {a}, {a, {a}}}.

3. (a) yes; (c) yes; (e) yes.

5. Hints: Let $a := \{\varnothing\}$, $b := \{\varnothing, \{\varnothing\}\} = \{\varnothing, a\}$, $c := \{\{\varnothing\}, \{\varnothing, \{\varnothing\}\}\} = \{a, b\}$, $d := \{\{\varnothing\}\} = \{a\}$, $e := \{\{\{\varnothing\}\}\} = \{d\}$. $\#a = 1$, therefore $a \neq \varnothing$. This implies $\#b = 2$,

$a \neq b$, $\#c = 2$. Then each of a, d, e is different from each of b, c (different cardinalities). $a \neq d$ because $\varnothing \neq a$, $d \neq e$ because $a \neq d$, $a \neq e$ because $\varnothing \neq d$. Also, $b \neq c$ because $b \neq \varnothing$ and $b \neq a$. The five sets are all distinct.
E.g. $f := \{\{\varnothing\}, \{\{\varnothing\}\}\} = \{a, d\}$; $\#f = 2$ because $a \neq d$; f is distinct from a, d, e, (different cardinality); $f \neq b$ ($a \neq d$, $\varnothing \neq d$); $f \neq c$ ($a \neq d$, $b \neq d$).

Exercises 5.3

1. (a) {A, B, C}, 3; (c) {A, B, C, D}, 4.

3. (a) 1, 2, 3, 4, 5; 5; (c) $-4, -3, -2, -1, 0, 1, 2, 3, 4, 5, 6, 7, 8$; 13; (e) 2, 3, 4, 5; 4; (g) 0; 1.

Complements 5.3

1. Hints. If $x \in A \cup (B \cup C)$, then $x \in A$ or $x \in B \cup C$, i.e. $x \in A$ or $x \in B$ or $x \in C$, i.e., $x \in X$ for at least one X. The argument can be reversed. Then it can be applied to $(A \cup B) \cup C$.

2, 3. Just apply the definitions.

Exercises 5.4

1. (a) B; 1; (c) C, D, E; 3.

3. (a) 1, 4, 6; (c) A, E; (e) 1, 4, 9, 16.

5. Just use the definitions.

Complements 5.4

1, 2. Just use the definitions.

Exercises 5.5

1. (a) {0, 1, 2, 3}; (c) $\{x \in \mathbb{N} \mid x \text{ is odd}\}$.

3. E.g., $\{x \in$ {printed lines in the text of problem 2} $\mid$ x is shorter than the regular lines$\}$, $U :=$ {printed lines in the text of that problem}, result $=$ {lines of regular length}.

5. Just apply the definitions.

Exercises 5.6

1. 0 if $m > n$, $n - m + 1$ if $m \leqslant n$.

3. [$j :=$ {journalists}, $A :=$ {Americans}, $Be :=$ {British engineers}]: $\#(j \cup A) = (\#j) + (\#A) - \#(A \cap j) = 10 + 30 - 6 = 34$; $\#Be = \#((j \cup A)') = 50 - 34 = 16$.

5. Hints: (1) If $A \subseteq B$ but $\#A > \#B$, then counting all the elements of B (and therefore all those in A) would not reach the count of A. Since this cannot be, $\#A > \#B$ is false; then $\#A \leqslant \#B$. (2) $\#A \leqslant \#B$ and $\#A \neq \#B$, etc. (3) Count the elements of A, then count up the elements of B. (4) $A \cup B = \{x \in A \mid x \notin B\} \cup \{x \in B \mid x \notin A\} \cup (A \cap B)$ (mutually disjoint). Apply result (3) twice.

Complements 5.6

1. Hints. If W satisfies the axiom, then $\omega \subseteq W$. If x and y are in ω and if $x \neq y$, then $x \subset y$ or $y \subset x$ (not both). Using ω, 0, and $^+$, we can define the elementary arithmetic operations and prove their basic properties. For instance, $x + 0 := x$, $x + (y^+) := (x + y)^+$ recursively; $x \times 0 := 0$, $x \times (y^+) := (x \times y) + x$, etc. (But the proofs require several tools we do not have now.)

Exercises 5.7

1. {{0,1}, {2,3,4,5}}, {{0}, {1}, {2,3,4,5}}.

3. {{00000}, {00001, 00010, 00100, 01000, 10000}, {00011, 00101, 00110, 01001, 01010, 01100, 10001, 10010, 10100, 11000}, {00111, 01011, 01101, 01110, 10011, 10101, 10110, 11001, 11010, 11100}, {01111, 10111, 11011, 11101, 11110}, {11111}}; 1, 5, 10, 10, 5, 1. In general, the numbers of elements in the blocks are $C(n, 0), C(n, 1), C(n, 2), \ldots, C(n, n)$.

5. Hints. The verification is immediate. The two partitions are equal if $\#\{A\} = \#\{\{x\} \mid x \in A\}$, i. e., if $\#A = 1$.

6, 7. Just apply the definitions.

Complements 5.7

1. See problem 3 in Complements 4.6.

3, 4. Just apply the definitions.

5. {{0,2,4}, {6,8}, {1,3} {5,7,9}}.

Exercises 5.8

1. (a) $\mathcal{P}\{a,b,c\} = \{\varnothing, \{a\}, \{b\}, \{c\}, \{b,c\}, \{a,c\}, \{a,b\}, \{a,b,c\}\}$; $2^3 = 8$; (c) $\mathcal{P}\{a\} = \{\varnothing, \{a\}\}$, $2^1 = 2$; $\mathcal{P}\mathcal{P}\{a\} = \mathcal{P}\{\varnothing, \{a\}\} = \{\varnothing, \{\varnothing\}, \{\{a\}\}, \{\varnothing, \{a\}\}\}$, $2^2 = 4, \ldots$.

3. (a) 3, (c) 4, (e) 8.

Complements 5.8

1, 2. Just apply the definitions.

Exercises 5.9

1. (a) {aA, aB, bA, bB, cA, cB}, $6 = 3 \cdot 2$; (c) $\varnothing$, $0 = 3 \cdot 0$; (e) $\{(\varnothing,\varnothing), (\varnothing, \{b\}), (\{a\},\varnothing), (\{a\}, \{b\}\}$, $4 = 2^1 \cdot 2^1$; (g) $\{(\varnothing,\varnothing)\}$, $1 = 2^0 \cdot 2^0$.

3. (a) {11, 12, 13, 21, 22, 23, 31, 32, 33}; (c) $\{\varnothing, \{a\}, \{b\}, \{a,b\}\}$; (e) 2^6.

5. Hint. Multiple application of Theorem 27, part 3.

6, 7. Just apply the definitions.

Complements 5.9

1. Hints for the proof that $(x, y) = (u, v)$ implies $x = u$ and $y = v$. Suppose $\{\{x\}, \{x,y\}\} = \{\{u\}, \{u,v\}\}$. Then: (1) $\{x\} = \{u\}$ or (2) $\{x\} = \{u,v\}$. In case (1), $x = u$, $\{x,y\} = \{u,v\}$, $y = v$. In case (2), $x = u = v$, $\{x,y\} = \{u\}$, $y = u = v$.

Chapter 6

Exercises 6.1

1. R S * RR RS R* SR SS S* *R *S ** RRR RRS RR* RSR RSS RS* R*R R*S R** SRR SRS SR* SSR SSS SS* S*R S*S S** *RR *RS *R* *SR *SS *S* **R **S ***.

3. (a) {Λ, a, b, aa, ab, ba, bb}; (c) {a, b, aa, ab, ba, bb, aaa, aab, aba, abb, baa, bab, bba, bbb}.

5. (a) 13; (c) 243; (e) 27; (g) 0.

7. (a) E.g., {Λ, x, y, xx, xy, yx, yy, xxx}; (c) E.g., {Λ, yxx, xyxx, yxxyxx, yxxxyxx, xyxxyxx, xyxxxyxx, yxxyxxyxx, yxxyxxxyxx, yxxxyxxyxx, xyxxyxxyxx}; (e) {y, xy, xx}; (g) {yy, yxy, xyy, xyxy}; (i) {xxy, xxxy}.

9. See hint in the book. $n=5$, $k=2$, AA AB AC AD AE BA BB ... EE; $n=2$, $k=5$, AAAAA AAAAB AAABA AAABB ... BBBBB.

Exercises 6.2

1. Theorem 6b. $U \cup \mathrm{b} = U$ [subst. $x = U$]; $U \cup \mathrm{b} = \mathrm{b}$ [by 4b]; $\mathrm{b} = U$.
Theorem 10. $\varnothing \cup U = U$ [by Thm. 7a], $\varnothing \cap U = \varnothing$ [by Thm. 7b]; by Thm. 8, $\varnothing' = U$, $U' = \varnothing$.
Theorem 12. Suppose $x \cap y = x$. Then $(x \cup y) \cup y' = x \cup (y \cup y')$ [by 2a] $=$ $x \cup U$ [by 5a] $= U$ [by Thm. 7a], $(x \cup y) \cap y' = (x \cap y') \cup (y \cap y')$ [by 1b, 3a] $=$ $(x \cap y') \cup \varnothing$ [by 5b] $= x \cap y'$ [by 4a] $= (x \cap y) \cap y'$ [subst.] $= x \cap (y \cup y')$ [by 2b] $=$ $x \cap \varnothing$ [by 5b] $= \varnothing$ [by Thm. 7b]. $(x \cup y) = (y')'$ [by Thm. 8] $= y$ [by Thm. 9]. Suppose $x \cup y = y$. Then $(x \cap y) \cap x' = (x \cap x') \cap y$ [by 1b, 2b] $= \varnothing \cap y$ [by 5b] $=$ $\varnothing$ [by Thm. 7b], $(x \cap y) \cup x' = (x \cup x') \cap (y \cup x')$ [by 3b] $= U \cap (y \cup x')$ [by 5a] $=$ $y \cup x'$ [by 4b] $= (x \cup y) \cup x'$ [subst.] $= (x \cup x') \cup y$ [by 1a, 2a] $= U \cup y$ [by 5a] $=$ U [by Thm. 7a]. $x \cap y = (x')'$ [by Thm. 8] $= x$ [by Thm. 9].

3. (e) $x + y = \{\mathrm{a, b, e}\}$.

Complements 6.2

1.

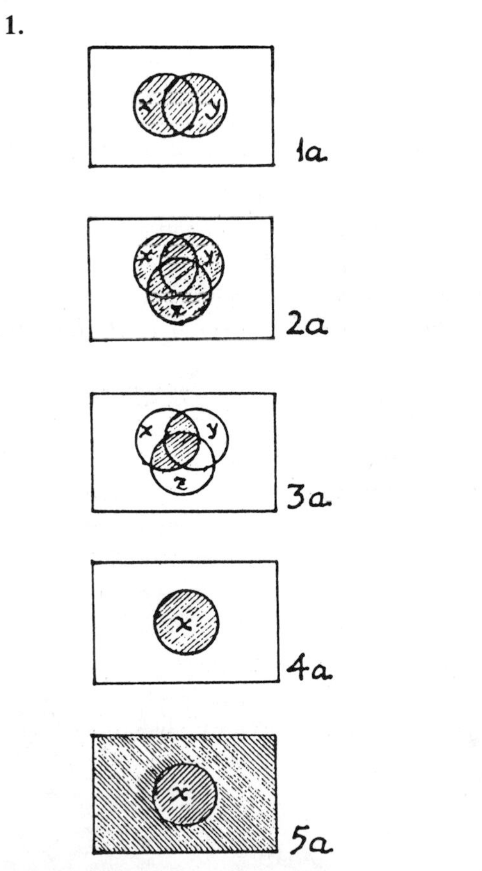

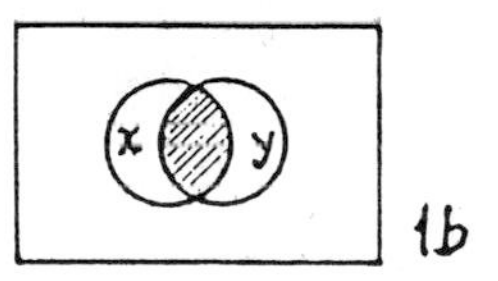

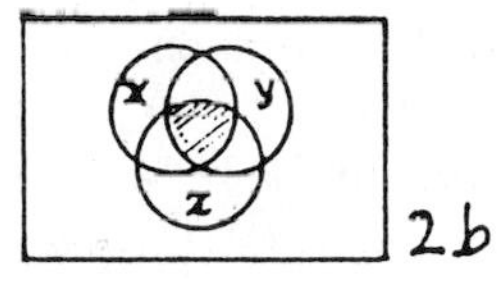

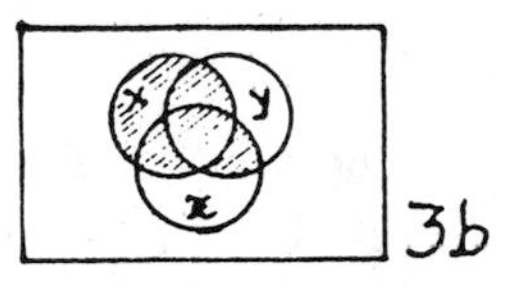

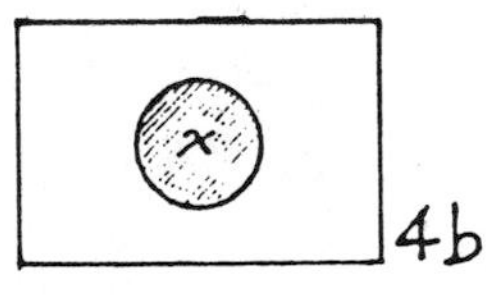

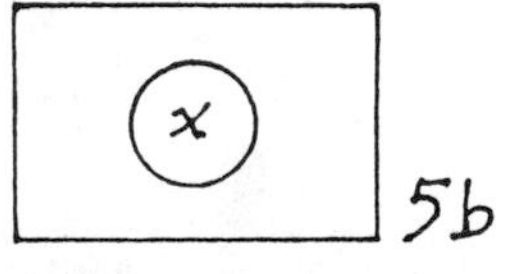

3. (a) $(x \cup y) - z = (x \cup y) \cap (z')$ [by def.] $= (x \cap (z')) \cup (y \cap (z'))$ [by 3a] $= (x - z) \cup (y - z)$ [by def.].

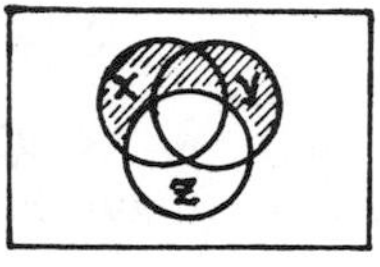

(c) $x + y = (x \cap y') \cup (x' \cap y)$ [by def.] $= (x \cup x') \cap (x \cup y) \cap (x' \cup y') \cap (y \cup y')$ [by 3b, 1a, 1b, 2b] $= (x \cup y) \cap (x' \cup y')$ [by 5a, 4b].

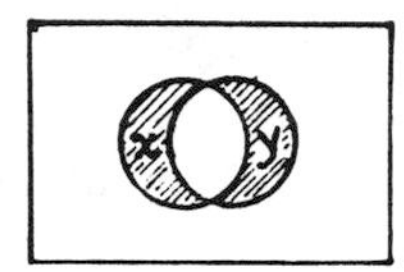

(e) $(x|x)|(y|y) = x'|y'$ [by part (d)] $= (x' \cap y')'$ [by def.] $= (x')' \cup (y')'$ [by DeMorgan] $= x \cup y$ [double complement].

Exercises 6.3

1. Hints. 1a, 1b, 4a, 4b are valid by construction. If $x = 0$, then $x' = 1$, $x \vee x' = 1$, $x \wedge x' = 0$; if $x = 1$, then $x' = 0$, $x \vee x' = 1$, $x \wedge x' = 1$; this proves 5a, 5b. Associativity and distributivity are verified directly.

3. Just apply the definitions and the properties of OR, AND, and NOT. For instance: $(i,j) \wedge ((k,l) \vee (m,n)) = (i,j) \wedge ((k\text{OR}m, l\text{OR}n))$ [by def.] $= (i\text{AND}(k\text{OR}m), j\text{AND}(l\text{OR}n))$ [by def.] $= ((i\text{AND}k)\text{OR}(i\text{AND}m), (j\text{AND}l)\text{OR}(j\text{AND}n))$ [by distrib.] $= (i\text{AND}k, j\text{AND}l) \vee (i\text{AND}m, j\text{AND}n)$ [by def.] $= ((i,j) \wedge (k,l)) \vee ((i,j) \wedge (m,n))$; $(i,j) \wedge 0 = (i,j) \wedge (0,0) = (i\text{AND}0, j\text{AND}0) = (0,0) = 0$.

5. (a) 011; (c) 010; (e) 010.

7. (a) if $x \leqslant y$ and if $y \leqslant z$, then $x \wedge y = x$ and $y \wedge z = y$ by definition, $x \wedge z = (x \wedge y) \wedge z$ [by substit.] $= x \wedge (y \wedge z)$ [by associat.] $= x \wedge y$ [subst.] $= x$ [by subst.]; then $x \leqslant z$ by definition. (c) $x \wedge x = x$ by idempotency; $x \leqslant x$ by definition.

Complements 6.3

1. E.g., $x - y = x \wedge (\sim y)$, $x + y = (x \wedge \sim y) \vee ((\sim x) \wedge y)$, $x|y = \sim(x \wedge y)$.

Exercises 6.4

1.

	p	q	$\sim p$	$\sim q$	$p \wedge q$	$p \vee q$	$\sim(p \wedge q)$	$\sim(p \vee q)$
(a)	T	T	F	F	T	T	F	F
(c)	F	T	T	F	F	T	T	F

3. (a) T; (c) F; (e) F; (f, h, j) T; (l) F; (n) T (in ordinary algebra).

5. F, T, T, F.

Exercises 6.5

1. E.g. idempotency: the disjunction or the conjunction of a statement with itself is equivalent to just the original statement.

3. E.g.: The truth value of a composite OR or AND statement is independent of the order of the components. The truth value of a triple composite OR or AND statement is independent of the interpretation of the double binary composition. A composite OR [AND] statement in which one of the components is false [true] is equivalent to the other component.

5.

p	q	$p \vee q$	$p \wedge q$	$\sim(p \wedge q)$	(1) $(p \vee q) \wedge \sim(p \wedge q)$	table p XOR q	$\sim p$	(2) $(\sim p) \vee (p \wedge q)$	table $p \Rightarrow q$
1	1	1	1	0	0	0	0	1	1
1	0	1	0	1	1	1	0	0	0
0	1	1	0	1	1	1	1	1	1
0	0	0	0	1	0	0	1	1	1

Columns (1) and "XOR table" coincide; columns (2) and "$\Rightarrow$ table" coincide.

7. $x = y$ ($x \subseteq y$ and $y \subseteq x$).

9. Idempotency: $(p \vee p) \Leftrightarrow p$, $(p \wedge p) \Leftrightarrow p$ by Theorem 16. Uniqueness of identities: if $(p \vee q) \Leftrightarrow p$ for all p, then $q \Leftrightarrow$F; if $(p \wedge q) \Leftrightarrow p$ for all p, then $q \Leftrightarrow$T; by Theorem 17. Annihilation: $(p \vee \text{T}) \Leftrightarrow$T, $(p \wedge \text{F}) \Leftrightarrow$F; by Theorem 18. Uniqueness of complement: if $(p \vee q) \Leftrightarrow$T and if $(p \wedge q) \Leftrightarrow$F, then $q \Leftrightarrow \sim p$ by Theorem 19. Double complement: $\sim(\sim p) \Leftrightarrow p$ by Theorem 20. Duality of identities: $\sim$F$\Leftrightarrow$T, $\sim$T$\Leftrightarrow$F by Theorem 21.

11. $(\sim q) \Rightarrow (\sim p)$ iff $(\sim\sim q) \vee (\sim q \wedge \sim p)$ [by def.] iff $(q \vee \sim q) \wedge (q \vee \sim p)$ [by double neg., distrib.] iff $q \vee \sim p$ [by ident.] iff $(\sim p \vee p) \wedge (\sim p \vee q)$ [by commut., ident.] iff $\sim p \vee (p \wedge q)$ [by distrib.] iff $p \Rightarrow q$ [by def.].

13. (a) True, take $q := 1$; $\forall\, p \in A \,|\, \exists\, q \in A \,|\, p \wedge q = p$. (c) True, take $p := 0$; $\exists\, p \in A \,|\, \forall\, q \in A \,|\, p \wedge q = 0$.

15. First two: direct verification. Third: x^2 is always nonnegative. Fourth: take $t := {}^-x$. Fifth: if $t + x = 0$ for every x, then $t = {}^-x$ for every x, that is, $t = {}^-1, {}^-2, \ldots$ for a given t; impossible.

17. (a) true; (c) true; (e) true; (g) false; (h) true; (j) E.g., let A be the set of the elements of a Boolean algebra; $\exists\, x \in A \mid \forall y \in A \mid x \wedge y = 0$, true.

19. (a) $\bigwedge_{x \in S} P(x)$ is true if and only if each $P(x)$ is true, i.e., $\forall x \in S \mid P(x)$.

Complements 6.5

1. See problem 3(d)(e)(f) in Complements 6.2.

3. Sixteen:

p	q	T	$p \vee q$	$q \Rightarrow p$	p	$p \Rightarrow q$	q	$p \Leftrightarrow q$	$p \wedge q$
T	T	T	T	T	T	T	T	T	T
T	F	T	T	T	T	F	F	F	F
F	T	T	T	F	F	T	T	F	F
F	F	T	F	T	F	T	F	T	F

p	q	$p \mid q$ $\sim(p \wedge q)$	$\sim(p \Leftrightarrow q)$	$\sim q$	$\sim(p \Rightarrow q)$	$\sim p$	$\sim(q \Rightarrow p)$	$p \dagger q$ $\sim(p \vee q)$	F
T	T	F	F	F	F	F	F	F	F
T	F	T	T	T	T	F	F	F	F
F	T	T	T	F	F	T	T	F	F
F	F	T	F	T	F	T	F	T	F

5.

p	q	$\sim p$	$\sim q$	$p \wedge q$	$p \wedge \sim q$	$\sim p \vee (p \wedge q)$	$(\sim p) \vee q$	$\sim(p \wedge \sim q)$	$\Rightarrow$
1	1	0	0	1	0	1	1	1	1
1	0	0	1	0	1	0	0	0	0
0	1	1	0	0	0	1	1	1	1
0	0	1	1	0	0	1	1	1	1

7. E.g. BAROKO: $((r \Rightarrow q) \wedge \sim(p \Rightarrow q)) \Rightarrow \sim(p \Rightarrow r)$ iff $((\sim r \vee q) \wedge \sim(\sim p \vee q)) \Rightarrow \sim(\sim p \vee r)$ [by the def. of $\Rightarrow$] iff $((\sim r \vee q) \wedge (p \wedge \sim q)) \Rightarrow (p \wedge \sim r)$ [by DeMorgan] iff $(\sim r \wedge p \wedge \sim q) \Rightarrow (p \wedge \sim r)$ [by distrib. and ident.] iff $(r \vee \sim p \vee q) \vee (p \wedge \sim r)$ [by the def. of $\Rightarrow$ and DeMorgan] iff $(r \vee \sim p \vee q \vee p) \wedge (r \vee \sim p \vee q \vee \sim r)$ [by distrib.] iff 1 [by duality, ident.].

Exercises 6.6

1. **(a1)** $p \vee \sim p$ **(b1)** $p \wedge \sim p$
   (a2) $p \vee 1$ **(b2)** $p \wedge 0$
   (a3) 1 **(b3)** 0

3. See text.

5. E.g.: Idempotency:

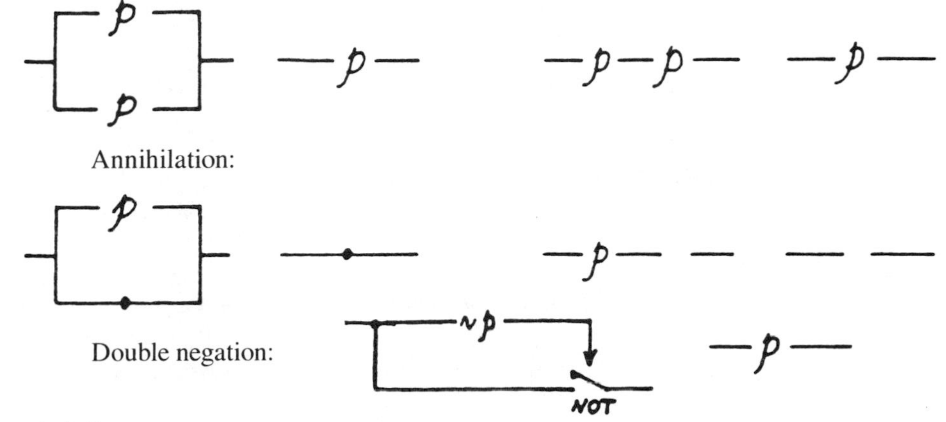

Complements 6.6

1. $(p \wedge q) \vee ((\sim p) \wedge (\sim q))$ gives the truth table of the circuit and is equivalent to $p \Leftrightarrow q$. A 2TSN is given by the figure on the left. A T-switch in terms of ordinary switches is given by the figure on the right.

3. A many-locations circuit, in terms of T- and X-switches:

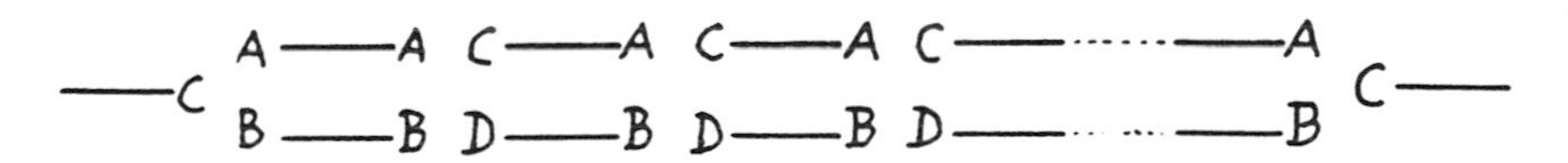

A many-locations circuit, as a 2TSN:

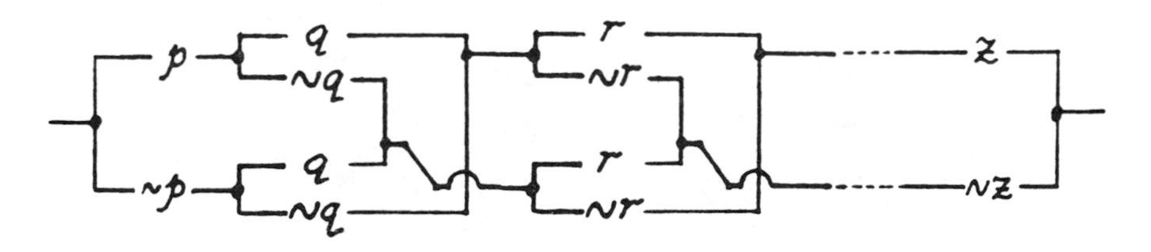

Exercises 6.7

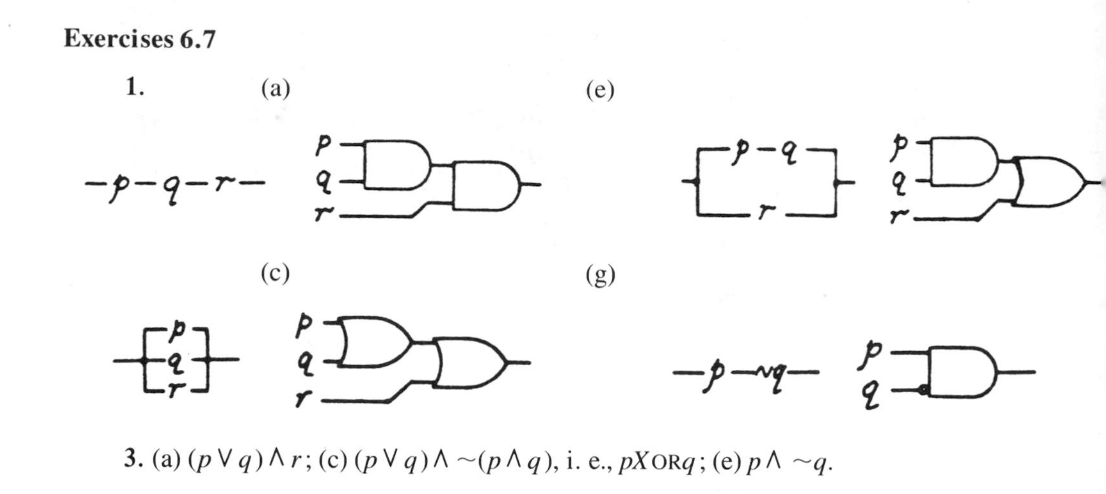

3. (a) $(p \vee q) \wedge r$; (c) $(p \vee q) \wedge \sim(p \wedge q)$, i. e., $pX\text{OR}q$; (e) $p \wedge \sim q$.

Complements 6.7

1. A gate realization of an X-switch is given below left. In all of the examples that follow, inputs A and B are negation of each other. In that case, a simpler solution to the realization of an X-switch in terms of gates is given below right.

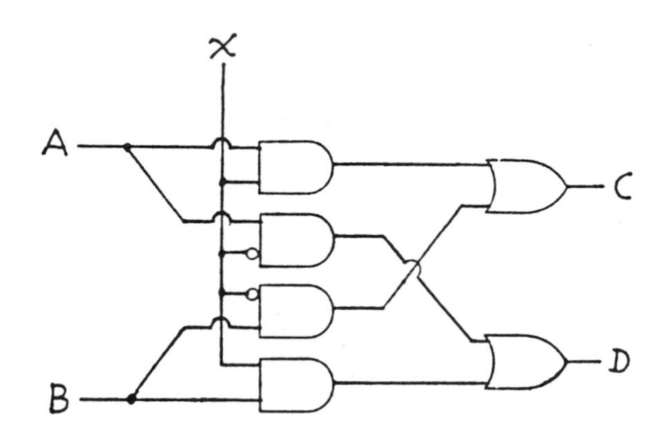

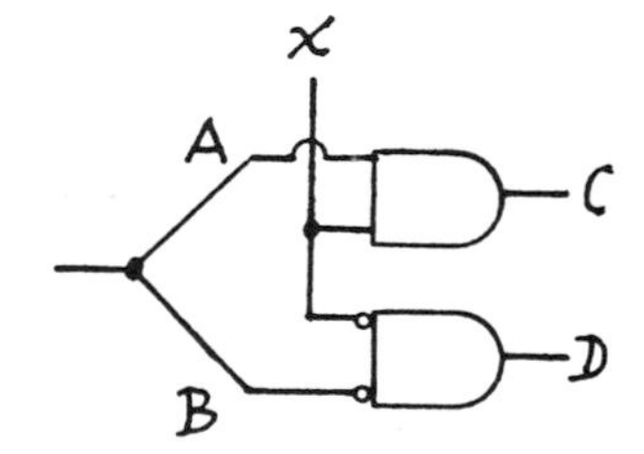

One- and two-location circuits are:

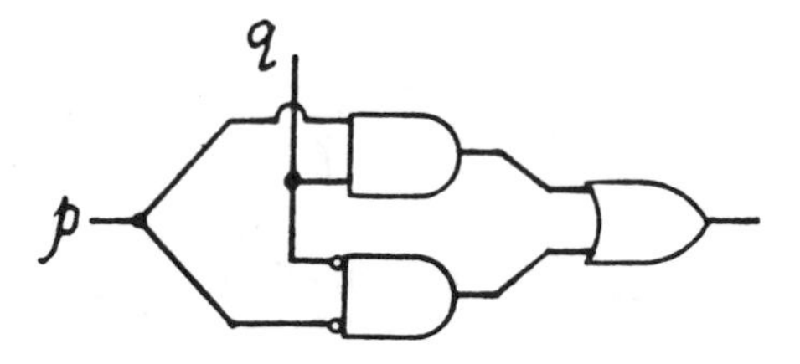

A circuit for three or more locations is:

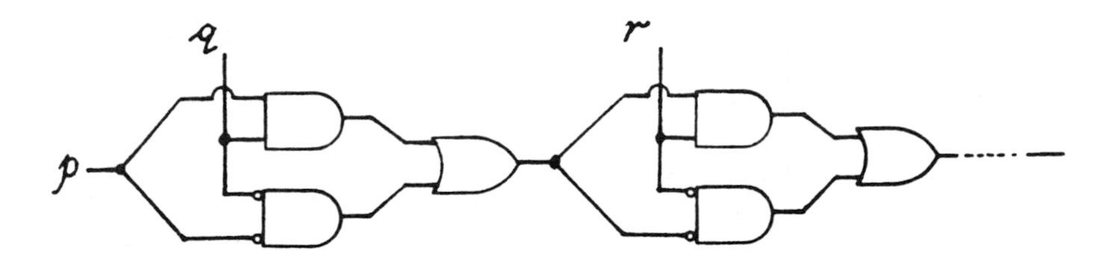

2, 3. E.g.: In the same order as in problem 3 of Complements 6.5 or problem 4 of Exercises 6.7, the postfix formulae are:

1	$pq\vee$	$pq{\sim}\vee$	p
$p{\sim}q\vee$	q	$pq\wedge p{\sim}q{\sim}\wedge\vee$	$pq\wedge$
$pq\wedge\ {\sim}$	$pq\wedge p{\sim}q{\sim}\wedge\vee\ {\sim}$	$q{\sim}$	$p{\sim}q\vee\ {\sim}$
$p{\sim}$	$pq{\sim}\vee{\sim}$	$pq\vee\ {\sim}$	0

Chapter 7

Exercises 7.1

1. Hint. By exhaustion:

x	0	0	0	0	1	1	1	2	2	4
y	1	2	4	7	2	4	7	4	7	7
$x+y$	1	2	4	7	3	5	8	6	9	11

Each of 1 to 9 is obtained precisely once.

Exercises 7.2

1. (a)

x	x^2 mod 5
0	0
1	1
2	4
3	4
4	1

(c) e.g. (d) e.g.

X	Y	$X\cup Y$	$X\cap Y$
$\{0,1\}$	$\{0\}$	$\{0,1\}$	$\{0\}$
$\{0,1\}$	$\{1\}$	$\{0,1\}$	$\{1\}$
$\{0,1\}$	$\{2\}$	$\{0,1,2\}$	$\varnothing$

3.

	total	1-1	surjection	bijection
(a)	yes	yes	no	no
(c)	yes	no	no	no
(e)	yes	yes	yes	yes
(g)	yes	no	no	no
(j)	no	no	yes	no

5. E.g.:

total	onto	*A*	*B*	*g*
yes	yes	{0}	{0}	{(0,0)}
yes	no	{0}	{0,1}	{(0,0)}
no	yes	{0,1}	{0}	{(0.0)}
no	no	{0,1}	{0,1}	{(0,0)}

7.

+	0	1	2
0	0	1	2
1	1	2	0
2	2	0	1

×	0	1	2
0	0	0	0
1	0	1	2
2	0	2	1

	+
(0, 0)	0
(0, 1)	1
(0, 2)	2
(1, 0)	1
(1, 1)	2
(1, 2)	0
(2, 0)	2
(2, 1)	0
(2, 2)	1

	×
(0, 0)	0
(0, 0)	0
(0, 2)	0
(1, 0)	0
(1, 1)	1
(1, 2)	2
(2, 0)	0
(2, 1)	2
(2, 2)	1

9. 4 unary

T	T	T	F	F
F	T	F	T	F

16 binary; see problem 3 in Complements 6.5.

11. $(\#A)!$; each represents a permutation of *A*.

13. Hint. *f* is a total function, because *fS* is uniquely defined for every *S*. *f* is 1-1 and onto, because each $(\#U)$-selection from $\{0, 1\}$ uniquely defines a subset. Therefore *f* is a bijection.

15. See Section 6.5.2.

Complements 7.2

1. (a) $C(237, 3)$.

3. (a) $f(f(x, y), z)=f(x, f(y, z))$ for all x, y, z. (c) $f(x, u)=x$ and $f(u, x)=x$ for every x. (e) $f(x, x)=x$ for every x.

Exercises 7.3

1. 7.1.1. $g=\{(x, x^2+2x+1) \mid x\in\mathbb{R}\}$, $g^{-1}=\{(x^2+2x+1, x) \mid x\in\mathbb{R}\}=\{(y, -1\pm\sqrt{y}) \mid y\in\mathbb{R}, y\geq 0\}$.
7.1.3. $g=\{(x, y) \mid x\in\{\text{names}\}, y\in\{0, \ldots, 9\}^7, (x, y) \text{ is in the book}\}$, $g^{-1}=\{(y, x) \mid x\in\{\text{names}\}, y\in\{0, \ldots, 9\}^7, (x, y) \text{ is in the book}\}$.

3. Hints. (1) If g is not total, then there is an x such that no pair (x, y) is in g. Then no pair (y, x) is in g^{-1}; that is, no y maps to x under g^{-1}; that is, g^{-1} is not onto. (2) Similar argument. (3) If g is not a function, then for some x there are distinct u and v such that (x, u) and (x, v) are in g. Then (u, x) and (v, x) are in g^{-1}; that is, two elements map to x under g^{-1}; that is, g^{-1} is many-to-one. (4) Similar argument. (5) If g is an isomorphism, then g is a total, one-to-one, surjective function. By the previous parts, so is g^{-1}; and conversely.

Complements 7.3

1. E.g.: 15260-1234 has check digit 6; 15260-1234-6 (correct) and 15206-1234-6 (one interchange) both check; 15260-1234-6 checks, but 15261-1234-6 (one mistyping) does not.

Exercises 7.4

1. $g\circ fx=x^3+1, f\circ gx=(x+1)^3, f\circ fx=(x^3)^3, g\circ gx=x+1+1$.

3. (a) $(x+5)^2+1$; (c) $(x^2+1)^2+1$; (e) $((x^2+1)^2+1)^2+1$; (g) $(x^2+1+5)^2+1$; (i) $x^2+1+5+5$.

5. $f^2x=x+2, f^3x=x+3, f^0x=x, f^{-1}x=x-1, f^{-2}x=x-2, f^nx=x+n\ (n\in\mathbb{Z})$.

Complements 7.4

1. Hints. (1) Direct verification. (2) Obvious.
(3a) If $(x, y)\in g$, then $(y, x)\in g^{-1}$ by definition, $(y, x)\in g$ by hypothesis, and $(x, y)\in g^{-1}$ by definition. Then $g\subseteq g^{-1}$ and $g=g^{-1}$.
(3b) If $(x_0, x_n)\in g^n$ and $n\geq 2$, then $(x_0, x_1)\in g, (x_1, x_2)\in g, \ldots, (x_{n-1}, x_n)\in g$ for some $x_1, x_2, \ldots$ Then $(x_0, x_2)\in g^2$, therefore $(x_0, x_2)\in g$; also, $(x_0, x_3)\in g^2$, therefore $(x_0, x_3)\in g$; etc. The case $n=1$ is obvious. The case $n=0$ depends on $i_A\subseteq g$.

Chapter 8

Exercises 8.1

1.

	total binary operation	associative	identity	inverse
8.1.1	by def.	no (see text)	irrelevant (yes, Λ)	irrelevant (no)
8.1.3	by def.	yes (see 6.1.2)	yes, Λ	no (if $x \neq \Lambda$, x:any $\neq \Lambda$)

3. Refer to Section 6.6. Let A satisfy the condition: if networks x and y are in A, then also network $x \vee y$ [$x \wedge y$] is in A. Then we have an abelian semigroup. If, in addition, also F (open circuit) [T (short circuit)] is in A, then we have an abelian monoid with identity F [T].

Complements 8.1

1. Hints. Cross-partition is a commutative and associative operation as a consequence of commutativity and associativity of union and intersection. The partition $\{A\}$ is the identity, as a consequence of the property $x \cap A = x$ (if x is a subset of A). The partition $\{\{x\} \mid x \in A\}$ is the annihilator, as a consequence of the fact that $\{x\} \cap B$ is $\varnothing$ or $\{x\}$.

3. $(A, \circ)$ and $(S^S, \circ)$ are non-abelian monoids. The identity is i_S in each case. The composition of a correspondence with its own inverse does not give the identity in every case, that is, the inverse does not exist in general. For any integer exponents, the two definitions of g^k (iteration of a correspondence and power in the algebraic sense) are consistent with each other and the rules of exponents are valid.

5. $S := \{1, 2, 3\}$. Define $i := \begin{pmatrix} 1 & 2 & 3 \\ 1 & 2 & 3 \end{pmatrix}$, $a := \begin{pmatrix} 1 & 2 & 3 \\ 1 & 3 & 2 \end{pmatrix}$, $b := \begin{pmatrix} 1 & 2 & 3 \\ 2 & 1 & 3 \end{pmatrix}$. Then $ba = \begin{pmatrix} 1 & 2 & 3 \\ 2 & 3 & 1 \end{pmatrix}$, $ab = \begin{pmatrix} 1 & 2 & 3 \\ 3 & 1 & 2 \end{pmatrix}$, $aba = \begin{pmatrix} 1 & 2 & 3 \\ 3 & 2 & 1 \end{pmatrix}$. The table is

i	a	b	ba	ab	aba
a	i	ab	aba	b	ba
b	ba	i	a	aba	ab
ba	b	aba	ab	i	a
ab	aba	a	i	ba	b
aba	ab	ba	b	a	i

$i^{-1}=i$, $a^{-1}=a$, $b^{-1}=b$, $(ba)^{-1}=ab$, $(ab)^{-1}=ba$, $(aba)^{-1}=aba$.

$S := \{1, 2\}$. Let $i := \begin{pmatrix}1 & 2\\ 1 & 2\end{pmatrix}$, $a := \begin{pmatrix}1 & 2\\ 2 & 1\end{pmatrix}$, Then the table is $\boxed{\begin{matrix} i & a \\ a & i\end{matrix}}$.

$S := \{1\}$. Let $i := \begin{pmatrix}1\\ 1\end{pmatrix}$. Then the table is just $\boxed{i}$.

7. E.g., hints for problem 4. If $\#S = n$, then enumerating the n-permutations of set $\{1, 2, \ldots, n\}$ provides the elements of A, say in the form $u(k)_{1 \leqslant k \leqslant n}$. If $u(k)_{1 \leqslant k \leqslant n}$ and $v(k)_{1 \leqslant k \leqslant n}$ are two such permutations (see Section 7.2.4.2), the $u(v(k))_{1 \leqslant k \leqslant n}$ is their composition. The inverse of $u(k)$ is (h such that $u(h)=k)_{1 \leqslant k \leqslant n}$, that is, u^{-1} as a function.

9. The group of the triangle is the same as the total group on $\{1, 2, 3\}$. See problem 5 above. For the square $\begin{matrix}4 & 3\\ 1 & 2\end{matrix}$: there are four destinations for vertex 1 and, after that, two possible destinations for vertex 2, for a total of eight permutations. Let $a := \begin{pmatrix}1 & 2 & 3 & 4\\ 2 & 3 & 4 & 1\end{pmatrix}$, $b := \begin{pmatrix}1 & 2 & 3 & 4\\ 2 & 1 & 4 & 3\end{pmatrix}$. Then the table is

i	a	a^2	a^3	b	ab	a^2b	a^3b
a	a^2	a^3	i	ab	a^2b	a^3b	b
a^2	a^3	i	a	a^2b	a^3b	b	ab
a^3	i	a	a^2	a^3b	b	ab	a^2b
b	a^3b	a^2b	ab	i	a^3	a^2	a
ab	b	a^3b	a^2b	a	i	a^3	a^2
a^2b	ab	b	a^3b	a^2	a	i	a^3
a^3b	a^2b	ab	b	a^3	a^2	a	i

A table of the group of the cube (left half)

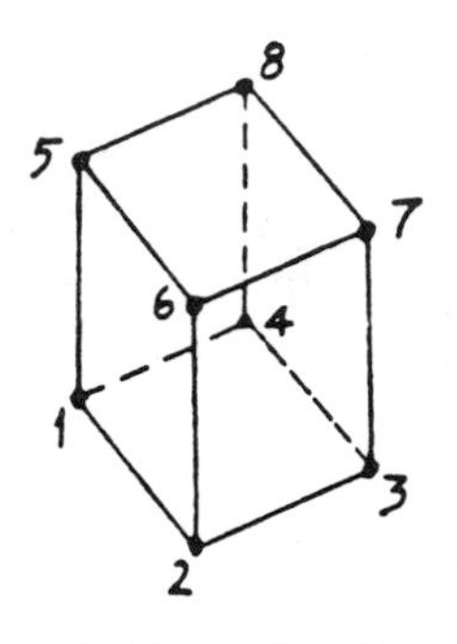

action on 1 2	12	23	34	41	15	26	37	48	14	21	32	43
permutation	i	a	a^2	a^3	c	ac	a^2c	a^3c	c^2	ac^2	a^2c^2	a^3c^2
	a	a^2	a^3	i	ac	a^2c	a^3c	c	ac^2	a^2c^2	a^3c^2	c^2
	a^2	a^3	i	a	a^2c	a^3c	c	ac	a^2c^2	a^3c^2	c^2	ac^2
	a^3	i	a	a^2	a^3c	c	ac	a^2c	a^3c^2	c^2	ac^2	a^2c^2
	c	ca	$acac^2$	ac^2	c^2	cac	aca	a	i	cac^2	$acac$	ac
	ac	aca	a^2cac^2	a^2c^2	ac^2	$acac$	a^2ca	a^2	a	$acac^2$	a^2cac	a^2c
	a^2c	a^2ca	a^3cac^2	a^3c^2	a^2c^2	a^2cac	a^3ca	a^3	a^2	a^2cac^2	a^3cac	a^3c
	a^3c	a^3ca	cac^2	c^2	a^3c^2	a^3cac	ca	i	a^3	a^3cac^2	cac	c
	c^2	a^3c	a^3ca	cac^2	i	a^3c^2	a^3cac	ca	c	a^3	a^3cac^2	cac
	ac^2	c	ca	$acac^2$	a	c^2	cac	aca	ac	i	cac^2	$acac$
	a^2c^2	ac	aca	a^2cac^2	a^2	ac^2	$acac$	a^2ca	a^2c	a	$acac^2$	a^2cac
	a^3c^2	a^2c	a^2ca	a^3cac^2	a^3	a^2c^2	a^2cac	a^3ca	a^3c	a^2	a^2cac^2	a^3cac
	ca	$acac^2$	ac^2	c	cac	aca	a	c^2	cac^2	$acac$	ac	i
	aca	a^2cac^2	a^2c^2	ac	$acac$	a^2ca	a^2	ac^2	$acac^2$	a^2cac	a^2c	a
	a^2ca	a^3cac^2	a^3c^2	a^2c	a^2cac	a^3ca	a^3	a^2c^2	a^2cac^2	a^3cac	a^3c	a^2
	a^3ca	cac^2	c^2	a^3c	a^3cac	ca	i	a^3c^2	a^3cac^2	cac	c	a^3
	cac	a^3cac	a^2cac	$acac$	cac^2	a^3cac^2	a^2cac^2	$acac^2$	ca	a^3ca	a^2ca	aca
	$acac$	cac	a^3cac	a^2cac	$acac^2$	cac^2	a^3cac^2	a^2cac^2	aca	ca	a^3ca	a^2ca
	a^2cac	$acac$	cac	a^3cac	a^2cac^2	$acac^2$	cac^2	a^3cac^2	a^2ca	aca	ca	a^3ca
	a^3cac	a^2cac	$acac$	cac	a^3cac^2	a^2cac^2	$acac^2$	cac^2	a^3ca	a^2ca	aca	ca
	cac^2	c^2	a^3c	a^3ca	ca	i	a^3c^2	a^3cac	cac	c	a^3	a^3cac^2
	$acac^2$	ac^2	c	ca	aca	a	c^2	cac	$acac$	ac	i	cac^2
	a^2cac^2	a^2c^2	ac	aca	a^2ca	a^2	ac^2	$acac$	a^2cac	a^2c	a	$acac^2$
	a^3cac^2	a^3c^2	a^2c	a^2ca	a^3ca	a^3	a^2c^2	a^2cac	a^3cac	a^3c	a^2	a^2cac^2

For the group of the cube, there are eight destinations for vertex 1, followed by three rotations about a diagonal, for a total of 24 permutations. To characterize each permutation, we need only its action on vertices 1 and 2; see the first row of the table above.

A table of the group of the cube (right half)

56	67	78	85	58	65	76	87	51	62	73	84
ca	aca	a^2ca	a^3ca	cac	$acac$	a^2cac	a^3cac	cac^2	$acac^2$	a^2cac^2	a^3cac^2
aca	a^2ca	a^3ca	ca	$acac$	a^2cac	a^3cac	cac	$acac^2$	a^2cac^2	a^3cac^2	cac^2
a^2ca	a^3ca	ca	aca	a^2cac	a^3cac	cac	$acac$	a^2cac^2	a^3cac^2	cac^2	$acac^2$
a^3ca	ca	aca	a^2ca	a^3cac	cac	$acac$	a^2cac	a^3cac^2	cac^2	$acac^2$	a^2cac^2
a^3c	a^3cac	a^2cac^2	a^2	a^3c^2	a^3cac^2	a^2ca	a^2c	a^3	a^3ca	a^2cac	a^2c^2
c	cac	a^3cac^2	a^3	c^2	cac^2	a^3ca	a^3c	i	ca	a^3cac	a^3c^2
ac	$acac$	cac^2	i	ac^2	$acac^2$	ca	c	a	aca	cac	c^2
a^2c	a^2cac	$acac^2$	a	a^2c^2	a^2cac^2	aca	ac	a^2	a^2ca	$acac$	ac^2
a	a^2c	a^2cac	$acac^2$	ac	a^2c^2	a^2cac^2	aca	ac^2	a^2	a^2ca	$acac$
a^2	a^3c	a^3cac	a^2cac^2	a^2c	a^3c^2	a^3cac^2	a^2ca	a^2c^2	a^3	a^3ca	a^2cac
a^3	c	cac	a^3cac^2	a^3c	c^2	cac^2	a^3ca	a^3c^2	i	ca	a^3cac
i	ac	$acac$	cac^2	c	ac^2	$acac^2$	ca	c^2	a	aca	cac
a^3cac	a^2cac^2	a^2	a^3c	a^3cac^2	a^2ca	a^2c	a^3c^2	a^3ca	a^2cac	a^2c^2	a^3
cac	a^3cac^2	a^3	c	cac^2	a^3ca	a^3c	c^2	ca	a^3cac	a^3c^2	i
$acac$	cac^2	i	ac	$acac^2$	ca	c	ac^2	aca	cac	c^2	a
a^2cac	$acac^2$	a	a^2c	a^2cac^2	aca	ac	a^2c^2	a^2ca	$acac$	ac^2	a^2
c^2	a^3c^2	a^2c^2	ac^2	i	a^3	a^2	a	c	a^3c	a^2c	ac
ac^2	c^2	a^3c^2	a^2c^2	a	i	a^3	a^2	ac	c	a^3c	a^2c
a^2c^2	ac^2	c^2	a^3c^2	a^2	a	i	a^3	a^2c	ac	c	a^3c
a^3c^2	a^2c^2	ac^2	c^2	a^3	a^2	a	i	a^3c	a^2c	ac	c
$acac^2$	a	a^2c	a^2cac	aca	ac	a^2c^2	a^2cac^2	$acac$	ac^2	a^2	a^2ca
a^2cac^2	a^2	a^3c	a^3cac	a^2ca	a^2c	a^3c^2	a^3cac^2	a^2cac	a^2c^2	a^3	a^3ca
a^3cac^2	a^3	c	cac	a^3ca	a^3c	c^2	cac^2	a^3cac	a^3c^2	i	ca
cac^2	i	ac	$acac$	ca	c	ac^2	$acac^2$	cac	c^2	a	aca

The vertices of the cube are labelled as indicated in the figure above the table. Let $a := \begin{pmatrix} 1 & 2 & 3 & 4 & 5 & 6 & 7 & 8 \\ 2 & 3 & 4 & 1 & 6 & 7 & 8 & 5 \end{pmatrix}$ and $c := \begin{pmatrix} 1 & 2 & 3 & 4 & 3 & 6 & 8 & 7 \\ 1 & 5 & 4 & 2 & 6 & 8 & 3 & 7 \end{pmatrix}$. Then see the resulting composition table above.

11. (a) $h(xfu)=(h(x))g(h(u))$; also, $h(xfu)=h(x)$; then $(h(x))g(h(u))=h(x)$, which implies that $h(u)$ is the identity for g. (c) Consequences of the rules of exponents.

13. Hint. See problem 11(c) above and delete the part of the codomain of h which contains the negative reals and 0.

Exercises 8.2

1. $\mathbb{N}$ gives a commutative semiring with identity 1; all of $\mathbb{Z}$, $\mathbb{Z}_m$ $(m>1)$, $\mathbb{Q}$, and $\mathbb{R}$ give a commutative ring with identity 1. Furthermore, the multiplication in each of $\mathbb{Z}_m-\{0\}$ (m prime), $\mathbb{Q}-\{0\}$, and $\mathbb{R}-\{0\}$ gives an abelian group.

3. (a) Commutative ring with identity 1. See problem 2(a) in Exercises 8.1. (c) Boolean algebra (and therefore semiring) with identity A. See Sections 6.2 and 6.3.4.

5. Yes if the ring is commutative, no if it is not. In any ring, $(x-y)(x+y)=x^2-yx+xy-y^2$; then $-yx+xy$ cancel if and only if $xy=yx$.

7. Direct verification.

9. Since $1+1=1$ would imply $1=0$, it must be that $1+1=0$, which (with the identity property) completes the addition table. Also, $1\cdot 1=1$ or $1\cdot 1=0$. The second case is excluded if 1 is required to be the multiplicative identity. The first case gives $\mathbb{Z}_2$.

11. Since $0'=1$ and $1'=0$, it must be that $a'=b$ and $b'=a$ (uniqueness of complement). Then $a\vee b=1$ and $a\wedge b=0$ (complement properties). Through commutativity, idempotency, and identity properties, the tables are so complete.

Complements 8.2

1. Hints. Using the symbols $+$ and $\times$ for the operations in the first semiring, and $\oplus$ and $*$ for those in the second, the conditions for a homomorphism h are: (1) h is a total function; (2) $h(x+y)=(hx)+(hy)$, $h(x\times y)=(hx)*(hy)$; (3) $h(0)=0$; (4) if u and v are the respective multiplicative identities, then $h(u)=v$. For an isomorphism, there is the additional condition of bijectivity. From $\mathbb{Z}$ to $\mathbb{Z}_m$ the natural homomorphism is reduction mod m. From a Boolean algebra to its own dual the natural isomorphism is complementation.

Exercises 8.3

1. See example in Section 8.3.

3, 5. Patient verification.

Complements 8.3

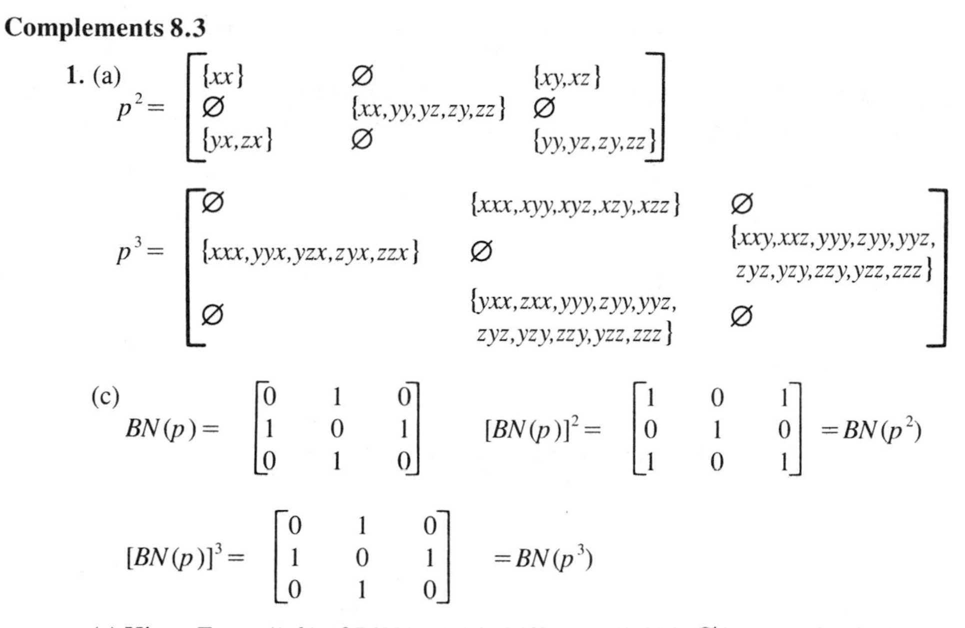

1. (a)

$$p^2 = \begin{bmatrix} \{xx\} & \varnothing & \{xy, xz\} \\ \varnothing & \{xx, yy, yz, zy, zz\} & \varnothing \\ \{yx, zx\} & \varnothing & \{yy, yz, zy, zz\} \end{bmatrix}$$

$$p^3 = \begin{bmatrix} \varnothing & \{xxx, xyy, xyz, xzy, xzz\} & \varnothing \\ \{xxx, yyx, yzx, zyx, zzx\} & \varnothing & \{xxy, xxz, yyy, zyy, yyz, zyz, yzy, zzy, yzz, zzz\} \\ \varnothing & \{yxx, zxx, yyy, zyy, yyz, zyz, yzy, zzy, yzz, zzz\} & \varnothing \end{bmatrix}$$

(c)

$$BN(p) = \begin{bmatrix} 0 & 1 & 0 \\ 1 & 0 & 1 \\ 0 & 1 & 0 \end{bmatrix} \qquad [BN(p)]^2 = \begin{bmatrix} 1 & 0 & 1 \\ 0 & 1 & 0 \\ 1 & 0 & 1 \end{bmatrix} = BN(p^2)$$

$$[BN(p)]^3 = \begin{bmatrix} 0 & 1 & 0 \\ 1 & 0 & 1 \\ 0 & 1 & 0 \end{bmatrix} = BN(p^3)$$

(e) Hints. Entry (i, k) of $B°N(x \vee y)$ is 0 iff entry (i, k) is $\varnothing$ in $x \vee y$, that is, in both x and y; entry (i, k) in $B°Nx \vee B°Ny$ is 0 iff both terms are 0, that is, iff entry (i, k) is $\varnothing$ in both x and y; this is the same condition as before. Entry (i, k) of $B°N(x{*}y)$ is 1 iff entry (i, k) in $x{*}y$ is nonempty, that is, iff, for some j, entries (i, j) in x and entry (j, k) in y are both nonempty; entry (i, k) of $(B°Nx) * (B°Ny)$ is 1 iff, for some j, entry (i, j) in $B°Nx$ and entry (j, k) in $B°Ny$ are both 1, that is, the same condition as before. (g) Counterexample: $n := 1$, $x := [\{a, aa\}]$, $y := [\{ab, b\}]$. Then $x{*}y = [\{aab, ab, aaab\}]$, $Nx = [2]$, $Ny = [2]$, $Nx{*}Ny = [4] \neq [3] = N(x{*}y)$.

3. Patient verification. Note that, at each bit, 1 OR 1 = 1, 1 XOR 1 = 0.

Chapter 9

Exercises 9.1

1. (a) 0, 1, 2, 3, 4, 5, 6, 7; (c) 5, 19, 61, 187, 565, 1699, 5101, 15307; (e) 1, 1, 3, 5, 9, 15, 25, 41.

3. $f0 = 1$, $f1 = \sim(11 \vee 10) = \sim 11 = 0$, $f2 = \sim(f1 \vee 10) = \sim 10 = 01$, $f3 = \sim(f2 \vee 10) = \sim 1 = 0, \ldots, fn = 01$ if n is even, $fn = 0$ if n is odd $(n > 1)$.

5. (a) $A^0 := \{\Lambda\}$, $A^{h+1} := A^h{:}\{(x) \mid x \in A\}$ $(h \geq 0)$. (c) $\bigvee_{i=1}^{0} x_i := 0$, $\bigvee_{i=1}^{h+1} x_i := (\bigvee_{i=1}^{h} x_i) \vee x_{h+1}$, $\bigwedge_{i=1}^{0} x_i := 1$, $\bigwedge_{i=1}^{h+1} x_i := (\bigwedge_{i=1}^{h} x_i) \wedge x_{h+1}$, $(h \geq 0)$.

7. Let) or (denote the 1-string containing) or (only. Let $(x_i)_{0 \leqslant i \leqslant n}$ be the given sequence. (1) Start from $h := 0$ and $E_0 := \Lambda$. (2) If $h = n$, then stop; string E_n is the result. If $0 \leqslant h < n$, define E_{h+1} as E_h:(when $x_{h+1} > x_h$ or as E_h:) when $x_{h+1} < x_h$. In the given example E_0 to E_8 are, respectively, Λ ((((() (()((()() (()()) (()())((()())().

9. Replace

	unary +, − with	binary + − */ with
(a)	~(complementation)	∪, ∩
(c)	~(negation)	∨, ∧

Complements 9.1

1. Remark. The definition implies $f0 = b, f1 = gb, f2 = g^2b, f^3 = g^3b$, ... Induction is needed to show that the process is valid for every fn ($n \in \mathbb{N}$).

Exercises 9.2

1. The cores of the proofs are the following.

(a) $\sum_{i=0}^{h+1} i^2 = (\sum_{i=0}^{h} i^2) + (h+1)^2 = h(h+1)(2h+1)/6 + 6(h+1)^2/6 = ((h+1)(h+2)(2(h+1)+1)/6$.

(c) $\sum_{i=1}^{h+1} (2i-1) = (\sum_{i=1}^{h} (2i-1)) + (2h+1) = h^2 + 2h + 1 = (h+1)^2$.

(e) $\sum_{k=0}^{h+1} C(h+1,\ k) = 1 + (\sum_{k=1}^{h} C(h+1,\ k)) + 1 = 1 + (\sum_{k=1}^{h} (C(h,\ k-1) + C(h,\ k))) + 1 = 1 + (\sum_{k=0}^{h-1} C(h,\ k)) + (\sum_{k=1}^{h} C(h,\ k)) + 1 = 2(\sum_{k=0}^{h} C(h,\ k)) = 2 \cdot 2^h = 2^{h+1}$.

3. # {permutations} $-\Sigma$ #{permutations with one fixed element} $+\Sigma$ #{permutations with two fixed elements} $-$ . . . $= n! - C(n,\ 1) \cdot (n-1)! + C(n, 2) \cdot (n-2)! - C(n,\ 3) \cdot (n-3)! + C(n,\ 4) \cdot (n-4)! -$ . . . $= n! - n(n-1)! + n(n-1)(n-2)!/2! - n(n-1)(n-2)(n-3)!/3! +$ . . . $= n! - n! + n!/2! - n!/3! + n!/4! -$ . . . $= n!(1/0! - 1/1! + 1/2! - 1/3! + 1/4! - \ldots)$.

5. (a)

q	5	2		
r	55	10	5	0;
			=	

(c)

q	2	87		
r	175	87	1	0.
			=	

7. (a) $x = -2, y = 1$; (c) $x = 3, y = -2$.

Complements 9.2

1. Remark: see Complement 5.6.

Exercises 9.3

In these exercises, the symbol "En" denotes "$\times 10^n$".

1. (a) $x_n = C \cdot 5^n$; $x_n = 5^n$; 1, 5, 25, 125, 625, 3125, 15625, 78125, 390625, 1953125; $\simeq$7.89E69. (c) $x_n = C \cdot 9^n$; $x_n = 0$; all 0.

3. (a) $C_1 \cdot 2^n + C_2 \cdot (-3)^n$; $2^n + (-3)^n$;

0	1	2	3	4	5	6	7	8	9	100
2	−1	13	−19	97	−211	793	−2059	6817	−19171	≃5.15E47

(c) $(C_1 + nC_2) \cdot 3^n$; $2(1+n) \cdot 3^n$;

0	1	2	3	4	5	6	7	8	9	100
2	12	54	216	810	2916	10206	34992	118098	393660	≃1.04E50

5. For the particular solution: in the general solution of Fibonacci's equation (see text), substitute:
(a) $C_1 = 1/\sqrt{5}$, $C_2 = -1/\sqrt{5}$;
(c) $C_1 = (-1+\sqrt{5})/\sqrt{5}$, $C_2 = (1+\sqrt{5})/\sqrt{5}$;
(e) $C_1 = 0$, $C_2 = 0$.

	0	1	2	3	4	5	6	7	8	9	50≃	100≃	150≃	200≃	250≃
(a)	0	1	1	2	3	5	8	13	21	34	1.26E10	3.54E20	9.97E30	2.81E41	7.90E51
(c)	2	0	2	2	4	6	10	16	26	42	1.56E10	4.38E20	1.23E31	3.47E41	9.76E51
(e)	all zero														

7. Hint. Immediate proof by induction.

9. (a) $B + C \cdot (-2)^n + D \cdot 3^n$. (c) $B + Cn + Dn^2$.

Chapter 10

Exercises 10.3

1. (a) [strict] = {(0, 1)}; [r] = {(0, 0), (0, 1), (1, 1)}. (c) [r] = given; [strict] = {(u, v) | first element of u = first element of v and $u \neq v$}.

Exercises 10.4

1. (a) Antisymmetric; [r], see Section 10.3; [s] = u (universal relation); [rs] = u. (c) Symmetric; [r] = given; [s] = given; [rs] = given.

3. (2) The definition implies $R^{-1} \subseteq R$. Theorem 7.17(3) implies $R^{-1} = R$. (3) Prove the contrapositives: $(R^{[\text{strict}]})^2 \cap i_A \neq \varnothing$ iff there are x, y such that xRy, yRx, and $x \neq y$; that is, iff R is not antisymmetric.

Exercises 10.5

1. (a) yes; (c) yes.

3. Hint for transitivity. If $u:z = v$ and if $v:t = w$, then $u:(z:t) = w$.

Complements 10.5

1. Hint. Obviously, $V = S*T$ implies $V \lessdot T$. Apply this when $V = S$ and obtain the "if" part. Suppose $S \lessdot T$. Let S' be a block of S; $\varnothing \neq S' \subseteq T'$ for some T'; $S' = S' \cap T' \in S*T$, $S \subseteq S*T$. Let S'' be a block of $S*T$; $S'' = S' \cap T'$ for some S', T'; $S'' \subseteq S' \subseteq T''$ for some T''; $\varnothing \neq S'' \subseteq T' \cap T''$; $T' = T''$, $S'' = S'$, $S*T \subseteq S$. $S*T = S$.

3. Hint. $u \lessdot v$ and $v \lessdot s$ imply $u*w = v$ and $v*t = s$ (w and t in H); then $u*w*t = s$, $w*t \in H$, which imply $u \lessdot s$.

5. (a) Intersections of subsets of U are subsets of U. Subset of. (c) Sums of nonnegative integers are nonnegative. $\leqslant$.

Exercises 10.6

1.

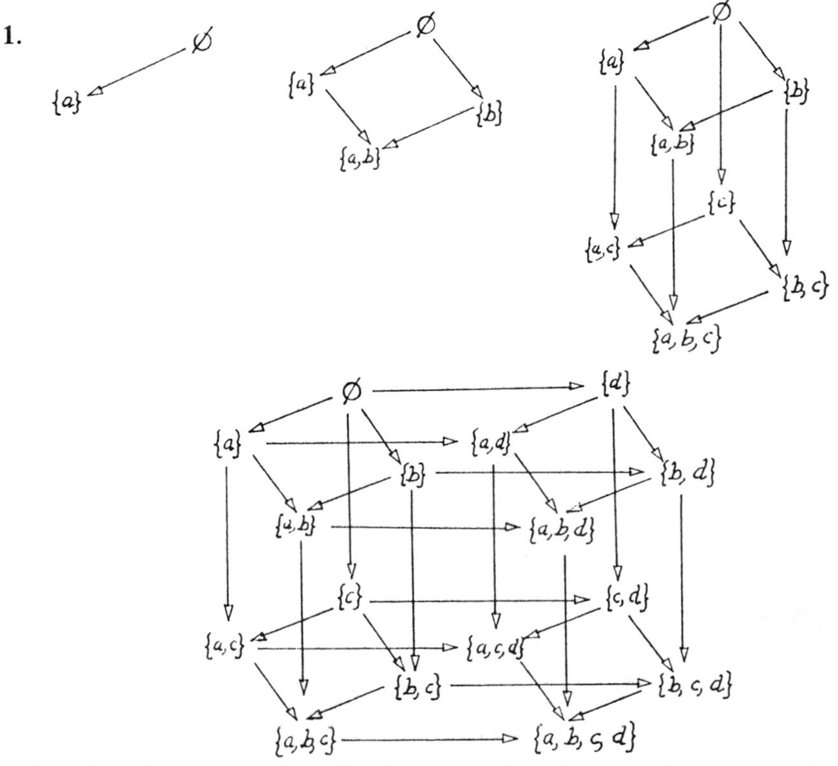

3. (a) Partial order, not total; $v \to x \quad w \to y \to z$. (c) Partial order, not total; $v \quad w \quad z$.

5. A set with one element.

7. 1%*%1 1%*%* 1%**1 **1%* **11* **1*1.

9. Hint. Define $f: \mathbb{Z} \to \{\text{odd integers}\}$, $fx := 2x+1$.

11. $0 \Rightarrow 1 \quad \varnothing \subseteq \{a\}$.

13. Hint. Let $S := R^{[\text{strict}]}$, $L := R^{[r]}$; then $L = S^{[r]}$, $S = L^{[\text{strict}]}$. L is reflexive, S is irreflexive. By problem 2 of Exercises 10.4, S is antisymmetric if and only if L is antisymmetric. Suppose L is a linear order. Then xLy or yLx. If both hold, then $x = y$ by antisymmetry; otherwise, xSy or ySx (not both); trichotomy is valid for S. Suppose S is trichotomic. Then only one of $x = y$, xSy, ySx is valid. Then xLy (by reflexivity) or yLx, and L is linear.

15. Interchanging the roles of the four elements 1, 2, 3, 4 in any possible structure will clearly produce an isomorphic structure. In the following diagrams, if the symbols a, b, c, d are replaced by a permutation of the symbols 1, 2, 3, 4, then isomorphic structures are obtained. Additional remarks on isomorphism are made in each case. Antiisomorphic cases are separated by ***.
$a \ b \ c \ d$ (two such structures are identical)
$a \to b \ c \ d$ (antiisomorphic structures result if a and b are interchanged)
$a \leftarrow b \to c \ d$ *** $a \to b \leftarrow c \ d$
$a \to b \ c \to d$ (antiisomorphic structures result if a and b are interchanged, and so are c and d)
$a \to b \to c \ d$ (same remark, for ac)
$a \to b \leftarrow c \to d$ (same remark, for ad and bc)
$a \to b \leftarrow c \leftarrow d$ *** $a \leftarrow b \to c \to d$
$a \to b \to c \to d$ (same remark, for ad and bc)

$$\begin{array}{ccc} a \to b \leftarrow c & *** & a \leftarrow b \to c \\ \uparrow & & \downarrow \\ d & & d \end{array}$$

$$\begin{array}{ccc} a \to b \to c & *** & a \leftarrow b \leftarrow c \\ \downarrow & & \uparrow \\ d & & d \end{array}$$

17. Hint, cont. Define xRy if x is even and y is odd, or if $x < y$ and x and y are both even or both odd.

Complements 10.6

1. Except for (g), see hints in the text. For (g): by the definition of $\Rightarrow$, (0,0), (0,1), and (1,1) are in R, and (1,0) is not. The strict relation is $\{(0,1)\}$, which is clearly irreflexive and trichotomic, and degenerately transitive.

Exercises 10.7

1. See Section 11.6.

Exercises 10.8

1. See text for hints.

3.

$\downarrow$	1	2	3	5	6	10	15	30
1	1	1	1	1	1	1	1	1
2	1	2	1	1	2	2	1	2
3	1	1	3	1	3	1	3	3
5	1	1	1	5	1	5	5	5
6	1	2	3	1	6	2	3	6
10	1	2	1	5	2	10	5	10
15	1	1	3	5	3	5	15	15
30	1	2	3	5	6	10	15	30

$\uparrow$	1	2	3	5	6	10	15	30
1	1	2	3	5	6	10	15	30
2	2	2	6	10	6	10	30	30
3	3	6	3	15	6	30	15	30
5	5	10	15	5	30	10	15	30
6	6	6	6	30	6	30	30	30
10	10	10	30	10	30	10	30	30
15	15	30	15	15	30	30	15	30
30	30	30	30	30	30	30	30	30

$\inf A = 1$, $\sup A = 30$.

5. Just apply the definitions.

7. See text for hints.

9. (a) For x^{-1} to be the inverse of x, A must have infimum, and the condition $x \downarrow x^{-1} = \inf A$ must determine x^{-1} uniquely. Since $x \downarrow \inf A = \inf A$, it must be that x^{-1} (if existent) is $\inf A$. If A contains only one element, then the condition is satisfied. If A contains more than one element and if $x = \inf A$, then the condition is not satisfied because $x \downarrow y = \inf A$ for every y. If A has more than one element and $x \neq \inf A$, it must be that, for every y other than $\inf A$, there is a z such that $z \lessdot x$, $z \lessdot y$, $z \neq \inf A$. This condition is easily verified to be sufficient, but the inverse is trivial and cannot exist for each x.

11. Just recall the definitions.

13. Hints. Given: $A = \{a_j\}_{1 \leq j \leq n}$, $a_j < a_{j+1}$ (and the consequences via reflexivity and transitivity). (a) $a_1 = \inf A$, $a_n = \sup A$. (c) If $a_h \downarrow a_k = a_1$ and $a_h \uparrow a_k = a_n$, then $\{h, k\} = \{1, n\}$ and $n \leq 2$. Etc.

15. (a) Boolean algebra; (c) Boolean algebra; (e) not a lattice; (g) bounded lattice; (i) bounded distributive lattice.

17. (a) $b \to z$, distributive and complemented.
(c) $b \to c \to d \to z$, $b \to d \to c \to z$, distributive;

$$\begin{array}{ccc} b & \to & c \\ \downarrow & & \downarrow \\ d & \to & z \end{array}$$
, distributive and complemented.

Exercises 10.9

1. {a, b, d}, {c, e}.

3. $S := R^{[r]}$, $x \perp y$ or $x = y$. Classes: $\{(x, y\} \mid x \perp y\}$.

5. i_A, $\{\{x\} \mid x \in A\}$. u_A, $\{A\}$. $\varnothing$ is an equivalence relation if $A = \varnothing$.

7. E.g.: $\subseteq$ on $\mathcal{P}\{a, b\}$,

	∅	{a}	{b}	{a,b}
∅	1	1	1	1
{a}	0	1	0	1
{b}	0	0	1	1
{a,b}	0	0	0	1

i on {a, b, c, d}

	a	b	c	d
a	1	0	0	0
b	0	1	0	0
c	0	0	1	0
d	0	0	0	1

"Visual" rules. REFLEXIVITY: the main diagonal contains all 1s. SYMMETRY: mirror symmetry about the main diagonal. TRANSITIVITY: if two 1s and a "vertex" • on the main diagonal are placed as in the figure, then the fourth vertex (?) of the "rectangle" is a 1.

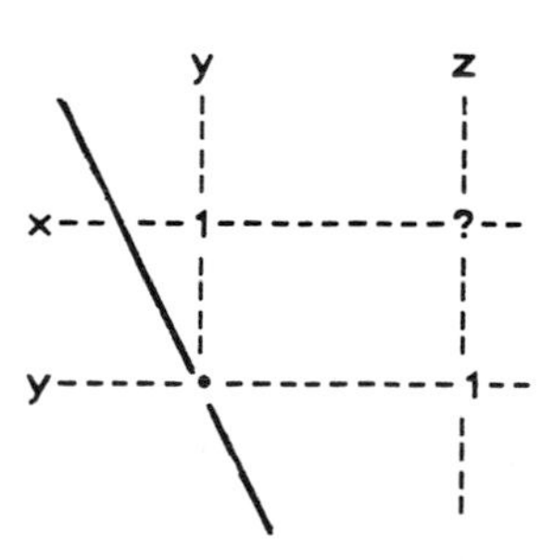

9. All closures equal R. $A/R = \{\{\text{even integers}\}, \{\text{odd integers}\}\}$.

11. Hint. A reasonable procedure is:
Start from a list of the elements of A, all unmarked.
Scan the pairs of R. For each of them: if the two elements of the pair are both unmarked, mark both of them with a new mark; if one is marked, mark the other with the same mark; if both are marked, but with different marks, change one of the two marks to the other in all elements of A.
At the end, mark with all different marks those elements that may have been left unmarked.
This method constructs the equivalence classes of the rst closure of the given relation; these are the equivalence classes if the relation was in fact an equivalence relation.
(c) The equivalence classes are the sets of elements with the same mark. (d) A Hasse diagram is obtained by drawing arrows joining, in any one order, the elements of each one of the equivalence classes. (b) $R^{[rst]}$ is the relation "x and y have the same mark." (a) R is an equivalence relation if all pairs with the same mark appear in R.

Complements 10.9

1. The matrix

of	is	which equals
$R^{[r]}$	$B \bigvee B^0$	same
$R^{[s]}$	$B \bigvee B^{-1}$	same
$B^{[t]}$	$\bigvee_{k>0} B^k$	$\bigvee_{0<k<\#A} B^k$
$R^{[rs]}$	$B \bigvee B^0 \bigvee B^{-1}$	same
$R^{[rt]}$	$\bigvee_{k\geq 0} B^k$	$\bigvee_{0\leq k<\#A} B^k$
$R^{[st]}$	$\bigvee_{k\in \mathbb{Z}-\{0\}} B^k$	$\bigvee_{0<\lvert k\rvert<\#A} B^k$
$R^{[rst]}$	$\bigvee_{k\in \mathbb{Z}} B^k$	$\bigvee_{\lvert k\rvert<\#A} B^k$

Exercises 10.10

1. (00000)
00001→00010→00100→01000→10000
00011→00101→00110→01001→01010→01100→10001→10010→10100→11000
00111→01011→01101→01110→10011→10101→10110→11001→11010→11100
01111→10111→11011→11101→11110
(11111).

3. Just apply the definitions.

Complements 10.10

1. Hints. Reflexivity and symmetry are clear. Transitivity follows from $\#(A \cup B) \leqslant \#A + \#B$. If $f \sim g$ and if f and g differ at $x_1, \ldots, x_m$, let $f_0 := f, f_{m+1} := g, f_k x := fx$ if $x > x_k$, $f_k x := gx$ if $x \leqslant x_k$. Then $f_k R f_{k+1}$, $f R^{m+1} g$, $\sim \subseteq R^{[t]}$. The other parts are immediate.

3. Patient verification.

Exercises 10.11

1. Hints, Reflexivity: $x+y=y+x$. Symmetry: if $x+v=y+u$, then $y+u=x+v$. Transitivity: if $x+v=y+u$ and if $u+b=v+a$, then $x+b=y+a$.

3-10. Just apply the definitions.

11. (a) $-3+5i$; (c) $(4/29)+(19/29)i$, (e) $-i$, (g) i, (j) $-i$, (l) $i^n = 1, i, -1, -i$ according as $n = 0, 1, 2, 3 \bmod 4$, respectively.

13. Direct verification.

Chapter 11

Exercises 11.1

1. Figure 11.1: $V = \{1, 2, 3, 4\}$, $A = \{a, \ldots, g\}$, $\mathrm{b} = \{$
$((1, 2), a)$, $((2, 1), a)$, $((1, 4), e)$, $((4, 1), e)$,
$((1, 2), b)$, $((2, 1), b)$, $((2, 4), f)$, $((4, 2), f)$,
$((1, 3), c)$, $((3, 1), c)$, $((3, 4), g)$, $((4, 3), g)$,
$((1, 3), d)$, $((3, 1), d)\}$.
Figure 11.3a: $V = \varnothing$, $A = \varnothing$, $\mathrm{b} = \varnothing$.
Figure 11.3c: $V = \{1\}$, $A = \{a, b\}$, $\mathrm{b} = \{((1, 1), a), ((1, 1), b)\}$.
Figure 11.4a: $V = \{1, 2, 3\}$, $A = \{a, b, c\}$, $\mathrm{b} = \{((1, 2), c), ((2, 1), c), ((2, 3), a), ((3, 2), a), ((3, 1), b), ((1, 3), b)\}$.
Figure 11.4c: $V = \{$A, B, C, D, E$\}$, $A = \{$2-combinations of $\{$A, B, C, D, E$\}\}$, $\mathrm{b} = \{$pairs (2-permutation, respective 2-combination)$\}$.
Figure 11.5c: $V = \{1, 2, 3, 4, 5, 6\}$, $A = \{$2-combinations $x \leqslant 4$ $y > 4\}$, $\mathrm{b} = \{$pairs (2-permutation $x \leqslant 4$ $y > 4$, respective 2-combination)$\}$.

3. Figure 11.3c: one loop after the other.
Figure 11.5c: e.g., start from node bottom left and—at each step—take the "highest" arc not yet taken.

Complements 11.1

1. Hints. In the first part, just use the definitions. An undirected (directed) graph is a relation on V if and only if each unordered (ordered) pair of nodes admits at most one arc.

Exercises 11.2

1. E.g.

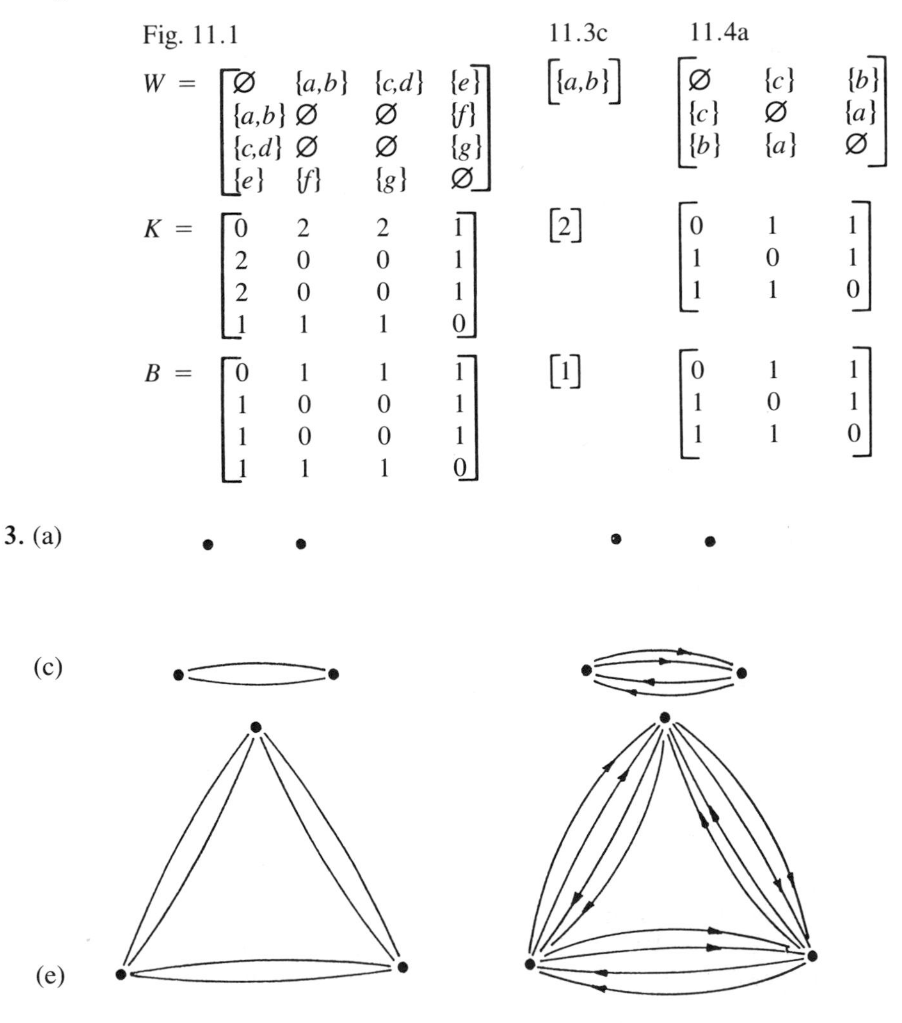

Fig. 11.1

$$W = \begin{bmatrix} \varnothing & \{a,b\} & \{c,d\} & \{e\} \\ \{a,b\} & \varnothing & \varnothing & \{f\} \\ \{c,d\} & \varnothing & \varnothing & \{g\} \\ \{e\} & \{f\} & \{g\} & \varnothing \end{bmatrix} \quad K = \begin{bmatrix} 0 & 2 & 2 & 1 \\ 2 & 0 & 0 & 1 \\ 2 & 0 & 0 & 1 \\ 1 & 1 & 1 & 0 \end{bmatrix} \quad B = \begin{bmatrix} 0 & 1 & 1 & 1 \\ 1 & 0 & 0 & 1 \\ 1 & 0 & 0 & 1 \\ 1 & 1 & 1 & 0 \end{bmatrix}$$

11.3c

$$W = \begin{bmatrix} \{a,b\} \end{bmatrix} \quad K = \begin{bmatrix} 2 \end{bmatrix} \quad B = \begin{bmatrix} 1 \end{bmatrix}$$

11.4a

$$W = \begin{bmatrix} \varnothing & \{c\} & \{b\} \\ \{c\} & \varnothing & \{a\} \\ \{b\} & \{a\} & \varnothing \end{bmatrix} \quad K = \begin{bmatrix} 0 & 1 & 1 \\ 1 & 0 & 1 \\ 1 & 1 & 0 \end{bmatrix} \quad B = \begin{bmatrix} 0 & 1 & 1 \\ 1 & 0 & 1 \\ 1 & 1 & 0 \end{bmatrix}$$

3. (a)

(c)

(e)

5. For an undirected graph, each element in entry (i, j) of the matrix appears also in entry (j, i) and only in this entry. For a directed graph, each element of entry (i, j) appears in that entry only.

Exercises 11.3

1. E.g. Figure 11.1, from node 1 to node 3: 1*afg*, 1*bfed*.

3. (a) (1) 1*bcda*, 4*cdab*, 3*dabc*, 2*abcd*; (2) 1*bcda*; (3) none. (c) (1) 1, 1 loop, 2 loop, 2 other loop; (2) 1 *b* loop loop *a*, 1 *b* loop other loop *a*, 1 *b* other loop loop *a*, 1 *b* other loop other loop *a*; (3) the last four paths and 1 *b* loop *a*, 1 *b* other loop *a*, 1 loop *b a*, 1 loop loop *b a*, 1 loop *b* loop *a*, 1 loop *b* other loop *a*, 1 loop up right, 1 loop loop up right.

Complements 11.3

1. Just recall the definitions.

3. Hints. (All indices are from 1 to n.) Let the temporary symbol $i[h]k$ denote the truth value of "there exists a path from node i to node k whose intermediate nodes have indices not exceeding h". Proceed by induction on j. Initially, $B^*(i,k)$ $(i \neq k)$ denotes the existence of an arc from i to k, that is, $B^*(i,k) = i[0]k$. Prove that, after column j is scanned, $B^*(i,k) = i[j]k$. By the hypothesis of induction, $B^*(i,k) = i[j-1]k$ before column j is scanned. If $B^*(i,j) = 1$ and if $B^*(j,k) = 1$, then $B^*(i,k)$ is set to be 1; this occurs if $i[j-1]j$ and $j[j-1]k$, which imply $i[j]k$; if $B^*(i,k)$ was 1, then it is unaltered, reflecting the fact that $i[j-1]k$ implies $i[j]k$. Conversely, if $i[j]k$, then there is a path from i to k that either (a) does not contain j as an intermediate node or (b) can be reduced (see the process before Theorem 7) to a path containing j just once. Then (a) $i[j-1]k$ or (b) $i[j-1]j$ and $j[j-1]k$. In either case, scanning column j sets $B^*(i,k)$ equal to $i[j]k$. The induction implies that $B^{\#}(i,k) = i[n-1]k$ $(i \neq k)$, which is the same as the existence of any path from i to k (see Theorem 7).

Exercises 11.4

1. E.g. Figure 11.4c: each node has degree 4.

Complements 11.4

1. Immediate. Then see complement 3.

3. Hints. Proving equivalence is a direct verification. Statements (a), (b), and (c) are consequences of the fact that free insertion and deletion do not change the nodes of degree other than 2 and do not essentially alter the paths that contain nodes of degree 2 (but not as end nodes). An additional invariant is, for instance, $\#V - \#A$ because an insertion increases by 1 both $\#V$ and $\#A$, and a deletion decreases both by 1.

Exercises 11.5

1. E.g. Each of the graphs in Figures 11.1a and 11.2a is connected, Figure 11.3d is not.

3.

		(i) number of		(ii) number of nodes of degree			
		nodes	arcs	0	1	3	4
(a)	(1)	5	5		2	2	
	(2)	5	5		2	2	
(c)	(1)	2	1		2		
	(2)	2	2				
	(3)	4	3		2		
	(4)	2	1		2		
	(5)	6	5		4	2	
	(6)	6	5		4	2	
	(7)	3	2		2		
(e)		6	9			6	

Exercises 11.6

1. (a) The third path is critical and gives 23 hours. (c) Three paths. The critical path gives 23 hours. (e) Three paths. The critical path gives 26 hours.

Exercises 11.7

1. E.g. the pentagram:

3. **ART**

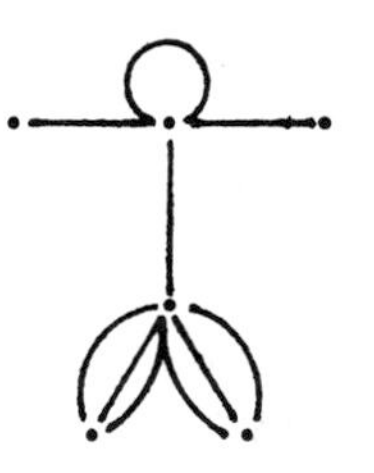

The least number of arcs is 3:

5. E.g. 00 01 11 12 22 23 33 34 44 45 55 56 66 60 02 24 46 61 13 35 50 03 36 62 25 51 14 40

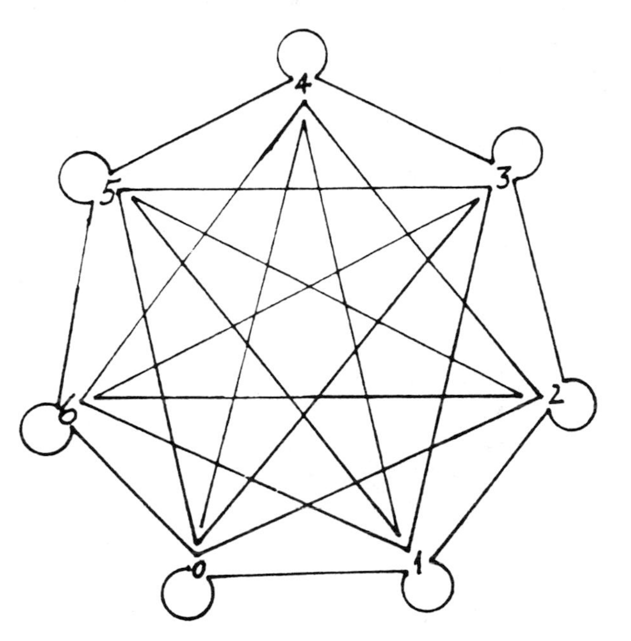

If the numbers are 1 to 6, then there is no solution; if 2 to 6, there is one.

7. Squares := nodes, separations := arcs. No for 8×8, because there are 24 odd nodes (the edge squares other than the corner squares have degree 3). To eliminate all or all but two of such edge squares we need sizes 2×3, 2×2, or $1 \times n$, all of which have a solution.

Complements 11.7

1. Hint. Adjoin m arcs initially. Then proceed as in the text (case $m = 0$). Then permute and delete the m extra arcs (no two of which are consecutive).

Chapter 12

Exercises 12.1.1

1. 2, 3, 4, 30, 10, 5; $2+3$; $2+3\times4$; $(2+3\times4)$; $30-(2+3\times4)$; $30-(2+3\times4)+10$, $30-(2+3\times4)-5$; $(30-(2+3\times4)+10)$, $(30-(2+3\times4)-5)$; $(30-(2+3\times4)+10)+(30-(2+3\times4)-5)$.

3. Hints. U = alphabet whose letters denote subsets, $S = \{\cup, \cap, \sim\}$. In Definition 9.5 (*) replace unary $+\ -$ with $\sim$, and replace binary $+\ -\ \cdot\ /$ with $\cup\cap$. Ambiguous expression: $x \cup y \cap z$. In Definition 1 replace as in (*) above.

Exercises 12.1.2

1. (a)

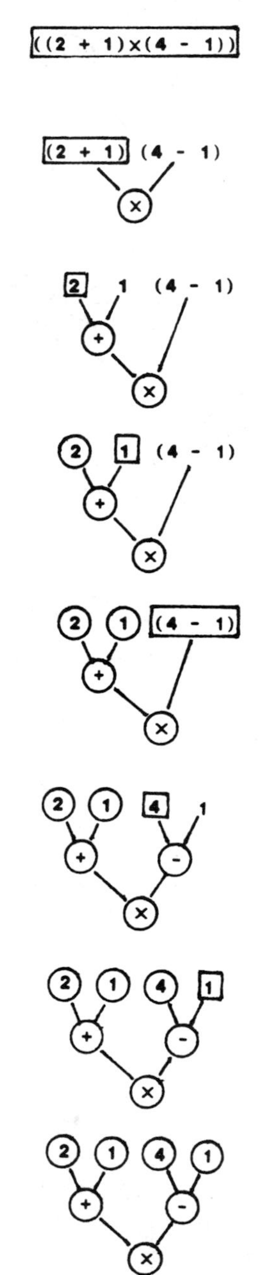

3. Hint. In Definition 3 and in the algorithm, interpret unary $\oplus$ as $\sim$ and binary $\oplus$ as $\cup\cap$.

Exercises 12.1.3

1. E.g. postfix:
   (i) 30 2 3+4×−10+30 2 3+4×−5−+;
   (iii) 30 2 3+4×−10+30 2 3+4×5−−+;
   (v) 30 2 3+4×10+−30 2 3+4×−5−+;
   (vii) 30 2 3+4×10+−30 2 3+4×5−−+.

3. Hints. In Theorem 2, in Definition 3, and in the algorithm, replace $(\oplus v_1)$ with $v_1\oplus$ and replace $(v_1\oplus v_2)$ with $v_1v_2\oplus$.

Exercises 12.1.4

1. E.g. see problem 1 in Exercises 12.1.3.

3. (a) 2 3−4−; (c) 2 3 4+−; (e) $xy \vee \sim 1 \wedge$; (g) $xy \vee 1 \wedge \sim$.

5. (a) 2 3+4−, ((2+3)−4); 2 3−4+; ((2−3)+4); 2 3 4−−, (2−(3−4)); 2 3 4+−, (2−(3+4)). (c) $xy+z/*$, $((x+y)*(/z))$; $xy+z*$, $((x+y)*z)$; $xy*z/*/$, $(/((x*y)*(/z)))$; $xy/z+*$, $(x*((/y)+z))$.

7. Hints. a. In Definition 1 add rule (2′) (similar to 2): if v is in L, then (v) is in L. In Theorem 2 add case (2′) (similar to 2): ... such that $v=(v_1)$. In Definition 3 add rule (2′) (similar to 2): if $v_1=(v_2)$, then ...; (and the second graphic symbol is ⓞ $\rightarrow v_2$). The infix part after Definition 6 requires a rule (2′) (similar to 2) and a slight graphical change: if label$(q)=()$, then $t(q):=(t(s_1))$.
   c. In case (ii), isomorphism is preserved (also with W, if nodes ⓞ are introduced). In case (i), we have a situation similar to homology in graphs (see Complements 11.4). Each expression or tree is replaced by a class of expressions or trees. The elements of the same class are obtained from each other by a finite number of deletions or insertions of: enclosure for prefix or postfix expressions, (...) for infix expressions, nodes ⓞ for trees. The insertion can be done at any position except: last for prefix expressions, first for postfix expressions, when (...) would not contain a subexpression for infix expressions, when ⓞ would be a leaf in a tree. There is still isomorphism between the respective sets of such equivalence classes.

Complements 12.1.4

1. Hints. An n-ary operation from $C_1, \ldots, C_n$ to C_k is a function of $C_1 \times \ldots \times C_n \rightarrow C_k$. Set Y must be the union of sets Y_1 ({symbols for elements of C_1}), ..., Y_n ({symbols for elements of C_n}), or—alternatively—Y is a set of pairs (symbol, set C_j to which the symbol refers); this is the "mode" used in computer languages. Whenever an n-ary operation has arguments $v_1, \ldots, v_n$, they must refer to the appropriate sets $C_1, \ldots, C_n$. Similar remarks are valid for operators; for instance, $x+y$ means different operations if x and y are numbers or sets.

Exercises 12.1.5

1. (a) −7, (c) 0. (e) 0, (g) 00, (i) 11.

Exercises 12.2

1. E.g. (c) Leaves: open dots.

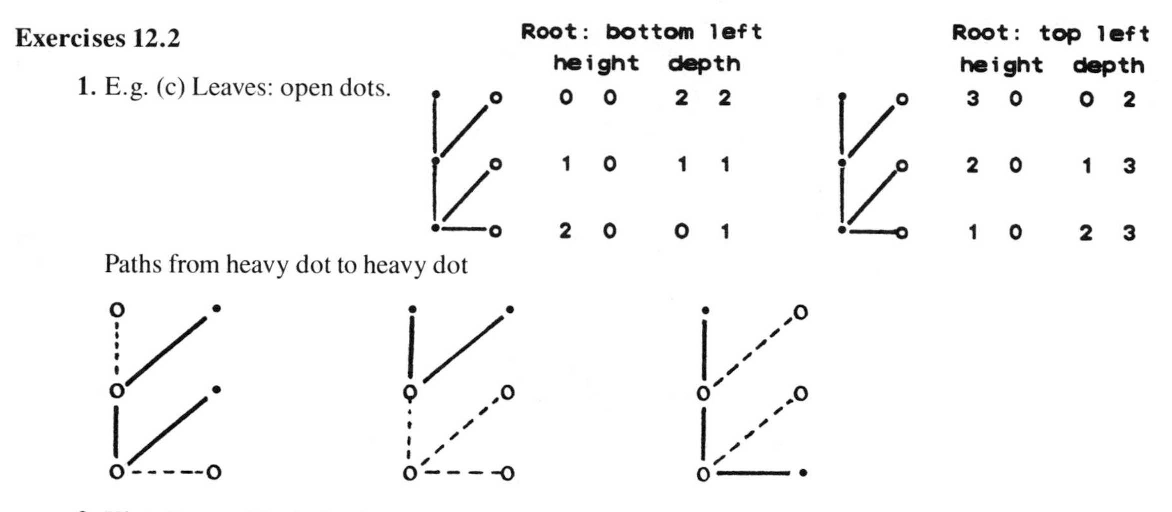

Paths from heavy dot to heavy dot

3. Hint. Proceed by induction.

Complements 12.2

1. Direct verification.

Exercises 12.3

1. E.g.: Given list: D1, S3, C4, F5, B6. Sort

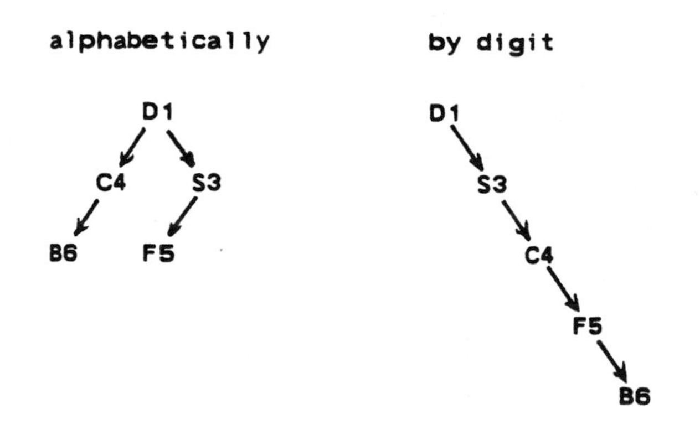

3. See second part of problem 1 above.

5.

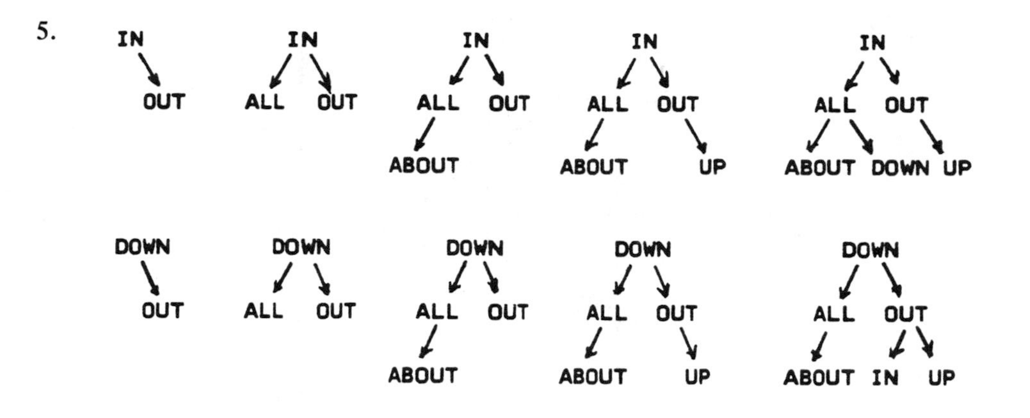

Exercises 12.4

1. For parts (i) and (v) see problem 3 in Exercises 11.5.

	(ii) c	(iii) to obtain a spanning tree	(iv) fundamental circuits
(a) (1)	1	skip the cross bar	the triangle
(2)	1	skip the curved arc	the upper part
(c) (1)	0	whole graph	∅
(2)	1	skip the curved arc	whole graph
(3)	0	whole graph	∅
(4)	0	whole graph	∅
(5)	0	whole graph	∅
(6)	0	whole graph	∅
(7)	0	whole graph	∅
(e)	4	path with nodes 123456	circuits with nodes 41234, 6123456, 52345, 63456

3.

	arcs of the component	(i) numbers of n a	(ii) c	(iii) arcs of a spanning tree	(iv) arcs of a set of fundamental circuits	(v) degree, # of nodes, $d =$	$\# =$
(a) (1)	1 2 3 4 5 6	5 6	2	1 2 4 5	1 3	1	2
					6	3	1
						5	1
(2)	7 8 9	4 3	0	7 8 9	∅	1	3
						3	1
(c) (1)	1	1 1	1	node only	1		
(2)	2 3	1 2	2	node only	2	4	1
					3		
(3)	4 5 6	1 3	3	node only	4	6	1
					5		
					6		
(4)	7 8 9 10	1 4	4	node only	7	8	1
					8		
					9		
					10		

5. Hint. If not arc-simple, there is a repeated arc and at least one of its endpoints is a repeated node.

Complements 12.4

1. Insertion: if a column has two 1s, replace it with two columns, each one containing one of the two 1s. If a column has a 2, replace it with two columns with 1s in the row of the 2. In either case, add a row with two 1s in the new columns. Deletion: delete a row that contains just two 1s; replace the columns of those two 1s with their sum.

Exercises 12.5

1. Hint. Refer to the figure below. The circuit 123451 results in a simple closed curve in the plane. Arc 35 must be inside or outside, arcs 14 and 24 must be on the other side, arc 13 must be on the same side as 35; then arc 25 must go from one side to the other with respect to the circuit 1341. The graph can be realized on a donut as in the figure below.

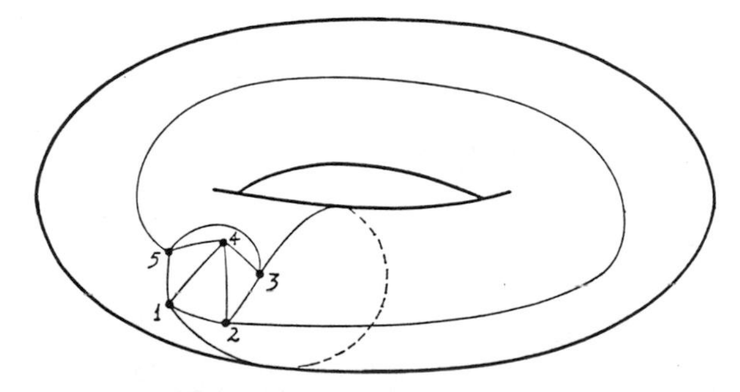

Complements 12.5

1. See hint in the text.

Chapter 13

Exercises 13.1

1. (a) 1 011 000 0 101 011 1 011 011 binary, 130 053 131 octal. (c) 1 100 001 1 101 110 1 111 001 0 100 000 1 110 000 1 101 000 1 110 010 1 100 001 1 110 011 1 100 101 binary, 141 156 171 040 160 150 162 141 163 145 octal.

3. E.g., in elevators, floors 1, 2, and 3 are coded, respectively:

Exercises 13.2

1. 111 000 000 111 111 111 111 111 000 000 111 111 111 000 111 000 000 000 111 000 111 000 111 000 000 000 000 000 111 000 000 000 000 111 000 111 000 000 111 000 000 111 111 000 111 000 111 000 000.

3. (a) If the received word contains five xs, decode it as x; otherwise, declare error. (c) If the received word contains at least three xs, decode it as x; otherwise, declare error. The decision will be correct if the channel introduces fewer than (a) 5 (c) 3 errors.

5. Just construct the code.

7. If a received word is not a code word, declare error. If there are no more than two errors, the decision is correct, because (see text) no word differs from a code word by more than one bit and therefore no two code words differ by less than two bits.

Complements 13.2

1. Hints. By brute force: $n=1$, $a=1$; $n=2$, $a=4$ (square); $n=3$, $a=12$ (cube). In general, repeat the previous graph, then draw 2^n connecting arcs: $f(n+1)=2f(n)+2^n$. Then: $n=4$, $a=12\cdot 2+8=32$; $n=5$, $a=32\cdot 2+16=80$; $n=6$, $a=80\cdot 2+32=192$; $n=7$, $a=192\cdot 2+64=448$.
   Or use the same idea and induction to prove the general formula $f(n)=n\cdot 2^{n-1}$ (Proof: $f(n+1)=2f(n)+2^n=2\cdot n\cdot 2^{n-1}+2^n=(n+1)2^n$).

Exercises 13.3

1.

	n	k	G	$\#C$	$\#(\{0, 1\}^n)$
(a)	3	1	11 1	2	8
(c)	7	4	011 1000 101 0100 110 0010 111 0001	16	128
(e)	4	1	111 1	2	16
(g)	4	2	01 10 01 01	4	16

3. $\#(\text{class of } 0) = \#C = 2^k$. Any one class is obtained from C by adding a given n-word v; that is, if x is in C, then $x + v$ is in the new class. Then $\mathbf{syn}(x + v) = (x + v)H = xH + vH = vH$, the same for the class. $\#(\text{any class}) = 2^k$. $\#\{\text{classes}\} = \#\text{range} = \#(\{0, 1\}^{n-k}) = 2^{n-k}$. Total number $= (\#$ any class$) \cdot (\#\{\text{classes}\}) = 2^n$, the same as $\#\{\text{all words}\}$.

Complements 13.3

1. The first $n - k$ rows of H form I_{n-k}. Any of the other rows can be obtained as a sum of rows among the first $n - k$. Then there are $n - k + 1$ rows whose sum is 0. Then $r \leqslant n - k + 1$, $r = d \leqslant n - k + 1$. If $d = n - k + 1$, then no $n - k$ (or fewer) rows have sum 0. Let v be a row of H beyond the first $n - k$. If v is other than $11\ldots1$, then it is the sum of fewer than $n - k$ rows of I; if we add 1, we have $n - k$ or fewer rows of H whose sum is 0, which cannot be, because $n - k < d$. Then there is at most one more row in H, and it must be all 1s. Then the only codes are those for which $H = I_{n-k}$ or H is obtained by adjoining to I_{n-k} a row of 1s. Let $f(h) := \sum_{j=0}^{h} C(n, j)$. Then $1 = f(0) < f(1) < f(2) < \ldots < f(n) = f(n+1) = \ldots = 2^n$. Then there is a specific value h for I, such that $f(I - 1) < 2^{n-k} \leqslant f(I)$; then $d/2$ cannot exceed h.

Exercises 13.5

1. Error-correcting: Select p such that $2p < d$. Construct V, L. If $\mathbf{syn}\,y \in V$, decide that the code word was $x := y + L^{-1}\mathbf{syn}\,y$. The decision is correct if there are fewer than $d/2$ errors. Error detecting: Select q such that $q < d$. If $\mathbf{syn}\,y = 0$, decide that the code word was $x := y$; otherwise, declare error. The decision is correct if there are fewer than d errors.

3. E.g.: $n = 4$, $k = 1$, $d = 4$, $G = [1\ 1\ 1\ 1]$, $K = [1\ 1\ 1]$,

$$H = \begin{bmatrix} 1 & 0 & 0 \\ 0 & 1 & 0 \\ 0 & 0 & 1 \\ 1 & 1 & 1 \end{bmatrix}$$

p	q	A	L →	V	info	x	y	syn$y$	decoded	check
1	1	1000		100	1	1111	1101	001	1111	y
		0100		010			1001	011	err	y
		0010		001			1000	100	0000	n
		0001		111	–	----	1010	101	err	–
		0000		000	–	----	1011	010	1111	–

Exercises 13.6

1. $m = 1$, K and G are 0×1, $H = [1]$, $C = \{0\}$ (the zero code of length 1), $d = 0$.
   $m = 2$, $K = [1\ 1]$, $G = [1\ 1\ 1]$,

   $$H = \begin{bmatrix} 1 & 0 \\ 0 & 1 \\ 1 & 1 \end{bmatrix}$$

   $C = \{000, 111\}$ (the repetition code of length 3), $d = 3$.

3. Guidelines. (1) Check $0 \leq m \leq ?$ (depending on the computer). (2) $n := 2^m - 1$, $k := n - m$. (3) The k rows of K are the m-digit binary numerals with at least two 1s, enumerated in any one order. (4) The first m rows of H contain the $m \times m$ identity matrix, rows from $m+1$ to $m+k$ are the same as rows 1 to k of K. (5) The first k columns of G are the same as K, columns $k+1$ to $k+m$ are the same as I_k.

Exercises 13.7

1. Parity check over Example 1. $C = \{0000, 1111\}$. See problem 3 of Exercises 13.5.
   Parity check over Example 3. $C = \{0000, 0011, 0110, 0101\}$.

n	k	d	G	K	H
					10
					01
2	2	2	01 10	01	01
			01 01	01	01

Parity check over Example 4. $C = \{0000, 0011, 0110, 0101\}$.

n	k	d	G	K	H
					1000
					0100
					0010
					0001
7	4	3	1011 1000	1011	1011
			1101 0100	1101	1101
			1110 0010	1110	1110
			0111 0001	0111	0111

Exercises 13.8

1. Example (4)

$$G = \begin{bmatrix} 0 & 1 & 1 & 1 & 1 & & \\ 1 & 0 & 1 & 1 & & 1 & \\ 1 & 1 & 0 & 1 & & & 1 \end{bmatrix} \qquad H = \begin{bmatrix} 1 & & & \\ & 1 & & \\ & & 1 & \\ & & & 1 \\ 0 & 1 & 1 & 1 \\ 1 & 0 & 1 & 1 \\ 1 & 1 & 0 & 1 \end{bmatrix}$$

Length $= 7$, dimension $= 3$, $d = 4$.

Example (5) G is $0 \times n$, $H = I_n$, length $= n$, dimension $= 0$, $d = 0$.

Example (10)

$$G = \begin{bmatrix} 00 & 10 \\ 11 & 01 \end{bmatrix} \qquad H = \begin{bmatrix} 10 \\ 01 \\ 00 \\ 11 \end{bmatrix}$$

Length $= 4$, dimension $= 2$, $d = 1$.

Complements 13.8

1. Hints. Let $y := [x_{n-k+1} \ldots x_n x_1 \ldots x_{n-k}] := [w\ t]$.
By part 2 of Theorem 4, applied to C^*, we have that y is in C^* iff $w = tK^{\mathrm{T}}$. These equations are equivalent to $x_{n-k+h} = \sum_{j=1}^{n-k} x_j g_{hj}$ $(1 \leqslant h \leqslant k)$ (recall that the entries of K are g_{hj} when $1 \leqslant h \leqslant k$ and $1 \leqslant j \leqslant n-k$, and that the superscript ^{T} denotes transposition). The last equations are equivalent to (*) $\sum_{i=1}^{n} x_i g_{hi} = 0$ $(1 \leqslant h \leqslant k)$ because the right part of G is the identity matrix. Adding together any number of words $(g_{hi})_{1 \leqslant i \leqslant n}$ gives all the words z of C; consequently, adding together any number of equations (*) gives equations $\Sigma_i x_i z_i = 0$, as we wanted to show.

Exercises 13.9

1. In Examples 4, 5, no reduction is possible. In Example 10, the first position can be deleted. In the results of problem 1 of Exercises 13.8 (the duals of Examples 4, 5, 10), no reduction of the type described in the text is possible (but note that the first two columns are equal in the dual of Example 10).

3. Hints. Since a deleted column must be one of the first $n-k$, the $k \times k$ identity matrix is unaltered and so is the dimension. Since the deletion of a 0 does not alter the weights, the distance is unaltered.

Exercises 13.10

1. $K^{**} = [11], G^{**} = [11\ \ 1], G^{*} = [111\ \ 1], G = \begin{bmatrix} 1\ 100 \\ 1\ 010 \\ 1\ 001 \end{bmatrix}$

   $C*$ is the 4-repetition code. C is the parity check code of length 4.

3.

$$H^{*} = \begin{bmatrix} 1 & & & & \\ & 1 & & & \\ & & 1 & & \\ & & & 1 & \\ & & & & 1 \\ 1 & 1 & 1 & 0 & 0 \\ 1 & 1 & 0 & 1 & 0 \\ 1 & 1 & 0 & 0 & 1 \\ 1 & 0 & 0 & 1 & 1 \\ 1 & 0 & 1 & 0 & 1 \\ 1 & 0 & 1 & 1 & 0 \\ 0 & 0 & 1 & 1 & 1 \\ 0 & 1 & 0 & 1 & 1 \\ 0 & 1 & 1 & 0 & 1 \\ 0 & 1 & 1 & 1 & 0 \\ 1 & 1 & 1 & 1 & 1 \end{bmatrix}$$

$$G^{*} = \begin{bmatrix} 1\ 1100 & 1 & & & & & & & & & & \\ 1\ 1010 & & 1 & & & & & & & & & \\ 1\ 1001 & & & 1 & & & & & & & & \\ 1\ 0011 & & & & 1 & & & & & & & \\ 1\ 0101 & & & & & 1 & & & & & & \\ 1\ 0110 & & & & & & 1 & & & & & \\ 0\ 0111 & & & & & & & 1 & & & & \\ 0\ 1011 & & & & & & & & 1 & & & \\ 0\ 1101 & & & & & & & & & 1 & & \\ 0\ 1110 & & & & & & & & & & 1 & \\ 1\ 1111 & & & & & & & & & & & 1 \end{bmatrix}$$

$k = 11, d = 4.$

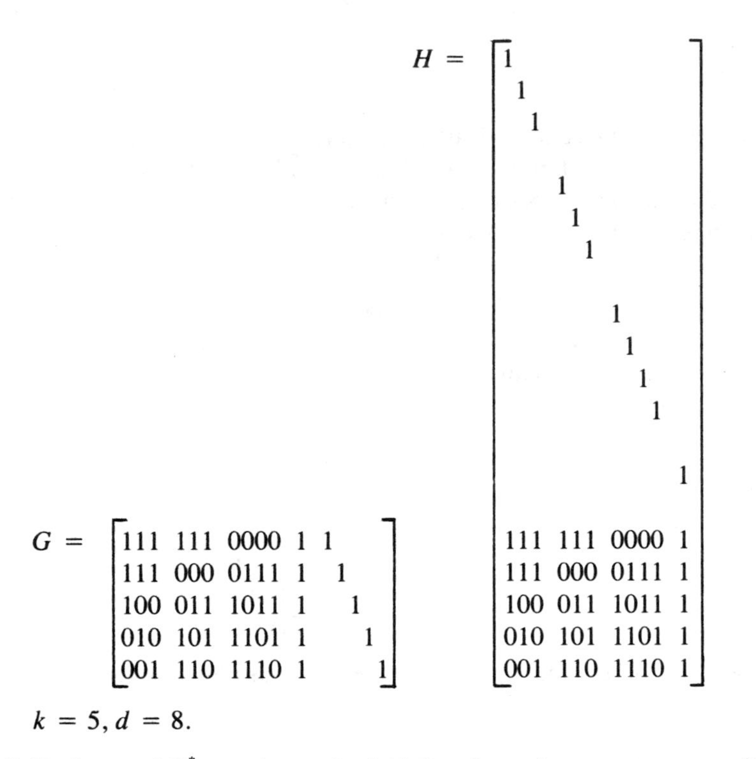

$k = 5, d = 8.$

5. Each row of G^* contains: a check bit in column 1, an m-sequence of h 1s and $m-h$ 0s $(2 \leq h \leq m)$ in columns 2 to $m+1$, a single 1 in the last k columns $(k = 2^m - m - 1)$. The weight of each row is even and is $h+1$ or $h+2$. When $h=2$, the weight is 4. Then $d \leq 4$. We have to exclude the possibility of a code word of weight less than 4; since code words are even, we have to exclude just weight 2.
   Adding three or more rows of G^* gives three or more 1s in the last k columns and the weight is more than 2. Adding two rows gives two 1s in the last k columns; since the two sequences in columns 2 to m are different from each other, their sum contributes at least one more 1. Then the weight is more than 2.

Exercises 13.11

1. E.g., Examples 4 and 10:

(4)
$$G = [11 \ \ 1] \qquad H = \begin{bmatrix} 10 \\ 01 \\ 11 \end{bmatrix} \qquad G^* = [11 \ \ 1] \qquad H^* = \begin{bmatrix} 01 \\ 10 \\ 11 \end{bmatrix}$$

(10)
$$G = \begin{bmatrix} 01 & 10 \\ 01 & 01 \end{bmatrix} \qquad H = \begin{bmatrix} 10 \\ 01 \\ 01 \\ 01 \end{bmatrix} \qquad G^* = \begin{bmatrix} 01 & 10 \\ 11 & 00 \end{bmatrix} \qquad H^* = \begin{bmatrix} 01 \\ 01 \\ 01 \\ 10 \end{bmatrix}$$

Chapter 14

Exercises 14.1

1. $M_3(p, q, r) = (p \wedge q) \vee (p \wedge r) \vee (q \wedge r)$, $M_5(p, q, r, s, t) = (r \wedge s \wedge t) \vee (q \wedge s \wedge t) \vee (q \wedge r \wedge t) \vee (q \wedge r \wedge s) \vee (p \wedge s \wedge t) \vee (p \wedge r \wedge t) \vee (p \wedge r \wedge s) \vee (p \wedge q \wedge t) \vee (p \wedge q \wedge s) \vee (p \wedge q \wedge r)$.
   n odd: construct the $C(n, (n+1)/2)$ elements of $\{0, 1\}^n$ that contain $(n+1)/2$ 1s; for each of these strings, AND together the arguments in the positions occupied by 1s, then OR together the terms so obtained.

3. $f(p, q, \ldots) := 0$ for every choice of $p, q, \ldots$.

5. E.g., see formulae 1 and 2; or $p \wedge q$ and $(p \wedge q \wedge r) \vee (p \wedge q \wedge \sim r)$.

Exercises 14.2

1.

p	q	$\sim p$	$\sim q$	$p \wedge \sim q$	$p \wedge q$	(a)	(b)	(c)	(d)
1	1	0	0	0	1	1	1	1	1
1	0	0	1	1	0	0	0	0	0
0	1	1	0	0	0	1	1	1	1
0	0	1	1	0	0	1	1	1	1

or: (b) $= (\sim p) \vee (\sim\sim q) =$ (a), (c) $= ((\sim p) \vee p) \wedge ((\sim p) \vee q) = 1 \wedge ((\sim p) \vee q) =$ (a), (d) $= (\sim p) \vee q$ by one of the definitions ($=$ (a)).

3. (a)

p	q	r	$q \Rightarrow r$	$p \Rightarrow q$	$(q \Rightarrow r) \wedge (p \Rightarrow q)$	$q \Rightarrow r$	(a)
1	1	1	1	1	1	1	1
1	1	0	0	0	0	0	1
1	0	1	1	0	0	1	1
1	0	0	1	0	0	1	1
0	1	1	1	1	1	1	1
0	1	0	0	1	0	0	1
0	0	1	1	1	1	1	1
0	0	0	1	1	1	1	1

5. $f = (p \wedge q \wedge r) \vee (p \wedge q \wedge \sim r) \vee (p \wedge \sim q \wedge \sim r)$, $g = (p \wedge q \wedge \sim r) \vee (p \wedge \sim q \wedge r) \vee (p \wedge \sim q \wedge \sim r) \vee (\sim p \wedge \sim q \wedge r) \vee (\sim p \wedge \sim q \wedge r)$, $h = (p \wedge q \wedge r) \vee (\sim p \wedge \sim q \wedge r)$.

7. (a) DNF: $p \wedge \sim q$,
$(p \wedge \sim q \wedge r) \vee (p \wedge \sim q \wedge \sim r)$,
$(p \wedge \sim q \wedge r \wedge s) \vee (p \wedge \sim q \wedge \sim r \wedge s) \vee (p \wedge \sim q \wedge r \wedge \sim s) \vee (p \wedge \sim q \wedge \sim r \wedge \sim s)$;
CNF: $(p \vee q) \wedge (p \vee \sim q) \wedge (\sim p \vee \sim q)$,
$(p \vee q \vee r) \wedge (p \vee \sim q \vee r) \wedge (\sim p \vee \sim q \vee r) \wedge (p \vee q \vee \sim r) \wedge (p \vee \sim q \vee \sim r) \wedge (\sim p \vee \sim q \vee \sim r)$,
$(p \vee q \vee r \vee s) \wedge (p \vee \sim q \vee r \quad s) \wedge (\sim p \vee \sim q \vee r \vee s) \wedge (p \vee q \vee \sim r \vee s) \wedge (p \vee \sim q \vee \sim r \vee s) \wedge (\sim p \vee \sim q \vee \sim r \vee s) \wedge (p \vee q \vee r \vee \sim s) \wedge (p \vee \sim q \vee r \vee \sim s) \wedge (\sim p \vee \sim q \vee r \vee \sim s) \wedge (p \vee q \vee \sim r \vee \sim s) \wedge (p \vee \sim q \vee \sim r \vee \sim s) \wedge (\sim p \vee \sim q \vee \sim r \vee \sim s)$.

(c) DNF: $(p \wedge q) \vee (\sim p \wedge q) \vee (\sim p \wedge \sim q)$,
$(p \wedge q \wedge r) \vee (\sim p \wedge q \wedge r) \vee (\sim p \wedge \sim q \wedge r) \vee (p \wedge q \wedge \sim r) \vee (\sim p \wedge q \wedge \sim r) \vee (\sim p \wedge \sim q \wedge \sim r)$,
$(p \wedge q \wedge r \wedge s) \vee (\sim p \wedge q \wedge r \wedge s) \vee (\sim p \wedge \sim q \wedge r \wedge s) \vee (p \wedge q \wedge \sim r \wedge s) \vee (\sim p \wedge q \wedge \sim r \wedge s) \vee (\sim p \wedge \sim q \wedge \sim r \wedge s) \vee (p \wedge q \wedge r \wedge \sim s) \vee (\sim p \wedge q \wedge r \wedge \sim s) \vee (\sim p \wedge \sim q \wedge r \wedge \sim s) \vee (p \wedge q \wedge \sim r \wedge \sim s) \vee (\sim p \wedge q \wedge \sim r \wedge \sim s) \vee (\sim p \wedge \sim q \wedge \sim r \wedge \sim s)$;
CNF: $\sim p \vee q$,
$(\sim p \vee q \vee r) \wedge (\sim p \vee q \vee \sim r)$,
$(\sim p \vee q \vee r \vee s) \wedge (\sim p \vee q \vee \sim r \vee s) \wedge (\sim p \vee q \vee r \vee \sim s) \wedge (\sim p \vee q \vee \sim r \vee \sim s)$.

9. Hint. Consider f^*: $\bigvee_J f(J) \wedge \min_J$. Then $f^*(I) = \bigvee_J f(J) \wedge \min_J I$. By Theorem 5, the only nonzero term is $f(I) \wedge \min_I I$, which equals $f(I)$. Therefore, $f^*(I) = f(I)$, and the normal form exists. Uniqueness is proved by the fact that $a_I = f(I)$ is a consequence of $\bigvee_J a_J \wedge \min_J(I) = \bigvee_J f(J) \wedge \min_J I$. Dually for conjunctive forms.

Exercises 14.3

1. Hints. Functions $\wedge$ (or $\vee$) and $\sim$ are obtained as shown in the problem. The constants are obtained as $1 = p|(p|p)$, $0 = 1|1$; $0 = p \dagger (p \dagger p)$, $1 = 0 \dagger 0$.

3. Hints. For $\{\sim\}$: $\sim 0 = 1$, $\sim 1 = 0$; a valid sequence of 0, 1, $\sim$, p, q produces the functions 0, 1, p, q, $\sim p$, $\sim q$.
For $\{\vee\}$: $p \vee p = p$, $p \vee 0 = p$, $p \vee 1 = 1$; any valid sequence of 0, 1, $\vee$, p, q produces the functions 0, 1, p, q, $p \vee q$.
For $\{\wedge\}$: By duality, we obtain only 0, 1, p, q, $p \wedge q$.
Then $\{\vee, \wedge\}$ can express only 0, 1, p, q, $p \vee q$, $p \wedge q$.

Complements 14.3

1. $(x \Rightarrow 0) = \sim x$, $((y \Rightarrow x) \Rightarrow x) = (x \vee y)$; then the alphabet $\{\Rightarrow, 0, p, q, \ldots\}$ is sufficient, but $\{\Rightarrow, p, q, \ldots\}$ alone is not, because $(x \Rightarrow x) = 1$, $(1 \Rightarrow x) = x$, $(x \Rightarrow 1) = 1$, and any valid sequence from $\{1, \Rightarrow, p, q, \ldots\}$ produces only the functions 1, p, q, $p \Rightarrow q$, $q \Rightarrow p$.

For the list of all total binary Boolean functions, refer to problem 3 of Complements 6.5. Consider their individual completeness. Functions 0, 1, p, q, $\sim p$, $\sim q$ are clearly not complete. Functions $\vee, \wedge$ are not complete (see problem 3 of Exercises 14.3). Functions $|$, $\dagger$ are strongly complete (see again problem 3 of Exercises 14.3). Function $p \Rightarrow q$ is weakly complete (see above). A similar argument is valid for $q \Rightarrow p$, $\sim(p \Rightarrow q)$, $\sim(q \Rightarrow p)$. Function $\Leftrightarrow$ is incomplete: first, it is commutative and associative; second, $(x \Leftrightarrow x) = 1$, $(x \Leftrightarrow 1) = x$, $(x \Leftrightarrow 0) = \sim x$; therefore, any valid sequence from $\{0, 1, \Leftrightarrow, p, q\}$ can be reduced to a sequence involving at most one p, then at most one q, then at most one 0 or one 1 only; such sequences are 1, 0, p, q, $p \Leftrightarrow 0$, $q \Leftrightarrow 0$, $p \Leftrightarrow q$, $p \Leftrightarrow q \Leftrightarrow 0$, and they produce only the functions 1, 0, p, q, $\sim p$, $\sim q$, $p \Leftrightarrow q$, $\sim(p \Leftrightarrow q)$. A similar argument is valid for the remaining operation, $\sim(p \Leftrightarrow q)$. In summary:

operation	complete	commutative	$(x \text{op} x) = \sim x$
0	no	yes	no
1	no	yes	no
p	no	no	no
q	no	no	no
$\sim p$	no	no	no
$\sim q$	no	no	no
$p \vee q$	no	yes	no
$p \wedge q$	no	yes	no
$p \mid q$	strongly	yes	yes
$p \dagger q$	strongly	yes	yes
$p \Rightarrow q$	weakly	no	no
$q \Rightarrow p$	weakly	no	no
$\sim(p \Rightarrow q)$	weakly	no	no
$\sim(q \Rightarrow p)$	weakly	no	no
$p \Leftrightarrow q$	no	yes	no
$\sim(p \Leftrightarrow q)$	no	yes	no

Exercises 14.4

1. (a) $(\sim p \wedge q) \vee (\sim p \wedge \sim q)$, (c) $(\sim p \wedge q) \vee (\sim p \wedge \sim q)$.

3. E.g., $f(p, q) = 1$ if $p = 01$ and $q = 10$, $f(p, q) = 0$ otherwise.

Exercises 14.5

1. By Stone's theorem, $a = 01$, $b = 10$, or $a = 10$, $b = 01$. In either case, the operations are

$\vee$				
	0	1	a	b
	1	1	1	1
	a	1	a	1
	b	1	1	b

$\wedge$				
	1	0	a	b
	0	0	0	0
	a	0	a	0
	b	0	0	b

	$\sim$
0	1
1	0
a	b
b	a

3. Same arguments as in problem 2.

Complements 14.5

1. If n is not a power of 2, then there is no Boolean algebra on set A. If $n = 2^k$, establish a bijection h: $A \to \{0, 1\}^k$, and establish operations on A by defining: $x \vee y := \mathrm{h}^{-1}$ (hx bitwise or hy), $x \wedge y := \mathrm{h}^{-1}$ (hx bitwise and hy), $\sim x := \mathrm{h}^{-1}$ (bitwise not hx).

Chapter 15

Exercises 15.1

1. (a) $\{\Lambda, x, y, z, xy, yz\}$; (c) $\{\Lambda, x, xy, yz, xx, xxy, xyz, xyx, xyxy, xyyz, yzx, yzxy, yzyz\}$; (e) B; (g) $\varnothing$; (i) $\{\Lambda\}$.

3. (a) $yxxxxx$; (c) $\{\Lambda, xyy, xyyxyy, \ldots\}$; (e) $\{\Lambda\}$.

5. They overlap at $\{\Lambda\}$ only.

Exercises 15.2

1. (a)

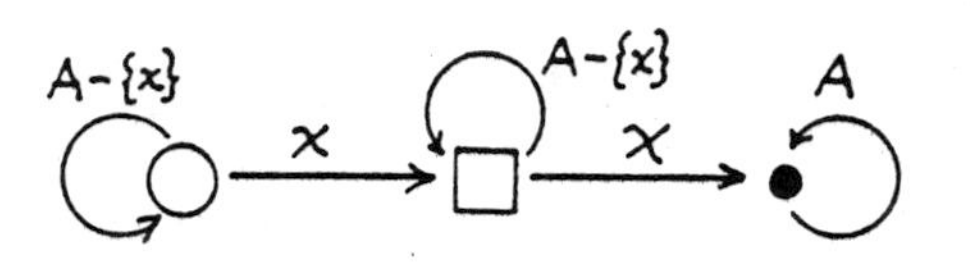

(e)

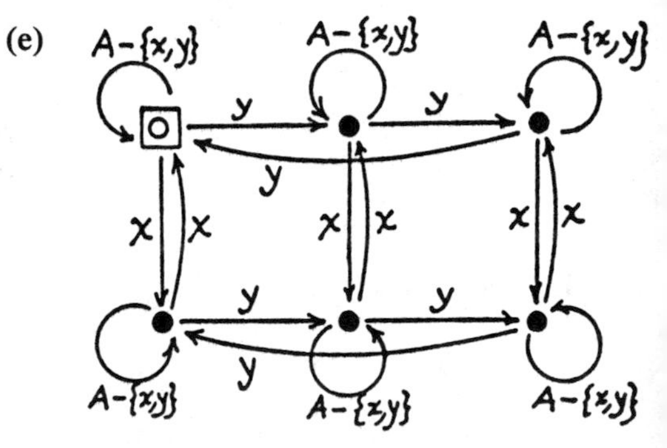

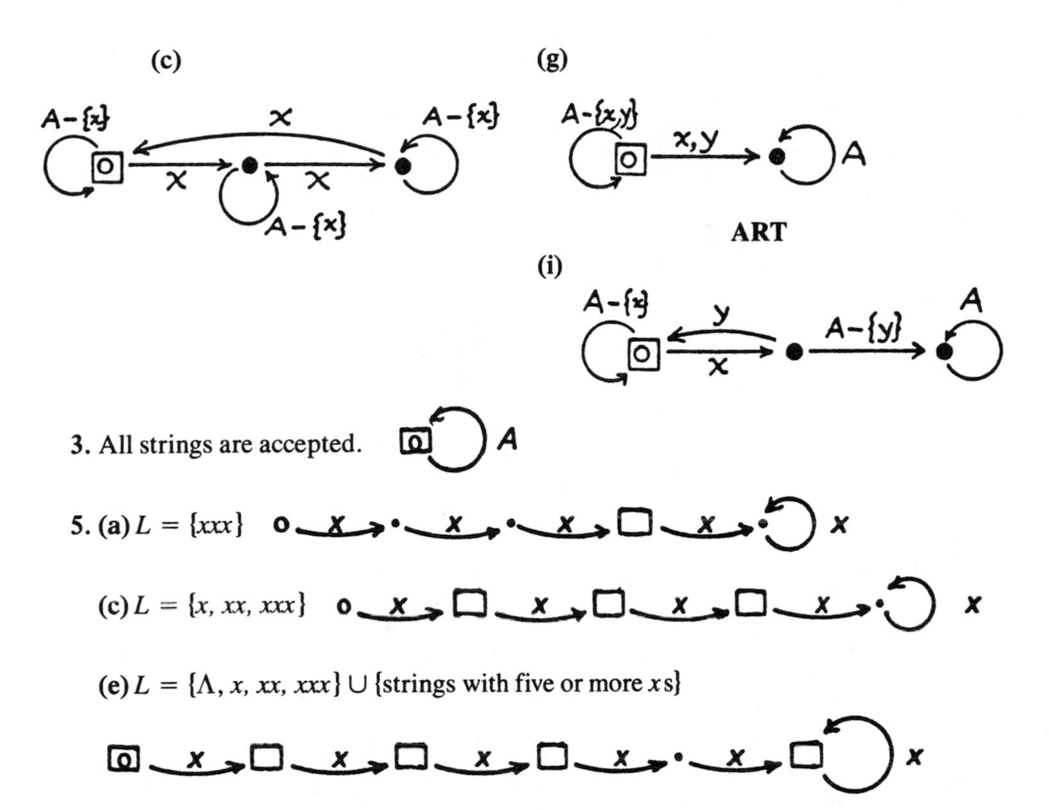

3. All strings are accepted.

5. (a) $L = \{xxx\}$

(c) $L = \{x, xx, xxx\}$

(e) $L = \{\Lambda, x, xx, xxx\} \cup$ {strings with five or more xs}

7. $\mathcal{L}(M_1)$ and $\mathcal{L}(M_2)$ are complement of each other within A^*. Hint for the proof: the processing of each finite string p "stops" at a state (say $z(p)$) which is uniquely determined by p. If $z(p) \in T_1$, then $p \in \mathcal{L}(M_1)$; if $z(p) \notin T_1$, then $p \notin \mathcal{L}(M_1)$; $z(p) \in V - T_1 = T_2$, and $p \in \mathcal{L}(M_2)$.

9. Numbering the states 1, 2, …, left to right:
(a)

15.2(a)

	x	y
1	1	1

15.2(b)

	x	y
1	1	1

15.2(c)

	x	y
1	2	2
2	2	2

15.3(a)

	x	y
1	2	3
2	3	3
3	3	3

15.3(b)

	x	y
1	1	2
2	2	2

(c)

15.1

	S	R
1	2	3
2	4	5
3	5	6
4	7	8
5	8	9
6	9	10
7	16	11
8	11	12
9	12	13
10	13	20
11	16	14
12	14	15
13	15	20
14	16	18
15	18	20
16	21	21
17	18	16
18	17	19
19	18	20
20	21	21
21	21	21

Exercises 15.3

1. (a) $L_1 = (x, y)^\star$, M_1: L_2 $\varnothing$, M_2:

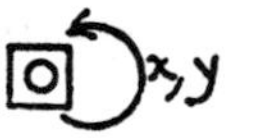

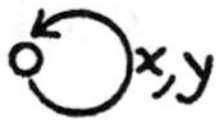

For $L_1 \cup L_2$: same as M_1. For $L_1{:}L_2$, same as M_2. For $L_1{}^\star$, same as M_1.

(c) $L_1 = \{\Lambda\}$, M_1: $L_2 = x^\star$, M_2:

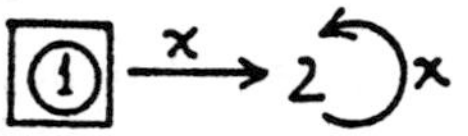

For $L_1 \cup L_2$:

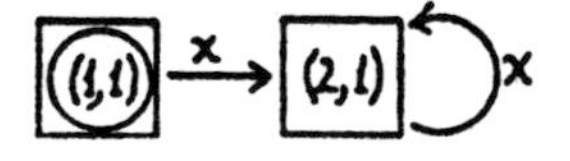

simplified to (same as M_2):

For $L_1{:}L_2$:

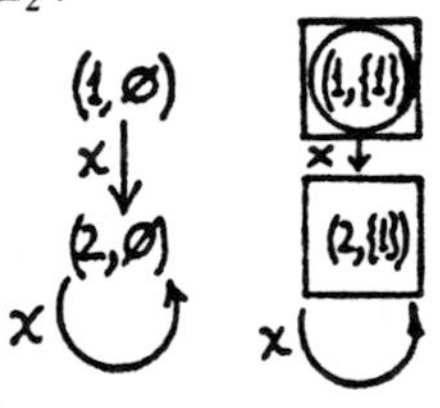

simplified to (same as M_2):

For $L_1{}^\star$:

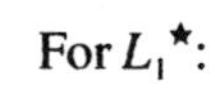

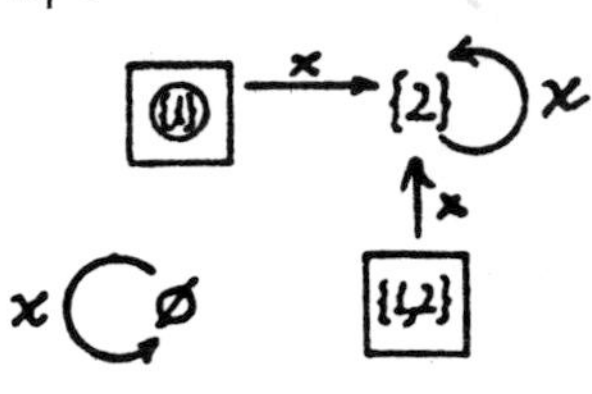

simplified to (same as M_1):

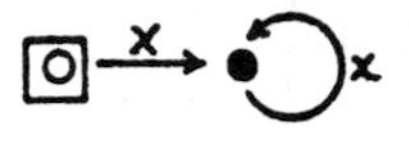

Complements 15.3

1. Hint. Since the operation set $\{\cup, \sim\}$ is functionally complete, every language that can be expressed in terms of $L_1, L_2, \cup, \cap, \sim$ can also be expressed in terms of $\cup$ and $\sim$. By Theorem 8, such languages are in $\mathcal{F}(A)$.

Exercises 15.4

1. Hints. **(a)** If $x_1, \ldots, x_h$ are in A, then $\{(x_i)\} \in \mathcal{R}(A)$ by definition, $(x_1 \ldots x_h) \in \mathcal{R}(A)$ by concatenation, $A^h \in \mathcal{R}(A)$ by finite union. **(b)** If $x \in A$, then $(x) \in \mathcal{R}(A)$ by definition, $A \in \mathcal{R}(A)$ by finite union, $A^{\star} \in \mathcal{R}(A)$ by iteration, $A^* = A^{\star}$ because A is the alphabet. **(c)** Same arguments as in (a) and (b).

3. **(a)** $\{xyx\}$; **(c)** $\{x, xx, xxx\}$; **(e)** $x^{\star}yy^{\star}$.

Complements 15.4

1. Hints (cont.) If x and y, or xy are involved, then the iteration will produce $xyxy \ldots$. When $A = \{x\}$, iteration is needed to obtain an infinite language. Iteration on $\varnothing$ and $\{\Lambda\}$ leads nowhere. We need an x^m for some fixed positive m. Then we obtain (at least) the strings x^{hm} (fixed m, arbitrary h). Further unions, concatenations (except with $\varnothing$), and iterations do not destroy the strings x^{hm} but may change them to x^{hm+c} (constant m and c, arbitrary h), and possibly add more strings. The language H so obtained contains the strings x^{hm+c}, whose lengths are m units apart. Language L contains only the strings x^{2^k}, whose lengths are 2^k units apart. Select k so that $2^k > 3m + c$ and then h so that $2^k < hm + c < 2^{k+1}$. Then there are strings in H that are not in L, $H \not\subseteq L$ for every infinite regular language H, $L \notin \mathcal{L}(A)$, $L \notin \mathcal{F}(A)$.

Appendix C
Table of Symbols

Abbreviations

- 2TSN (two-terminal switching network), 6.6. *See also* network
- A. P. R. (annual percentage rate), Compl. 9.3
- ASCII (American Standard Code for Information Interchange), 13.1.2
- CNF. *See:* form, conjunctive normal
- CPM (critical path method), 11.6
- DNF. *See:* form, disjunctive normal
- FSA. *See:* acceptor, finite state
- GCD (greatest common divisor), Ex. 9.2
- glb (greatest lower bound). *See* infimum
- inf. *See* infimum
- ISBN (International Standard Book Number), 3.3
- lub (least upper bound). *See* supremum
- max. *See* maxterm
- min. *See* minterm
- PERT (Project Evaluation and Review Technique), 10.7, 11.6
- Q. E. D. (end of proof), 3.4.2
- RM. *See:* code, Reed-Mueller
- RPN (reverse Polish notation), 12.1.3
- sup. *See* supremum
- syn. *See* syndrome
- UCF. *See:* form, universal conjunctive
- UDF. *See:* form, universal disjunctive
- w. *See* weight
- WFF. *See:* formula, well-formed

Number Systems

- $\mathbb{C}$ (complex numbers). *See* number
- $\mathbb{N}$ (natural numbers). *See* number
- $\mathbb{Q}$ (rationals). *See* number
- $\mathbb{R}$ (reals). *See* number
- $\mathbb{Z}$ (integers). *See* number
- $\mathbb{Z}_m$ (integers mod m), 3.1

Operations and Functions

- Σ (summation), 2.1.1
- $\cdot$ (multiplication), 2.1.1

Relations

Set theory

Special objects

Index

References are to section numbers. When there is a main reference, it appears first.

A 7
B 8
C 9
D 0
E 1
F 2
G 3